MARKOV PROCESSES

AN INTRODUCTION FOR PHYSICAL SCIENTISTS

MARKOV PROCESSES

AN INTRODUCTION FOR PHYSICAL SCIENTISTS

Daniel T. Gillespie

Research Department
Naval Weapons Center
China Lake, California

ACADEMIC PRESS, INC.
An Imprint of Elsevier
Boston   San Diego   New York
London   Sydney   Tokyo   Toronto

Cover design by Elizabeth E. Tustian and June S. Deatherage

ACADEMIC PRESS, INC.
An Imprint of Elsevier
1250 Sixth Avenue, San Diego, CA 92101

United Kingdom Edition published by
ACADEMIC PRESS LIMITED
24–28 Oval Road, London NW1 7DX

Library of Congress Cataloging-in-Publication Data

Gillespie, Daniel T.
    Markov processes : an introduction for physical scientists /
Daniel T. Gillespie.
        p.   cm.
    Includes bibliographical references and index.

        ISBN-13: 978-0-12-283955-9   ISBN-10: 0-12-283955-2 (acid-free paper)
    1. Markov processes. I. Title.
QA274.7.G55   1992
519.2'33—dc20   91-25738
                                                           CIP

ISBN-13: 978-0-12-283955-9
ISBN-10: 0-12-283955-2
Transferred to Digital Printing 2008

CONTENTS

† This section or subsection can be omitted without impairment to the sequel.

PREFACE

Markov process theory can be broadly described as a generalization of ordinary calculus to accommodate mathematical functions that are imbued with a certain kind of unpredictability or "randomness." That brief description is correct so far as it goes, but it must leave many unanswered questions in the mind of a prospective reader who has never knowingly encountered a Markov process. And since no one should be asked to begin a long book without having a fair idea of what the book is supposed to be about, I feel obliged to begin this preface with a few paragraphs that explain in simple terms just what a "Markov process" is.

By a "process" in general, we mean any function X of time t that can be regarded as specifying the instantaneous state of some real or hypothetical system. The process $X(t)$ might be the vertical velocity at time t of a tossed ball, or the electrical current at time t in some circuit, or the number of customers at time t waiting in a grocery checker's line. Processes that are amenable to analysis can generally be classified as being either "deterministic" or "stochastic."

We say that a process is *deterministic* if a knowledge of its values up to and including time t allows us to unambiguously predict its value at any infinitesimally later time $t + dt$. In symbols, the values $X(t')$ for $t' \leq t$ uniquely determine the value of $X(t + dt)$ for any positive infinitesimal dt. An important subclass of deterministic processes is comprised of those that are *memoryless*. For them, the value of $X(t)$ *alone* uniquely determines $X(t + dt)$, so the process can advance in time without having to recall its past values. An example of a memoryless deterministic process is one for which the value $X(t + dt)$ is obtained from the value $X(t)$ through a formula of the form

$$X(t + dt) = X(t) + f(X(t), t)\, dt,$$

where f is some ordinary function. Subtracting $X(t)$ from both sides and then dividing through by dt, we see that this process is simply the solution to the ordinary differential equation

$$\frac{dX}{dt} = f(X, t),$$

subject to some prescribed initial condition $X(t_0) = x_0$. Deterministic processes of this particular kind provide the mathematical framework for many traditional analytical disciplines.

A broader class of processes, which actually includes deterministic processes as a special case, is the class of stochastic processes. We say that a process is *stochastic* if a knowledge of its values up to and including time t allows us to *probabilistically* predict its value at any infinitesimally later time $t + dt$. More precisely, the values $X(t')$ for $t' \leq t$ determine the *probability* that $X(t + dt)$ will equal any particular value v for any given positive infinitesimal dt. If that probability happens always to be zero for all v values but one, then we are dealing once again with a deterministic process. But excepting that special circumstance, we see that a knowledge of all the values of a stochastic process prior to and including time t will only allow us to make "probabilistic guesses" about the value of the process at time $t + dt$. As in the case of deterministic processes, an important subclass of stochastic processes is comprised of those that are *memoryless*. For a memoryless stochastic process, the probabilities assigned to the possible values of $X(t + dt)$ on the basis of the value $X(t)$ *alone* cannot be "sharpened" by taking cognizance of any values $X(t')$ for $t' < t$; so the process just "forgets" those past values.

A *memoryless stochastic process* is called a *Markov process*. Thus, a Markov process $X(t)$ is the state function of some system whose state value at time $t + dt$ can be probabilistically predicted from its state value at time t, but in a way that cannot be improved upon by taking account of the state values prior to time t. The name given to memoryless stochastic processes is that of the Russian mathematician A. A. Markov (1856–1922), one of several people who first envisioned the concept around the turn of the Twentieth Century. Since that time, many people from many math-related disciplines have been instrumental in developing an analytical theory for Markov processes — a theory that obviously must somehow go beyond ordinary calculus. In this book I have tried to set forth an introductory, deductive exposition of Markov process theory.

The book assumes on the part of the reader mainly a working knowledge of differential and integral calculus. Its "introductory" nature is reflected in two ways: First, it treats only *univariate* processes; i.e., it assumes that $X(t)$ is a scalar function, and not a vector function with several component functions. This is actually a rather severe restriction, because most real-world processes are best treated as multivariate. But since multivariate stochastic processes are notationally cumbersome and often analytically intractable, the univariate restriction seems quite a reasonable one for an introductory work.

The second introductory concession of this book is that *it assumes no prior knowledge of probability and random variables*. Since such knowledge is nevertheless essential to a meaningful understanding of Markov process theory — for this is the vehicle that Markov process theory uses to "go beyond" ordinary calculus — the book begins in Chapter 1 with a self-contained, pragmatic exposition of random variable theory. I have tried to make that exposition as brief as possible by limiting its scope to only those concepts and results that will be needed in the following chapters. But I cannot overemphasize the importance of Chapter 1 for understanding the rest of the book.

I value rigor, and I have striven to achieve it in this book. But I must forewarn the reader that I am at heart a physicist, not a mathematician. And although I do take certain precautions, I frequently and deliberately treat "dt" as "a little bit of t" (as in the early paragraphs of this preface), and I also make frequent use of the "Dirac delta function" (which will be introduced in Chapter 1 and Appendix B). These practices, although frowned upon by many mathematicians, seem to me to be indispensable if one is to develop an *intuitive feeling* for Markov processes, and that in turn seems to be essential if one is to be able to creatively apply Markov process theory to interesting problems. My "rigor" thus lies only in my conscientious attempt to be very clear about things that I am assuming or defining, and very convincing about things that I am deriving. My goal is to convey *insight* and *perspective* in as direct a way as possible; completeness, generality and formal mathematical rigor are employed only to the extent that they support this goal.

The organization of the book is best described with reference to the diagram on the next page. In Chapter 2 we use the random variable theory introduced in Chapter 1 to define a Markov process and develop some of its general concepts and formulas. We discover in Chapter 2 that a Markov process can be fully specified only if we specify its "propagator density function" Π. This function Π is, in the jargon of Chapter 1, a conditioned density function for the random variable $X(t+dt) - X(t)$, and it plays a pivotal role in our exposition: In Chapter 2 we do everything we can without assuming a specific form for Π. In Chapter 3 we show how one way of specifying Π leads to the class of so-called "continuous" Markov processes. And in Chapters 4 and 5 we show how another way of specifying Π leads to the class of so-called "jump" Markov processes. We find that jump Markov processes come in two categories, namely those with "continuum" states and those with "discrete" states, and these are discussed separately in Chapters 4 and 5. In the last and longest chapter, 6, we consider in detail a subclass of discrete state jump Markov processes

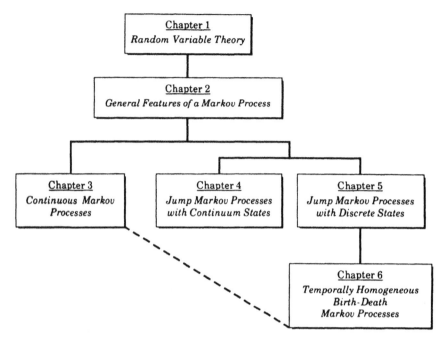

called "temporally homogeneous birth-death" Markov processes; they are in many ways the simplest and most widely applicable kind of Markov process. The story as thus told is a bit long for a typical university course, so some strategies for shortening it will be suggested at the end of this preface.

The "top-down" organization just described has the unfortunate consequence that many of the simpler results of Markov process theory do not emerge until relatively late in the book. But the usual teaching strategy of starting simply and then generalizing would have made it very difficult to convey a picture of the *overall structure* of Markov process theory — to show the forest and not just some trees — and as mentioned earlier the establishment of such a perspective is an overriding goal of this book. For example, the dashed line in the diagram signifies the many interesting parallels and connections that exist between *continuous* Markov processes and *birth-death* Markov processes. But those features can be properly appreciated only in the context of the hierarchical development depicted in the figure.

I have come to believe that one's knowledge of any dynamical system is deficient unless one knows a valid way to *numerically simulate* that system on a computer. Accordingly, I have devoted considerable attention throughout this book to deriving correct numerical simulation procedures for Markov processes, and to exhibiting and discussing the results of many actual numerical simulations. I also firmly believe that the *reasoning* used to cast any specific phenomenon as a Markov process should be very carefully done: Randomness in a physical phenomenon is not a license for sloppy logic in the construction of a mathematical model for that phenomenon. To give substance to this view, I have exhibited in some detail the reasoning used to construct Markov process models of Brownian motion, molecular diffusion and chemical reactions. These particular examples reflect my own past research interests, and should not be regarded as exhausting the practical applications of Markov process theory.

In writing this book, I have naturally drawn heavily on the hard won insights and results of many people past and present. But those who are familiar with Markov process theory will perceive that I have organized and presented most of the standard material in a rather novel way, and also that I have presented some material that is apparently new. Some examples that partake, in widely varying degrees, of both these categories are as follows: the pragmatic development of random variable theory in Chapter 1; the focus on the process propagator in Chapter 2, and the derivations of the forms of the continuous and jump propagator density functions in Sections 3.1 and 4.1; the definition of the time-integral of a Markov process in Section 2.6, and the derivation of its moment evolution equations in Section 2.7; the derivation of a quadrature solution for completely homogeneous Markov processes in Subsection 2.8.B and its several applications in later chapters; the discussion of the different forms of the Langevin equation in Section 3.4; the definition and use of the potential and barrier functions in Chapters 3 and 6; the discussion of noise-enhanced and noise-induced stability in Section 3.6; the discussion of the continuous simulation algorithm in Subsection 3.9.A and Appendix E; the general method for approximately solving open moment evolution equations in Appendix C; the introduction of the next-jump density function in Chapters 4, 5 and 6; the analysis of one-dimensional Brownian motion and self-diffusion in Sections 4.5 and 4.6; the development in Subsection 6.6.B of a tractable set of recursion relations for numerically calculating all the moments of a birth-death first passage time; the analysis in Subsection 6.7.A of birth-death splitting probabilities and mean first exit times; and the simple approximate formulas derived in Section 6.8 for characterizing stable state fluctuations and transitions in a bistable birth-death Markov process.

Those who are familiar with Markov process theory may also perceive that I have omitted some things that they would not have, even at the price of lengthening an already long book. Some such omissions may simply be due to ignorance on my part. Three omissions, however, were deliberate and deserve comment. First, I have not made use in this book of the "stochastic calculus" of Ito or Stratonovich. I am presently of the opinion that, at least from the standpoint of Markov process theory, the investment of effort in setting up those admittedly interesting calculi is not adequately repaid in either convenience or insight.

My second conscious omission is more discomfiting. Since this book is an introduction to, and not a review of, Markov process theory, I have not annotated it with references detailing who deserves credit for what result or insight. I feel that such references would have interrupted and interfered with the heuristic flow of the book. Furthermore, I frankly had neither the time nor the mandate to research the *history* of Markov process theory. Considering that this subject has been developed over many years through the parallel efforts of mathematicians, physicists, chemists and electrical engineers, who typically hold differing views as to which results are most significant and how those results ought to be conveyed, it would surely be a monumental task to assign credit accurately. I have not risen to that task, but I hope that this shortcoming will be somewhat ameliorated by the remarks in the Acknowledgements and Bibliography.

My third deliberate omission in this book is formal exercises, which will probably make the book less attractive to teachers. But I would suggest to teachers that there are many opportunities in this book for the assignment of *computer-oriented* exercises: Many numerical results are quoted in the text and displayed in tables and graphs, and the comprehending student should be able to reproduce almost all of those results (modulo in many cases statistical fluctuations) by properly programming an ordinary PC-type computer. My experience with traditional exercises suggests that they are often more of a distraction than an aid to understanding. Had I been forced to include exercises, I would have made them the "fill-in-the-steps" type, integrating them with the text and thereby shortening the narrative. But I have not done that, and have elected instead to supply nearly all of the steps in order to enhance the self-study value of the book. Accordingly, the reader should regard *any* part of this book that seems to be grindingly tedious as merely a "worked exercise," supplied by a benevolent author, which the reader can either read through for enlightenment, or try to work through with the book closed as a challenge, or even defer by skipping over.

Finally, since this book is too long to be covered completely in a normal one-year university course, I feel obliged to make some suggestions to teachers on how coverage of the book might feasibly be abbreviated.

The severest possible abbreviation of the book would be to simply omit everything except Chapter 1. One would then have a textbook for a short course on random variable theory, rather than a long course on Markov process theory. But I submit that such a short course, which with the addition of a few supplemental topics and applications could easily fill a semester, would be worthwhile for *all* students of science, even those who do not intend to study Markov process theory. And those students who do intend to study Markov process theory would, after such a course, be in a position to cover nearly all of Chapters 2 through 6 in a subsequent one-year effort.

For a stand-alone one-year course on Markov process theory proper, one should start by covering all of Chapters 1 and 2, inasmuch as they lay the necessary foundation. Next, because it is so basic and so beautiful, the theory of *continuous* Markov processes in Chapter 3 should be covered, pretty much in its entirety. Proceeding then to the theory of *jump* Markov processes, I would suggest that one make a hard choice between treating jump processes with *continuum* states or jump processes with *discrete* states. If one chooses to treat jump Markov processes with *continuum* states, then one should cover Sections 4.1 through 4.4 and the first half of the simulation examples in Section 4.6; the material thus omitted in Chapter 4 is a rather detailed analysis of one-dimensional Brownian motion and self-diffusion, and it could be used as the course finale if time permitted. If on the other hand one chooses to treat jump Markov processes with *discrete* states, then one should skip Chapter 4 altogether and proceed with Chapters 5 and 6. In fact, since Chapter 5 begins by quickly redeveloping the Chapter 2 fundamentals in integer variable form, it would be possible to assimilate most of Chapters 5 and 6 *without* first reading Chapters 2 and 3; this however would occasion some loss of perspective, especially as regards appreciating the interesting and often useful connections with the results derived in Chapter 3. In any case, Chapter 6 can be shortened if desired by omitting Section 6.5 and Subsection 6.6.C.

China Lake, California   DANIEL T. GILLESPIE

ACKNOWLEDGEMENTS

I learned most of what I know about stochastic processes from the writings of S. Chandrasekhar and N. G. van Kampen, and from the writings of and discussions with Crispin Gardiner, Dan Walls and Marc Mangel. Any egregious errors that I have committed in this book are no doubt due to my not having paid careful enough attention to their words. Other individuals whose writings in various related areas have substantively contributed to my understanding of the subject at hand are: P. D. Drummond, W. Feller, R. Görtz, S. Karlin, V. M. Kenkre, L. Kleinrock, R. Landauer, K. Lindenberg, M. Malek-Mansour, I. Matheson, K. J. McNeil, D. A. McQuarrie, E. W. Montroll, G. Nicolis, A. Nitzan, R. D. Present, I. Procaccia, F. Rasetti, L. E. Reichl, F. Reif, J. Ross, J. V. Sengers, V. Seshadri, M. F. Shlesinger, L. Takács J. S. Turner, G. H. Weiss, B. J. West and R. Zwanzig.

I began work on this book in 1986 after a series of informal study sessions with John Cannon, and I am grateful to him for encouraging me to take up this project. I thereafter benefited from many helpful discussions with my coworkers at the Naval Weapons Center, especially Warren Willman, Jorge Martin, Charles Kenney, William Alltop and Gary Hewer.

The development of this book over the past five years, and some assorted gestational studies during the dozen years preceding that, were all accomplished as specific elements of the Independent Research Program of the Naval Weapons Center. I especially thank Ron Derr, Head of the Naval Weapons Center's Research Department, for his explicit and positive endorsement of this project since its inception. I also thank the Office of Naval Research for some supplemental funding through its Mathematical Sciences Division and its Navy Laboratory Participation Program. From a broader perspective, this book has been made possible by the continuous financial support of the U. S. Taxpayer, to whom I am genuinely and humbly grateful. In that connection, I should note that publication of this book is controlled by a Cooperative Research and Development Agreement between Academic Press and the Naval Weapons Center, in accordance with the provisions of the Technology Transfer Act of 1986.

I composed all of the book's text, and a few of its figures, on a Xerox 6085 wordprocessor running Xerox's ViewPoint software. Since that wordprocessing system lightened the logistical burden of creating this book by at least ten-fold, I cannot refrain from expressing my gratitude to the people of Xerox for producing their marvelous "supertypewriter." And I want to thank Debbie Ayers of Xerox for her efforts in helping me to realize the full potential of the ViewPoint system.

The number-based figures in this book were all created using CA-DISSPLA software running on a VAX 8600 computer and a Talaris 1590 printer, all provided by the Naval Weapons Center's Information Systems Group. I want to especially thank Claire McGraedy of that group for her help in using those graphics capabilities.

Lastly, but mostly, it gives me great pleasure to publicly thank my wife Carol, not only for amiably putting up with my obsessive pursuit of this book over the past five years, but also for giving of her professional talents as a mathematician and computer scientist to help me in that pursuit. All the numerical computations reported in this book, and many exploratory computations that were not reported, were carried out by her on her home computer (a Zenith Z-120) over the course of many evenings and weekends. Many of those computations held unexpected lessons for me, lessons that materially affected the course of this book, and she was an active participant in my efforts to understand those lessons. The benefit of her help and emotional support to me in writing this book is beyond quantifying, and so is my gratitude to her.

BIBLIOGRAPHY

As is the case with many physicists, my interest in stochastic processes was originally kindled many years ago by S. Chandrasekhar's masterpiece article "Stochastic Problems in Physics and Astronomy." That article originally appeared in the January 1943 issue of *Reviews of Modern Physics*, but it is more readily available today, along with a number of other classics in the field, in the book

- N. Wax, ed., *Selected Papers on Noise and Stochastic Processes* (Dover, New York, 1954).

For the writing of this book, however, I relied for help and guidance more frequently on the following two works:

- N. G. van Kampen, *Stochastic Processes in Physics and Chemistry* (North-Holland, Amsterdam, 1981);

- C. W. Gardiner, *Handbook of Stochastic Methods for Physics, Chemistry and the Natural Sciences* (Springer-Verlag, Berlin, 1985).

These two books are generally more advanced and comprehensive than mine, and both provide through their bibliographies an extensive guide to the literature on stochastic processes which I cannot improve upon here.

MARKOV PROCESSES

An Introduction for Physical Scientists

- 1 -

RANDOM VARIABLE THEORY

In this chapter we shall develop those definitions, concepts and theorems of random variable theory that will be required for our development of Markov process theory in the following chapters. Our discussion of random variable theory will, for the most part, be as brief and narrowly circumscribed as its limited purpose here will allow.

1.1 THE LAWS OF PROBABILITY

For practically all of its applications in the physical sciences, probability can be defined and understood through the so-called *frequency interpretation*. Basic to that interpretation is some kind of experiment that can be repeatedly performed or tried, under ostensibly identical conditions, to yield a series of outcomes. Each trial of the experiment may have any number of specific outcomes 1, 2, 3, ..., and each outcome i either occurs or does not occur on each trial. Also legitimate outcomes of a trial are: "$\underline{i}$" (read "not i"), the nonoccurrence of outcome i; "$i{\wedge}j$" (read "i And j"), the occurrence of both outcomes i and j; and "$i{\vee}j$" (read "i or j"), the occurrence of either or both of outcomes i and j. The compounding of outcomes using $\wedge$ and $\vee$ can be extended to three or more outcomes, and we note in particular that

$$i{\wedge}(j{\wedge}k) = (j{\wedge}i){\wedge}k = \ldots \equiv i{\wedge}j{\wedge}k,$$

$$i{\vee}(j{\vee}k) = (j{\vee}i){\vee}k = \ldots \equiv i{\vee}j{\vee}k.$$

Combinations of outcomes that *mix* $\wedge$ and $\vee$ do *not* behave so simply, but we shall not have occasion to deal with such combinations here.

If in n trials of the experiment a particular outcome i is found to occur $m_n(i)$ times, then

$$p(i) \equiv \lim_{n \to \infty} \frac{m_n(i)}{n} , \qquad (1.1\text{-}1)$$

if it exists, is called the **probability** of outcome i with respect to that experiment. We say that on any future trial of the experiment, "the probability that outcome i will occur is $p(i)$." Of course, outcome i in this definition can be replaced by any compound outcome, such as $i \vee j$ or $i \wedge j \wedge k$.

An allied definition is that of the **conditional probability**: If, in the same series of n trials, two specific outcomes i and j are found to occur together $m_n(i \wedge j)$ times, then

$$p(j \mid i) \equiv \lim_{n \to \infty} \frac{m_n(i \wedge j)}{m_n(i)} , \qquad (1.1\text{-}2)$$

if it exists, is called the probability of outcome j given, or conditioned on, outcome i. The symbol "$\mid$" is read "given" or "conditioned on." We say that on any future trial of the experiment, "the probability that outcome j will occur given that outcome i occurs is $p(j \mid i)$." If outcome i never occurs, so that $m_n(i) = 0$ for all n, then $p(j \mid i)$ is simply regarded as undefined. As before, the two outcomes i and j in Eq. (1.1-2) can be separately replaced by any combination of outcomes.

A set of outcomes 1, 2, ..., N is said to be **mutually exclusive** if and only if not more than one of them can occur on the same trial; in that case $m_n(i \wedge j) = 0$ for all $1 \le i < j \le N$ and for any number of trials n, so the definition (1.1-1) gives

Outcomes 1, ..., N are mutually exclusive

$$\Leftrightarrow \quad p(i \wedge j) = 0 \text{ for all } 1 \le i < j \le N. \quad (1.1\text{-}3)$$

A set of outcomes 1, 2, ..., N is said to be **collectively exhaustive** if and only if at least one of them occurs on every trial; in that case, $m_n(1 \vee 2 \vee ... \vee N) = n$ for all numbers of trials n, so definition (1.1-1) gives

Outcomes 1, ..., N are collectively exhaustive

$$\Leftrightarrow \quad p(1 \vee 2 \vee ... \vee N) = 1. \qquad (1.1\text{-}4)$$

A set of outcomes might very well be *both* mutually exclusive and collectively exhaustive, in that on each trial *some* one, but *only* one, of those outcomes occurs; a simple example of a mutually exclusive and collectively exhaustive set of outcomes is the pair of outcomes i and $\underline{i}$.

From the foregoing definitions we can now deduce the following three *laws of probability.*[†]

Range Law. The probability $p(i)$ of any outcome i is a real number satisfying

$$0 \le p(i) \le 1, \tag{1.1-5}$$

with $p(i) = 0$ corresponding to the circumstance that i *never* occurs, and $p(i) = 1$ corresponding to the circumstance that i *always* occurs.

Proof. Since $0 \le m_n(i) \le n$, then $0 \le m_n(i)/n \le 1$, which with definition (1.1-1) gives Eq. (1.1-5). Outcome i is "impossible" if and only if $m_n(i) = 0$ for all n, giving $p(i) = 0$; outcome i is "certain" if and only if $m_n(i) = n$ for all n, giving $p(i) = 1$. QED

Addition Law. For any N *mutually exclusive* outcomes $1, ..., N$ we have

$$p(1 \vee 2 \vee ... \vee N) = p(1) + p(2) + ... + p(N). \tag{1.1-6}$$

Proof. Mutual exclusivity implies that no more than one of the outcomes can occur on any one trial, so $m_n(1 \vee ... \vee N) = m_n(1) + ... + m_n(N)$ for all n. Dividing through by n and taking the limit $n \to \infty$, we obtain Eq. (1.1-6).
QED

Multiplication Law. For any two outcomes i and j we have

$$p(i \wedge j) = p(i)\, p(j \mid i) = p(j)\, p(i \mid j). \tag{1.1-7}$$

Proof. Dividing both the numerator and denominator in Eq. (1.1-2) by n gives, in the limit $n \to \infty$, $p(j \mid i) = p(i \wedge j)/p(i)$; this establishes the left equality in Eq. (1.1-7). The right equality then follows from the fact that outcomes $i \wedge j$ and $j \wedge i$ are identical. QED

Some comments on these three laws are in order. First, we note that the structure of the definitions (1.1-1) and (1.1-2) of $p(i)$ and $p(j \mid i)$ is such that Eqs. (1.1-5) – (1.1-7) remain valid if all probabilities therein are conditioned on some other outcome o. So Eqs. (1.1-5) and (1.1-6) can respectively be written more generally as

---

[†] Mathematicians typically prefer to avoid the "frequency interpretation" of probability, and to simply regard the following three laws as *axioms*, from which many theorems about probability can then be rigorously derived. But in order to apply those theorems to physical phenomena, it is necessary to tie the axioms to the physical world through some kind of frequency interpretation. Since we shall be exclusively interested here in physical applications of probability theory, then we may as well embrace the frequency interpretation at the outset and regard the three laws as its consequences.

$$0 \leq p(i\,|\,o) \leq 1$$

and

$$p(1\vee \ldots \vee N\,|\,o) = p(1\,|\,o) + \ldots + p(N\,|\,o),$$

with the latter relation assuming that events 1, ..., N are mutually exclusive when o occurs. And Eq. (1.1-7) can similarly be generalized to

$$p(i\wedge j\,|\,o) = p(i\,|\,o)\,p(j\,|\,i\wedge o) = p(j\,|\,o)\,p(i\,|\,j\wedge o).$$

Secondly, by combining the definition (1.1-3) of mutual exclusivity with Eq. (1.1-7), and assuming that neither $p(i)$ nor $p(j)$ vanishes [so that $p(i\,|\,j)$ and $p(j\,|\,i)$ are both well defined] we deduce that outcomes i and j are mutually exclusive if and only if

$$p(i\wedge j) = 0 \ \Leftrightarrow \ p(i\,|\,j) = 0 \ \Leftrightarrow \ p(j\,|\,i) = 0.$$

Thirdly, we note that the addition law, unlike the multiplication law, deals *only* with outcomes that are mutually exclusive. A generalization of the addition law that does *not* require mutual exclusivity can be framed, but its form becomes complicated, and its usefulness correspondingly limited, as the number N of outcomes considered is increased. Although we shall not require this generalized addition law in our work here, it is nevertheless instructive to derive that law for $N=2$. We reason as follows: Since outcome 1 is equivalent to either outcome $1\wedge 2$ or outcome $1\wedge \underline{2}$, which two outcomes are obviously mutually exclusive, then we have by the addition law,

$$p(1) = p((1\wedge 2) \vee (1\wedge \underline{2})) = p(1\wedge 2) + p(1\wedge \underline{2}).$$

Similarly, we may deduce that

$$p(2) = p(1\wedge 2) + p(\underline{1}\wedge 2).$$

Now, the outcome $1\vee 2$ is defined to be *either* of the three outcomes $1\wedge 2$ or $1\wedge \underline{2}$ or $\underline{1}\wedge 2$. Since these three outcomes are evidently mutually exclusive, then the addition law gives

$$p(1\vee 2) = p(1\wedge 2) + p(1\wedge \underline{2}) + p(\underline{1}\wedge 2).$$

By eliminating $p(1\wedge \underline{2})$ and $p(\underline{1}\wedge 2)$ on the right hand side of this last equation by means of the two preceding equations, we get

$$p(1\vee 2) = p(1) + p(2) - p(1\wedge 2). \qquad (1.1\text{-}8)$$

This is the generalized addition law for a set of any two outcomes. Notice that if outcomes 1 and 2 happen to be mutually exclusive, so that

$p(1\wedge2)=0$, then Eq. (1.1-8) reduces to the $N=2$ version of Eq. (1.1-6), just as we should expect.

For later reference we shall now enumerate all the implications of the multiplication law for a set of *three* outcomes 1, 2 and 3 (which need not be mutually exclusive). In addition to the pairwise relations (1.1-7), we also have relations of the form

$$p(1\wedge2\wedge3) = p((1\wedge2)\wedge3) = p(1\wedge2)\,p(3\mid 1\wedge2),$$

as well as relations of the form

$$p(1\wedge2\wedge3) = p(1\wedge(2\wedge3)) = p(1)\,p(2\wedge3\mid 1).$$

In general, then, for any three outcomes 1, 2 and 3 we have the following relations, valid for (i,j,k) any permutation of $(1,2,3)$:

$$p(i\wedge j) = p(i)\,p(j\mid i), \tag{1.1-9a}$$

$$p(1\wedge2\wedge3) = p(i\wedge j)\,p(k\mid i\wedge j), \tag{1.1-9b}$$

$$p(1\wedge2\wedge3) = p(i)\,p(j\wedge k\mid i). \tag{1.1-9c}$$

Also, by combining the first two of these relations, we have

$$p(1\wedge2\wedge3) = p(i)\,p(j\mid i)\,p(k\mid i\wedge j), \tag{1.1-10}$$

which again is valid for (i,j,k) any permutation of $(1,2,3)$.

Finally, a set of N outcomes is said to be **statistically independent** if and only if the probability of any one of those outcomes i, conditioned on any one or any "anded" combination of the other outcomes, is equal to $p(i)$. Thus, for a set of three outcomes we have, by definition,

Outcomes 1, 2, 3 are statistically independent
$$\Leftrightarrow\ p(i) = p(i\mid j) = p(i\mid j\wedge k), \tag{1.1-11a}$$

for (i,j,k) all permutations of $(1,2,3)$. There is some redundancy in relations (1.1-11a), but not as much as one might at first suppose. For instance, since Eq. (1.1-7) implies that $p(1)p(2\mid1)=p(2)p(1\mid2)$, then if it is true that $p(2\mid1)=p(2)$ it must also be true that $p(1\mid2)=p(1)$. On the other hand, if it is true that both $p(1\mid2)$ and $p(1\mid3)$ are equal to $p(1)$, it is *not* necessarily true that $p(1\mid2\wedge3)=p(1)$. By using Eqs. (1.1-9), it is possible to prove from (1.1-11a) that

Outcomes 1, 2, 3 are statistically independent
$$\Leftrightarrow\ p(i\wedge j) = p(i)\,p(j)\ \text{and}\ p(1\wedge2\wedge3) = p(1)\,p(2)\,p(3), \tag{1.1-11b}$$

for (i,j) all pairs from $(1,2,3)$. Equations $(1.1\text{-}11b)$ are only four in number; they do *not* contain any redundancy, and thus provide a more compact criterion than $(1.1\text{-}11a)$ for the statistical independence of three outcomes. More generally, a set of N outcomes is statistically independent if and only if the probability of every "anded" combination of those outcomes is equal to the product of their separate probabilities.

1.2 DEFINITION OF A RANDOM VARIABLE

From an applications point of view, a **variable** is simply any entity that always has a *value* which we can measure or "sample" at will. Our main concern here will be with real variables, whose possible values are by definition real numbers. The value of a variable generally depends upon the context in which it is sampled. Normally, we try to define the sampling context in such a way that the value obtained on any sampling is uniquely determined. For example, if the variable X is the position at time t of a simple harmonic oscillator of intrinsic frequency v, then the value of X in the circumstance that the initial position was A and the initial velocity was 0 will always be found to be $A\cos 2\pi v t$. In general, when a variable X is such that its sampling context uniquely determines its value, we say that X is a **sure variable**. The mathematical variables used in elementary expositions of science and engineering are nearly always tacitly assumed to be sure variables.

If the sampling context of a variable does *not* uniquely determine its value, then we might be tempted to call that variable "random." However, the definition of a random variable requires a bit more than mere value uncertainty in contextual sampling. Specifically, X is said to be a (real) **random variable** if and only if there exists a function P of a (real) variable x such that $P(x)\Delta x$ equals, to first order in Δx, the probability of finding the value of X in the interval $[x, x+\Delta x)$. In symbols,

$$\text{Prob}\{X \in [x, x+\Delta x)\} = P(x)\Delta x + o(\Delta x), \qquad (1.2\text{-}1a)$$

where $o(\Delta x)$ denotes terms that satisfy $o(\Delta x)/\Delta x \to 0$ as $\Delta x \to 0$. Therefore, if we divide through by Δx, and then let Δx shrink to some *infinitesimal* (vanishingly small) positive value dx, we get

$$P(x) = \text{Prob}\{X \in [x, x+dx)\} / dx, \qquad (1.2\text{-}1b)$$

an equation that is often also written as

$$P(x)\mathrm{d}x = \mathrm{Prob}\{X \in [x, x + \mathrm{d}x)\}. \tag{1.2-1c}$$

Equation (1.2-1b) says that $P(x)$ is essentially the "density of probability" of the random variable X at the value x, a circumstance that leads us to call P the **density function** of X. Any one of the three equivalent equations (1.2-1) can be regarded as the *definition* of P. We say that "X is randomly distributed according to the density function P." It is to be emphasized that the functional form of P completely defines the random variable X, and vice versa. So if two random variables X and Y have the same density function, then we write $X = Y$. Notice that to say that $X = Y$ is *not* to say that X and Y will necessarily be found, upon sampling, to have equal values.

To calculate the probability that a sample value of X will be found inside some *finite* interval $[a, b)$, imagine that interval to be subdivided into a set of nonoverlapping infinitesimal subintervals $[x, x + \mathrm{d}x)$. Since the outcome $X \in [a, b)$ can be realized if and only if X is found inside some one of those subintervals, and since the nonoverlapping nature of the subintervals guarantees that X cannot be found in more than one of them, then the addition law of probability allows us to calculate $\mathrm{Prob}\{X \in [a, b)\}$ by *summing* $\mathrm{Prob}\{X \in [x, x + \mathrm{d}x)\}$ over all the subintervals. Thus we deduce from Eq. (1.2-1c) that

$$\mathrm{Prob}\{X \in [a, b)\} = \int_a^b P(x)\,\mathrm{d}x. \tag{1.2-2}$$

Since any sample value of X will surely lie *somewhere* in the interval $(-\infty, \infty)$, then it follows from Eq. (1.2-2) that P must satisfy

$$\int_{-\infty}^{\infty} P(x)\,\mathrm{d}x = 1, \tag{1.2-3}$$

an equation that is called the "normalization condition." Furthermore, since $\mathrm{d}x$ in Eq. (1.2-1c) is intrinsically positive, then the mere fact that $P(x)\mathrm{d}x$ is a probability implies by the range law that P must satisfy

$$P(x) \geq 0 \quad \text{for all } x. \tag{1.2-4}$$

Indeed, *any* function P that satisfies both conditions (1.2-3) and (1.2-4) can be regarded as a *density* function, defining some random variable X.

It should be clear from the foregoing that the density function brings some semblance of *order* to the otherwise haphazard behavior of a random variable: Even though we cannot say for sure what value X will have on any *individual* sampling, Eq. (1.2-1c) implies that, in the limit of *infinitely many* samplings, a fraction $P(x)\mathrm{d}x$ thereof will yield values in the interval $[x, x + dx)$; therefore, a normalized frequency histogram of the sample values of X will always approach the curve $P(x)$-versus-x as the number of samplings approaches infinity. So P can be said to describe the "bias" in X, and in that sense of the word *all* random variables are biased; random variables differ from one another only in the *forms* of their biases. It is the "tension" between the unpredictability of a single sampling and the bias evinced over very many samplings that makes random variables such fascinating objects of study.

The **Dirac delta function** $\delta(x - x_0)$ may be defined *either* by the statement that

$$\int_{-\infty}^{\infty} f(x)\, \delta(x - x_0)\, dx = f(x_0) \tag{1.2-5}$$

for any function f of x, *or* by the *pair* of equations

$$\delta(x - x_0) = 0, \text{ if } x \neq x_0, \tag{1.2-6a}$$

$$\int_{-\infty}^{\infty} \delta(x - x_0)\, dx = 1. \tag{1.2-6b}$$

Comparing Eqs. (1.2-6) with conditions (1.2-4) and (1.2-3), we see that the function $\delta(x - x_0)$ satisfies all the requirements of a *density* function, and therefore defines some random variable X. Taking account of Eq. (1.2-2), we see that this random variable X has the property that

$$\text{Prob}\{X \in [a,b)\} = \int_{a}^{b} \delta(x - x_0)\, dx = \begin{cases} 1, & \text{if } x_0 \in [a,b), \\ 0, & \text{if } x_0 \notin [a,b), \end{cases}$$

where the second equality follows from Eqs. (1.2-6). In words, the probability of finding the value of X inside any given interval is either 1 or 0 according to whether that interval does or does not contain x_0. Clearly, this implies that the value of X must be precisely x_0, and so we conclude that

$$X \text{ has density function } \delta(x - x_0) \Leftrightarrow X \text{ is the sure variable } x_0. \tag{1.2-7}$$

The reverse implication in (1.2-7) follows from the fact that $\delta(x - x_0)$ is the *only* function of x whose integral from $x = a$ to $x = b$ is unity if $x_0 \in [a,b)$ and zero otherwise. From this point of view, a sure variable is just a special type of random variable.

Closely associated with the density function P of a random variable X is the **distribution function** F of X. It is defined by

$$F(x) \equiv \text{Prob}\{X < x\}. \tag{1.2-8}$$

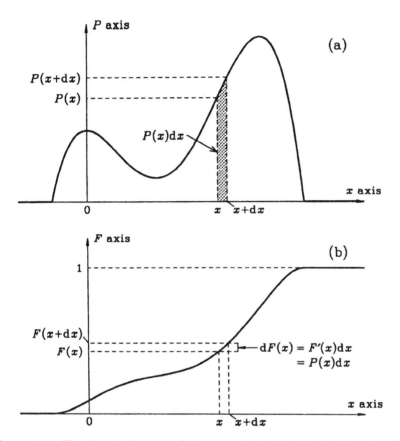

Figure 1-1. The *density* function P in (a), and the *distribution* function F in (b), for a hypothetical random variable X. Equations (1.2-9) show that P and F are related as derivative and integral. The probability that a sample value of X will fall between x and $x + dx$ is equal to the shaded area $P(x)dx$ in (a), and to the vertical distance $dF(x)$ in (b). The probability that a sample value of X will be less than x is equal to the ordinate value $F(x)$ of the (b) curve, and to the area under the (a) curve to the left of the abscissa value x.

Combining this definition with Eq. (1.2-2) shows that

$$F(x) = \int_{-\infty}^{x} P(x')\,dx', \qquad (1.2\text{-}9a)$$

and upon differentiating this equation with respect to x, we get

$$P(x) = F'(x). \qquad (1.2\text{-}9b)$$

This derivative-integral relation between the two functions P and F is illustrated in Fig. 1-1. If we know P we can calculate F from Eq. (1.2-9a), and if we know F we can calculate P from Eq. (1.2-9b); therefore, the random variable X can be defined by specifying *either* its density function P or its distribution function F. With Eqs. (1.2-9a), (1.2-3) and (1.2-4), it is easy to see that $F(x)$ rises nondecreasingly from 0 at $x = -\infty$ to 1 at $x = +\infty$; indeed, *any* function with this property can be regarded as a distribution function F that defines some random variable X.

In Section 1.6 we shall consider a way of defining a random variable that does not entail the overt specification of its density or distribution functions, although of course both will be implicitly specified.

1.3 AVERAGES AND MOMENTS

If h is any univariate function, then the **average** of h with respect to the random variable X is denoted by $\langle h(X) \rangle$, and is defined by

$$\langle h(X) \rangle \equiv \lim_{N \to \infty} \frac{1}{N} \sum_{i=1}^{N} h(x^{(i)}), \qquad (1.3\text{-}1)$$

where $x^{(1)}, x^{(2)}, ..., x^{(N)}$ are the values assumed by X in N independent samplings (i.e., samplings that have not been screened for acceptance on the basis of their results). Now, according to the definitions (1.2-1c) and (1.1-1), $P(x)dx$ gives the *approximate fraction* of any N sample values of X that will fall inside $[x, x+dx)$, the approximation becoming exact as $N \to \infty$; thus, the contribution to the sum in Eq. (1.3-1) coming from X values in the interval $[x, x+dx)$ is approximately equal to $NP(x)dx \times h(x)$, so we can approximate that sum by

$$\sum_{i=1}^{N} h(x^{(i)}) \approx \int_{x=-\infty}^{\infty} [N P(x)dx \times h(x)] = N \int_{-\infty}^{\infty} h(x) P(x) \, dx.$$

Dividing through by N and then taking the limit $N \to \infty$, so that the approximate equality becomes an exact equality, we conclude that

$$\langle h(X) \rangle = \int_{-\infty}^{\infty} h(x) P(x) \, dx. \qquad (1.3\text{-}2)$$

This formula, which gives $\langle h(X) \rangle$ in terms of the function h and the density function of X, is analytically more convenient that the formula (1.3-1), and is sometimes regarded as an alternate definition of $\langle h(X) \rangle$.

Especially useful are averages of the form

$$\langle X^n \rangle \equiv \int_{-\infty}^{\infty} x^n P(x)\,dx \quad (n=0, 1, 2, ...). \tag{1.3-3}$$

$\langle X^n \rangle$ is called the nth **moment of X**, or sometimes also the nth moment of P. In view of the normalization condition (1.2-3), the zeroth moment of X always exists and has the value unity:

$$\langle X^0 \rangle = 1. \tag{1.3-4}$$

Higher moments may or may not exist, depending upon how rapidly $P(x) \to 0$ as $|x| \to \infty$. For most random variables of physical interest, at least some of the higher moments will exist. If it happens that *all* the moments of X exist, then they will collectively determine the average with respect to X of any analytic (infinitely differentiable) function h; because, using Taylor's theorem, we have from Eqs. (1.3-2) and (1.3-3) that

$$\langle h(X) \rangle = \int_{-\infty}^{\infty} \left(\sum_{n=0}^{\infty} \frac{h^{(n)}(0)}{n!} x^n \right) P(x)\,dx = \sum_{n=0}^{\infty} \frac{h^{(n)}(0)}{n!} \langle X^n \rangle,$$

where $h^{(n)}(0)$ is the nth derivative of h evaluated at $x=0$. So in many cases (there are exceptions), a knowledge of all the moments of X is tantamount to a complete knowledge of the density function of X.

Notice that if X is not identically zero, so that $P(x)$ is other than $\delta(x)$ [see (1.2-7)], then the integral in the definition (1.3-3) is strictly positive when n is even; therefore,

$$X \text{ not} \equiv 0 \quad \Rightarrow \quad \langle X^{2n} \rangle > 0 \quad (n=1, 2, ...). \tag{1.3-5}$$

The **mean of X** is defined to be the first moment of X:

$$\text{mean}\{X\} \equiv \langle X \rangle = \int_{-\infty}^{\infty} x P(x)\,dx. \tag{1.3-6}$$

The **variance of X** is defined to be

$$\text{var}\{X\} \equiv \langle (X - \langle X \rangle)^2 \rangle = \int_{-\infty}^{\infty} (x - \langle X \rangle)^2 P(x)\,dx. \tag{1.3-7}$$

Since the integral on the right can be evaluated, using Eqs. (1.3-3) and (1.2-3), as

$$\int_{-\infty}^{\infty} \left(x^2 - 2\langle X \rangle x + \langle X \rangle^2 \right) P(x)\,dx = \langle X^2 \rangle - 2\langle X \rangle\langle X \rangle + \langle X \rangle^2 = \langle X^2 \rangle - \langle X \rangle^2,$$

then the variance of X can be expressed rather simply in terms of the first and second moments of X as

$$\text{var}\{X\} = \langle X^2 \rangle - \langle X \rangle^2. \tag{1.3-8}$$

The non-negativity of the integrand in Eq. (1.3-7) guarantees that $\text{var}\{X\} \geq 0$, so it follows from Eq. (1.3-8) that

$$\langle X^2 \rangle \geq \langle X \rangle^2, \tag{1.3-9}$$

with equality obtaining if and only if $\text{var}\{X\}=0$. The circumstance of a vanishing variance actually has a profound significance, in that

$$\text{var}\{X\} = 0 \quad \Leftrightarrow \quad X \text{ is the sure variable } \langle X \rangle. \tag{1.3-10}$$

To prove the forward implication here we observe that, since $(x-\langle X \rangle)^2$ is strictly positive everywhere except at $x=\langle X \rangle$ while $P(x)$ is never negative, then the only way for the integral in Eq. (1.3-7) to vanish is for $P(x)$ to be zero everywhere except possibly at $x=\langle X \rangle$; in that case, normalization would demand that $P(x)=\delta(x-\langle X \rangle)$, which by (1.2-7) would imply that X is the sure variable $\langle X \rangle$. To prove the reverse implication in (1.3-10) we observe from (1.2-7) that we have $P(x)=\delta(x-\langle X \rangle)$, and when that is substituted into Eq. (1.3-7) the x-integration gives, because of Eq. (1.2-5), $\text{var}\{X\}=0$. In light of (1.3-10), we can now see that equality obtains in (1.3-9) if and only if X is a sure variable.

The square root of the variance of X is called the **standard deviation of X**:

$$\text{sdev}\{X\} \equiv [\text{var}\{X\}]^{1/2} \equiv \langle (X - \langle X \rangle)^2 \rangle^{1/2} = [\langle X^2 \rangle - \langle X \rangle^2]^{1/2}. \tag{1.3-11}$$

Since $\text{sdev}\{X\}$ is by definition the square root of the average of the square of the difference between X and its mean $\langle X \rangle$, then $\text{sdev}\{X\}$ measures the size of the expected difference between the sample values of X and mean of X, or as we shall say, the size of the expected "dispersion of" or "fluctuations in" X about $\langle X \rangle$. For *most* random variables X, a sampling will yield a result that, more often than not, lies somewhere between $\langle X \rangle - \text{sdev}\{X\}$ and $\langle X \rangle + \text{sdev}\{X\}$.

Although a complete knowledge of any random variable X requires a knowledge of its density function P, the partial knowledge afforded by just the pair of values $\langle X \rangle$ and $\text{sdev}\{X\}$ often suffices for many practical purposes. The value of $\langle X \rangle$ is usually the best possible "sure number approximation" to the random variable X, while the value of $\text{sdev}\{X\}$, or more precisely the smallness thereof, is a measure of how adequate that approximation is. So the first two moments of X are often sufficient to characterize X for practical purposes, even though such a characterization is neither unique nor complete.

1.4 FOUR IMPORTANT RANDOM VARIABLES

In practical applications of random variable theory, four particular species of random variables occur so often that they merit special attention. These are the "uniform," the "exponential," the "normal," and the "Cauchy" random variables.

The **uniform random variable** is defined by the density function [see Fig. 1-2a]

$$P(x) = \begin{cases} 1/(b-a), & \text{if } a \leq x < b, \\ 0, & \text{if } x < a \text{ or } x \geq b, \end{cases} \tag{1.4-1}$$

where a and b are any real numbers satisfying $-\infty < a < b < \infty$. This function obviously satisfies the normalization condition (1.2-3) and the nonnegativity condition (1.2-4). We say that the random variable X defined by this density function is "uniformly distributed on $[a,b)$," and we write $X = U(a,b)$. If we insert the density function (1.4-1) into the moment definition (1.3-3) and then invoke the integral identity (A-1), we obtain

$$\langle X^n \rangle = \frac{b^{n+1} - a^{n+1}}{(n+1)(b-a)} = \frac{1}{n+1} \sum_{j=0}^{n} a^j b^{n-j}. \quad [X = U(a,b)] \tag{1.4-2}$$

Putting $n=1$ and 2 in Eq. (1.4-2) and then invoking Eqs. (1.3-6) and (1.3-11), we find that

$$\text{mean}\{X\} = \frac{a+b}{2}, \quad \text{sdev}\{X\} = \frac{b-a}{2\sqrt{3}}. \quad [X = U(a,b)] \tag{1.4-3}$$

If the parameter b is allowed to approach a from above, then it is clear from these last equations that $\text{mean}\{X\} \to a$ and $\text{var}\{X\} \to 0$; hence, we conclude from (1.3-10) that

$$\lim_{b \to a} U(a,b) = \text{the sure variable } a. \tag{1.4-4}$$

A popular but erroneous notion is that the uniform random variable is unbiased and therefore "truly random." The fallacy of this notion is seen by noting that the rectangularly shaped density function in Eq. (1.4-1) clearly favors values *inside* $[a,b)$ over values *outside* $[a,b)$; furthermore, any attempt to expand the interval $[a,b)$ to cover the entire x-axis while keeping $P(x)$ a positive constant would necessarily run afoul

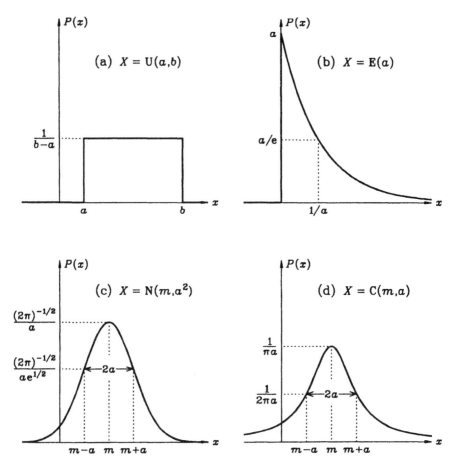

Figure 1-2. The density functions of four frequently encountered random variables. Plot (a) shows the density function (1.4-1) for X *uniformly* distributed on $[a,b)$, for which $\langle X \rangle = (a+b)/2$ and sdev$\{X\} = (b-a)/(12)^{1/2}$. Plot (b) shows the density function (1.4-5) for X *exponentially* distributed with decay constant a, for which $\langle X \rangle = $sdev$\{X\} = 1/a$. Plot (c) shows the density function (1.4-9) for X *normally* distributed with mean m and variance a^2, for which $\langle X \rangle = m$ and sdev$\{X\} = a$. And plot (d) shows the density function (1.4-13) for X *Cauchy* distributed about m with half-width a, for which $\langle X \rangle$ and sdev$\{X\}$ do not exist. It can be seen from the identically scaled plots (c) and (d) that the density function of $X = C(m,a)$ is slightly more sharply peaked, but very much heavier tailed, than the density function of $X = N(m,a^2)$.

of the normalization condition (1.2-3). To repeat an earlier observation, *all random variables are biased* in the sense that some sampling outcomes are more likely than others.

The **exponential random variable** is defined by the density function [see Fig. 1-2b]

$$P(x) = \begin{cases} a \exp(-ax), & \text{if } x \geq 0, \\ 0, & \text{if } x < 0, \end{cases} \qquad (1.4\text{-}5)$$

where a is any positive real number. This function is clearly nonnegative, and as we shall see shortly it satisfies the normalization condition. We say that the random variable X defined by this density function is "exponentially distributed with decay constant a," and we write $X = \mathbf{E}(a)$. If we insert the density function (1.4-5) into the moment definition (1.3-3) and then invoke the integral identity (A-2), we find that

$$\langle X^n \rangle = n! \, / \, a^n . \qquad [X = \mathbf{E}(a)] \qquad (1.4\text{-}6)$$

Putting $n = 0$ in Eq. (1.4-6) gives $\langle X^0 \rangle = 1$, showing that P does indeed satisfy the normalization condition. Putting $n = 1$ and 2 in Eq. (1.4-6) and then invoking Eqs. (1.3-6) and (1.3-11), we find that

$$\text{mean}\{X\} = \text{sdev}\{X\} = 1/a. \quad [X = \mathbf{E}(a)] \qquad (1.4\text{-}7)$$

If the parameter a is allowed to approach ∞, then it is clear from these last results that mean$\{X\} \to 0$ and var$\{X\} \to 0$; hence we conclude from (1.3-10) that

$$\lim_{a \to \infty} \mathbf{E}(a) = \text{the sure number } 0. \qquad (1.4\text{-}8)$$

The **normal random variable** (sometimes also called *Gaussian* random variable) is defined by the density function [see Fig. 1-2c]

$$P(x) = \frac{1}{(2\pi a^2)^{1/2}} \exp\left(-\frac{(x-m)^2}{2a^2} \right), \qquad (1.4\text{-}9)$$

where a and m are any real numbers satisfying $0 < a < \infty$ and $-\infty < m < \infty$. This function is clearly nonnegative, and as we shall see shortly it satisfies the normalization condition. We say that the random variable X defined by this density function is "normally distributed with mean m and variance a^2," and we write $X = \mathbf{N}(m, a^2)$. If we insert Eq. (1.4-9) into the moment definition (1.3-3), change the integration variable from x to $z = (x-m)/a\sqrt{2}$, and then use the binomial formula

$$(x+y)^n = \sum_{k=0}^{n} \frac{n!}{k!\,(n-k)!}\, x^{n-k}\, y^k,$$

we can evaluate the integral using the identity (A-3); in this way we find that

$$\langle X^n \rangle = n! \sum_{\substack{k=0 \\ (k\,even)}}^{n} \frac{m^{n-k}\,(a^2)^{k/2}}{(n-k)!\,(k/2)!\,2^{k/2}}. \quad [X=\mathrm{N}(m,a^2)] \quad (1.4\text{-}10)$$

Putting $n=0$ in Eq. (1.4-10) gives $\langle X^0 \rangle = 1$, showing that P does indeed satisfy the normalization condition. Putting $n=1$ and 2 in Eq. (1.4-10) and then invoking Eqs. (1.3-6) and (1.3-11), we find that

$$\mathrm{mean}\{X\} = m, \quad \mathrm{sdev}\{X\} = a, \quad [X=\mathrm{N}(m,a^2)] \quad (1.4\text{-}11)$$

a result that we anticipated earlier. If the parameter a is allowed to approach zero, then it is clear from Eqs. (1.4-11) that mean$\{X\}\to m$ and var$\{X\}\to 0$; hence we conclude from (1.3-10) that

$$\lim_{a\to 0} \mathrm{N}(m,a^2) = \text{the sure variable } m. \quad (1.4\text{-}12)$$

A popular but erroneous notion concerning the normal random variable is that it is the only random variable whose mean and variance uniquely parametrize its density function. In truth, the uniform, the exponential, and a good many other random variables also have this property. Another common but obviously incorrect notion is that any random variable that has a variance or a standard deviation is necessarily normally distributed.

Finally, we mention the **Cauchy** random variable, which is defined by the density function [see Fig. 1-2d]

$$P(x) = \frac{a/\pi}{(x-m)^2 + a^2}, \quad (1.4\text{-}13)$$

where a and m are any real numbers satisfying $0<a<\infty$ and $-\infty<m<\infty$. We say that the random variable X defined by this density function is "Cauchy distributed about m with half-width a," and we write $X=\mathrm{C}(m,a)$. As shown in Fig. 1-2d, the Cauchy density function has a single symmetric peak at $x=m$ whose height is $1/\pi a$ and whose width at half-maximum is $2a$. It is easy to see from the integral identity (A-5) that the density function (1.4-13) satisfies the normalization condition (1.2-3). However, the Cauchy density function is such that $\langle X^n \rangle$ is *undefined* for *all* $n \geq 1$. The reason is that the tails of the function $P(x)$ in Eq. (1.4-13) do not go to zero quickly enough to permit the moment integral in Eq. (1.3-3) to converge for any $n \geq 1$. Nor will the limit in the definition (1.3-1) converge for $h(X) = X^n$ for any $n \geq 1$ when X is $C(m,a)$. As was intimated earlier, it is *not* a mark of illegitimacy for a random variable to have no moments. (But of course, it's no great honor either.)

As can be seen from Figs. 1-2c and 1-2d, the Cauchy random variable $C(m,a)$ has some obvious points of similarity with the normal random variable $N(m,a^2)$. In particular, the density function of each of these two random variables has a peak at m of "width" $2a$. Furthermore, since the limit as $a \rightarrow 0$ of the function (1.4-13) is one of the many "representations" of the Dirac delta function $\delta(x - m)$, then it follows that

$$\lim_{a \rightarrow 0} C(m,a) = \text{the sure variable } m, \qquad (1.4\text{-}14)$$

in close analogy with the result (1.4-12) for $N(m,a^2)$. On the other hand, whereas $N(m,a^2)$ has *all* of its moments defined, $C(m,a)$ has *none* of its moments defined. We shall touch further on the relation between the normal and Cauchy random variables in Section 1.6.

The four random variables just defined by no means exhaust all those for which the density function has a simple, analytic, easily parametrized form. There are many other random variables (e.g., the log-normal, the chi-squared, the Student's-t, the gamma, the Rayleigh, etc.) that also have those convenient features, and which respectively play important roles in physical applications of random variable theory. But a knowledge of only the four random variables discussed above will suffice for the limited purposes of this book. And of the many distinct random variables subsumed under these four classes, two particular ones will be so frequently invoked that we shall give them special symbols, apart from the generic "X": We shall denote the uniform random variable $U(a,b)$ with $a = 0$ and $b = 1$ by the symbol U,

$$U \equiv U(0,1), \qquad (1.4\text{-}15)$$

and call it the **unit uniform** random variable. And we shall denote the normal random variable $N(m,a^2)$ with $m = 0$ and $a = 1$ by the symbol N,

$$N \equiv N(0,1), \qquad (1.4\text{-}16)$$

and call it the **unit normal** random variable. The density functions of the random variables U and N can easily be found by making the appropriate substitutions in Eqs. (1.4-1) and (1.4-9), respectively.

1.5 JOINT RANDOM VARIABLES

A set of n variables $X_1,...,X_n$ is said to be a set of **joint random variables** if and only if (i) all n variables can be sampled simultaneously, and (ii) there exists an n-variate function P such that $P(x_1,...,x_n)dx_1\cdots dx_n$ equals the probability that such a simultaneous sampling will find X_i inside the infinitesimal interval $[x_i,x_i+dx_i)$ for each $i=1$ to n. For simplicity we shall focus our discussion here on the case $n=3$, the extension of our results to other values of n being straightforward. Thus we consider three random variables, X_1, X_2 and X_3, for which in any simultaneous sampling[†]

$$P(x_1,x_2,x_3)\,dx_1 dx_2 dx_3 = \text{Prob}\{X_i \in [x_i,x_i+dx_i) \text{ for } i=1, 2 \text{ and } 3\}. \quad (1.5\text{-}1)$$

The function P is called the **joint density function** of X_1, X_2 and X_3, and Eq. (1.5-1) can be regarded as its definition. We say that X_1, X_2 and X_3 are "randomly distributed according to the joint density function P."

Implicit in the hypothesis that X_1, X_2 and X_3 can be simultaneously sampled is the *assumption* that individually sampling any *one* variable is equivalent to simultaneously sampling *all three* variables and then ignoring the values found for the other two. This assumption will be invoked repeatedly in our subsequent analysis.

By appealing to the addition law of probability in the same way as we did in deriving Eq. (1.2-2), we can show that

$$\int_{a_1}^{b_1} dx_1 \int_{a_2}^{b_2} dx_2 \int_{a_3}^{b_3} dx_3 \, P(x_1,x_2,x_3)$$

$$= \text{Prob}\{X_i \in [a_i,b_i) \text{ for } i=1, 2 \text{ and } 3\}. \quad (1.5\text{-}2)$$

In particular, since each of the random variables X_i is *certain* to be found in the interval $(-\infty,\infty)$, then P satisfies

$$\int_{-\infty}^{\infty} dx_1 \int_{-\infty}^{\infty} dx_2 \int_{-\infty}^{\infty} dx_3 \, P(x_1,x_2,x_3) = 1. \quad (1.5\text{-}3)$$

---

† Equation (1.5-1) is assumed to be valid only to first order in each of the positive infinitesimals dx_1, dx_2 and dx_3, just as Eq. (1.2-1c) is valid only to first order in dx. The omitted higher order correction terms will not affect the results of any subsequent calculations that we shall do here.

This equation, like Eq. (1.2-3), is called the "normalization condition." And since the infinitesimals dx_i in Eq. (1.5-1) are intrinsically positive, then the function P also satisfies [see Eq. (1.2-4)]

$$P(x_1, x_2, x_3) \geq 0. \qquad (1.5\text{-}4)$$

Indeed, *any* three-variate function P satisfying both of conditions (1.5-3) and (1.5-4) can be regarded as a joint density function, defining a set of joint random variables X_1, X_2 and X_3.

In analogy with the distribution function defined in Eq. (1.2-8), we can also define a **joint distribution function** for X_1, X_2 and X_3 by

$$F(x_1, x_2, x_3) \equiv \text{Prob}\{X_i < x_i \text{ for } i = 1, 2 \text{ and } 3\}, \qquad (1.5\text{-}5a)$$

$$= \int_{-\infty}^{x_1} dx_1' \int_{-\infty}^{x_2} dx_2' \int_{-\infty}^{x_3} dx_3' \, P(x_1', x_2', x_3'), \qquad (1.5\text{-}5b)$$

where the last equality follows from Eq. (1.5-2). However, the joint distribution function F does not turn out to be very useful in practice.

But the joint density function P spawns a number of "subordinate" density functions, and they turn out to be extremely useful. These subordinate density functions can be grouped into two classes as follows, wherein (i,j,k) denotes any permutation of $(1,2,3)$: The **marginal density functions** are defined by

$$P_i(x_i) \, dx_i \equiv \text{Prob}\{X_i \in [x_i, x_i + dx_i), \text{regardless of } X_j \text{ and } X_k\}, \qquad (1.5\text{-}6a)$$

$$P_{ij}(x_i, x_j) \, dx_i \, dx_j \equiv \text{Prob}\{X_i \in [x_i, x_i + dx_i) \text{ and}$$
$$X_j \in [x_j, x_j + dx_j), \text{regardless of } X_k\}. \qquad (1.5\text{-}6b)$$

And the **conditional density functions** are defined by

$$P_i^{(j)}(x_i \mid x_j) \, dx_i \equiv \text{Prob}\{X_i \in [x_i, x_i + dx_i), \text{given}$$
$$X_j = x_j, \text{regardless of } X_k\}, \qquad (1.5\text{-}7a)$$

$$P_i^{(j,k)}(x_i \mid x_j, x_k) \, dx_i \equiv \text{Prob}\{X_i \in [x_i, x_i + dx_i),$$
$$\text{given } X_j = x_j \text{ and } X_k = x_k\}, \qquad (1.5\text{-}7b)$$

$$P_{ij}^{(k)}(x_i, x_j \mid x_k) \, dx_i \, dx_j \equiv \text{Prob}\{X_i \in [x_i, x_i + dx_i) \text{ and}$$
$$X_j \in [x_j, x_j + dx_j), \text{given } X_k = x_k\}. \qquad (1.5\text{-}7c)$$

In connection with the conditional density functions, we note that the infinitesimal nature of dx_i is presumed to render the conditions "given $X_i \in [x_i, x_i + dx_i]$" and "given $X_i = x_i$" equivalent and interchangeable.

Comparing the definitions of the subordinate density functions with the definitions (1.2-1c) and (1.5-1), we see that all these density functions must be nonnegative and normalized, and that the specific normalization conditions are as follows:

$$\int_{-\infty}^{\infty} dx_i \, P_i(x_i) = \int_{-\infty}^{\infty} dx_i \, P_i^{(j)}(x_i \mid x_j) = \int_{-\infty}^{\infty} dx_i \, P_i^{(j,k)}(x_i \mid x_j, x_k) = 1,$$

$$\tag{1.5-8a}$$

$$\int_{-\infty}^{\infty} dx_i \int_{-\infty}^{\infty} dx_j \, P_{i,j}(x_i, x_j) = \int_{-\infty}^{\infty} dx_i \int_{-\infty}^{\infty} dx_j \, P_{i,j}^{(k)}(x_i, x_j \mid x_k) = 1.$$

$$\tag{1.5-8b}$$

The joint, marginal and conditional density functions introduced above are intimately interrelated. To elucidate the interrelations, we recall Eqs. (1.1-9) for the probabilities of three general outcomes 1, 2 and 3. If we identify outcome i with the finding "$X_i \in [x_i, x_i + dx_i]$," then in view of the definitions (1.5-6) and (1.5-7) we may write Eqs. (1.1-9) as follows:

$$P_{i,j}(x_i, x_j) \, dx_i dx_j = P_i(x_i) \, dx_i \cdot P_j^{(i)}(x_j \mid x_i) \, dx_j,$$

$$P(x_1, x_2, x_3) \, dx_1 dx_2 dx_3 = P_{i,j}(x_i, x_j) \, dx_i dx_j \cdot P_k^{(i,j)}(x_k \mid x_i, x_j) \, dx_k,$$

$$P(x_1, x_2, x_3) \, dx_1 dx_2 dx_3 = P_i(x_i) \, dx_i \cdot P_{j,k}^{(i)}(x_j, x_k \mid x_i) \, dx_j dx_k.$$

Upon canceling the differentials, we obtain the following equations, valid for (i,j,k) any permutation of $(1,2,3)$:

$$P_{i,j}(x_i, x_j) = P_i(x_i) P_j^{(i)}(x_j \mid x_i), \tag{1.5-9a}$$

$$P(x_1, x_2, x_3) = P_{i,j}(x_i, x_j) P_k^{(i,j)}(x_k \mid x_i, x_j), \tag{1.5-9b}$$

$$P(x_1, x_2, x_3) = P_i(x_i) P_{j,k}^{(i)}(x_j, x_k \mid x_i). \tag{1.5-9c}$$

These equations, when supplemented by the equation

$$P_{i,j}^{(k)}(x_i, x_j \mid x_k) = P_i^{(k)}(x_i \mid x_k) P_j^{(i,k)}(x_j \mid x_i, x_k), \tag{1.5-9d}$$

which follows by simply subjecting the arguments leading to the first equation to the *condition* $X_k = x_k$, embody all relevant interrelations among the joint, marginal and conditional density functions for three joint random variables.

For example, we can use Eqs. (1.5-9), along with the normalization conditions (1.5-8), to derive formulas for calculating every subordinate density function from the joint density function. Thus, by integrating Eq. (1.5-9c) over x_j and x_k and then using Eq. (1.5-8b), we deduce

$$P_i(x_i) = \int_{-\infty}^{\infty} dx_j \int_{-\infty}^{\infty} dx_k \, P(x_1, x_2, x_3). \qquad (1.5\text{-}10a)$$

By integrating Eq. (1.5-9b) over x_k and then using Eq. (1.5-8a), we deduce

$$P_{i,j}(x_i, x_j) = \int_{-\infty}^{\infty} dx_k \, P(x_1, x_2, x_3). \qquad (1.5\text{-}10b)$$

By dividing Eq. (1.5-9a) by $P_i(x_i)$ and then using the preceding two results, we deduce

$$P_j^{(i)}(x_j \mid x_i) = \frac{\int_{-\infty}^{\infty} dx_k \, P(x_1, x_2, x_3)}{\int_{-\infty}^{\infty} dx_j \int_{-\infty}^{\infty} dx_k \, P(x_1, x_2, x_3)}. \qquad (1.5\text{-}10c)$$

By dividing Eq. (1.5-9b) by $P_{i,j}(x_i, x_j)$ and then using Eq. (1.5-10b), we deduce

$$P_k^{(i,j)}(x_k \mid x_i, x_j) = \frac{P(x_1, x_2, x_3)}{\int_{-\infty}^{\infty} dx_k \, P(x_1, x_2, x_3)}. \qquad (1.5\text{-}10d)$$

And finally, by dividing Eq. (1.5-9c) by $P_i(x_i)$ and then using Eq. (1.5-10a), we deduce

$$P_{j,k}^{(i)}(x_j, x_k \mid x_i) = \frac{P(x_1, x_2, x_3)}{\int_{-\infty}^{\infty} dx_j \int_{-\infty}^{\infty} dx_k \, P(x_1, x_2, x_3)}. \qquad (1.5\text{-}10e)$$

As the first two equations above indicate, each *marginal* density function is obtained by integrating the joint density function over all the ignored variables. And as the last three equations indicate, each *conditional* density function is obtained by taking the ratio of two integrals of the joint density function, with the integral in the numerator being taken over all the ignored variables, and the integral in the denominator being taken over all variables *except* the conditioning variables.

It follows from Eqs. (1.5-10) that the joint density function uniquely determines all the marginal and conditional density functions. Conversely, Eqs. (1.5-9b) and (1.5-9c) show that *certain subsets* of the marginal and conditional density functions uniquely determine the joint density function. Eqs. (1.5-9b) and (1.5-9c) are sometimes referred to as *partial conditionings* of the joint density function P. By substituting Eq. (1.5-9a) into Eq. (1.5-9b), we obtain

$$P(x_1,x_2,x_3) = P_i(x_i) P_j^{(i)}(x_j | x_i) P_k^{(i,j)}(x_k | x_i,x_j). \qquad (1.5\text{-}11)$$

This important relation, which holds for (i,j,k) any permutation of $(1,2,3)$, is called a *full conditioning* of the joint density function P. Similarly, Eq. (1.5-9a) can be regarded as a full conditioning of $P_{i,j}$, and Eq. (1.5-9d) can be regarded as a full conditioning of $P_{i,j}^{(k)}$.

There also exists a variety of *composition formulas* that express each subordinate density function as an integral involving other subordinate density functions. Thus, for example, by integrating Eq. (1.5-9d) over x_j and then invoking the normalization condition (1.5-8a), we get

$$P_i^{(k)}(x_i | x_k) = \int_{-\infty}^{\infty} dx_j\, P_{i,j}^{(k)}(x_i,x_j | x_k). \qquad (1.5\text{-}12a)$$

By noting from the index permutability of Eq. (1.5-9a) that

$$P_i(x_i) P_j^{(i)}(x_j | x_i) = P_j(x_j) P_i^{(j)}(x_i | x_j),$$

and then integrating over x_j using the normalization condition (1.5-8a), we get

$$P_i(x_i) = \int_{-\infty}^{\infty} dx_j\, P_j(x_j) P_i^{(j)}(x_i | x_j), \qquad (1.5\text{-}12b)$$

By integrating Eq. (1.5-9b) over both x_i and x_j and then using Eq. (1.5-10a), we get

$$P_k(x_k) = \int_{-\infty}^{\infty} dx_i \int_{-\infty}^{\infty} dx_j\, P_{i,j}(x_i,x_j) P_k^{(i,j)}(x_k | x_i,x_j). \qquad (1.5\text{-}12c)$$

And by integrating Eq. (1.5-11) over x_i and then using Eq. (1.5-10b), we get

$$P_{j,k}(x_j,x_k) = \int_{-\infty}^{\infty} dx_i\, P_i(x_i) P_j^{(i)}(x_j | x_i) P_k^{(i,j)}(x_k | x_i,x_j). \qquad (1.5\text{-}12d)$$

Of course, all of these composition formulas (1.5-12) hold for (i,j,k) any permutation of $(1,2,3)$.

In analogy with the definition of statistically independent *outcomes* in Eq. (1.1-11a), we make the following definition of statistically independent *random variables*:

X_1, X_2, X_3 are **statistically independent**

$$\Leftrightarrow P_i(x_i) = P_i^{(j)}(x_i \mid x_j) = P_i^{(j,k)}(x_i \mid x_j, x_k), \quad (1.5\text{-}13a)$$

where the equations are understood to hold for (i,j,k) *every* permutation of (1,2,3). We can easily prove that a necessary and sufficient condition for X_1, X_2 and X_3 to be statistically independent is that their joint density function be always equal to the product of their univariate marginal density functions; i.e.,

X_1, X_2, X_3 are statistically independent

$$\Leftrightarrow P(x_1, x_2, x_3) = P_1(x_1) P_2(x_2) P_3(x_3). \quad (1.5\text{-}13b)$$

That Eqs. (1.5-13a) imply Eq. (1.5-13b) follows immediately from the full conditioning formula (1.5-11). And that Eq. (1.5-13b) implies Eqs. (1.5-13a) follows by substituting the former into Eqs. (1.5-10c) and (1.5-10d), and then invoking the normalization of P_i. If we compare Eq. (1.5-13b) with the equations in (1.1-11b) that establish the statistical independence of any three "outcomes," we see that the statistical independence of three random variables is somewhat easier to ensure: For the three random variables we need to establish only the *one* equation in (1.5-13b), but for any three outcomes we need to establish *all four* of equations (1.1-11b). The reason for this difference is that, whereas the function $P(x_1, x_2, x_3)$ completely characterizes the statistics of the three random variables X_1, X_2, X_3, the probability $p(1 \wedge 2 \wedge 3)$ does *not* completely characterize the statistics of the three outcomes 1,2,3. For example, the function $P(x_1, x_2, x_3)$ completely determines the function $P_1(x_1)$ through Eq. (1.5-10a), but the single probability $p(1 \wedge 2 \wedge 3)$ does not in general determine the probability $p(1)$.

The **average** of any three-variate function h with respect to the set of random variables X_1, X_2, X_3 is defined, in direct analogy with Eq. (1.3-1), to be

$$\langle h(X_1, X_2, X_3) \rangle \equiv \lim_{N \to \infty} \frac{1}{N} \sum_{n=1}^{N} h(x_1^{(n)}, x_2^{(n)}, x_3^{(n)}), \quad (1.5\text{-}14)$$

where $x_j^{(n)}$ is the value found for X_j in the nth simultaneous sampling of the three random variables. Just as we deduced Eq. (1.3-2) from Eq.

(1.3-1), so can we deduce from Eq. (1.5-14) the analytically more convenient formula

$$\langle h(X_1,X_2,X_3)\rangle = \int_{-\infty}^{\infty} dx_1 \int_{-\infty}^{\infty} dx_2 \int_{-\infty}^{\infty} dx_3\, h(x_1,x_2,x_3)\, P(x_1,x_2,x_3). \quad (1.5\text{-}15)$$

If h happens to be independent of X_3, then Eq. (1.5-15) reads

$$\langle h(X_1,X_2)\rangle = \int_{-\infty}^{\infty} dx_1 \int_{-\infty}^{\infty} dx_2\, h(x_1,x_2)\left[\int_{-\infty}^{\infty} dx_3\, P(x_1,x_2,x_3)\right],$$

and the integral in brackets can be replaced with $P_{1,2}(x_1,x_2)$ in accordance with Eq. (1.5-10b). More generally, we have for (i,j,k) any permutation of $(1,2,3)$,

$$\langle h(X_i,X_j)\rangle = \int_{-\infty}^{\infty} dx_i \int_{-\infty}^{\infty} dx_j\, h(x_i,x_j)\, P_{i,j}(x_i,x_j), \quad (1.5\text{-}16a)$$

$$\langle h(X_i)\rangle = \int_{-\infty}^{\infty} dx_i\, h(x_i)\, P_i(x_i). \quad (1.5\text{-}16b)$$

Notice that Eq. (1.5-16b) has the same form as Eq. (1.3-2). This means that $\langle X_i^n\rangle$, mean$\{X_i\}$, var$\{X_i\}$ and sdev$\{X_i\}$ are all as defined in Section 1.3, except that P is everywhere replaced by P_i.

In addition to mean$\{X_i\}$ and var$\{X_i\}$, another useful quantity pertaining to sets of joint random variables is the **covariance** of X_i and X_j. It is defined by

$$\text{cov}\{X_i,X_j\} = \langle\, (X_i-\langle X_i\rangle)\,(X_j-\langle X_j\rangle)\,\rangle, \quad (1.5\text{-}17a)$$

$$= \langle X_i X_j\rangle - \langle X_i\rangle\langle X_j\rangle. \quad (1.5\text{-}17b)$$

The equivalence of these two expressions for cov$\{X_i,X_j\}$ can be easily proved using Eqs. (1.5-16). We shall examine the significance of the covariance in the next section. For now it is sufficient to note, by comparing Eq. (1.5-17b) with Eqs. (1.3-8), that

$$\text{cov}\{X_i,X_i\} = \text{var}\{X_i\}. \quad (1.5\text{-}18)$$

Finally, we note that it is possible to define a variety of **conditioned averages**, such as for example

$$\langle h(X_i,X_j)\,|\,X_j=x_j\rangle \equiv \int_{-\infty}^{\infty} dx_i\, h(x_i,x_j)\, P_i^{(j)}(x_i\,|\,x_j). \quad (1.5\text{-}19)$$

From the "sampling" point of view of Eq. (1.5-14), this conditioned average is calculated by ignoring all samplings in which the value of X_j is not found to be equal (or infinitesimally close) to x_j. Conditioned averages lead, in an obvious way, to conditioned means, conditioned variances, and conditioned covariances.

1.6 SOME USEFUL THEOREMS

We saw in Section 1.2 that a random variable X is generally defined by specifying its density function P, which characterizes X according to Eqs. (1.2-1). Another way to define a random variable X would be to specify some mathematical procedure by which the sample values of X are to be produced. Both of these approaches allow one to calculate averages with respect to X, the first approach utilizing Eq. (1.3-2), and the second approach utilizing Eq. (1.3-1).

An especially useful application of the second method of defining a random variable is the following: Given some random variable X_1 with density function P_1, let the random variable X_2 be *defined* by the rule that its sample value $x_2^{(i)}$ is produced by first producing a sample value $x_1^{(i)}$ of X_1 and then taking $x_2^{(i)} = f(x_1^{(i)})$, where f is some given (ordinary) function. We shall call the random variable X_2 thus defined "$f(X_1)$," and we shall write

$$X_2 = f(X_1). \tag{1.6-1a}$$

This in fact is what we shall *mean* by a "function of a random variable." But what is the density function P_2 of the random variable X_2 thus defined? To find out, let X_1 and X_2 be regarded as *joint* random variables. Then the composition formula (1.5-12b) gives

$$P_2(x_2) = \int_{-\infty}^{\infty} dx_1 \, P_1(x_1) \, P_2^{(1)}(x_2 \mid x_1).$$

Now, if we are *given* that X_1 has the value x_1, then the definition (1.6-1a) implies that X_2 will surely have the value $f(x_1)$; hence, by (1.2-7), it follows that

$$P_2^{(1)}(x_2 \mid x_1) = \delta(x_2 - f(x_1)).$$

Substituting this into the previous equation, we conclude that if the random variable X_1 has density function P_1 then the density function of the random variable $X_2 = f(X_1)$ is

$$P_2(x_2) = \int_{-\infty}^{\infty} dx_1 \, P_1(x_1) \, \delta(x_2 - f(x_1)). \tag{1.6-1b}$$

This equation evidently provides the desired bridge between the two alternate methods described above for specifying a random variable.

To see how Eqs. (1.6-1) generalize to higher numbers of random variables, let's consider the case of three random variables X_1, X_2 and X_3. First let's suppose that X_3 is defined functionally in terms of X_1 and X_2 by

$$X_3 = f(X_1, X_2), \tag{1.6-2a}$$

with X_1 and X_2 in turn being defined through their joint density function $P_{1,2}$. We know from the composition formula (1.5-12c) that the density function of X_3 is given by

$$P_3(x_3) = \int_{-\infty}^{\infty} dx_1 \int_{-\infty}^{\infty} dx_2 P_{1,2}(x_1, x_2) P_3^{(1,2)}(x_3 \mid x_1, x_2).$$

Now, if we are *given* that $X_1 = x_1$ and $X_2 = x_2$, then the random variable X_3 is by definition the *sure* variable $f(x_1, x_2)$; therefore, by the result (1.2-7), the second factor in the integrand here is $\delta(x_3 - f(x_1, x_2))$, and so we conclude that the density function of the random variable X_3 defined in (1.6-2a) is

$$P_3(x_3) = \int_{-\infty}^{\infty} dx_1 \int_{-\infty}^{\infty} dx_2 P_{1,2}(x_1, x_2) \delta(x_3 - f(x_1, x_2)). \tag{1.6-2b}$$

Alternatively, suppose that X_1 and X_2 are both defined functionally in terms of X_3 by

$$X_1 = f_1(X_3), \quad X_2 = f_2(X_3), \tag{1.6-3a}$$

with X_3 in turn being defined through its density function P_3. We know from the composition formula (1.5-12d) that the joint density function of the two random variables X_1 and X_2 is given by

$$P_{1,2}(x_1, x_2) = \int_{-\infty}^{\infty} dx_3 P_3(x_3) P_1^{(3)}(x_1 \mid x_3) P_2^{(3,1)}(x_2 \mid x_3, x_1).$$

Now, if we are *given* that $X_3 = x_3$, then the random variables X_1 and X_2 are by definition the *sure* variables $f_1(x_3)$ and $f_2(x_3)$ respectively; therefore, the last two factors in the integrand here are respectively $\delta(x_1 - f_1(x_3))$ and $\delta(x_2 - f_2(x_3))$, and so we conclude that the joint density function of the random variables X_1 and X_2 defined in Eqs. (1.6-3a) is

$$P_{1,2}(x_1, x_2) = \int_{-\infty}^{\infty} dx_3 P_3(x_3) \delta(x_1 - f_1(x_3)) \delta(x_2 - f_2(x_3)). \tag{1.6-3b}$$

By straightforwardly extending these arguments to more than three variables, we arrive at the following general theorem:

Random Variable Transformation (RVT) Theorem. If the n random variables X_1, ..., X_n have the joint density function $P(x_1,...,x_n)$, and if the m random variables Y_1, ..., Y_m are defined by $Y_i = f_i(X_1,...,X_n)$ [$i=1$ to m], then the joint density function $Q(y_1,...,y_m)$ of Y_1, ..., Y_m is given by

$$Q(y_1,...,y_m) = \int_{-\infty}^{\infty} dx_1 \cdots \int_{-\infty}^{\infty} dx_n \, P(x_1,...,x_n) \prod_{i=1}^{m} \delta(y_i - f_i(x_1,...,x_n)).$$

$$(1.6\text{-}4)$$

A useful special case of the RVT theorem is the following result:

RVT Corollary. If $m=n$, and if further the transformation $y_i = f_i(x_1,...,x_n)$ [$i=1$ to n] is one-to-one and smooth, so that there exists a well-behaved inverse transformation $x_i = g_i(y_1,...,y_n)$ [$i=1$ to n], then the joint density function of the random variables $Y_i = f_i(X_1,...,X_n)$ [$i=1$ to n] is

$$Q(y_1,...,y_n) = P(x_1,...,x_n) \left| \frac{\partial(x_1,...,x_n)}{\partial(y_1,...,y_n)} \right|, \qquad (1.6\text{-}5)$$

where x_i on the right is understood to be the *function* $g_i(y_1,...,y_n)$.

Proof. First change the integration variables in Eq. (1.6-4) from $\{x_i\}$ to $\{z_i \equiv f_i(x_1,...,x_n)\}$. This evidently entails putting $x_i = g_i(z_1,...,z_n)$ in the arguments of P, putting $f_i(x_1,...,x_n) = z_i$ in the arguments of the delta functions, and of course putting

$$dx_1 \cdots dx_n = \left| \frac{\partial(x_1,...,x_n)}{\partial(z_1,...,z_n)} \right| dz_1 \cdots dz_n.$$

Since the delta functions in the thusly transformed integral (1.6-4) will have the form $\delta(y_i - z_i)$, then the z_i-integrations can be easily performed by applying the rule (1.2-5). The delta functions are thereby eliminated and z_i is everywhere replaced by y_i, giving the result (1.6-5).   QED

As much of the remainder of this section will demonstrate, many important results can be derived by using the RVT theorem and its corollary. Perhaps the simplest such result is the following theorem:

Linear Transformation Theorem. If the random variable X has density function $P(x)$, and if β and λ are any two constants, then the random variable $Y = \beta X + \lambda$ has density function

$$Q(y) = P(\beta^{-1}(y - \lambda)) \, |\beta|^{-1}. \qquad (1.6\text{-}6)$$

Proof. It follows from the RVT corollary that the density function of Y is $Q(y) = P(x)|dx/dy|$, where x is the inverse of $y = \beta x + \lambda$. Solving the latter equation for x gives $x = \beta^{-1}(y - \lambda)$, and hence $|dx/dy| = |\beta|^{-1}$. Substituting these results into the previous expression for $Q(y)$ gives Eq. (1.6-6).   QED

A useful application of the linear transformation theorem is for the case in which $X = N(m, a^2)$. Substituting the normal density function (1.4-9) into the right hand side of Eq. (1.6-6) yields

$$Q(y) = |\beta|^{-1}(2\pi a^2)^{-1/2} \exp\{-[\beta^{-1}(y - \lambda) - m]^2/2a^2\},$$
$$= (2\pi \beta^2 a^2)^{-1/2} \exp\{-[y - (\beta m + \lambda)]^2/2\beta^2 a^2\}.$$

Thus we see, by comparing with Eq. (1.4-9), that

$$X = N(m, a^2) \implies \beta X + \lambda = N(\beta m + \lambda, \beta^2 a^2). \tag{1.6-7}$$

This result will be invoked several times in our later work.

We shall now use the $m = 1, n = 2$ version of the RVT theorem (1.6-4) to establish some interesting facts concerning the normal and Cauchy random variables. Consider any two *statistically independent* random variables X_1 and X_2 with respective density functions $P_1(x_1)$ and $P_2(x_2)$. Since X_1 and X_2 are statistically independent, we have from theorem (1.5-13b) that their joint density function $P(x_1, x_2) = P_1(x_1)P_2(x_2)$. It then follows from the RVT theorem that, for any bivariate function f, the random variable $Y = f(X_1, X_2)$ will have the density function

$$Q(y) = \int_{-\infty}^{\infty} dx_1 \int_{-\infty}^{\infty} dx_2\, P_1(x_1)\, P_2(x_2)\, \delta(y - f(x_1, x_2)). \tag{1.6-8}$$

Using this general result, we now establish the following three theorems:

Normal Sum Theorem. If $X_1 = N(m_1, a_1^2)$ and $X_2 = N(m_2, a_2^2)$, with X_1 and X_2 statistically independent of each other, then $Y = X_1 + X_2$ is the random variable $N(m_1 + m_2, a_1^2 + a_2^2)$.

Proof (outlined). In Eq. (1.6-8) put $f(x_1, x_2) = x_1 + x_2$, and also put, in accordance with Eq. (1.4-9),

$$P_j(x_j) = (2\pi a_j^2)^{-1/2} \exp[-(x_j - m_j)^2/2a_j^2].$$

Then write the Dirac delta function in its so-called *integral representation* [see Appendix B, Eqs. (B-6) and (B-7)]:

$$\delta(y - x_1 - x_2) = (2\pi)^{-1} \oint_{-\infty}^{\infty} du \, \exp[iu(y - x_1 - x_2)]. \tag{1.6-9}$$

All that done, change the integration variables (x_1, x_2) to (z_1, z_2), where $z_j \equiv (x_j - m_j)/2a_j$, and then integrate over z_1, z_2 and u in that order with the help of the integral identity (A-6). Note that the final u-integral converges as an ordinary integral, so that, as discussed in Appendix B, its special nature can be ignored. The result of the three integrations is

$$Q(y) = [2\pi(a_1{}^2 + a_2{}^2)]^{-1/2} \exp\{-[y - (m_1 + m_2)]^2/2(a_1{}^2 + a_2{}^2)\},$$

and this, by Eq. (1.4-9), proves the theorem. QED

Cauchy Sum Theorem. If $X_1 = C(m_1, a_1)$ and $X_2 = C(m_2, a_2)$, with X_1 and X_2 statistically independent of each other, then $Y = X_1 + X_2$ is the random variable $C(m_1 + m_2, a_1 + a_2)$.

Proof (outlined). In Eq. (1.6-8) put $f(x_1, x_2) = x_1 + x_2$, and also put, in accordance with Eq. (1.4-13),

$$P_j(x_j) = (a_j/\pi)[(x_j - m_j)^2 + a_j{}^2]^{-1}.$$

Then write the Dirac delta function in its integral representation (1.6-9). All that done, change the integration variables (x_1, x_2) to (z_1, z_2), where $z_j = (x_j - m_j)/a_j$, and then integrate over z_1 and z_2 with the help of the integral identity (A-8). The remaining u-integral converges as an ordinary integral, and so is evaluated by applying the integral identity (A-9). The result is

$$Q(y) = [(a_1 + a_2)/\pi]\{[y - (m_1 + m_2)]^2 + (a_1 + a_2)^2\}^{-1},$$

and this, by Eq. (1.4-13), proves the theorem. QED

Comparing these two "sum" theorems, we see additional evidence of the similarity between the random variables $N(m, a^2)$ and $C(m, a)$, which was noted earlier in connection with Figs. 1-2c and 1-2d. Both of these random variables preserve their "class" under statistically independent summation, a feature that is *not* shared by either the uniform or the exponential random variables. Notice that, when summing independent normal random variables, it is their *variances* that add and not their standard deviations. We shall see shortly that the variance of the sum of *any* two *statistically independent* random variables is equal to the sum of their variances.

The fact that the variance of the normal random variable $N(0,1)$ is finite while the variance of the Cauchy random variable $C(0,1)$ is infinite suggests that large sample values of the random variable $C(0,1)$ should be much more commonplace that large sample values of the random variable $N(0,1)$. Our next theorem helps us to appreciate why this is so.

Normal Ratio Theorem. If X_1 and X_2 are both $N(0,1)$ and statistically independent of each other, then $Y = X_1/X_2$ is the random variable $C(0,1)$.

Proof (outlined). In Eq. (1.6-8), put $f(x_1,x_2) = x_1/x_2$, and also put, in accordance with Eq. (1.4-9),

$$P_j(x_j) = (2\pi)^{-1/2} \exp[-x_j^2/2].$$

The change of variable $x_1 \to w \equiv x_1/x_2$, with $dx_1 = |x_2|dw$, allows the delta function to be easily eliminated by integrating over w. Then the change of variable $x_2 \to u \equiv x_2^2$ brings the remaining integral into the form

$$Q(y) = \frac{1}{2\pi} \int_0^\infty du \, \exp\left[-\frac{1}{2}\left(1 + y^2\right)u\right] = \frac{1/\pi}{y^2 + 1},$$

where the last equality follows from the integral identity (A-2). This result, by Eq. (1.4-13), proves the theorem.   QED

A curious but, as we shall see later, very useful consequence of the RVT corollary is the

Distribution Function Theorem. If the random variable X has distribution function $F(x)$, then

$$F(X) = U. \tag{1.6-10}$$

where U is the unit uniform random variable, $U(0,1)$.

Proof. Since the distribution function F rises monotonically from 0 at $x = -\infty$ to 1 at $x = +\infty$, then the transformation $y = F(x)$ maps the interval $-\infty < x < \infty$ onto the interval $0 < y < 1$; this means that the random variable $Y \equiv F(X)$ can have sample values only inside the unit interval. The density function $Q(y)$ of Y must therefore vanish for $y \notin [0,1]$. For $y \in [0,1]$, the RVT corollary gives

$$Q(y) = P(x) \left| \frac{dx}{dy} \right| = \frac{P(x)}{|dy/dx|} = \frac{P(x)}{F'(x)} = 1,$$

where the last equality follows from the fundamental relation (1.2-9b) between the density function and the distribution function. Thus, the density function of Y is 0 outside the unit interval and 1 inside the unit interval, which means that Y must be the random variable $U(0,1)$.   QED

In proving the distribution function theorem by invoking the RVT corollary, we implicitly assumed that the transformation $y = F(x)$ is *invertible* everywhere that $P(x) > 0$. This property is guaranteed by the relation $F''(x) = P(x)$, which implies that $F(x)$ will be strictly increasing for such x values. So the distribution function theorem is essentially the statement that, if F is any differentiable function of x that increases monotonically from 0 at $x = -\infty$ to 1 at $x = +\infty$, then

$$F^{-1}(U) \text{ is the random variable with density function } F'', \quad (1.6\text{-}11)$$

where U is the unit uniform random variable. This fact, as we shall see later in this chapter, is the practical springboard of the so-called "Monte Carlo" computation method.

If X_1, X_2 and X_3 have joint density function P, and if for any three-variate function h we define the random variable Y to be $h(X_1,X_2,X_3)$, then the RVT theorem implies that the density function of Y is

$$Q(y) = \int_{-\infty}^{\infty} dx_1 \int_{-\infty}^{\infty} dx_2 \int_{-\infty}^{\infty} dx_3 \, P(x_1,x_2,x_3) \, \delta(y - h(x_1,x_2,x_3)).$$

It follows that the mean of the random variable Y is

$$\langle Y \rangle = \int_{-\infty}^{\infty} dy \, y \, Q(y),$$

$$= \int_{-\infty}^{\infty} dy \int_{-\infty}^{\infty} dx_1 \int_{-\infty}^{\infty} dx_2 \int_{-\infty}^{\infty} dx_3 \, y \, P(x_1,x_2,x_3) \, \delta(y - h(x_1,x_2,x_3)).$$

Integrating over y with the help of the delta function property (1.2-5), and recalling that Y on the left side here is by definition the random variable $h(X_1,X_2,X_3)$, we conclude that

$$\langle h(X_1,X_2,X_3) \rangle = \int_{-\infty}^{\infty} dx_1 \int_{-\infty}^{\infty} dx_2 \int_{-\infty}^{\infty} dx_3 \, h(x_1,x_2,x_3) \, P(x_1,x_2,x_3). \quad (1.6\text{-}12)$$

Although Eq. (1.6-12) is identical in form to Eq. (1.5-15), the left hand sides of those two equations in fact represent formally different entities: The left hand side of Eq. (1.6-12) is the mean of the random variable $h(X_1,X_2,X_3)$, while the left hand side of Eq. (1.5-15) is the average of the function h with respect to the random variables X_1, X_2 and X_3. The agreement between Eqs. (1.6-12) and (1.5-15) signifies that {the mean of the random variable $h(X_1,X_2,X_3)$} *equals* {the average of the function h with respect to the random variables X_1, X_2 and X_3}. The use of a common symbol for those two quantities is therefore justified. For example, the nth moment of the random variable X, defined in Eq. (1.3-3), may also be regarded as the mean of the random variable $Y = X^n$. And the average $\langle X_i X_j \rangle$ that occurs in the definition (1.5-17b) of the covariance of two random variables X_i and X_j may also be regarded as the mean of the random variable $Y = X_i X_j$.

A convenient and often used property of the bracket notation for the mean or average is its *linearity property*: For any two constants β and λ, and any two functions h and g, we have

$$\langle \beta h(X_1,X_2,X_3) + \lambda g(X_1,X_2,X_3) \rangle$$
$$= \beta \langle h(X_1,X_2,X_3) \rangle + \lambda \langle g(X_1,X_2,X_3) \rangle, \qquad (1.6\text{-}13)$$

as may easily be proved from Eq. (1.6-12). By using this property of the bracket notation, we can greatly simplify many discussions and calculations in random variable theory. For example, consider the proof of the following useful theorem:

Linear Combination Theorem. For any set of random variables $X_1, ..., X_n$ and any set of constants $a_1, ..., a_n$, we have

$$\text{mean}\left\{ \sum_{i=1}^{n} a_i X_i \right\} = \sum_{i=1}^{n} a_i \,\text{mean}\{X_i\}, \qquad (1.6\text{-}14)$$

$$\text{var}\left\{ \sum_{i=1}^{n} a_i X_i \right\} = \sum_{i=1}^{n} a_i^2 \,\text{var}\{X_i\} + 2 \sum_{i=1}^{n-1} \sum_{j=i+1}^{n} a_i a_j \,\text{cov}\{X_i,X_j\}. \quad (1.6\text{-}15)$$

Proof. Eq. (1.6-14) is just $\langle \Sigma_i a_i X_i \rangle = \Sigma_i a_i \langle X_i \rangle$, which is an immediate consequence of the linearity property (1.6-13). To prove Eq. (1.6-15), we use Eqs. (1.3-8) and (1.6-13) to write

$\text{var}\{\Sigma_i a_i X_i\}$

$= \langle (\Sigma_i a_i X_i)^2 \rangle - (\langle \Sigma_i a_i X_i \rangle)^2 = \langle \Sigma_i \Sigma_j a_i a_j X_i X_j \rangle - (\Sigma_i a_i \langle X_i \rangle)^2$

$= \Sigma_i \Sigma_j a_i a_j \langle X_i X_j \rangle - \Sigma_i \Sigma_j a_i a_j \langle X_i \rangle \langle X_j \rangle$

$= \Sigma_i a_i^2 \langle X_i^2 \rangle + \Sigma\Sigma_{i \neq j} a_i a_j \langle X_i X_j \rangle - \Sigma_i a_i^2 \langle X_i \rangle^2 - \Sigma\Sigma_{i \neq j} a_i a_j \langle X_i \rangle \langle X_j \rangle$

$= \Sigma_i a_i^2 [\langle X_i^2 \rangle - \langle X_i \rangle^2] + \Sigma\Sigma_{i \neq j} a_i a_j [\langle X_i X_j \rangle - \langle X_i \rangle \langle X_j \rangle].$

With Eqs. (1.3-8) and (1.5-17b), this last equation evidently reduces to Eq. (1.6-15).   QED

In particular, the linear combination theorem implies that

$$\text{mean}\{aX\} = a \,\text{mean}\{X\}, \qquad (1.6\text{-}16a)$$

$$\text{var}\{aX\} = a^2 \,\text{var}\{X\}, \qquad (1.6\text{-}16b)$$

and also that

$$\text{mean}\{X_1 + X_2\} = \text{mean}\{X_1\} + \text{mean}\{X_2\}, \qquad (1.6\text{-}17a)$$

$$\text{var}\{X_1 + X_2\} = \text{var}\{X_1\} + \text{var}\{X_2\} + 2\,\text{cov}\{X_1,X_2\}. \qquad (1.6\text{-}17b)$$

The result (1.6-17b), that var$\{X_1+X_2\}$ differs from var$\{X_1\}$+var$\{X_2\}$ by twice cov$\{X_1,X_2\}$, motivates us to consider the covariance in more detail. For that, it is useful to begin by proving the following theorem:

Covariance Range Theorem. For any two random variables X_1 and X_2, we have

$$|\text{cov}\{X_1,X_2\}| \leq \text{sdev}\{X_1\} \cdot \text{sdev}\{X_2\}. \tag{1.6-18}$$

Proof. Consider the random variable $a_1X_1 - a_2X_2$, where a_1 and a_2 are any constants. Since the variance of this random variable, like the variance of any random variable, must be nonnegative, then Eq. (1.6-15) gives

$$0 \leq \text{var}\{a_1X_1 - a_2X_2\} = a_1{}^2\,\text{var}\{X_1\} + a_2{}^2\,\text{var}\{X_2\} - 2a_1a_2\,\text{cov}\{X_1,X_2\}.$$

Now taking $a_1 = \text{cov}\{X_1,X_2\}$ and $a_2 = \text{var}\{X_1\}$, and then dividing through by var$\{X_1\}$ [the case var$\{X_1\}$=0 is rather uninteresting since it implies that both sides of (1.6-18) vanish], we get

$$0 \leq \text{var}\{X_1\}\,\text{var}\{X_2\} - \text{cov}^2\{X_1,X_2\}.$$

Rearranging, and recalling that the standard deviation is just the square root of the variance, we obtain Eq. (1.6-18).   QED

The **correlation coefficient** of X_1 and X_2 is defined by

$$\text{corr}\{X_1,X_2\} \equiv \frac{\text{cov}\{X_1,X_2\}}{\text{sdev}\{X_1\} \cdot \text{sdev}\{X_2\}}. \tag{1.6-19}$$

This definition is motivated by the covariance range theorem (1.6-18), which evidently implies that

$$-1 \leq \text{corr}\{X_1,X_2\} \leq +1. \tag{1.6-20}$$

Note that it is possible for either equality in Eq. (1.6-20) to hold. Thus, if $X_2 = X_1$ we have

$$\text{cov}\{X_1,X_2\} = \text{cov}\{X_1,X_1\} = \langle X_1X_1\rangle - \langle X_1\rangle\langle X_1\rangle$$
$$= \text{var}\{X_1\} = (\text{sdev}\{X_1\})^2,$$

and the right hand side of the definition (1.6-19) reduces to +1; we say that X_1 is "maximally correlated" with with itself. On the other hand, if $X_2 = -X_1$ we have

$$\text{cov}\{X_1,X_2\} = \text{cov}\{X_1,-X_1\} = \langle X_1(-X_1)\rangle - \langle X_1\rangle\langle -X_1\rangle$$
$$= -\text{var}\{X_1\} = -(\text{sdev}\{X_1\})^2,$$

and the right hand side of the definition (1.6-19) reduces to -1; we say that X_1 and $-X_1$ are "maximally anti-correlated."

We say that two random variables X_1 and X_2 are "uncorrelated" if and only if $\text{cov}\{X_1,X_2\}=0$. More generally, we say for any *set* of random variables $X_1, ..., X_n$ that

$X_1, ..., X_n$ are **uncorrelated**

$$\Leftrightarrow \ \text{cov}\{X_i,X_j\} = 0 \ \text{ for all } 1 \leq i < j \leq n. \quad (1.6\text{-}21)$$

For a set of uncorrelated random variables, all the covariance terms in the variance sum formula (1.6-15) vanish, thereby making that formula more symmetric with the mean sum formula (1.6-14). In other words, for any set of *uncorrelated* random variables, not only is the mean of the sum equal to the sum of the means, but the variance of the sum is equal to the sum of the variances.

A sufficient (but not necessary) condition for $X_1, ..., X_n$ to be uncorrelated is that they be statistically independent. To prove this, we simply observe that for any pair X_i and X_j from a statistically independent set we have

$$\langle X_i X_j \rangle = \int_{-\infty}^{\infty} dx_i \int_{-\infty}^{\infty} dx_j \, (x_i x_j) \, P_{i,j}(x_i,x_j) \qquad \text{[by (1.5-16a)]}$$

$$= \int_{-\infty}^{\infty} dx_i \int_{-\infty}^{\infty} dx_j \, (x_i x_j) \, P_i(x_i) \, P_j(x_j) \qquad \text{[by (1.5-13b)]}$$

$$= \int_{-\infty}^{\infty} dx_i \, x_i \, P_i(x_i) \int_{-\infty}^{\infty} dx_j \, x_j \, P_j(x_j) = \langle X_i \rangle \langle X_j \rangle.$$

This implies that $\text{cov}\{X_i,X_j\}=0$, and hence that X_i and X_j are uncorrelated. But statistical independence is *not* a *necessary* condition for uncorrelatedness, because the numerical values of $\langle X_i X_j \rangle$ and $\langle X_i \rangle \langle X_j \rangle$ might just *happen* to be equal *without* the function $P_{i,j}(x_i,x_j)$ being everywhere equal to the product of the two functions $P_i(x_i)$ and $P_j(x_j)$.

Perhaps the most often used property of statistically independent random variables is the simple theorem that

$X_1, ..., X_n$ are statistically independent

$$\Rightarrow \ \text{mean\&var} \left\{ \textstyle\sum_i X_i \right\} = \textstyle\sum_i \text{mean\&var} \{X_i\}. \quad (1.6\text{-}22)$$

Of course, the "mean" relation here holds even without statistical independence [see Eq. (1.6-14)]. The "variance" relation is proved simply by observing that, as just noted, statistical independence implies that the variables are uncorrelated, which by definition implies that all the covariances vanish, which by Eq. (1.6-15) implies the conclusion.

The final theorem that we shall state and prove here is perhaps the most famous of all theorems in statistics, and a theorem that will be invoked at several critical junctures in later chapters of this book:

Central Limit Theorem. Let X_1, ..., X_n be n statistically independent random variables, each distributed according to a common density function P which has a finite mean μ and a finite variance σ^2. Let the two random variables S_n and A_n be defined in terms of X_1, ..., X_n by

$$S_n \equiv \sum_{j=1}^{n} X_j, \quad A_n \equiv \frac{1}{n} \sum_{j=1}^{n} X_j. \qquad (1.6\text{-}23)$$

Then, regardless of the form of the common density function P,

$$S_n \rightarrow N(n\mu, n\sigma^2) \quad \text{as} \quad n \rightarrow \infty, \qquad (1.6\text{-}24)$$

$$A_n \rightarrow N(\mu, \sigma^2/n) \quad \text{as} \quad n \rightarrow \infty. \qquad (1.6\text{-}25)$$

Before proving the central limit theorem, let us consider its import. Suppose we regard the random variable X_j as "the outcome of the jth independent sampling of X," where X is the random variable with density function P, mean μ and variance σ^2. Then from the definitions (1.6-23), we see that a sample value of S_n can be regarded as the *sum* of n independent sample values of X, while a sample value of A_n can be regarded as the *average* of n independent sample values of X. So we shall call the random variable S_n the "n-sample sum of X," and the random variable A_n the "n-sample average of X." The major assertion of the central limit theorem is that, *if n is sufficiently large*, then both the n-sample sum and the n-sample average of *any* random variable X will be essentially *normally* distributed. Notice that the values asserted for the means and variances of S_n and A_n in Eqs. (1.6-24) and (1.6-25) are actually simple consequences of the linear combination theorem [see Eqs. (1.6-14) and (1.6-15)]; it is the *asymptotic normality* of S_n and A_n, independently of the density function P of X, that is the remarkable message of the central limit theorem.

Since, for n sufficiently large, the n-sample average A_n behaves as though it were a normal random variable with mean μ and variance σ^2/n, it follows that the probability that A_n will be found to be within some positive distance ε of μ can be calculated by simply integrating the normal density function (1.4-9), with $m = \mu$ and $a^2 = \sigma^2/n$, from $x = \mu - \varepsilon$ to $x = \mu + \varepsilon$. Taking $\varepsilon = \gamma \sigma n^{-1/2}$, and letting "~" denote asymptotic equality as $n \rightarrow \infty$, we thus have

$$\text{Prob}\{|A_n - \mu| < \gamma\sigma n^{-1/2}\} \sim \int_{\mu-\gamma\sigma/\sqrt{n}}^{\mu+\gamma\sigma/\sqrt{n}} [2\pi(\sigma^2/n)]^{-1/2} \exp\left\{-\frac{(x-\mu)^2}{2(\sigma^2/n)}\right\} dx$$

$$= \frac{1}{\sqrt{\pi}} \int_{-\gamma/\sqrt{2}}^{+\gamma/\sqrt{2}} \exp(-t^2) dt \equiv \text{erf}(\gamma/\sqrt{2}),$$

where erf is the "error function." From tables of values of the error function we find in particular that

$$\text{Prob}\{|A_n - \mu| < \gamma\sigma n^{-1/2}\} \sim \begin{cases} 0.683 & \text{for } \gamma=1, \\ 0.954 & \text{for } \gamma=2, \\ 0.997 & \text{for } \gamma=3, \end{cases} \qquad (1.6\text{-}26)$$

where again "$\sim$" denotes asymptotic equality as $n \to \infty$. Rule (1.6-26) allows us to assign numerical *confidence limits* to our best estimate of the value of a physical quantity that has been measured experimentally, or (as will be discussed in Section 1.8) the value of a definite integral that has been calculated by Monte Carlo methods. Notice that the confidence interval size in (1.6-26) is directly proportional to σ and inversely proportional to $\sqrt{n}$; the latter dependence implies that, as n becomes larger and larger, the sample values of A_n will tend to cluster closer and closer about μ.

The central limit theorem says nothing about *how rapidly* S_n and A_n approach normality with increasing n. Presumably, that rate of approach will depend upon the form of the density function P. Now, it follows from a simple generalization of the previously discussed normal sum theorem that, if the X_i's in the central limit theorem are themselves normal, then S_n and A_n will be *exactly* normal for *any* $n>1$. Thus, the convergence to normality alleged by the central limit theorem is *immediate* if the common density function P is normal. We may therefore expect that the *more* normal-like P is the *faster* will be the approach of S_n and A_n to normality, while the *less* normal-like P is the *slower* will that approach be.

Notice that the central limit theorem does *not* apply if P is the *Cauchy* density function; because that density function does not have a (finite) variance. In fact, by generalizing the arguments used earlier in our proof of the Cauchy sum theorem, we can show that if $X_1,...,X_n$ are statistically independent and each is $C(m,a)$, then the random variable A_n defined in Eq. (1.6-23) will *also* be $C(m,a)$. This implies that if we try to estimate the pseudo-mean m of $C(m,a)$ by averaging n sample values of that random variable, then using $n = 10^{10}$ sample values will give us a no more accurate estimate than using $n = 1$! The reason for this can be traced to the heavy tails of the Cauchy density function; they give rise to frequent extreme sample values that keep spoiling any "sharpening" of the A_n distribution in the manner of (1.6-26).

Proof of the Central Limit Theorem. Our proof will focus not on the random variables S_n and A_n directly, but rather on a third random variable Z_n, defined by

$$Z_n \equiv n^{-1/2} \sum_{j=1}^{n} (X_j - \mu). \tag{1.6-27}$$

By using the definitions (1.6-23), it is straightforward to show that the random variables S_n and A_n are functionally related to the random variable Z_n by

$$S_n = n^{1/2} Z_n + n\mu \quad \text{and} \quad A_n = n^{-1/2} Z_n + \mu. \tag{1.6-28}$$

The object of our proof will be to establish that

$$Z_n \to N(0, \sigma^2) \quad \text{as} \quad n \to \infty. \tag{1.6-29}$$

The two assertions (1.6-24) and (1.6-25) of the central limit theorem will then follow by applying the theorem (1.6-7): It implies that if Z_n does indeed approach $N(0, \sigma^2)$, then by Eqs. (1.6-28), S_n will approach

$$N(n^{1/2} \cdot 0 + n\mu, (n^{1/2})^2 \cdot \sigma^2) = N(n\mu, n\sigma^2),$$

while A_n will approach

$$N(n^{-1/2} \cdot 0 + \mu, (n^{-1/2})^2 \cdot \sigma^2) = N(\mu, \sigma^2/n),$$

precisely as claimed in Eqs. (1.6-24) and (1.6-25).

To prove that Z_n becomes $N(0, \sigma^2)$ as n becomes large, we reason as follows. Since $X_1, ..., X_n$ are by hypothesis statistically independent, each with the density function $P(x)$, then their joint density function is $P(x_1)P(x_2)\cdots P(x_n)$. Therefore, since Z_n is defined in terms of the X_j's by Eq. (1.6-27), then the RVT theorem (1.6-4) implies that the density function of Z_n is

$$Q(z_n) = \int_{-\infty}^{\infty} dx_1 \cdots \int_{-\infty}^{\infty} dx_n \left[\prod_{i=1}^{n} P(x_i) \right] \delta\left(z_n - n^{-1/2} \sum_{j=1}^{n} (x_j - \mu) \right).$$

Writing the Dirac delta function here using the integral representation formula (B-7),

$$\delta\left(z_n - n^{-1/2} \sum_{j=1}^{n} (x_j - \mu)\right)$$

$$= (2\pi)^{-1} \oint_{-\infty}^{\infty} du \exp\left[iu\left(z_n - n^{-1/2} \sum_{j=1}^{n} (x_j - \mu)\right)\right]$$

$$= (2\pi)^{-1} \oint_{-\infty}^{\infty} du \exp(iuz_n) \prod_{j=1}^{n} \exp\left[-iu n^{-1/2}(x_j - \mu)\right],$$

we get

$$Q(z_n) = (2\pi)^{-1} \oint_{-\infty}^{\infty} du \exp(iuz_n)$$

$$\times \prod_{j=1}^{n} \left\{\int_{-\infty}^{\infty} dx_j\, P(x_j) \exp\left[-iu n^{-1/2}(x_j - \mu)\right]\right\}.$$

But this is evidently the same as

$$Q(z_n) = (2\pi)^{-1} \oint_{-\infty}^{\infty} du \exp(iuz_n) \left[G(u n^{-1/2})\right]^n,$$

where we have defined the function G by

$$G(\xi) \equiv \int_{-\infty}^{\infty} dx\, P(x) \exp[-i\xi(x - \mu)].$$

Now we make a Taylor series expansion of $G(\xi)$ about $\xi = 0$,

$$G(\xi) = G(0) + G'(0)\xi + G''(0)\xi^2/2 + o(\xi^2),$$

where $o(t)/t \to 0$ as $t \to 0$. The above definition of the function G implies that its nth derivative at ξ is

$$G^{(n)}(\xi) = \int_{-\infty}^{\infty} dx\, P(x) [-i(x - \mu)]^n \exp[-i\xi(x - \mu)],$$

and for $\xi = 0$ this reduces to

$$G^{(n)}(0) = (-i)^n \int_{-\infty}^{\infty} dx\, (x - \mu)^n P(x) = (-i)^n \langle (X - \mu)^n \rangle.$$

In particular, since P has mean μ and variance σ^2, then we have $G(0) = 1$, $G'(0) = 0$ and $G''(0) = -\sigma^2$; therefore, the above expansion of $G(\xi)$ about $\xi = 0$ becomes

$$G(\xi) = 1 - \sigma^2 \xi^2/2 + o(\xi^2).$$

Substituting this formula for G into our last expression for $Q(z_n)$ gives

$$Q(z_n) = (2\pi)^{-1} \oint_{-\infty}^{\infty} du \exp(iuz_n) \left[1 - \frac{\sigma^2 u^2}{2n} + o(u^2 n^{-1}) \right]^n.$$

Now since $(1 - a/n)^n \to \exp(-a)$ as $n \to \infty$, then for n sufficiently large the last factor in the integrand can be approximated by $\exp(-\sigma^2 u^2/2)$. Thus we have

$$Q(z_n) \sim (2\pi)^{-1} \oint_{-\infty}^{\infty} du \exp(iuz_n) \exp(-\sigma^2 u^2/2),$$

where "$\sim$" denotes asymptotic equality as $n \to \infty$. The integral on the right converges as an ordinary integral, so its special nature can be ignored. Evaluating it with the help of the integral identity (A-6), we get

$$Q(z_n) \sim (2\pi\sigma^2)^{-1/2} \exp(-z_n^2/2\sigma^2).$$

This says that the random variable Z_n is asymptotically $N(0, \sigma^2)$; therefore, by the remarks immediately following Eq. (1.6-29), the central limit theorem is proved.

1.7 INTEGER RANDOM VARIABLES

It is sometimes convenient to use random variables that take on *only* *integer* values. We say that X is an **integer random variable** if and only if there exists a function P of an integer variable n such that $P(n)$ equals the probability of finding the value of X to be n. In symbols [cf. Eq. (1.2-1)],

$$P(n) \equiv \text{Prob}\{X = n\}. \tag{1.7-1}$$

We again call P the *density function* of X, and Eq. (1.7-1) may be regarded as its definition. Using the addition law of probability, we easily deduce that the probability of finding a value for X between n_1 and n_2 inclusively is [cf. Eq. (1.2-2)]

$$\text{Prob}\{X \in [n_1, n_2]\} = \sum_{n=n_1}^{n_2} P(n). \tag{1.7-2}$$

In particular, since X will surely take some integer value between $-\infty$ and $+\infty$, then [cf. Eq. (1.2-3)]

$$\sum_{n=-\infty}^{\infty} P(n) = 1, \tag{1.7-3}$$

an equation that is called the "normalization condition" for P. Also, since $P(n)$ is a probability, then the range law implies that [cf. Eq. (1.2-4)]

$$0 \le P(n) \le 1. \tag{1.7-4}$$

Indeed, *any* function P that satisfies both conditions (1.7-3) and (1.7-4) can be regarded as a *density* function, defining some integer random variable X according to Eq. (1.7-1).

If the density function $P(n)$ is the **Kronecker delta function,**

$$\delta(n,n_0) \equiv \begin{cases} 1, & \text{if } n=n_0, \\ 0, & \text{if } n \ne n_0, \end{cases} \tag{1.7-5}$$

which function clearly satisfies conditions (1.7-3) and (1.7-4), then X will evidently *always* be found to have the value n_0. And since the converse is also true, then we conclude that [cf. (1.2-7)]

$$X \text{ has density function } \delta(n,n_0) \Leftrightarrow X \text{ is the sure variable } n_0. \tag{1.7-6}$$

The *distribution function* F for an integer random variable X is defined by [cf. Eq. (1.2-8)]

$$F(n) \equiv \text{Prob}\{X \le n\}. \tag{1.7-7}$$

Equations (1.7-7) and (1.7-2) together imply that [cf. Eq. (1.2-9a)]

$$F(n) = \sum_{n'=-\infty}^{n} P(n'), \tag{1.7-8a}$$

and this in turn implies that [cf. Eq. (1.2-9b)]

$$P(n) = F(n) - F(n-1). \tag{1.7-8b}$$

If we know P we can calculate F from Eq. (1.7-8a), and if we know F we can calculate P from Eq. (1.7-8b); thus, an integer random variable X can be defined by specifying *either* its density function or its distribution function. It is easy to see from Eqs. (1.7-3), (1.7-4) and (1.7-8a) that $F(n)$ rises nondecreasingly from 0 at $n = -\infty$ to 1 at $n = +\infty$.

Since the integers are a subset of the reals, it might be suspected that integer random variables can be viewed as a special class of real random variables. This is indeed the case. Consider the *real* random variable Y defined by the density function

$$Q(y) = \sum_{n=-\infty}^{\infty} P(n)\,\delta(y-n), \tag{1.7-9}$$

where P is the density function of the *integer* random variable X as defined above, and $\delta(y-n)$ is the Dirac delta function of the real variable y. It follows from Eq. (1.2-2) that

$$\text{Prob}\{Y \in [a,b]\} = \int_a^b Q(y)\,dy = \sum_{n=-\infty}^{\infty} P(n)\int_a^b \delta(y-n)\,dy.$$

From this result it is easy to show that the probability of finding Y inside any interval that does not contain at least one integer is zero, while the probability of finding Y inside any interval that contains only one integer n' is $P(n')$. This implies that the *real* random variable Y with density function Q in Eq. (1.7-9) is in fact identical to the *integer* random variable X with density function P.

Since integer random variables are thus a subset of the real random variables, we may expect that the theory of integer random variables will not be substantially different from the theory of real random variables as outlined in the earlier sections of this chapter. Indeed, a comparison of Eqs. (1.7-1) – (1.7-8) with their counterparts in Section 1.2 shows that the basic relations for integer random variables can be obtained from the basic relations for real random variables by simply replacing *integrations* over x by *sums* over n, and replacing *Dirac* delta functions by *Kronecker* delta functions. By way of illustration, let us see how the fundamental equation (1.3-2) adapts to the integer variable case. Let Y be the real random variable whose density function is Q in Eq. (1.7-9), and which is thus identical to the integer random variable X whose density function is P. Then we have for the average of any univariate function h with respect to Y,

$$\langle h(Y)\rangle = \int_{-\infty}^{\infty} h(y)\,Q(y)\,dy \qquad \text{[by (1.3-2)]}$$

$$= \sum_{n=-\infty}^{\infty} \int_{-\infty}^{\infty} h(y)\,P(n)\,\delta(y-n)\,dy \qquad \text{[by (1.7-9)]}$$

$$= \sum_{n=-\infty}^{\infty} h(n)\,P(n). \qquad \text{[by (1.2-5)]}$$

Therefore, since the random variables Y and X are one and the same,

$$\langle h(X) \rangle = \sum_{n=-\infty}^{\infty} h(n) \, P(n) . \qquad (1.7\text{-}10)$$

This is the integer variable version of Eq. (1.3-2). It could also have been derived from Eq. (1.3-1), by arguing from the definitions (1.7-1) and (1.1-1) in much the same way that we did when we derived Eq. (1.3-2).

The kth *moment* of the integer random variable X is defined by

$$\langle X^k \rangle \equiv \sum_{n=-\infty}^{\infty} n^k \, P(n) . \qquad (1.7\text{-}11)$$

Using this general formula, we can calculate as before the mean and variance of X. And just as in the real variable case [cf. (1.3-10)], we find that the vanishing of var$\{X\}$ is a necessary and sufficient condition for X to be the *sure* variable $\langle X \rangle$.

Three specific integer random variables occur so often as to merit special mention. The **discrete uniform** random variable X is defined by the density function

$$P(n) = \begin{cases} (n_2 - n_1 + 1)^{-1}, & \text{if } n_1 \le n \le n_2, \\ 0, & \text{if } n < n_1 \text{ or } n > n_2, \end{cases} \qquad (1.7\text{-}12a)$$

where n_1 and n_2 are any two integers satisfying $n_1 \le n_2$. Evidently, for this random variable all integer values between n_1 and n_2 inclusively are equally likely. The results

$$\langle X \rangle = \frac{(n_1 + n_2)}{2}, \quad \text{var}\{X\} = \frac{(n_2 - n_1)(n_2 - n_1 + 2)}{12} \qquad (1.7\text{-}12b)$$

follow by substituting Eq. (1.7-12a) into Eq. (1.7-11) for $k=1$ and 2, and then evaluating the sums with the aid of the two algebraic identities

$$\sum_{n=1}^{N} n = \frac{N(N+1)}{2}, \quad \sum_{n=1}^{N} n^2 = \frac{N(N+1)(2N+1)}{6} .$$

Notice from Eqs. (1.7-12b) that, if $n_2 = n_1$, then $\langle X \rangle = n_1$ and var$\{X\}=0$, implying that X is the sure variable n_1.

The **binomial** random variable X is defined by the density function

$$P(n) = \begin{cases} \dfrac{N!}{n! \, (N-n)!} \, p^n \, (1-p)^{N-n}, & \text{if } 0 \le n \le N, \\ 0, & \text{if } n < 0 \text{ or } n > N, \end{cases} \qquad (1.7\text{-}13a)$$

where N is any positive integer and p is any real number satisfying $0 \le p \le 1$. If p is interpreted as the probability that the toss of a certain coin will yield heads, then the three-factor product on the right hand side of Eq. (1.7-13a) can be interpreted as the probability of tossing exactly n heads, in any order, in N tosses. The results

$$\langle X \rangle = Np, \quad \text{var}\{X\} = Np(1-p) \tag{1.7-13b}$$

follow by calculating $\langle X \rangle$ and $\langle X(X-1) \rangle \equiv \langle X^2 \rangle - \langle X \rangle$ for the density function in Eq. (1.7-13a), these calculations being made from Eq. (1.7-10) with the help of the algebraic identity

$$1 = [p + (1-p)]^n = \sum_{n=0}^{N} \frac{N!}{n!\,(N-n)!}\, p^n (1-p)^{N-n}.$$

Notice from Eqs. (1.7-13b) that if $p=0$ then X will be the sure number 0, while if $p=1$ then X will be the sure variable N.

Finally, the **Poisson** random variable X is defined by the density function

$$P(n) = \begin{cases} \dfrac{e^{-a} a^n}{n!}, & \text{if } n \ge 0, \\[2mm] 0, & \text{if } n < 0, \end{cases} \tag{1.7-14a}$$

where a is any positive real number. The results

$$\langle X \rangle = \text{var}\{X\} = a \tag{1.7-14b}$$

follow by calculating $\langle X \rangle$ and $\langle X(X-1) \rangle \equiv \langle X^2 \rangle - \langle X \rangle$ for the density function in Eq. (1.7-14a), these calculations being made from Eq. (1.7-10) with the help of the algebraic identity

$$e^a = \sum_{n=0}^{\infty} \frac{a^n}{n!}.$$

It is interesting to note that the Poisson random variable can be viewed as a special case of the binomial random variable, namely the case in which the binomial parameters N and p are respectively very large and very small. To prove this statement, we simply observe that when N is sufficiently large we can approximate in Eq. (1.7-13a)

$$\frac{N!}{(N-n)!}\, p^n = N(N-1)\cdots(N-n+1)\, p^n \approx N^n p^n = (Np)^n$$

and

$$(1-p)^{N-n} \approx (1-p)^{N} = \left(1 - \frac{Np}{N}\right)^{N} \approx e^{-Np};$$

thus it follows that

$$\frac{N!}{n!\,(N-n)!}\,p^{n}\,(1-p)^{N-n} \xrightarrow[p=a/N]{N\to\infty} \frac{e^{-a}\,a^{n}}{n!}. \qquad (1.7\text{-}15)$$

In words, if $N\to\infty$ and $p\to0$ with Np held constant, then the binomial random variable with parameters N and p becomes the Poisson random variable with mean and variance Np.

It is obvious from Eq. (1.7-14b) that, if the Poisson parameter $a\to0$, then the Poisson random variable X approaches the sure number 0. A more interesting limit is that in which the Poisson parameter a becomes very large: Observe from Eqs. (1.7-14b) that, when $a\gg1$, the Poisson density function will be appreciably different from zero only when n is relatively near a. For such large n we can use Stirling's approximation,

$$n! \approx (2\pi n)^{1/2}\,n^{n}\,e^{-n} \quad \text{for } n\gg1,$$

to establish that

$$\ln\left[\frac{e^{-a}\,a^{n}}{n!}\,(2\pi a)^{1/2}\right] \approx (n-a) - n\ln\left(\frac{n}{a}\right), \quad \text{for } n\gg1.$$

Then we can use the standard logarithm approximation,

$$\ln(1+\varepsilon) \approx \varepsilon - \varepsilon^{2}/2 \quad \text{for } |\varepsilon|\ll1,$$

with $\varepsilon=(n-a)/a$, to show that when n is near a the right hand side of the previous equation is approximately equal to $-a\varepsilon^{2}/2 = -(n-a)^{2}/2a$. Thus we can conclude that

$$\frac{e^{-a}\,a^{n}}{n!} \approx \frac{1}{(2\pi a)^{1/2}}\,\exp\left[-\frac{(n-a)^{2}}{2a}\right] \qquad (a\gg1). \qquad (1.7\text{-}16)$$

In words, when a is very large, the *Poisson* random variable with mean and variance a resembles the *normal* random variable with mean and variance a. And by combining this result with (1.7-15), we may further deduce that

$$\frac{N!}{n!\,(N-n)!}\,p^{n}\,(1-p)^{N-n} \xrightarrow[p=a/N,\,a\gg1]{N\to\infty} \frac{1}{(2\pi a)^{1/2}}\,\exp\left[-\frac{(n-a)^{2}}{2a}\right].$$

$$(1.7\text{-}17)$$

In words, if $N \to \infty$ and $p \to 0$ with Np a *large constant*, then the binomial random variable with parameters N and p begins to look like a normal random variable with mean and variance Np.

For *sets* of integer random variables, we can define joint, marginal and conditional density functions in much the same way as we did for real random variables in Section 1.5. The relations among these various density functions may be obtained from the real variable formulas by simply replacing integrals over x_i by sums over n_i. The definitions (1.5-13) of statistical independence hold also for integer random variables, and the average formulas in Eqs. (1.5-15) and (1.5-16) hold as well if the integrals therein are replaced by summations. The covariance of two integer random variables X_i and X_j is defined by the same formula (1.5-17) as for two real random variables.

If X is an integer random variable and f is some function that maps integers onto integers, then we define the integer random variable $f(X)$ as that random variable whose sample values are $f(n^{(i)})$, where $n^{(i)}$ is a sample value of X. We can easily derive an integer version of the RVT theorem (1.6-4), in which integrals are replaced by sums and Dirac delta functions are replaced by Kronecker delta functions. The foregoing definition of a function of a random variable is again such that Eq. (1.7-10), which gives the "average of the function h with respect to the random variable X," may also be interpreted as the "mean of the random variable $h(X)$."

1.8 RANDOM NUMBER GENERATING PROCEDURES

It is often advantageous to be able to numerically construct sample values of a random variable with a prescribed density function, or more generally, simultaneous sample values of several random variables with a prescribed joint density function. Procedures for making such constructions are called *random number generating algorithms*. And any numerical calculation that makes explicit use of random variable sample values, or "random numbers," is called a *Monte Carlo calculation*.

Two kinds of Monte Carlo calculation are especially useful in practice. One is the *numerical simulation of random processes*, such as the evolving population profile of a collection of radioactive atoms, or the trajectory of an electron passing through a cloud of randomly positioned scatterers. In later chapters of this book we shall see how certain random number generating algorithms can be employed to numerically simulate random processes of the "Markov" type.

Another important kind of Monte Carlo calculation is the *numerical evaluation of definite integrals*. The justification for this seemingly incongruous use of random numbers may be found in the two fundamental formulas for $\langle h(X) \rangle$ in Eqs. (1.3-1) and (1.3-2); together, those two equations imply that

$$\int_{-\infty}^{\infty} h(x)\,P(x)\,dx = \lim_{N \to \infty} \frac{1}{N} \sum_{i=1}^{N} h(x^{(i)}), \qquad (1.8\text{-}1)$$

where $x^{(1)}$, ..., $x^{(N)}$ are N independent sample values of the random variable whose density function is P. Now, in the discipline of statistical mechanics, the usual strategy is to calculate the average on the right by actually computing the integral on the left. The basic idea behind "Monte Carlo integration" is to simply turn this procedure around, and to calculate the integral on the left by actually computing the average on the right, albeit for a *finite* value of N. In taking N to be finite, however, we unavoidably incur some uncertainty in our estimate of the integral. But if N is taken large enough, we can get a quantitative estimate of this uncertainty by appealing to the central limit theorem corollary (1.6-26). Specifically, if we compute for "reasonably large" N the two N-sample averages

$$\langle h \rangle_N \equiv \frac{1}{N} \sum_{i=1}^{N} h(x^{(i)}) \quad \text{and} \quad \langle h^2 \rangle_N \equiv \frac{1}{N} \sum_{i=1}^{N} [h(x^{(i)})]^2, \qquad (1.8\text{-}2)$$

then we may assert that

$$\int_{-\infty}^{\infty} h(x)\,P(x)\,dx \approx \langle h \rangle_N \pm \gamma \left[\langle h^2 \rangle_N - \langle h \rangle_N^2 \right]^{1/2} N^{-1/2}. \qquad (1.8\text{-}3)$$

To explain the rationale for formulas (1.8-2) and (1.8-3), we begin by writing the *left* hand side of Eq. (1.6-26), with $n = N$, in the form

$$\mu \approx A_N \pm \gamma \sigma N^{-1/2}.$$

Next, we *replace* the random variable X by the random variable $h(X)$: The mean μ of X is thereby replaced by the mean $\int h(x)P(x)dx$ of $h(X)$; the N-sample average A_N of X is replaced by the N-sample average $\langle h \rangle_N$ of $h(X)$; and the standard deviation σ of X is replaced by the standard deviation $[\langle h^2 \rangle_\infty - \langle h \rangle_\infty^2]^{1/2}$ of $h(X)$, which however we *approximate* as $[\langle h^2 \rangle_N - \langle h \rangle_N^2]^{1/2}$. Recalling now the *right* hand side of Eq. (1.6-26), we see that Eq. (1.8-3) is to be interpreted as follows: *Provided N is sufficiently large*, the integral on the left will lie in the indicated $\pm$-interval about $\langle h \rangle_N$ with probabilities 0.683 for $\gamma = 1$, 0.954 for $\gamma = 2$, and 0.997 for $\gamma = 3$. Notice that, as the number N of sample points is made larger, the interval of uncertainty in Eq. (1.8-3) decreases like $1/\sqrt{N}$. And in the limit $N \to \infty$, Eq. (1.8-3) properly reduces to Eq. (1.8-1).

The Monte Carlo integration method proves to be most useful in practice when the number m of integration variables exceeds 1. In that case, the Monte Carlo integration scheme is still as prescribed in Eqs. (1.8-2) and (1.8-3), except that x is now regarded as the m-component *vector* variable $(x_1,...,x_m)$. This means that the computation of the two sums in Eqs. (1.8-2) will require the generation of N simultaneous sample values $x_1^{(i)}$, ..., $x_m^{(i)}$ of m random variables X_1, ..., X_m whose joint density function is $P(x_1,...,x_m)$.

The feasibility of Monte Carlo calculations in general hinges on the availability of fast, reliable algorithms for generating random numbers according to arbitrarily specified density functions. As we shall see shortly, most such algorithms designed for use on digital computers require a "subprogram" that generates sample values of the unit uniform random variable, $U \equiv U(0,1)$. A sample value of U is customarily refered to as a *unit uniform random number*, and denoted generically by r.[†]

One common procedure for computing a sequence $r^{(1)}$, $r^{(2)}$, ... of unit uniform random numbers is *Lehmer's method*: With N_0, C and M being three "suitably chosen" integers, one first generates a sequence of integers $N_1, N_2, ...$ through the recursion relation

$$N_i = C N_{i-1} \text{ (modulo } M\text{)}, \tag{1.8-4a}$$

which means simply that N_i is the *remainder* obtained when CN_{i-1} is divided by M; one then takes

$$r^{(i)} = N_i / M. \tag{1.8-4b}$$

That the numbers $r^{(1)}$, $r^{(2)}$, ... produced by this simple recursive procedure can approximately mimic independent samplings of the random variable U was originally viewed with a combination of awe and skepticism. Although the awe still seems warranted, the skepticism has largely abated with the recent discovery that even very simple *nonlinear*

[†] An apt name for a computer subprogram that generates *unit uniform random numbers* is URN, not only for the obvious acronymic reason, but also because of the homage thus paid to the vessel traditionally used by classically attuned probabilists to hold colored balls for "random drawings." Other names, such as JAR or CAN, could certainly be used instead, but they may well call into question the taste and breeding of the user.

recursion relations can produce apparently "chaotic" sequences. Notice that the recursion relation in Eq. (1.8-4a) is in fact quite nonlinear, owing to the modulo operation.

A "minimal standard" unit uniform random number generator currently favored by many experts in this area† is the foregoing Lehmer algorithm with $M = 2^{31} - 1 = 2147483647$, $C = 7^5 = 16807$, and N_0 (called the "seed") any integer between 1 and $M - 1$ inclusively. This generator has a nonrepetition period of $M - 1$, it never produces the two values $r = 0$ and $r = 1$, and it gives scores on statistical tests for randomness that, while not perfect, are adequate for most applications. This generator can also be easily coded for most digital computers. It should be noted, however, that the calculations in Eqs. (1.8-4) usually must be performed in a "high precision" mode, because *it is crucial that the integer arithmetic in Eq. (1.8-4a) be done exactly.* In any case, *this* is the unit uniform random number generator that has been used for all the Monte Carlo calculations reported in later chapters of this book.

But our goal here is not to delve into the mysteries of unit uniform random number generators;† we shall simply take their availability as a given fact. Our chief concern here will be with the problem of using the output of such generators, i.e., unit uniform random numbers, to construct sample values of *arbitrarily specified* random variables.

Consider first the problem of generating a sample value x of a real random variable X with a prescribed density function P. The most elegant procedure for accomplishing this is as follows:

Inversion Generating Method. Let X be a real random variable with density function P and distribution function F. Then if r is a unit uniform random number, the number x obtained by solving the equation $F(x) = r$, i.e., the number

$$x = F^{-1}(r), \tag{1.8-5}$$

can be regarded as a sample value of X.

Proof. The fact that $F(x)$ is unambiguously invertible at all points for which $P(x) > 0$ follows from the fundamental relation $F'(x) = P(x)$, which ensures that $F(x)$ is strictly increasing at such points. The inversion formula (1.8-5) then becomes a simple consequence of the distribution function theorem: $F^{-1}(r)$, being by definition a sample value of the random variable $F^{-1}(U)$, is by (1.6-11) a sample value of the random

---

† See the article "Random Number Generators: Good Ones Are Hard To Find" by S. Park and K. Miller in *Communications of the ACM*, Vol. 31, pp 1192–1201 (1988), and references contained therein.

variable whose density function is F'. Since $F'=P$, then that random variable is the hypothesized random variable X.   QED

An intuitive understanding of how and why the inversion method works can be gained by refering to the schematic graph of the distribution function F in Fig. 1-1b: In essence, the inversion method tells us to randomly scatter points along the vertical axis of that graph uniformly between 0 and 1, and to then project those points individually onto the horizontal axis through the curve $F(x)$-versus-x. Clearly, points on the vertical axis lying in the interval $[F(x), F(x+dx))$ will project into the horizontal axis interval $[x, x+dx)$. Now, for a given interval size $[F(x), F(x+dx))$, a *large* value of the local slope $F'(x) \equiv P(x)$ of the projection curve will correspond to a relatively small interval $[x, x+dx)$, and will thus give rise to a *compression* of the projected points; conversely, a *small* value of the local slope $P(x)$ will correspond to a relatively large interval $[x, x+dx)$, and so will give rise to a *rarefaction* of the projected points. Since this compression or rarefaction will always be in direct proportion to the slope $P(x)$, it follows that the local density of the projected points on the horizontal axis will tend to be directly proportional to the local value of the function P. And that is essentially the defining property of the density function of X.

Let us illustrate the inversion generating method by using it to derive random number generating formulas for the three random variables $\mathbf{U}(a,b)$, $\mathbf{E}(a)$ and $\mathbf{C}(m,a)$.

The general uniform random variable $\mathbf{U}(a,b)$, as defined by the density function in Eq. (1.4-1), has the distribution function

$$F(x) = \int_a^x (b-a)^{-1}\,dx' = \frac{x-a}{b-a} \quad (a \le x \le b).$$

Setting this equal to a unit-interval uniform random number r and then solving for x, we obtain the generating formula

$$x = a + (b-a)\,r \quad [x \text{ from } \mathbf{U}(a,b)]. \tag{1.8-6}$$

The exponential random variable $\mathbf{E}(a)$, as defined by the density function in Eq. (1.4-5), has the distribution function

$$F(x) = \int_0^x a\exp(-ax')\,dx' = 1 - \exp(-ax) \quad (0 \le x < \infty).$$

Setting this equal to $(1-r)$, which like r is a unit-interval uniform random number, and then solving for x, we obtain the generating formula

$$x = (1/a)\ln(1/r) \quad [x \text{ from } \mathbf{E}(a)]. \tag{1.8-7}$$

And finally, the Cauchy random variable $C(m,a)$, as defined by the density function in Eq. (1.4-13), has the distribution function

$$F(x) = \int_{-\infty}^{x} \frac{(a/\pi)}{(x'-m)^2 + a^2}\, dx' = \frac{1}{\pi}\left[\arctan\!\left(\frac{x-m}{a}\right) + \frac{\pi}{2}\right].$$

Setting this equal to a unit-interval uniform random number r and then solving for x, we obtain the generating formula

$$x = m + a\tan[(r-\tfrac{1}{2})\pi] \quad [x \text{ from } C(m,a)].$$ (1.8-8)

A direct application of the inversion generating method to produce sample values of the normal random variable $N(m,a^2)$, as defined by the density function in Eq. (1.4-9), is complicated by the fact that the distribution function for that random variable can be inverted only by numerical means. At the end of this section we shall develop a somewhat indirect but nonetheless wholly analytic random number generating algorithm for $N(m,a^2)$, based on a multivariable generalization of the inversion method. But it is not at all uncommon to encounter a density function $P(x)$ whose inverse distribution function is so complicated that the inversion method is impractical. In some of those instances, it may be more convenient to use the following generating procedure.

Rejection Generating Method. If the density function P of the random variable X has the form $P(x) = Ch(x)$, where C is a normalizing constant and the function h satisfies

$$0 \leq h(x) \leq B \quad \text{for } x \in [a,b],$$ (1.8-9a)

$$h(x) = 0 \quad \text{for } x \notin [a,b],$$ (1.8-9b)

then the following procedure will produce a sample value of X:

1° Draw two independent unit uniform random numbers r_1 and r_2, and calculate the "trial number"

$$x_t = a + (b-a)\, r_1.$$ (1.8-10)

2° If $h(x_t)/B \geq r_2$, then *accept* x_t as a sample value of X; otherwise, *reject* x_t (along with the two values r_1 and r_2) and return to step 1°.

Furthermore, the average acceptance ratio or *efficiency* of this generating procedure is given by

$$E = \frac{\displaystyle\int_a^b h(x)\, dx}{B(b-a)}.$$ (1.8-11)

Proof. Let $P_A(x)$ denote the density function of the *accepted* numbers produced by this procedure, and let p denote the probability that step 1° will produce an accepted number that lies between x and $x + dx$. We can write the probability p in two different ways. First, we can write p as the product {the probability that the trial number produced in step 1° will be accepted in step 2°} times {the probability that an accepted number in step 2° will lie between x and $x + dx$}. So, by our definitions of E and P_A, our first expression for p is

$$p = [E] \times [P_A(x)dx].$$

But we can also write p as the product {the probability that the trial number produced in step 1° will lie between x and $x + dx$} times {the probability that a trial number with the value x will be accepted in step 2°}. Observing from Eq. (1.8-6) that the trial number in Eq. (1.8-10) is just a sample value of the random variable $U(a,b)$, then the first factor here is $(b-a)^{-1}dx$. The second factor, being the probability that a sample of the random variable $U(0,1)$ will be less than or equal to $h(x)/B$, is just the value $h(x)/B$ itself. So our second expression for p is

$$p = [(b-a)^{-1}dx] \times [h(x)/B].$$

If we now equate the above two expressions for p, we find that

$$P_A(x) = \frac{1}{EB(b-a)} h(x).$$

This proves that the density function of the numbers produced by steps 1° and 2° is indeed some constant times $h(x)$. And since this density function is by definition normalized, then we have

$$1 = \int_a^b P_A(x)\,dx = \frac{1}{EB(b-a)} \int_a^b h(x)\,dx,$$

which evidently implies Eq. (1.8-11).†   QED

Notice that the rejection method allows ignorance of both the distribution function F and the normalizing constant C. But the rejection method does require that the density function bound B and the supporting interval $[a,b]$ both be *known* and *finite*. Since an average of $1/E$ trial numbers x_t must be generated in order to get one accepted value, then the practical feasibility of the rejection method ultimately hinges on

---

† It is easy to modify this proof to establish the following *generalized rejection generating method*: If x_t is picked randomly according to some density function $p(x)$, instead of according to Eq. (1.8-10), then the density function of the accepted numbers will be some constant times $p(x)h(x)$, and the generating efficiency will be $E = B^{-1} \int p(x)h(x)dx$.

E not being too small. In that regard, although B can be taken to be *any* upper bound on $h(x)$ in $[a,b]$, Eq. (1.8-11) shows that E will be maximized if we take B to be the *least* upper bound. Eq. (1.8-11) also shows that E can be geometrically interpreted as the ratio of the area under $h(x)$ in $[a,b]$ to the area under the horizontal line $y=B$ over that same interval. Therefore, if $h(x)$ has a very narrow, high peak in $[a,b]$, then the efficiency of the rejection generating procedure will be very low.

   Both the inversion and the rejection generating methods can be adapted to the case of an *integer* random variable. In our work here we shall require only the former adaptation:

Integer Inversion Generating Method. Let X be an integer random variable with density function P. Then if r is a unit uniform random number, the integer n that satisfies the double inequality

$$\sum_{n'=-\infty}^{n-1} P(n') \le r < \sum_{n'=-\infty}^{n} P(n') \qquad (1.8\text{-}12)$$

can be regarded as a sample value of X.

Proof. The probability of generating a value r satisfying $a \le r < b$, for any $0 \le a < b \le 1$, is by Eqs. (1.2-2) and (1.4-1)

$$\int_{a}^{b} (1-0)^{-1} dx = b-a.$$

Making the appropriate substitutions for a and b, we thus see that the probability of inequality (1.8-12) being satisfied for a given value of n is

$$\left(\sum_{n'=-\infty}^{n} P(n') \right) - \left(\sum_{n'=-\infty}^{n-1} P(n') \right) = P(n).$$

So $P(n)$ is the probability that a given integer n will cause the inequality (1.8-12) to be satisfied; therefore, any integer n that is selected by requiring satisfaction of (1.8-12) must be a sample value of the integer random variable whose density function is P.   QED

   The reason for calling (1.8-12) the integer version of Eq. (1.8-5) is that (1.8-12) is just $F(n-1) \le r < F(n)$, and finding the n-value that satisfies this double inequality is analogous to finding the x-value that satisfies $F(x)=r$. Notice that the right inequality in (1.8-12) cannot be satisfied if $r=1$ because of the normalization condition (1.7-3). In fact, *we should always disallow the value* $r=1$. The formal reason for this is that the interval from a to b in Eq. (1.4-1) does not include the point b, so the random variable $U(0,1)$ should never assume the value unity.

To illustrate the integer inversion generating method, let us derive the generating algorithm for the integer uniform random variable whose density function is given by Eq. (1.7-12a). The inequality (1.8-12) reads for that density function

$$\sum_{n'=n_1}^{n-1} (n_2 - n_1 + 1)^{-1} \le r < \sum_{n'=n_1}^{n} (n_2 - n_1 + 1)^{-1}.$$

Multiplying through by $(n_2 - n_1 + 1)$ and evaluating the two sums, we get

$$(n-1) - n_1 + 1 \le r(n_2 - n_1 + 1) < n - n_1 + 1.$$

By adding n_1 to each member of this double inequality, we can immediately deduce that it implies

$$n = \text{greatest-integer-in}\{ n_1 + r(n_2 - n_1 + 1) \}, \qquad (1.8\text{-}13)$$

so this is the rule for generating a random integer uniformly in $[n_1, n_2]$. Notice by the way that if r were to assume the unallowed value 1, then Eq. (1.8-13) would give the erroneous value $n_2 + 1$ for n.

Now let us return to the *real* variable case, and ask how we might go about generating simultaneous sample values of several random variables with a specified *joint* density function P. There are two methods for doing this, which respectively generalize the inversion and rejection methods described earlier. The *joint rejection generating method* is basically the same as the simple rejection generating method [see Eqs. (1.8-9) – (1.8-11)], except that P is now regarded as a joint density function and all scalar variables are regarded as vector variables. Although the joint rejection method is straightforward enough, its efficiency in practical applications is usually quite low. A better approach to generating joint random numbers is the inversion method:

Joint Inversion Generating Method. Let X_1, X_2 and X_3 be three random variables with joint density function P. Define the functions F_1, F_2 and F_3 by

$$F_1(x_1) \equiv \int_{-\infty}^{x_1} P_1(x_1') \, dx_1', \qquad (1.8\text{-}14a)$$

$$F_2(x_2; x_1) \equiv \int_{-\infty}^{x_2} P_2^{(1)}(x_2' | x_1) \, dx_2', \qquad (1.8\text{-}14b)$$

$$F_3(x_3; x_1, x_2) \equiv \int_{-\infty}^{x_3} P_3^{(1,2)}(x_3' | x_1, x_2) \, dx_3', \qquad (1.8\text{-}14c)$$

where the integrands on the right are subordinate density functions obtained from P through Eqs. (1.5-10). Then if r_1, r_2 and r_3 are three independent unit uniform random numbers, the values x_1, x_2 and x_3 obtained by successively solving the three equations

$$\left\{ \begin{array}{ll} F_1(x_1) = r_1, & \text{(1.8-15a)} \\[2mm] F_2(x_2;x_1) = r_2, & \text{(1.8-15b)} \\[2mm] F_3(x_3;x_1,x_2) = r_3, & \text{(1.8-15c)} \end{array} \right.$$

can be regarded as simultaneous sample values of X_1, X_2 and X_3.

Proof. First observe that the non-negativity of the integrands in Eqs. (1.8-14) and the normalization equations (1.5-8a) together ensure that each function F_i increases monotonically from 0 to 1 as x_i increases from $-\infty$ to ∞; hence, each equation of the form $F_i(x_i) = r_i$ in Eqs. (1.8-15) can indeed be solved for x_i for any given r_i in the unit interval. Since each r_i in Eqs. (1.8-15) can be regarded as a sample of an independent unit uniform random variable U_i, then the values x_1, x_2 and x_3 found by solving Eqs. (1.8-15) can be regarded as samplings of three random variables Y_1, Y_2 and Y_3 that are functionally related to U_1, U_2 and U_3 by

$$F_1(Y_1) = U_1, \quad F_2(Y_2;Y_1) = U_2, \quad F_3(Y_3;Y_1,Y_2) = U_3.$$

We shall prove that the three random variables Y_1, Y_2 and Y_3 defined by this functional transformation have the same joint density function P as X_1, X_2 and X_3. This will imply that $Y_i = X_i$ for $i = 1$ to 3, and hence that the Y_i sample values (x_i) obtained by inverting Eqs. (1.8-15) are in fact sample values of the random variables X_i, as asserted by the theorem.

Applying the RVT corollary (1.6-5) to the above random variable transformation, we see that the joint density function P_Y of Y_1, Y_2 and Y_3 is related to the joint density function P_U of U_1, U_2 and U_3 by

$$P_Y(y_1,y_2,y_3) = P_U(u_1,u_2,u_3) \left| \frac{\partial(u_1,u_2,u_3)}{\partial(y_1,y_2,y_3)} \right|,$$

where the u_i's on the right are the *functions*

$$u_1 = F_1(y_1), \quad u_2 = F_2(y_2;y_1), \quad u_3 = F_3(y_3;y_1,y_2).$$

Now since U_1, U_2 and U_3 are statistically independent unit uniform random variables, then their joint density function is given by

$$P_U(u_1,u_2,u_3) = Q(u_1)\,Q(u_2)\,Q(u_3),$$

where $Q(u)$ is unity if $0 \leq u < 1$ and zero otherwise. Because the range of the functions $u_i = F_i$ is precisely the unit interval, each of the three factors $Q(u_i)$ in the above equation is in fact *always* unity; therefore $P_U = 1$, and the preceding formula for P_Y reduces to

$$P_Y(y_1, y_2, y_3) = \left| \frac{\partial(u_1, u_2, u_3)}{\partial(y_1, y_2, y_3)} \right| .$$

Since the function u_1 is independent of y_2 and y_3, and the function u_2 is independent of y_3, then the Jacobian determinant here has all zero entries everywhere on one side of its main diagonal; therefore, the determinant is simply equal to the product of its diagonal elements:

$$P_Y(y_1, y_2, y_3) = \left| \frac{\partial u_1}{\partial y_1} \cdot \frac{\partial u_2}{\partial y_2} \cdot \frac{\partial u_3}{\partial y_3} \right|$$

$$= \left| \frac{\partial F_1(y_1)}{\partial y_1} \cdot \frac{\partial F_2(y_2; y_1)}{\partial y_2} \cdot \frac{\partial F_3(y_3; y_1, y_2)}{\partial y_3} \right|$$

$$= P_1(y_1) \cdot P_2^{(1)}(y_2 | y_1) \cdot P_3^{(1,2)}(y_3 | y_1, y_2)$$

$$= P(y_1, y_2, y_3) .$$

Here, the third equality follows from the definitions (1.8-14) of the functions F_i, and the last equality follows from the full conditioning formula (1.5-11) for the joint density function P. Thus we see that P_Y, the joint density function of the three random variables whose simultaneous sample values x_1, x_2 and x_3 are obtained by inverting Eqs. (1.8-15), coincides with the joint density function of the three random variables X_1, X_2 and X_3. This means that x_1, x_2 and x_3 are in fact simultaneous sample values of X_1, X_2 and X_3.   QED

In a sense, the joint inversion method merely consists of successive applications of the simple inversion method: Solving Eq. (1.8-15a) generates a value x_1 for X_1 independently of X_2 and X_3; solving Eq. (1.8-15b) then generates a value x_2 for X_2, given that $X_1 = x_1$ but independently of X_3; and solving Eq. (1.8-15c) finally generates a value x_3 for X_3, given that $X_1 = x_1$ and $X_2 = x_2$. But from the point of view of the RVT theorem, the joint inversion method works because the r_i-to-x_i transformation defined by Eqs. (1.8-15) has the special property that its Jacobian is precisely the joint density function $P(x_1, x_2, x_3)$.

The joint inversion generating method obviously entails a lot of analytical work — first in the calculation of the subordinate density functions, then in the calculation of their integrals according to Eqs. (1.8-14), and finally in the inversion of the transformation formulas (1.8-15). Therefore, in any application of the method, one should make some effort to determine which of the 3! possible ways of conditioning the joint density function $P(x_1,x_2,x_3)$ will lead to the simplest implementation of the method. We should also note that, in generating any of the three x_i-values in Eq. (1.8-15), it is permissible to use the simple *rejection* method instead of the simple *inversion* method; e.g., if Eq. (1.8-15b) is too complicated to solve for x_2, then we could use the single-variable rejection method to generate x_2 according to the density function $P_2^{(1)}(x_2 \mid x_1)$.

We shall conclude this section by using the joint inversion generating method to derive a generating algorithm for the normal random variable $N(m,a^2)$. This generating algorithm will be used often in our later work. And the derivation of the algorithm given below will serve to illustrate many of the key random variable concepts and techniques that have been presented in this chapter.

The distribution function corresponding to the normal density function (1.4-9) cannot be computed analytically for essentially the same reason that the integral

$$I = \int_{-\infty}^{\infty} dx \exp(-x^2)$$

cannot be calculated straightforwardly. It may be recalled that the usual trick for evaluating this integral is to first write

$$I^2 = \left(\int_{-\infty}^{\infty} dx_1 \exp(-x_1^2) \right) \left(\int_{-\infty}^{\infty} dx_2 \exp(-x_2^2) \right)$$

$$= \int_{-\infty}^{\infty} dx_1 \int_{-\infty}^{\infty} dx_2 \exp[-(x_1^2+x_2^2)],$$

and to then transform from the Cartesian integration variables (x_1,x_2) to the polar variables (s,θ): Since $s^2=x_1^2+x_2^2$ and $dx_1dx_2=sdsd\theta$, then the double integral becomes under this transformation

$$I^2 = \int_0^{\infty} ds \int_0^{2\pi} d\theta \, s \exp(-s^2) = 2\pi \int_0^{\infty} (1/2) \, du \exp(-u) = \pi,$$

from which we conclude that $I=\sqrt{\pi}$.

Keeping this integration trick in mind, let us consider now the problem of generating simultaneous sample values of *two* random variables X_1 and X_2 whose *joint* density function is

$$P(x_1,x_2) = \prod_{i=1}^{2} (2\pi a^2)^{-1/2} \exp[-(x_i - m)^2 / 2a^2]$$

$$= (2\pi a^2)^{-1} \exp\{-[(x_1 - m)^2 + (x_2 - m)^2] / 2a^2\}.$$

From the form of this joint density function, it is clear that X_1 and X_2 are *statistically independent* random variables, and *each* is $N(m,a^2)$. Now, we obviously cannot generate simultaneous sample values of X_1 and X_2 by a direct application of the joint inversion method. However, consider the two random variables S and Θ, which are functionally related to the random variables X_1 and X_2 by the polar transformation formulas

$$X_1 = m + S\cos\Theta, \quad X_2 = m + S\sin\Theta .$$

It follows from the RVT corollary (1.6-5) that the joint density function of S and Θ is

$$Q(s,\theta) = P(x_1,x_2) \left| \frac{\partial(x_1,x_2)}{\partial(s,\theta)} \right|$$

$$= (2\pi a^2)^{-1} \exp\{-[(x_1 - m)^2 + (x_2 - m)^2] / 2a^2\} \left| \frac{\partial(x_1,x_2)}{\partial(s,\theta)} \right| ,$$

where x_1 and x_2 on the right are now regarded as the *functions*

$$x_1 = m + s\cos\theta, \quad x_2 = m + s\sin\theta .$$

It is easy to show from these last relations that

$$(x_1 - m)^2 + (x_2 - m)^2 = s^2 \quad \text{and} \quad \left| \frac{\partial(x_1,x_2)}{\partial(s,\theta)} \right| = s .$$

So we conclude that the joint density function Q of S and Θ is

$$Q(s,\theta) = (2\pi a^2)^{-1} s \exp(-s^2 / 2a^2) \quad (0 \le s < \infty, \ 0 \le \theta < 2\pi).$$

Now this joint density function Q, *unlike* the joint density function P, is very amenable to the joint inversion generating method: The relevant subordinate density functions are

$$Q_1(s) = \int_0^{2\pi} Q(s,\theta')\,d\theta' = a^{-2} s \exp(-s^2 / 2a^2),$$

$$Q_2^{(1)}(\theta \mid s) = \frac{Q(s,\theta)}{\displaystyle\int_0^{2\pi} Q(s,\theta')\,d\theta'} = (2\pi)^{-1}.$$

So the corresponding F_i functions in Eqs. (1.8-14) are

$$F_1(s) \equiv \int_0^s Q_1(s') \, ds' = 1 - \exp(-s^2/2a^2),$$

$$F_2(\theta;s) \equiv \int_0^\theta Q_2^{(1)}(\theta' \mid s) \, d\theta' = \frac{\theta}{2\pi}.$$

Now we are in a position to make use of Eqs. (1.8-15). Specifically, letting r_1 and r_2 be two independent unit uniform random numbers, we first set $F_1(s) = r_1$ and solve for s, and we then set $F_2(\theta;s) = r_2$ and solve for θ. The results, after replacing r_1 by the equally serviceable unit uniform random number $1 - r_1$, are

$$s = a[2 \ln(1/r_1)]^{1/2}, \quad \theta = 2\pi r_2. \tag{1.8-16a}$$

Then taking

$$x_1 = m + s \cos\theta, \quad x_2 = m + s \sin\theta, \tag{1.8-16b}$$

in accordance with the formulas relating X_1 and X_2 to S and Θ, we finally obtain the desired simultaneous sample values x_1 and x_2 of the statistically independent normal random variables X_1 and X_2. So, although we cannot analytically transform a *single* unit uniform random number r into a sample value x of $N(m,a^2)$, Eqs. (1.8-16) provide a simple recipe for analytically transforming a statistically independent *pair* of unit uniform random numbers r_1 and r_2 into a statistically independent *pair* of sample values x_1 and x_2 of $N(m,a^2)$. We shall find frequent use for the generating algorithm (1.8-16) in later chapters of this book.

- 2 -

GENERAL FEATURES OF A MARKOV
PROCESS

In this chapter we shall use the concepts of random variable theory set forth in Chapter 1 to define a *Markov process* and broadly frame its fundamental properties. This will involve introducing the key functions that are used to describe Markov processes, as well as deriving some general equations which those functions must obey. We shall quickly discover that any substantive characterization of a Markov process requires that we specify the form of what is called the *propagator density function*. In Chapters 3 and 4 we shall consider two different ways of specifying that critical function, which specifications lead to the two principal classes of Markov processes called "continuous" and "jump." In the present chapter we shall see how the propagator density function comes to play its pivotal role in Markov process theory, and we shall develop that theory as fully as we can without committing ourselves to a specific form for the propagator density function.

2.1 THE MARKOV STATE DENSITY FUNCTION

We consider a time-evolving or "dynamical" system whose possible states can be represented by points on the real axis, and we let

$$X(t) \equiv \text{the state point, or \textbf{state}, of the system at time } t. \quad (2.1\text{-}1)$$

We shall assume that the value of X at some initial time t_0 is fixed,

$$X(t_0) = x_0, \quad (2.1\text{-}2)$$

but that $X(t)$ for any $t > t_0$ can be predicted only probabilistically; more specifically, we assume that $X(t)$ for any given $t > t_0$ is a *random variable*, as defined in Section 1.2. Since it makes sense to inquire about the state of the system at successive instants t_1, t_2, ..., t_n, where $t_0 < t_1 < t_2 < \ ... \ < t_n$, then we can ascribe to the corresponding n random variables $X(t_1)$, $X(t_2)$, ..., $X(t_n)$ a *joint density function* $P_n^{(1)}$, which is defined as follows:

$$P_n^{(1)}(x_n, t_n; x_{n-1}, t_{n-1}; \ ... \ ; x_1, t_1 \mid x_0, t_0) \, dx_n \, dx_{n-1} \cdots dx_1$$

$$\equiv \text{Prob}\{ X(t_i) \in [x_i, x_i + dx_i) \text{ for } i = 1, 2, ..., n,$$

$$\text{given that } X(t_0) = x_0, \text{ with } t_0 \leq t_1 \leq \ ... \ \leq t_n \}. \quad (2.1\text{-}3)$$

If all these assumptions are satisfied, then we say that $X(t)$ is a **stochastic process**.

It is evident that a stochastic process $X(t)$ has infinitely many joint density functions $P_n^{(1)}$, corresponding to $n = 1, 2, \ ...$. And associated with each of these joint density functions is a plethora of subordinate density functions; for example,

$$P_{n-j}^{(j+1)}(x_n, t_n; \ ...; x_{j+1}, t_{j+1} \mid x_j, t_j; \ ... \ ; x_1, t_1; x_0, t_0),$$

is defined to be the joint density function of the $n - j$ random variables $X(t_{j+1})$, ..., $X(t_n)$ given the $j + 1$ conditions $X(t_0) = x_0$, $X(t_1) = x_1$, ..., $X(t_j) = x_j$. Notice that the *subscript* on the density function $P_k^{(j)}$ refers to the number of (x, t) pairs to the *left* of the "given" bar, while the *superscript* refers to the number of (x, t) pairs to the *right* of the "given" bar; thus, $P_k^{(j)}$ is a k-variate joint density function with j conditionings.

It is always possible to calculate the function $P_{n-1}^{(1)}$ from the function $P_n^{(1)}$ by simply integrating the latter over any one of the variables x_1, ..., x_n. However, it is *not* in general possible to deduce the function $P_{n+1}^{(1)}$ from the function $P_n^{(1)}$. This "open-ended" nature of the density functions for a general stochastic process usually makes any substantive analysis extremely difficult. But we shall be concerned here with only a very restricted subclass of stochastic processes, namely those that have the "past-forgetting" property that, for all $j \geq 2$ and $t_{i-1} \leq t_i$,

$$P_1^{(j)}(x_j, t_j \mid x_{j-1}, t_{j-1}; \ ... \ ; x_1, t_1; x_0, t_0)$$

$$= P_1^{(1)}(x_j, t_j \mid x_{j-1}, t_{j-1}) \equiv P(x_j, t_j \mid x_{j-1}, t_{j-1}). \quad (2.1\text{-}4)$$

This is called the **Markov property**, and it says that *only the most recent conditioning matters*: Given that $X(t') = x'$, then our ability to predict $X(t)$ for any $t > t'$ will not be enhanced by a knowledge of any values of the process earlier than t'. Any stochastic process $X(t)$ that has this past-forgetting property is called a *Markovian* stochastic process, or more simply, a **Markov process**. In what follows it may always be assumed, unless explicitly stated otherwise, that the stochastic process $X(t)$ under consideration is a Markov process.

The Markov property (2.1-4) breaks the open-endedness of the hierarchy of joint state density functions in a dramatic way. For the joint density function $P_2^{(1)}$ we have

$$P_2^{(1)}(x_2,t_2; x_1,t_1 \mid x_0,t_0)$$

$$= P_1^{(1)}(x_1,t_1 \mid x_0,t_0)\, P_1^{(2)}(x_2,t_2 \mid x_1,t_1; x_0,t_0) \qquad \text{[by (1.5-9d)]}$$

$$= P_1^{(1)}(x_1,t_1 \mid x_0,t_0)\, P_1^{(1)}(x_2,t_2 \mid x_1,t_1). \qquad \text{[by (2.1-4)]}$$

Hence, writing $P_1^{(1)} \equiv P$ in accordance with the notation suggested in Eq. (2.1-4), we have

$$P_2^{(1)}(x_2,t_2; x_1,t_1 \mid x_0,t_0) = P(x_2,t_2 \mid x_1,t_1)\, P(x_1,t_1 \mid x_0,t_0). \qquad (2.1\text{-}5)$$

The same kind of reasoning shows that

$$P_3^{(1)}(x_3,t_3; x_2,t_2; x_1,t_1 \mid x_0,t_0) = P(x_3,t_3 \mid x_2,t_2)\, P(x_2,t_2 \mid x_1,t_1)\, P(x_1,t_1 \mid x_0,t_0),$$

and more generally, for any set of times $t_n \geq t_{n-1} \geq \cdots \geq t_0$,

$$P_n^{(1)}(x_n,t_n; \ldots; x_1,t_1 \mid x_0,t_0) = \prod_{i=1}^{n} P(x_i,t_i \mid x_{i-1},t_{i-1}). \qquad (2.1\text{-}6)$$

So for a *Markov* process, every conditioned state density function $P_n^{(1)}$ can be written solely in terms of the particular conditioned state density function $P_1^{(1)} \equiv P$. The function $P_1^{(1)} \equiv P$ thus becomes the principle focus of our study, and we shall henceforth refer to it as the **Markov state density function**. For future reference, the formal definition of the Markov state density function is [cf. Eq. (2.1-3)]

$$P(x_2,t_2 \mid x_1,t_1)\, dx_2$$

$$\equiv \text{Prob}\{\, X(t_2) \in [x_2, x_2+dx_2), \text{ given } X(t_1) = x_1, \text{ with } t_2 \geq t_1 \,\}. \qquad (2.1\text{-}7)$$

2.2 THE CHAPMAN-KOLMOGOROV EQUATION

Since $P(x_2,t_2 \mid x_1,t_1)$ is a density function with respect to its argument x_2, it must satisfy conditions analogous to Eqs. (1.2-3) and (1.2-4), namely,

$$P(x_2,t_2 \mid x_1,t_1) \geq 0, \tag{2.2-1}$$

$$\int_{-\infty}^{\infty} dx_2\, P(x_2,t_2 \mid x_1,t_1) = 1. \tag{2.2-2}$$

Also, if we let t_2 *equal* its minimum value t_1 then the condition $X(t_1)=x_1$ obviously implies that $X(t_2)=x_1$, or that $X(t_2)$ is the *sure* variable x_1; hence, by (1.2-7), we have the requirement

$$P(x_2, t_2 = t_1 \mid x_1,t_1) = \delta(x_2 - x_1). \tag{2.2-3}$$

Of course, the function P would have to satisfy the preceding three equations even if the process $X(t)$ were not Markovian. A condition upon P that arises specifically because of the Markov property may be deduced as follows: For any three times $t_1 \leq t_2 \leq t_3$, we have

$$P_1^{(1)}(x_3,t_3 \mid x_1,t_1)$$

$$= \int_{-\infty}^{\infty} dx_2\, P_2^{(1)}(x_3,t_3; x_2,t_2 \mid x_1,t_1) \qquad \text{[by (1.5-12a)]}$$

$$= \int_{-\infty}^{\infty} dx_2\, P_1^{(1)}(x_2,t_2 \mid x_1,t_1)\, P_1^{(2)}(x_3,t_3 \mid x_2,t_2; x_1,t_1) \quad \text{[by (1.5-9d)]}$$

$$= \int_{-\infty}^{\infty} dx_2\, P_1^{(1)}(x_2,t_2 \mid x_1,t_1)\, P_1^{(1)}(x_3,t_3 \mid x_2,t_2), \qquad \text{[by (2.1-4)]}$$

where the last step explicitly invokes the assumed Markovian nature of $X(t)$. Interchanging the two factors in the last integrand and then abbreviating $P_1^{(1)}$ everywhere by P, we obtain what is known as the **Chapman-Kolmogorov equation:**

$$P(x_3,t_3 \mid x_1,t_1) = \int_{-\infty}^{\infty} P(x_3,t_3 \mid x_2,t_2)\, P(x_2,t_2 \mid x_1,t_1)\, dx_2$$

$$(t_1 \leq t_2 \leq t_3). \tag{2.2-4}$$

This integral equation, a graphical interpretation of which is given in Fig. 2-1, is essentially a *consistency condition* on the Markov state density function P for any Markov process $X(t)$. As we shall see later, it severely limits the range of acceptable functional forms of P for any Markov process.

Because of the initial condition (2.1-2), the quantity $P(x,t \mid x_0,t_0)$ is of special importance. As we shall see later, time-evolution equations for $P(x,t \mid x_0,t_0)$ can be derived from two specially phrased versions of the Chapman-Kolmogorov equation, obtained by relabeling the variables x_i

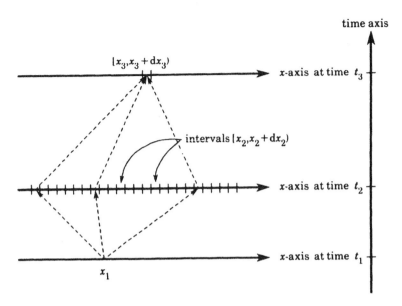

Figure 2-1. Graphical interpretation of the Chapman-Kolmogorov equation (2.2-4). The probability $P(x_3,t_3 \mid x_1,t_1)dx_3$ of going from x_1 at time t_1 to the interval $[x_3,x_3+dx_3)$ at time t_3 can always be written as the sum of the probabilities of this occurring via all possible intervals $[x_2,x_2+dx_2)$ at any fixed intermediate time t_2. When the Markov property (2.1-4) holds, the summand takes the form $[P(x_2,t_2 \mid x_1,t_1)dx_2] \times [P(x_3,t_3 \mid x_2,t_2)dx_3]$, whence the integral equation (2.2-4) for the function P.

and t_i according to the two diagrams in Fig. 2-2. These two "utility" versions of the Chapman-Kolmogorov equation are as follows:

$$P(x,t+\Delta t \mid x_0,t_0) = \int_{-\infty}^{\infty} d\xi \, P(x,t+\Delta t \mid x-\xi,t) \, P(x-\xi,t \mid x_0,t_0)$$

$$(t_0 < t < t+\Delta t); \quad (2.2\text{-}5a)$$

$$P(x,t \mid x_0,t_0) = \int_{-\infty}^{\infty} d\xi \, P(x,t \mid x_0+\xi,t_0+\Delta t_0) \, P(x_0+\xi,t_0+\Delta t_0 \mid x_0,t_0)$$

$$(t_0 < t_0+\Delta t_0 < t). \quad (2.2\text{-}5b)$$

Equation (2.2-5a) will give rise to equations that govern the x and t behavior of $P(x,t \mid x_0,t_0)$ for fixed x_0 and t_0, the so-called "forward" time-evolution equations. And Eq. (2.2-5b) will give rise to equations that govern the x_0 and t_0 behavior of $P(x,t \mid x_0,t_0)$ for fixed x and t, the so-called "backward" time-evolution equations. Notice that Δt and Δt_0 here need *not* be infinitesimally small.

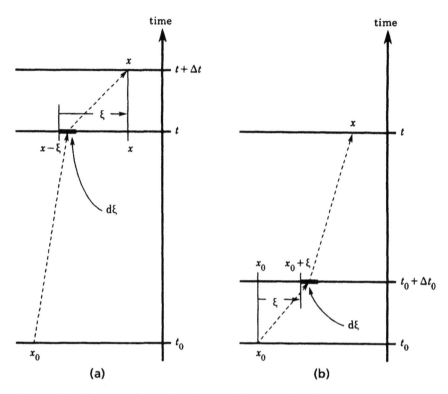

Figure 2-2. Showing the initial, intermediate and final state variables for the two special Chapman-Kolmogorov equations (2.2-5a) and (2.2-5b).

Another version of the Chapman-Kolmogorov equation that occasionally proves useful for determining $P(x,t \mid x_0,t_0)$ is obtained as follows: By an obvious extension of Eq. (1.5-12a), we have for any set of times $t_0 < t_1 < \cdots < t_n$,

$$P(x_n,t_n \mid x_0,t_0) = \int_{-\infty}^{\infty} dx_{n-1} \cdots \int_{-\infty}^{\infty} dx_1 \, P_n^{(1)}(x_n,t_n; \ldots; x_1,t_1 \mid x_0,t_0).$$

Substituting from Eq. (2.1-6) on the right then gives the result

$$P(x_n,t_n \mid x_0,t_0) = \int_{-\infty}^{\infty} dx_1 \cdots \int_{-\infty}^{\infty} dx_{n-1} \prod_{i=1}^{n} P(x_i,t_i \mid x_{i-1},t_{i-1})$$

$$(t_0 < t_1 < \ldots < t_n). \quad (2.2\text{-}6)$$

This equation, which evidently reduces to Eq. (2.2-4) for $n=2$, will be referred to as the **compounded Chapman-Kolmogorov equation.**

2.3 FUNCTIONS OF STATE AND THEIR AVERAGES

Any univariate function g that takes a value $g(x)$ for any possible system state x can be regarded as a **function of state**. We can also think of a function of state as a *random variable* $g(X(t))$, which is *defined* to be "the function g of the random variable $X(t)$" [see the discussion of Eqs. (1.6-1)]. The **conditional state average**,

$$\langle g(X(t)) \mid X(t') = x' \rangle = \int_{-\infty}^{\infty} dx\, g(x)\, P(x,t \mid x',t') \quad (t' \leq t), \quad (2.3\text{-}1)$$

may accordingly be viewed *either* as (i) the average of the function g with respect to the random variable $X(t)$ given that $X(t') = x'$, *or* as (ii) the mean of the random variable $g(X(t))$ given that $X(t') = x'$ [see the discussion following Eq. (1.6-12)].

Similarly, any bivariate function g can be used to define a **two-time function of state** $g(X(t_1), X(t_2))$ for $t_1 \leq t_2$, and its average or mean, given $X(t') = x'$ for some $t' \leq t_1$, is

$$\langle g(X(t_1), X(t_2)) \mid X(t') = x' \rangle = \int_{-\infty}^{\infty} dx_1 \int_{-\infty}^{\infty} dx_2\, g(x_1, x_2)\, P_2^{(1)}(x_2,t_2;\, x_1,t_1 \mid x',t').$$

With the Markov simplification (2.1-5), this is

$$\langle g(X(t_1), X(t_2)) \mid X(t') = x' \rangle$$
$$= \int_{-\infty}^{\infty} dx_1 \int_{-\infty}^{\infty} dx_2\, g(x_1, x_2)\, P(x_2,t_2 \mid x_1,t_1)\, P(x_1,t_1 \mid x',t')$$
$$(t' \leq t_1 \leq t_2) \quad (2.3\text{-}2)$$

If t' in Eqs. (2.3-1) and (2.3-2) is taken to be the initial time t_0, then we shall simply omit the conditioning notation; thus, the **initially conditioned averages** will be denoted by

$$\langle g(X(t)) \mid X(t_0) = x_0 \rangle \equiv \langle g(X(t)) \rangle$$
$$= \int_{-\infty}^{\infty} dx\, g(x)\, P(x,t \mid x_0,t_0) \quad (t_0 \leq t), \quad (2.3\text{-}3)$$

and

$$\langle g(X(t_1), X(t_2)) \mid X(t_0) = x_0 \rangle \equiv \langle g(X(t_1), X(t_2)) \rangle$$
$$= \int_{-\infty}^{\infty} dx_1 \int_{-\infty}^{\infty} dx_2\, g(x_1, x_2)\, P(x_2,t_2 \mid x_1,t_1)\, P(x_1,t_1 \mid x_0,t_0)$$
$$(t_0 \leq t_1 \leq t_2). \quad (2.3\text{-}4)$$

The most important applications of the univariate average (2.3-3) are to the two functions $g(x)=x$ and $g(x)=x^2$. This is because [see Eqs. (1.3-6) and (1.3-8)] the initially conditioned **mean** and **variance** of the process $X(t)$ are given respectively by

$$\text{mean}\{X(t)\} = \langle X(t)\rangle \quad (t_0 \leq t) \tag{2.3-5}$$

and

$$\text{var}\{X(t)\} = \langle X^2(t)\rangle - \langle X(t)\rangle^2 \quad (t_0 \leq t). \tag{2.3-6}$$

And the initially conditioned **standard deviation** of $X(t)$ is given by

$$\text{sdev}\{X(t)\} = [\text{var}\{X(t)\}]^{1/2} = [\langle X^2(t)\rangle - \langle X(t)\rangle^2]^{1/2} \quad (t_0 \leq t). \tag{2.3-7}$$

Given that $X(t_0)=x_0$, then at any time $t>t_0$ we can "usually expect" to find $X(t)$ to be within roughly $\text{sdev}\{X(t)\}$ of the value $\langle X(t)\rangle$.

The most important application of the bivariate average (2.3-4) is to the function $g(x_1,x_2)=x_1 x_2$. This is because [see Eq. (1.5-17)] the **covariance** of $X(t_1)$ and $X(t_2)$ for $t_0 \leq t_1 \leq t_2$, given that $X(t_0)=x_0$, is defined by

$$\text{cov}\{X(t_1),X(t_2)\} \equiv \langle X(t_1)X(t_2)\rangle - \langle X(t_1)\rangle\langle X(t_2)\rangle \quad (t_0 \leq t_1 \leq t_2). \tag{2.3-8}$$

As shown in Eq. (1.6-18), $\text{cov}\{X(t_1),X(t_2)\}$ is always bounded in absolute value by $\text{sdev}\{X(t_1)\}\cdot\text{sdev}\{X(t_2)\}$. If $\text{cov}\{X(t_1),X(t_2)\}$ assumes its positive bound then $X(t_1)$ and $X(t_2)$ are said to be *maximally correlated*, while if $\text{cov}\{X(t_1),X(t_2)\}$ assumes its negative bound then $X(t_1)$ and $X(t_2)$ are said to be *maximally anti-correlated*. If $\text{cov}\{X(t_1),X(t_2)\}=0$, then $X(t_1)$ and $X(t_2)$ are said to be *uncorrelated*. One way in which $X(t_1)$ and $X(t_2)$ can be uncorrelated is for them to be *statistically independent*, in the sense that $P(x_2,t_2 \,|\, x_1,t_1)$ is independent of x_1 (and hence also independent of t_1); because in that case we have from Eq. (2.3-4),

$$\langle X(t_1)X(t_2)\rangle = \int_{-\infty}^{\infty} dx_1 \int_{-\infty}^{\infty} dx_2 \, x_1 x_2 P(x_2,t_2|x_1,t_1) P(x_1,t_1|x_0,t_0)$$

$$= \int_{-\infty}^{\infty} dx_1 \int_{-\infty}^{\infty} dx_2 \, x_1 x_2 P(x_2,t_2) P(x_1,t_1)$$

$$= \int_{-\infty}^{\infty} dx_1 \, x_1 P(x_1,t_1) \int_{-\infty}^{\infty} dx_2 \, x_2 P(x_2,t_2)$$

$$= \langle X(t_1)\rangle \langle X(t_2)\rangle.$$

So if $X(t_1)$ and $X(t_2)$ are statistically independent, then $\text{cov}\{X(t_1),X(t_2)\}$ vanishes, implying by definition that $X(t_1)$ and $X(t_2)$ are uncorrelated. But notice that $X(t_1)$ and $X(t_2)$ could be uncorrelated (i.e., have a vanishing covariance) *without* being statistically independent.

It is an instructive exercise to show directly for a Markov process that

$$\text{cov}\{X(t_1),X(t_2=t_1)\} = \text{var}\{X(t_1)\}, \qquad (2.3\text{-}9)$$

as we should expect on the basis of Eq. (1.5-18). We have from Eq. (2.3-4),

$$\langle X(t_1)X(t_2=t_1)\rangle = \int_{-\infty}^{\infty} dx_1 \int_{-\infty}^{\infty} dx_2 \, x_1 x_2 \, P(x_2,t_2=t_1 \,|\, x_1,t_1) \, P(x_1,t_1 \,|\, x_0,t_0)$$

$$= \int_{-\infty}^{\infty} dx_1 \int_{-\infty}^{\infty} dx_2 \, x_1 x_2 \, \delta(x_2-x_1) \, P(x_1,t_1 \,|\, x_0,t_0)$$

$$= \int_{-\infty}^{\infty} dx_1 \, x_1 x_1 \, P(x_1,t_1 \,|\, x_0,t_0) = \langle X^2(t_1)\rangle.$$

where the second equality follows from Eq. (2.2-3). Substituting this result into Eq. (2.3-8) with $t_2=t_1$, and then recalling the definition of the variance in Eq. (2.3-6), we obtain Eq. (2.3-9).

2.4 THE MARKOV PROPAGATOR

Suppose that our system — our Markov process — is in the state x at time t; i.e., suppose that $X(t)=x$. Then by the infinitesimally later time $t+dt$, the system will have evolved to some new state that is *displaced* from x by the amount

$$\Xi(dt; x,t) \equiv X(t+dt) - X(t), \text{ given } X(t)=x. \qquad (2.4\text{-}1)$$

Notice that the condition "$X(t)=x$" in this definition affects the *first* term on the right side as well as the second. This state displacement $\Xi(dt; x,t)$ from state x during time $[t,t+dt)$ is clearly a *random variable*; we shall call it the **propagator** of the process $X(t)$. Like any random variable, the propagator is completely specified by its density function. We shall denote the density function of $\Xi(dt; x,t)$ by $\Pi(\xi \,|\, dt; x,t)$, and refer to it as the **propagator density function**; thus we have, by definition,

$$\Pi(\xi \,|\, dt; x,t) \, d\xi = \text{Prob}\{\, \Xi(dt; x,t) \in [\xi, \xi+d\xi) \,\}. \qquad (2.4\text{-}2)$$

Evidently, the propagator $\Xi(dt; x,t)$ tells us where the process, in state x at time t, will be at the infinitesimally later time $t+dt$; specifically, the process will be in the state $x+\Xi(dt; x,t)$. Here it is perhaps appropriate to remark that we are regarding dt as a real variable whose allowed range is the open interval $(0,\varepsilon)$, where ε is positive but "arbitrarily close to zero." Although we shall always take $\varepsilon \ll 1$, so that $(dt)^2$ is negligibly small compared to dt, the precise value of ε will depend upon the situation. For example, if we are considering a specific differentiable function h of the variable t, then we can always find a positive number ε (which may depend upon t) such that $h(t+dt)$ can for most purposes be equated to $h(t) + h'(t)dt$ for all dt in $(0,\varepsilon)$. In our work here we shall never need to know the actual value of ε, but merely that such a value exists.

So the random variable $\Xi(dt; x,t)$, and its associated density function $\Pi(\xi \mid dt; x,t)$, depend parametrically on the three real variables dt, x and t. We have chosen to notationally separate the parameter dt from the parameters x and t because, as we shall see later, it is possible for $\Pi(\xi \mid dt; x,t)$ to be explicitly independent of x or t, but *not* of dt. In fact, it usually turns out that the *moments* of $\Xi(dt; x,t)$ have the analytical structure

$$\langle \Xi^n(dt; x,t)\rangle \equiv \int_{-\infty}^{\infty} d\xi \, \xi^n \, \Pi(\xi \mid dt; x,t) = B_n(x,t) \, dt + o(dt)$$

$$(n=1,2,\ldots), \quad (2.4\text{-}3a)$$

where $B_1(x,t)$, $B_2(x,t)$, ... are all well-behaved functions of x and t, and $o(dt)$ denotes terms that go to zero with dt faster than dt:

$$o(dt)/dt \to 0 \quad \text{as} \quad dt \to 0.$$

Equation (2.4-3a) essentially *defines* the functions $B_n(x,t)$, as can be seen a little more clearly by dividing through by dt and then noting that $o(dt)/dt$ can be made arbitrarily small by taking dt small enough; in other words, for vanishingly small dt we have

$$B_n(x,t) = \frac{1}{dt} \langle \Xi^n(dt; x,t)\rangle = \frac{1}{dt} \int_{-\infty}^{\infty} d\xi \, \xi^n \, \Pi(\xi \mid dt; x,t)$$

$$(n=1,2,\ldots). \quad (2.4\text{-}3b)$$

We shall call $B_n(x,t)$ the nth **propagator moment function** of the Markov process $X(t)$. Much of what follows in this chapter will be devoted to deriving some fundamental equations involving these propagator moment functions. Obviously, those equations will have meaning *only* if the propagator density function $\Pi(\xi \mid dt; x,t)$ is such that the above definition of $B_n(x,t)$ truly makes sense. As we shall discover in the following chapters, the propagator moment functions $B_n(x,t)$ are nearly always well defined for Markov processes of practical interest.

Notice that the propagator $\Xi(dt; x,t)$, its density function $\Pi(\xi \mid dt; x,t)$ and its moments $\langle \Xi^n(dt; x,t)\rangle$ all need be specified *only to lowest order in* dt. This is because we can always choose the allowed range $(0,\varepsilon)$ of the variable dt to be arbitrarily small.

There is a very important connection between the propagator density function Π and the Markov state density function P. To expose that connection, we observe that the definition (2.1-7) of P implies that, for dt a positive infinitesimal,

$$P(x+\xi,t+dt \mid x,t)d\xi$$
$$= \text{Prob}\{ X(t+dt) \in [x+\xi, x+\xi+d\xi) \text{ given that } X(t)=x \}$$
$$= \text{Prob}\{ [X(t+dt)-x] \in [\xi,\xi+d\xi) \text{ given that } X(t)=x \}$$
$$= \text{Prob}\{ [X(t+dt)-X(t)] \in [\xi,\xi+d\xi) \text{ given that } X(t)=x \}$$
$$= \text{Prob}\{ \Xi(dt; x,t) \in [\xi,\xi+d\xi) \},$$

where the last equality follows from the definition (2.4-1) of the propagator $\Xi(dt; x,t)$. Hence, by the definition (2.4-2) of the propagator density function $\Pi(\xi \mid dt; x,t)$, we conclude that [see Fig. 2-3]

$$\Pi(\xi \mid dt; x,t) = P(x+\xi,t+dt \mid x,t). \tag{2.4-4}$$

So we see that the propagator density function $\Pi(\xi \mid dt; x,t)$ is just the Markov state density function $P(x',t' \mid x,t)$ in which x' exceeds x by the amount ξ, while t' exceeds t by the *infinitesimal* amount dt. Notice that ξ, unlike dt, is *not* assumed to be an infinitesimal; ξ is an unrestricted real variable.

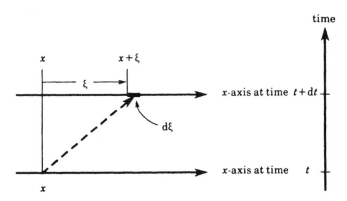

Figure 2-3. Illustrating the transition of the process from state x at time t to the interval $[x+\xi, x+\xi+d\xi)$ at time $t+dt$. The infinitesimal nature of $d\xi$ allows us to write the probability for this transition to occur as $P(x+\xi,t+dt \mid x,t)d\xi$. And because dt is likewise an infinitesimal, we can also express this transition probability as $\Pi(\xi \mid dt; x,t)d\xi$.

Because $\Pi(\xi \,|\, dt;\, x,t)$ is a *density* function with respect to its argument ξ, then it must satisfy the two conditions

$$\Pi(\xi \,|\, dt;\, x,t) \;\geq\; 0 \tag{2.4-5}$$

and

$$\int_{-\infty}^{\infty} d\xi \; \Pi(\xi \,|\, dt;\, x,t) \;=\; 1. \tag{2.4-6}$$

Furthermore, since it is clear from the definition (2.4-1) that $\Xi(dt=0;\, x,t)$ must be the *sure* number zero, then we must have [see (1.2-7)]

$$\Pi(\xi \,|\, dt=0;\, x,t) \;=\; \delta(\xi). \tag{2.4-7}$$

In addition to these three basic conditions on the propagator density function Π, there is one more important requirement that follows from the fundamental relation (2.4-4) and the Chapman-Kolmogorov equation (2.2-4). According to the latter equation, P must satisfy for any $\Delta t > 0$ and any value of a between 0 and 1,

$$P(x+\xi, t+\Delta t \,|\, x,t)$$
$$= \int_{-\infty}^{\infty} d\xi_1 \; P(x+\xi, t+\Delta t \,|\, x+\xi_1, t+a\Delta t) \; P(x+\xi_1, t+a\Delta t \,|\, x,t)$$
$$= \int_{-\infty}^{\infty} d\xi_1 \; P(x+\xi_1+\xi-\xi_1,\, t+a\Delta t+(1-a)\Delta t \,|\, x+\xi_1, t+a\Delta t)$$
$$\times P(x+\xi_1, t+a\Delta t \,|\, x,t).$$

Then taking Δt to be the *infinitesimal* dt, it follows on substituting from Eq. (2.4-4) that

$$\Pi(\xi \,|\, dt;\, x,t) \;=\; \int_{-\infty}^{\infty} d\xi_1 \, \Pi(\xi-\xi_1 \,|\, (1-a)dt;\, x+\xi_1, t+a\,dt) \, \Pi(\xi_1 \,|\, a\,dt;\, x,t). \tag{2.4-8}$$

This condition, which must hold to lowest order in dt for all a between 0 and 1, will be called the **Chapman-Kolmogorov condition** on the propagator density function. It and conditions (2.4-5) – (2.4-7) serve to restrict the possible functional forms that can be ascribed to the propagator density function $\Pi(\xi \,|\, dt;\, x,t)$.

We can gain a little more insight into the Chapman-Kolmogorov condition (2.4-8) by writing it in the equivalent form

$$\Pi(\xi \,|\, dt;\, x,t) \;=\; \int_{-\infty}^{\infty} d\xi_1 \int_{-\infty}^{\infty} d\xi_2 \; \Pi(\xi_1 \,|\, a\,dt;\, x,t)$$
$$\times \Pi(\xi_2 \,|\, (1-a)dt;\, x+\xi_1, t+a\,dt) \, \delta(\xi-\xi_1-\xi_2),$$

from which Eq. (2.4-8) readily follows by integrating over ξ_2 with the help of the delta function. Viewing this last equation in light of the RVT theorem (1.6-4), we see that it implies the following functional relation among three random variables corresponding to ξ, ξ_1 and ξ_2:

$$\varXi(dt; X(t),t) = \varXi(\alpha dt; X(t),t) + \varXi((1-\alpha)dt; X(t)+\varXi(\alpha dt;X(t),t), t+\alpha dt).$$
$$(2.4-9)$$

Although this relation may at first glance appear to be very complicated, it is merely the statement that {the change in the process over $(t,t+dt)$} must be equal to {the change in the process over $(t,t+\alpha dt)$} plus {the *subsequent* change in the process over $(t+\alpha dt, t+dt)$}. The *random variable* relation (2.4-9), which again must hold to lowest order in dt for all α between 0 and 1, is entirely equivalent to the *density function* relation (2.4-8). *Either* may be referred to as the "Chapman-Kolmogorov condition."

The importance of the propagator density function Π lies in the fact that it completely determines the Markov state density function P, from which of course everything knowable about the (Markov) process $X(t)$ can be computed. This fundamental fact is not readily apparent from Eq. (2.4-4), which shows rather that a knowledge of P implies a knowledge of Π; however, it can be proved by reasoning as follows.

In the compounded Chapman-Kolmogorov equation (2.2-6), let $t_n = t$ be any value greater that t_0, and let the points $t_1, t_2, ..., t_{n-1}$ divide the interval $[t_0, t]$ into n subintervals of equal length $(t-t_0)/n$. Further, change the integration variables in that equation according to

$$x_i \rightarrow \xi_i \equiv x_i - x_{i-1} \quad (i=1,...,n-1).$$

Notice that the Jacobian determinant of this transformation has zeros everywhere on one side of its main diagonal, and ones everywhere on its main diagonal, so $dx_1 \cdots dx_{n-1} = d\xi_1 \cdots d\xi_{n-1}$. Notice also that this transformation implies that

$$x_i = x_{i-1} + \xi_i = [x_{i-2} + \xi_{i-1}] + \xi_i = ... = x_0 + \xi_1 + ... + \xi_i.$$

Finally, relabel $x_n = x$, and *define* $\xi_n \equiv x - x_{n-1}$. With all these changes, the compounded Chapman-Kolmogorov equation (2.2-6) becomes

$$P(x,t|x_0,t_0) = \int_{-\infty}^{\infty} d\xi_1 \cdots \int_{-\infty}^{\infty} d\xi_{n-1}$$
$$\times \prod_{i=1}^{n} P(x_{i-1}+\xi_i, t_{i-1}+(t-t_0)/n \,|\, x_{i-1},t_{i-1}), \quad (2.4-10)$$

wherein

$$t_i = t_{i-1} + (t-t_0)/n \qquad (i=1,...,n-1), \qquad \text{(2.4-11a)}$$

$$x_i = x_0 + \xi_1 + ... + \xi_i \qquad (i=1,...,n-1), \qquad \text{(2.4-11b)}$$

$$\xi_n \equiv x - x_0 - \xi_1 - ... - \xi_{n-1}. \qquad \text{(2.4-11c)}$$

We emphasize that Eq. (2.4-10) is merely the compounded Chapman-Kolmogorov equation (2.2-6) in which the intermediate times t_1, t_2, ..., t_{n-1} divide the interval $(t-t_0)$ *evenly*, and the integrations are taken with respect to the state *displacement* variables ξ_i instead of the state variables x_i. Figure 2-4 schematizes the relations among the variables in the above equations for the particular case $n=4$. Now suppose we choose n *so large* that

$$(t-t_0)/n = dt, \text{ an infinitesimal.} \qquad \text{(2.4-11d)}$$

In that case, the P-factors in the integrand of Eq. (2.4-10) all become Π-factors by virtue of Eq. (2.4-4), and we obtain

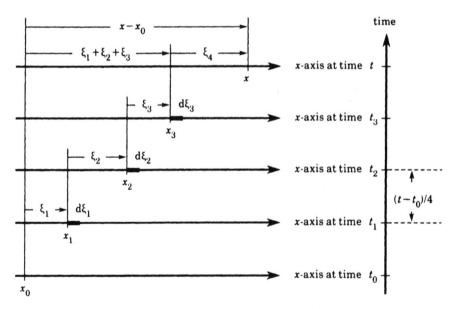

Figure 2-4. Illustrating the relations among the variables in Eq. (2.4-10) for the case $n=4$. Notice that ξ_1, ξ_2 and ξ_3 are integration variables, while ξ_4 is simply defined to be the difference between $(x-x_0)$ and $(\xi_1+\xi_2+\xi_3)$. This figure is a graphical encapsulation of relations (2.4-11a) – (2.4-11c) in which, for the sake of clarity, all ξ_i's have been taken positive.

$$P(x,t \mid x_0,t_0) = \int_{-\infty}^{\infty} d\xi_1 \cdots \int_{-\infty}^{\infty} d\xi_{n-1} \prod_{i=1}^{n} \Pi(\xi_i \mid dt; x_{i-1},t_{i-1}), \quad (2.4\text{-}12)$$

where now *all four* of Eqs. (2.4-11) apply. This result shows that if we specify $\Pi(\xi \mid dt; x',t')$ as a function of ξ for all x', all $t' \in [t_0,t)$, and all infinitesimally small dt, then $P(x,t \mid x_0,t_0)$ is uniquely determined for all x.

Although Eq. (2.4-12) gives proof to the assertion that the propagator density function Π completely determines the Markov state density function P, that formula would not seem to afford a universally simple means of actually computing P from Π. Notice in particular the potentially complicated dependence of the integrand in Eq. (2.4-12) on the integration variables through the functions x_i defined in Eq. (2.4-11b). In the next section we shall consider a different way of using the Chapman-Kolmogorov equation and the fundamental relation (2.4-4) to compute P from a knowledge of Π. But first it might be appropriate to make some general observations about the rationale for *any* method that seeks to calculate P from Π.

To say that we must specify Π in order to determine P is, because of Eq. (2.4-4), to say that we must specify $P(x',t' \mid x,t)$ for t' *infinitesimally* larger than t in order to calculate $P(x',t' \mid x,t)$ for *any* t' larger than t. This may seem like circular reasoning, but it does have an obvious precedent in mathematical physics in the elementary problem of a particle of mass m moving along the x-axis. The "state" of that simple system at time t can be represented by the ordered pair $[x_t,p_t]$, the position and momentum of the particle at time t. Now, it is well known that in order to calculate the particle's state at any time $t > t_0$, given the state at time t_0, it is necessary to specify the force function, $F(x,p;t)$. But the ultimate purpose of this force function is to specify the "state at time $t + dt$ given the state at time t," namely

$$[x_{t+dt}, p_{t+dt} \mid x_t, p_t] = [x_t + (p_t/m)dt, \, p_t + F(x_t,p_t;t)dt]. \quad (2.4\text{-}13)$$

Once this "conditional state" has been specified, the mathematical machinery of integral calculus may then be used to deduce an explicit formula for $[x_t,p_t \mid x_0, p_0]$ for *any* $t > t_0$. Of course, in the case of a Markov process, it is not generally possible to specify with certainty the value of $X(t+dt)$ given the value of $X(t)$. The most we can do toward that end is to specify the probability distribution for $X(t+dt)$ given the value of $X(t)$. And that is precisely what the propagator density function Π does [recall the definitions (2.4-1) and (2.4-2)].

Equation (2.4-13) also makes it clear that our freedom to specify the form of a particle's state at time $t+dt$ given its state at time t is *not* unlimited; because, even though we have considerable latitude in specifying the form of the force function $F(x,p;t)$, that function can be used in the conditional state formula (2.4-13) in only a very particular way. Similarly, for a Markov process we do not have unlimited freedom to invent propagator density functions $\Pi(\xi \mid dt; x,t)$. We must make sure that any such function satisfies conditions (2.4-5) – (2.4-8), of which the last is especially restrictive. And when those four conditions are supplemented by even rather modest additional requirements, the form of Π can become surprisingly rigid. In Chapters 3 and 4, we shall present two different sets of "reasonable additional requirements" on Π that lead to two particularly interesting and useful classes of Markov processes. More specifically, in Chapter 3 we shall derive the functional form that $\Pi(\xi \mid dt; x,t)$ must have if the random variable $\Xi(dt; x,t)$ is to have the character of a *well behaved infinitesimal*; this will give rise to the class of so-called *continuous* Markov processes. And then in Chapter 4 we shall derive the functional form that $\Pi(\xi \mid dt; x,t)$ must have if the random variable $\Xi(dt; x,t)$ is to be *usually zero but occasionally finite*; this will give rise to the class of so-called *jump* Markov processes. In the present chapter we shall develop some general results, involving either the propagator density function $\Pi(\xi \mid dt; x,t)$ or the propagator moment functions $\{B_n(x,t)\}$, that will be applicable to *both* of those classes of Markov processes.

2.5 THE KRAMERS-MOYAL EQUATIONS

Equation (2.4-12) is essentially an infinite order *integral* equation for the Markov state density function P — infinite because Eq. (2.4-11d) evidently requires n to be infinitely large. That equation is evidently completely determined if the propagator density function Π is specified. Now we shall derive two infinite order *differential* equations for P, which are likewise completely determined if Π is specified. These two infinite order differential equations are called the *Kramers-Moyal equations*, and they, like Eq. (2.4-12), are consequences of the Chapman-Kolmogorov equation; in particular, the "forward" Kramers-Moyal equation follows from Eq. (2.2-5a), and the "backward" Kramers-Moyal equation follows from Eq. (2.2-5b).

The derivation of the forward Kramers-Moyal equation starts by observing that if we define the function f by

$$f(x) \equiv P(x+\xi,t+\Delta t \mid x,t) \, P(x,t \mid x_0,t_0),$$

then the integrand of Eq. (2.2-5a) can be written $f(x-\xi)$. Assuming f to be infinitely differentiable, we have from Taylor's theorem that

$$f(x-\xi) = f(x) + \sum_{n=1}^{\infty} \frac{(-\xi)^n}{n!} \frac{\partial^n}{\partial x^n} f(x).$$

Substituting this for the integrand in Eq. (2.2-5a), we get

$$P(x,t+\Delta t \mid x_0,t_0)$$

$$= \int_{-\infty}^{\infty} d\xi \, P(x+\xi,t+\Delta t \mid x,t) \, P(x,t \mid x_0,t_0)$$

$$+ \sum_{n=1}^{\infty} \frac{(-1)^n}{n!} \frac{\partial^n}{\partial x^n} \left[\int_{-\infty}^{\infty} d\xi \, \xi^n \, P(x+\xi,t+\Delta t \mid x,t) \, P(x,t \mid x_0,t_0) \right],$$

where we have assumed that the order in which the n-summation and ξ-integration are performed is not important. The first term on the right, because of Eq. (2.2-2), integrates to $P(x,t \mid x_0,t_0)$. Subtracting that quantity from both sides and dividing by Δt gives

$$[P(x,t+\Delta t \mid x_0,t_0) - P(x,t \mid x_0,t_0)] / \Delta t$$

$$= \sum_{n=1}^{\infty} \frac{(-1)^n}{n!} \frac{\partial^n}{\partial x^n} \left[\left\{ \frac{1}{\Delta t} \int_{-\infty}^{\infty} d\xi \, \xi^n \, P(x+\xi,t+\Delta t \mid x,t) \right\} P(x,t \mid x_0,t_0) \right].$$

Now taking the limit $\Delta t \to 0$, and noting that

$$\lim_{\Delta t \to 0} \frac{1}{\Delta t} \int_{-\infty}^{\infty} d\xi \, \xi^n \, P(x+\xi,t+\Delta t \mid x,t)$$

$$= \frac{1}{dt} \int_{-\infty}^{\infty} d\xi \, \xi^n \, P(x+\xi,t+dt \mid x,t)$$

$$= \frac{1}{dt} \int_{-\infty}^{\infty} d\xi \, \xi^n \, \Pi(\xi \mid dt; x,t) \qquad \text{[by (2.4-4)]}$$

$$\equiv B_n(x,t), \qquad\qquad\qquad \text{[by (2.4-3b)]}$$

we obtain the equation

$$\frac{\partial}{\partial t} P(x,t \mid x_0, t_0) = \sum_{n=1}^{\infty} \frac{(-1)^n}{n!} \frac{\partial^n}{\partial x^n} \left[B_n(x,t) \, P(x,t \mid x_0, t_0) \right]. \quad (2.5\text{-}1)$$

This is the **forward Kramers-Moyal equation**. If all the functions $B_n(x,t)$ are known — i.e., if all the propagator moments $\langle \Xi^n(dt; x,t) \rangle$ are known *and* have the analytical form assumed in Eqs. (2.4-3) — then the forward Kramers-Moyal equation (2.5-1) constitutes a self-contained *t*-evolution equation for $P(x,t \mid x_0, t_0)$ for fixed values of x_0 and t_0.

The derivation of the backward Kramers-Moyal equation starts by observing that if we define the function h by

$$h(x_0) \equiv P(x,t \mid x_0, t_0 + \Delta t_0),$$

then the *first factor* in the integrand of Eq. (2.2-5b) can be written as $h(x_0 + \xi)$. Assuming h to be infinitely differentiable, we have from Taylor's theorem that

$$h(x_0 + \xi) = h(x_0) + \sum_{n=1}^{\infty} \frac{\xi^n}{n!} \frac{\partial^n}{\partial x_0^n} h(x_0).$$

Substituting this for the first factor in the integrand of Eq. (2.2-5b), we get

$$P(x,t \mid x_0, t_0)$$

$$= \int_{-\infty}^{\infty} d\xi \, P(x,t \mid x_0, t_0 + \Delta t_0) \, P(x_0 + \xi, t_0 + \Delta t_0 \mid x_0, t_0)$$

$$+ \sum_{n=1}^{\infty} \frac{1}{n!} \int_{-\infty}^{\infty} d\xi \, \xi^n \left[\frac{\partial^n}{\partial x_0^n} P(x,t \mid x_0, t_0 + \Delta t_0) \right] P(x_0 + \xi, t_0 + \Delta t_0 \mid x_0, t_0),$$

where again we have assumed that the order of *n*-summation and ξ-integration can be changed. The first term on the right, because of Eq. (2.2-2), integrates to $P(x,t \mid x_0, t_0 + \Delta t_0)$. Subtracting that quantity from both sides and dividing through by Δt_0 gives

$$[P(x,t \mid x_0, t_0) - P(x,t \mid x_0, t_0 + \Delta t_0)] / \Delta t_0$$

$$= \sum_{n=1}^{\infty} \frac{1}{n!} \left[\frac{1}{\Delta t_0} \int_{-\infty}^{\infty} d\xi \, \xi^n \, P(x_0 + \xi, t_0 + \Delta t_0 \mid x_0, t_0) \right] \frac{\partial^n}{\partial x_0^n} P(x,t \mid x_0, t_0 + \Delta t_0).$$

Now taking the limit $\Delta t_0 \to 0$, and noting that

$$\lim_{\Delta t_0 \to 0} \frac{1}{\Delta t_0} \int_{-\infty}^{\infty} d\xi\, \xi^n\, P(x_0+\xi, t_0+\Delta t_0 \,|\, x_0, t_0)$$

$$= \frac{1}{dt_0} \int_{-\infty}^{\infty} d\xi\, \xi^n\, P(x_0+\xi, t_0+dt_0 \,|\, x_0, t_0)$$

$$= \frac{1}{dt_0} \int_{-\infty}^{\infty} d\xi\, \xi^n\, \Pi(\xi \,|\, dt_0; x_0, t_0) \qquad \text{[by (2.4-4)]}$$

$$\equiv B_n(x_0, t_0), \qquad\qquad\qquad \text{[by (2.4-3b)]}$$

we obtain the equation

$$-\frac{\partial}{\partial t_0} P(x,t \,|\, x_0, t_0) = \sum_{n=1}^{\infty} \frac{1}{n!} B_n(x_0, t_0) \frac{\partial^n}{\partial x_0^n}\Big[P(x,t \,|\, x_0, t_0)\Big]. \qquad (2.5\text{-}2)$$

This is the **backward Kramers-Moyal equation**. If all the functions $B_n(x,t)$ are known — i.e., if all the propagator moments $\langle \Xi^n(dt; x,t)\rangle$ are known *and* have the analytical form assumed in Eqs. (2.4-3) — then the backward Kramers-Moyal equation (2.5-2) constitutes a self-contained t_0-evolution equation for $P(x,t \,|\, x_0, t_0)$ for fixed values of x and t.

It follows from Eq. (2.2-3) that the *forward* Kramers-Moyal equation (2.5-1), being a t-evolution equation equation for fixed x_0 and t_0, is to be solved subject to the *initial condition*

$$P(x, t=t_0 \,|\, x_0, t_0) = \delta(x-x_0), \qquad (2.5\text{-}3)$$

whereas the *backward* Kramers-Moyal equation (2.5-2), being a t_0-evolution equation for fixed x and t, is to be solved subject to the *final condition*

$$P(x,t \,|\, x_0, t_0=t) = \delta(x-x_0). \qquad (2.5\text{-}4)$$

The fact that the Kramers-Moyal equations are *infinite* order partial differential equations obviously makes their solution rather problematic. In any event, like the infinite order *integral* equation (2.4-12), the Kramers-Moyal equations show clearly that a determination of the Markov state density function $P(x,t \,|\, x_0, t_0)$ requires that the process propagator $\Xi(dt; x,t)$ be specified. In the case of the integral equation (2.4-12), $\Xi(dt; x,t)$ is specified by giving its density function $\Pi(\xi \,|\, dt; x,t)$; in the case of the Kramers-Moyal equations (2.5-1) and (2.5-2), $\Xi(dt; x,t)$ is specified by giving all of its moments $\langle \Xi^n(dt; x,t)\rangle = B_n(x,t)dt$.

It often happens that *all* these equations for the density function $P(x,t \mid x_0,t_0)$ of $X(t)$ are just too difficult for one to solve, and one would gladly settle for a knowledge of the behavior of just a few low order moments of $X(t)$, such as its mean, variance and covariance. In Section 2.7 we shall derive time-evolution equations for these moments in terms of the propagator moment functions $B_n(x,t)$. But before doing that, we want to explore the possibility of *differentiating* and *integrating* a Markov process $X(t)$ with respect to it argument t.

2.6 THE TIME-INTEGRAL OF A MARKOV PROCESS

In this section we shall consider whether it is possible to sensibly define the *derivative* and *integral* of a Markov process $X(t)$ with respect to t. We shall discover that the time-integral of $X(t)$ can indeed be well defined, but that the time-derivative of $X(t)$ does not exist unless $X(t)$ happens to be a completely deterministic process.

Let us consider first the matter of the *time-derivative* of $X(t)$. By the definition (2.4-1) of the process propagator, if $X(t) = x$ then in the next infinitesimal time interval dt the process will change by the amount

$$X(t+dt) - X(t) = \Xi(dt; x,t).$$

It follows that a *typical value* of this process increment in time dt will be

$$X(t+dt) - X(t) \sim \mathrm{mean}\{\Xi(dt; x,t)\} \pm \mathrm{sdev}\{\Xi(dt; x,t)\}. \qquad (2.6\text{-}1)$$

To estimate this typical value, we are going to *assume* that the first two propagator moment functions B_1 and B_2, as defined in Eq. (2.4-3b), both exist as well-behaved functions. We shall find in Chapters 3 and 4 that this is not a very restrictive assumption, as it is valid for nearly all Markov processes of practical interest; certainly it is a benign assumption for any Markov process for which the Kramers-Moyal equations (2.5-1) and (2.5-2) have meaning. Now, by the definition (2.4-3a), we have

$$\mathrm{mean}\{\Xi(dt; x,t)\} = \langle \Xi(dt; x,t) \rangle = B_1(x,t)\, dt + o(dt),$$

and

$$\mathrm{sdev}\{\Xi(dt; x,t)\} = \{\langle \Xi^2(dt; x,t) \rangle - \langle \Xi(dt; x,t) \rangle^2\}^{1/2}$$

$$= \{[B_2(x,t)\, dt + o(dt)] - [B_1(x,t)\, dt + o(dt)]^2\}^{1/2}$$

$$= B_2^{1/2}(x,t)\, (dt)^{1/2} \{1 + o(dt)/dt\}^{1/2} + o(dt).$$

Substituting these typical values into Eq. (2.6-1), we thus see that if $X(t)=x$ then we will typically have

$$X(t+dt) - X(t)$$

$$\sim B_1(x,t)\,dt + o(dt) \pm \left[B_2^{1/2}(x,t)\,(dt)^{1/2} \left\{ 1 + o(dt)/dt \right\}^{1/2} + o(dt) \right].$$

Dividing through by dt, we get

$$[X(t+dt) - X(t)]\,/\,dt$$

$$\sim B_1(x,t) + \frac{o(dt)}{dt} \pm \frac{B_2^{1/2}(x,t)}{(dt)^{1/2}} \left\{ 1 + \frac{o(dt)}{dt} \right\}^{1/2} \pm \frac{o(dt)}{dt}. \qquad (2.6\text{-}2)$$

Now taking the limit $dt \to 0$, keeping in mind that $o(dt)/dt \to 0$ in that limit, we get

$$\lim_{dt \to 0} \frac{X(t+dt) - X(t)}{dt} \sim \begin{cases} B_1(x,t), & \text{if } B_2(x,t)=0; \\ \pm \infty, & \text{if } B_2(x,t) \neq 0. \end{cases} \qquad (2.6\text{-}3)$$

So we see that the limit on the left, which obviously defines the derivative of the process at time t in state x, does *not* exist *except* in the special circumstance that $B_2(x,t)=0$. We shall discover in Chapters 3 and 4 [see especially Section 3.3A] that the only circumstance in which $B_2(x,t)$ vanishes identically is when the process $X(t)$ is a completely deterministic process. We thus conclude that *a genuinely stochastic Markov process does not have a derivative with respect to time.*

Now let us consider the possibility of defining a *time-integral* of the Markov process $X(t)$. We begin by noting that if the stochastic process

$$S(t) \equiv \int_{t_0}^{t} X(t')\,dt' \qquad (2.6\text{-}4)$$

exists at all, it *cannot* be a *Markov* process! This is because $S(t)$ would by definition have a derivative with respect to t, namely

$$dS(t)/dt = X(t), \qquad (2.6\text{-}5)$$

and this is something that, as we have just demonstrated, a genuinely stochastic Markov process does *not* have. Nevertheless, it may still be possible to define $S(t)$ as a viable *non-Markovian* process.

So, how might we define the process $S(t)$ which we have denoted *symbolically* by Eqs. (2.6-4) and (2.6-5)? If $X(t)$ in those representations were a *sure* function of t, then $S(t)$ would of course be that sure function for which

$$\frac{S(t+dt) - S(t)}{dt} = X(t) + \frac{o(dt)}{dt},$$

or equivalently

$$S(t+dt) = S(t) + X(t)\,dt + o(dt).$$

So if $X(t)$ is a Markov process instead of a sure function, let us *define* its **integral** to be that process $S(t)$ such that

$$S(t+dt) = s + x\,dt + o(dt), \text{ given } S(t)=s \text{ and } X(t)=x. \qquad (2.6\text{-}6)$$

With this definition and the *initial condition*

$$S(t_0) = 0, \qquad (2.6\text{-}7)$$

then $S(t)$ should indeed represent the "area under the graph" of the Markov process $X(t')$ between times $t' = t_0$ and $t' = t$, as is suggested by Eq. (2.6-4).

It is easy to see from the definition (2.6-6) that $S(t)$ is not Markovian: A knowledge of its state s at time t is not *by itself* sufficient to predict probabilistically its state at time $t+dt$. For such a prediction, we evidently also need to know the state x of the process $X(t)$ at time t, and x simply cannot be predicted, even probabilistically, from a knowledge of s alone. However, it is clear that if we specify the states of *both* processes X and S at time t, then it *is* possible to predict probabilistically the states of *both* processes at time $t+dt$. Specifically,

$$\text{Prob}\{\, X(t+dt)\in[x+\xi,\, x+\xi+d\xi)\mid X(t)=x \text{ and } S(t)=s\,\}$$
$$= \Pi(\xi\mid dt;\, x,t)\, d\xi, \qquad (2.6\text{-}8a)$$

and

$$\text{Prob}\{\, S(t+dt)\in[s+\eta,\, s+\eta+d\eta)\mid X(t)=x \text{ and } S(t)=s\,\}$$
$$= \delta(\eta - x\,dt)\, d\eta. \qquad (2.6\text{-}8b)$$

The first relation here follows of course from the definitions (2.4-1) and (2.4-2) of the propagator and its density function. The second relation follows from the definition (2.6-6) and the fact that $\delta(\eta - x\,dt)d\eta$ is unity if $\eta = x\,dt$ and zero otherwise. The results (2.6-8) imply that $X(t)$ and $S(t)$ *together* constitute a *bivariate* Markov process. The treatment of bivariate Markov processes lies outside the intended scope of our work here. Nevertheless, as we shall see in the next section, we will at least be able to derive from the definition (2.6-6) time-evolution equations for *all the moments* of the integral process $S(t)$.

2.7 TIME-EVOLUTION OF THE MOMENTS

Often our interest in a Markov process $X(t)$ is confined to a few of its lowest order moments, such as its mean, variance and covariance. One way to compute those moments would be to first compute $P(x,t \mid x_0,t_0)$, say by solving either the integral equation (2.4-12) or the forward Kramers-Moyal equation (2.5-1), and to then carry out the integrations prescribed in Eqs. (2.3-3) – (2.3-8). But it is usually much easier to circumvent the function $P(x,t \mid x_0,t_0)$ and work instead with explicit time-evolution equations for the moments themselves. In this section we shall derive those "moment evolution equations," not only for the process $X(t)$ but also for its time-integral $S(t)$, and we shall develop explicit solutions to those equations for two simple but frequently encountered cases.

2.7.A THE GENERAL MOMENT EVOLUTION EQUATIONS

Perhaps the most obvious way to derive an equation for the time derivative of the moment $\langle X^n(t) \rangle$ is to multiply the forward Kramers-Moyal equation (2.5-1) through by x^n, integrate the resulting equation over all x, and then reduce each term on the right side by repeated integrations-by-parts. But we shall take here a different approach which, besides being slightly simpler, will allow us to also deduce time-evolution equations for the moments of the (non-Markovian) integral process $S(t)$.

The definition (2.4-1) of the process propagator implies that the state of the process at time $t+dt$, namely $X(t+dt)$, is related to the state at the infinitesimally earlier time t, namely $X(t)$, by

$$X(t+dt) = X(t) + \varXi(dt; X(t),t). \qquad (2.7\text{-}1)$$

Raising this equation to the power n for any integer $n \geq 1$,

$$X^n(t+dt) = \left[X(t) + \varXi(dt; X(t),t) \right]^n,$$

and then expanding the right side using the binomial formula, we get

$$X^n(t+dt) = X^n(t) + \sum_{k=1}^{n} \binom{n}{k} X^{n-k}(t)\, \varXi^k(dt; X(t),t). \qquad (2.7\text{-}2)$$

Here, and frequently in the sequel, we use the common abbreviation for the binomial coefficient,

$$\binom{n}{k} \equiv \frac{n!}{k!\,(n-k)!} \quad (k=0,\ldots,n).$$

If we now *average* Eq. (2.7-2), invoking the inherent linearity of that operation, we get

$$\langle X^n(t+dt)\rangle = \langle X^n(t)\rangle + \sum_{k=1}^{n} \binom{n}{k} \langle X^{n-k}(t)\, \Xi^k(dt;\, X(t),t)\rangle. \quad (2.7\text{-}3)$$

The various averages appearing in this equation could in principle be calculated by sampling the values of the random variables $X(t)$, $X(t+dt)$ and $\Xi(dt;\, X(t),t) \equiv X(t+dt) - X(t)$ over many independent realizations of the Markov process, and then evaluating the appropriate sample averages in the manner of Eq. (1.5-14). Although that procedure helps us to understand the meanings of the averages in Eq. (2.7-3), for an *analytical* calculation of those averages we would want to use density function formulas like Eq. (1.5-15). But the precise form of the density function formula for the average that appears under the summation sign in Eq. (2.7-3) is not immediately obvious, owing to the parametric dependence of the random variable Ξ on the random variable X.

How we should calculate an average of the form $\langle g(X(t), \Xi(dt;\, X(t),t))\rangle$ where g is some given function? If the random variable $X(t)$ had a *definite* value x, then this average could be calculated simply as

$$\overline{g(x, \Xi(dt;\, x,t))} = \int_{-\infty}^{\infty} d\xi\, g(x,\xi)\, \Pi(\xi\,|\,dt;\, x,t),$$

because $\Pi(\xi\,|\,dt;\, x,t)$ is by definition the density function of the random variable $\Xi(dt;\, x,t)$. But $X(t)$ actually assumes a value in the interval $[x, x+dx)$ only with probability $P(x,t\,|\,x_0,t_0)dx$; therefore, according to the fundamental rule (2.3-3), the *overall* average should be given by

$$\langle g(X(t), \Xi(dt;\, X(t),t))\rangle = \int_{-\infty}^{\infty} dx\, \overline{g(x, \Xi(dt;\, x,t))}\, P(x,t\,|\,x_0,t_0).$$

Substituting from the preceding equation, we thus conclude that the average $\langle g(X(t), \Xi(dt;\, X(t),t))\rangle$ is to be calculated from the formula

$$\langle g(X(t), \Xi(dt;\, X(t),t))\rangle = \int_{-\infty}^{\infty} dx \int_{-\infty}^{\infty} d\xi\, g(x,\xi)\, \Pi(\xi\,|\,dt;\, x,t)\, P(x,t\,|\,x_0,t_0).$$

$$(2.7\text{-}4)$$

Another way to obtain this result is to recognize, with the help of the conditioning formula (1.5-9a), that $P(x,t\,|\,x_0,t_0)\,\Pi(\xi\,|\,dt;\, x,t)$ is the *joint* density function of the two random variables $X(t)$ and $\Xi(dt;\, X(t),t)$.

As a specific application of the general formula (2.7-4), let us take the case offered by Eq. (2.7-3) in which $g(x,\xi) = x^j \xi^k$ with $j \geq 0$ and $k \geq 1$:

$$\langle X^j(t)\, \Xi^k(dt;\, X(t),t)\rangle$$

$$= \int_{-\infty}^{\infty} dx \int_{-\infty}^{\infty} d\xi\, (x^j \xi^k)\, \Pi(\xi\,|\,dt;\, x,t)\, P(x,t\,|\,x_0,t_0) \qquad \text{[by (2.7-4)]}$$

$$= \int_{-\infty}^{\infty} dx\, x^j \left[\int_{-\infty}^{\infty} d\xi\, \xi^k\, \Pi(\xi\,|\,dt;\, x,t)\right] P(x,t\,|\,x_0,t_0)$$

$$= \int_{-\infty}^{\infty} dx\, x^j \left[B_k(x,t)\, dt + o(dt)\right] P(x,t\,|\,x_0,t_0) \qquad \text{[by (2.4-3a)]}$$

$$= \int_{-\infty}^{\infty} dx \left[x^j B_k(x,t)\right] P(x,t\,|\,x_0,t_0)\, dt + o(dt).$$

Therefore, appealing once again to the rule (2.3-3), we conclude that

$$\langle X^j(t)\, \Xi^k(dt;\, X(t),t)\rangle = \langle X^j(t)\, B_k(X(t),t)\rangle\, dt + o(dt)$$

$$(j \geq 0,\, k \geq 1). \quad (2.7\text{-}5)$$

In light of the result (2.7-5), Eq. (2.7-3) can evidently be written

$$\langle X^n(t+dt)\rangle = \langle X^n(t)\rangle + \sum_{k=1}^{n} \binom{n}{k} \langle X^{n-k}(t)\, B_k(X(t),t)\rangle\, dt + o(dt).$$

Transposing the first term on the right, dividing through by dt, and then taking the limit $dt \to 0$, we thus arrive at the following set of **moment evolution equations** for $X(t)$:

$$\frac{d}{dt}\langle X^n(t)\rangle = \sum_{k=1}^{n} \binom{n}{k} \langle X^{n-k}(t)\, B_k(X(t),t)\rangle \quad (t_0 \leq t,\, n = 1,2,\ldots). \quad (2.7\text{-}6)$$

Since $X(t_0)$ is by hypothesis the sure number x_0, then this set of first order differential equations is to be solved subject to the initial conditions

$$\langle X^n(t_0)\rangle = x_0^n \qquad (n = 1,2,\ldots). \qquad (2.7\text{-}7)$$

If the propagator moment functions $B_k(x,t)$ can be expressed as polynomials in x (for instance by a Taylor series expansion), then Eqs. (2.7-6) evidently become a set of coupled, ordinary differential equations for the moments $\langle X(t)\rangle$, $\langle X^2(t)\rangle$, $\langle X^3(t)\rangle$, ... Notice, however, that the coupling may — and in fact often does — pose severe problems for solving these equations. In particular, the only circumstance in which Eqs. (2.7-6) will be *closed*, and hence in principle exactly solvable, is when

$B_k(x,t)$ is a polynomial in x of degree $\leq k$; because only then will the right side of the nth equation contain only moments of order less than or equal to n, thus allowing the equations to be solved successively for $n = 1, 2, \ldots$. If $B_k(x,t)$ is *not* a polynomial in x of degree $\leq k$, then the most one can hope for is some kind of approximate solution. A systematic but somewhat laborious procedure for developing *approximate numerical* solutions for $\langle X(t) \rangle$ and $\langle X^2(t) \rangle$ is outlined in Appendix C. Alternatively, one can try to find in the specifics of the problem being considered some justification for artificially closing Eqs. (2.7-6); this might be accomplished either by directly approximating $B_k(x,t)$, or else by approximating a few higher order moments in terms of lower order moments.

Now let us consider the time-integral $S(t)$ of the Markov process $X(t)$. The definition (2.6-6) of the integral process implies that $S(t+dt)$ is related to the infinitesimally earlier random variables $S(t)$ and $X(t)$ according to [compare Eq. (2.7-1)]

$$S(t+dt) = S(t) + X(t)\,dt + o(dt). \tag{2.7-8}$$

Raising this equation to the power m for any integer $m \geq 1$,

$$S^m(t+dt) = \Big[S(t) + X(t)\,dt + o(dt) \Big]^m,$$

and then using the binomial formula to expand the right side to first order in dt, we get

$$S^m(t+dt) = S^m(t) + m\,S^{m-1}(t)\,X(t)\,dt + o(dt). \tag{2.7-9}$$

Averaging this equation gives

$$\langle S^m(t+dt) \rangle = \langle S^m(t) \rangle + m \langle S^{m-1}(t)\,X(t) \rangle\,dt + o(dt).$$

Transposing the first term on the right, dividing through by dt, and then taking the limit $dt \to 0$, we thus arrive at the following set of equations for the time derivatives of the moments of $S(t)$:

$$\frac{d}{dt}\langle S^m(t) \rangle = m \langle S^{m-1}(t)\,X(t) \rangle \quad (t_0 \leq t,\ m = 1, 2, \ldots). \tag{2.7-10}$$

Since $S(t_0)$ is the sure number 0 [by Eq. (2.6-7)], then these differential equations have the initial conditions

$$\langle S^m(t_0) \rangle = 0 \qquad (m = 1, 2, \ldots). \tag{2.7-11}$$

But we are obviously not yet done, because the right side of Eq. (2.7-10) involves the as yet unknown "cross moments" $\langle S^{m-1}(t)\,X(t) \rangle$.

To derive equations from which the cross moments in Eqs. (2.7-10) can be calculated, we begin by multiplying Eq. (2.7-9) by Eq. (2.7-2):

$$S^m(t+dt)\,X^n(t+dt)$$

$$= \left[S^m(t) + m\,S^{m-1}(t)\,X(t)\,dt + o(dt) \right]$$

$$\times \left[X^n(t) + \sum_{k=1}^{n} \binom{n}{k} X^{n-k}(t)\,\Xi^k(dt; X(t),t) \right]$$

$$= S^m(t)\,X^n(t) + m\,S^{m-1}(t)\,X^{n+1}(t)\,dt$$

$$+ \sum_{k=1}^{n} \binom{n}{k} S^m(t)\,X^{n-k}(t)\,\Xi^k(dt; X(t),t)$$

$$+ m\,dt \sum_{k=1}^{n} \binom{n}{k} S^{m-1}(t)\,X^{n-k+1}(t)\,\Xi^k(dt; X(t),t) + o(dt).$$

Averaging the last equation, and applying to the third and fourth terms a straightforward generalization of the rule (2.7-5), we get

$$\langle S^m(t+dt)\,X^n(t+dt) \rangle = \langle S^m(t)\,X^n(t) \rangle + m\,\langle S^{m-1}(t)\,X^{n+1}(t) \rangle\,dt$$

$$+ \sum_{k=1}^{n} \binom{n}{k} \langle S^m(t)\,X^{n-k}(t)\,B_k(X(t),t) \rangle\,dt + o(dt),$$

where we have noted that the average of the fourth term in the preceding equation turns out to be proportional to $(dt)^2$. Finally, transposing the first term on the right, dividing through by dt, and then taking the limit $dt \to 0$, we conclude that the cross moments between $X(t)$ and $S(t)$ are governed by the coupled set of first order differential equations

$$\frac{d}{dt} \langle S^m(t)\,X^n(t) \rangle = m\,\langle S^{m-1}(t)\,X^{n+1}(t) \rangle$$

$$+ \sum_{k=1}^{n} \binom{n}{k} \langle S^m(t)\,X^{n-k}(t)\,B_k(X(t),t) \rangle$$

$$(t_0 \le t,\, m \ge 1;\, n \ge 1). \quad (2.7\text{-}12)$$

Since $S(t_0) = 0$, then the initial conditions for this set of equations are

$$\langle S^m(t_0)\,X^n(t_0) \rangle = 0 \qquad (m \ge 1;\, n \ge 1). \quad (2.7\text{-}13)$$

The two sets of differential equations (2.7-12) and (2.7-10), along with their respective initial conditions (2.7-13) and (2.7-11), provide a basis for calculating all the moments of $S(t)$. But notice that the structure of the second term on the right side of Eq. (2.7-12) implies that that equation will be closed, and hence solvable in principle, if and only if $B_k(x,t)$ is a polynomial in x of degree $\leq k$. This closure condition is the *same* as that found for the time-evolution equations (2.7-6) for $\langle X^n(t) \rangle$. In fact, a detailed examination of Eqs. (2.7-10) and (2.7-12) will reveal that solving those equations jointly for the first N moments of $S(t)$ requires that we *already know* the first N moments of $X(t)$. Therefore, the general solution procedure is to first solve Eqs. (2.7-6) for the moments of $X(t)$, and to then solve Eqs. (2.7-10) and (2.7-12) together to get the moments of $S(t)$.

2.7.B MEAN, VARIANCE AND COVARIANCE EVOLUTION EQUATIONS

Now we shall deduce, from the general moment evolution equations derived in the Subsection 2.7.A, specific time-evolution equations for the means, variances and covariances of both $X(t)$ and $S(t)$. We shall find that these particular time-evolution equations involve only the first two propagator moment functions, $B_1(x,t)$ and $B_2(x,t)$. And we shall also find that these equations will be *closed* if and only if $B_1(x,t)$ is a polynomial in x of degree ≤ 1 and $B_2(x,t)$ is a polynomial in x of degree ≤ 2. If those rather stringent conditions on $B_1(x,t)$ and $B_2(x,t)$ are not satisfied, then the most we can hope for is approximate solutions, obtained either through the procedure outlined in Appendix C, or else by suitably approximating the two functions $B_1(x,t)$ and $B_2(x,t)$.

First we shall establish that the mean, variance and covariance of the Markov process $X(t)$ satisfy the differential equations

$$\frac{d}{dt}\langle X(t) \rangle = \langle B_1(X(t),t) \rangle \qquad (t_0 \leq t), \qquad (2.7\text{-}14)$$

$$\frac{d}{dt}\operatorname{var}\{X(t)\} = 2\Big(\langle X(t)\, B_1(X(t),t) \rangle - \langle X(t) \rangle \langle B_1(X(t),t) \rangle \Big) + \langle B_2(X(t),t) \rangle$$

$$(t_0 \leq t), \qquad (2.7\text{-}15)$$

$$\frac{d}{dt_2}\operatorname{cov}\{X(t_1),X(t_2)\} = \langle X(t_1)\, B_1(X(t_2),t_2) \rangle - \langle X(t_1) \rangle \langle B_1(X(t_2),t_2) \rangle$$

$$(t_0 \leq t_1 \leq t_2), \qquad (2.7\text{-}16)$$

these equations being subject to the respective initial conditions

$$\langle X(t=t_0)\rangle = x_0, \tag{2.7-17}$$

$$\text{var}\{X(t=t_0)\} = 0, \tag{2.7-18}$$

$$\text{cov}\{X(t_1), X(t_2=t_1)\} = \text{var}\{X(t_1)\}. \tag{2.7-19}$$

The first two initial conditions follow from Eq. (2.1-2), while the third follows from Eq. (2.3-9). Notice that a knowledge of the initial condition for the covariance evolution equation will normally require that we have already solved the variance evolution equation.

Equation (2.7-14) is an immediate consequence of the moment evolution equation (2.7-6) for $n=1$. To prove Eq. (2.7-15), we note that

$$\frac{d}{dt}\text{var}\{X(t)\} = \frac{d}{dt}\left[\langle X^2(t)\rangle - \langle X(t)\rangle^2\right]$$

$$= \frac{d}{dt}\langle X^2(t)\rangle - 2\langle X(t)\rangle\frac{d}{dt}\langle X(t)\rangle$$

$$= \left[2\langle X(t)\,B_1(X(t),t)\rangle + \langle B_2(X(t),t)\rangle\right] - 2\langle X(t)\rangle\left[\langle B_1(X(t),t)\rangle\right],$$

where the last step has invoked Eq. (2.7-6) for $n=1$ and $n=2$. A simple rearrangement of terms then yields Eq. (2.7-15).

To prove the covariance equation (2.7-16), we start by observing that

$$\frac{d}{dt_2}\text{cov}\{X(t_1), X(t_2)\} = \frac{d}{dt_2}\left[\langle X(t_1)\,X(t_2)\rangle - \langle X(t_1)\rangle\langle X(t_2)\rangle\right]$$

$$= \frac{d}{dt_2}\langle X(t_1)\,X(t_2)\rangle - \langle X(t_1)\rangle\frac{d}{dt_2}\langle X(t_2)\rangle$$

$$\frac{d}{dt_2}\text{cov}\{X(t_1), X(t_2)\} = \frac{d}{dt_2}\langle X(t_1)\,X(t_2)\rangle - \langle X(t_1)\rangle\langle B_1(X(t_2),t_2)\rangle,$$

where the last step follows from Eq. (2.7-14). To evaluate the first term on the right, we first replace t in Eq. (2.7-1) everywhere by t_2, and then multiply the resulting equation through by $X(t_1)$:

$$X(t_1)\,X(t_2+dt_2) = X(t_1)\,X(t_2) + X(t_1)\,\Xi(dt_2; X(t_2),t_2).$$

Averaging this equation gives

$$\langle X(t_1)\,X(t_2+dt_2)\rangle = \langle X(t_1)\,X(t_2)\rangle + \langle X(t_1)\,\Xi(dt_2; X(t_2),t_2)\rangle.$$

The first average on the right is of course to be computed with respect to the joint density function $P(x_2,t_2 \mid x_1,t_1) P(x_1,t_1 \mid x_0,t_0)$ of $X(t_1)$ and $X(t_2)$, as in Eq. (2.3-4). And by a straightforward extension of rule (2.7-4), the second average on the right is to be computed by first averaging as though $X(t_1)=x_1$ and $X(t_2)=x_2$, and then averaging *that* average jointly over x_1 and x_2:

$$\langle\, X(t_1)\, \Xi(dt_2; X(t_2),t_2)\,\rangle$$

$$= \int_{-\infty}^{\infty} dx_1 \int_{-\infty}^{\infty} dx_2 \int_{-\infty}^{\infty} d\xi\, (x_1\xi)\, \Pi(\xi \mid dt_2; x_2,t_2)$$
$$\times P(x_2,t_2 \mid x_1,t_1)\, P(x_1,t_1 \mid x_0,t_0)$$

$$= \int_{-\infty}^{\infty} dx_1 \int_{-\infty}^{\infty} dx_2\, x_1 \left[\int_{-\infty}^{\infty} d\xi\, \xi\, \Pi(\xi \mid dt_2; x_2,t_2) \right]$$
$$\times P(x_2,t_2 \mid x_1,t_1)\, P(x_1,t_1 \mid x_0,t_0)$$

$$= \int_{-\infty}^{\infty} dx_1 \int_{-\infty}^{\infty} dx_2\, x_1 \left[B_1(x_2,t_2)dt_2 + o(dt_2) \right]$$
$$\times P(x_2,t_2 \mid x_1,t_1)\, P(x_1,t_1 \mid x_0,t_0) \qquad \text{[by (2.4-3a)]}$$

$$= \langle\, X(t_1)\, B_1(X(t_2),t_2)\,\rangle\, dt_2 + o(dt_2). \qquad \text{[by (2.3-4)]}$$

Substituting this result into the preceding formula, we get

$$\langle\, X(t_1)\, X(t_2+dt_2)\,\rangle - \langle\, X(t_1)\, X(t_2)\,\rangle = \langle\, X(t_1)\, B_1(X(t_2),t_2)\,\rangle\, dt_2 + o(dt_2).$$

Dividing through by dt_2 and passing to the limit $dt_2 \to 0$, we get

$$\frac{d}{dt_2} \langle\, X(t_1)\, X(t_2)\,\rangle = \langle\, X(t_1)\, B_1(X(t_2),t_2)\,\rangle.$$

Substituting this result into the previous equation for the t_2-derivative of the $\text{cov}\{X(t_1),X(t_2)\}$, we finally obtain Eq. (2.7-16).

Notice that if $B_1(x,t)$ is independent of x, then the right side of Eq. (2.7-16) will be zero, thus implying that $\text{cov}\{X(t_1),X(t_2)\}$ will be independent of t_2. Therefore, taking account of the initial condition (2.7-19), we have the theorem

$$B_1(x,t) \text{ is independent of } x \;\Rightarrow\; \text{cov}\{X(t_1),X(t_2)\} = \text{var}\{X(t_1)\}$$
$$(t_0 \le t_1 \le t_2). \quad (2.7\text{-}20)$$

In the following two subsections we shall obtain solutions to Eqs. (2.7-14) – (2.7-16) for two explicit pairs of functions B_1 and B_2. But first we want to derive the time-evolution equations for the mean, variance and covariance of the integral process $S(t)$. We shall show that these equations are as follows: For the *mean* of $S(t)$ we have, not surprisingly,

$$\langle S(t) \rangle = \int_{t_0}^{t} \langle X(t') \rangle \, dt'. \tag{2.7-21}$$

For the *variance* of $S(t)$ we have

$$\text{var}\{S(t)\} = 2 \int_{t_0}^{t} \text{cov}\{S(t'), X(t')\} \, dt', \tag{2.7-22}$$

wherein the integrand is the solution of the differential equation

$$\frac{d}{dt} \text{cov}\{S(t), X(t)\} = \text{var}\{X(t)\} + \langle S(t) B_1(X(t), t) \rangle - \langle S(t) \rangle \langle B_1(X(t), t) \rangle$$

$$(t_0 \le t) \quad (2.7\text{-}23a)$$

that satisfies the initial condition

$$\text{cov}\{S(t = t_0), X(t = t_0)\} = 0. \tag{2.7-23b}$$

And finally, for the *covariance* of $S(t)$ we have

$$\text{cov}\{S(t_1), S(t_2)\} = \text{var}\{S(t_1)\} + \int_{t_1}^{t_2} \text{cov}\{S(t_1), X(t_2')\} \, dt_2'$$

$$(t_0 \le t_1 \le t_2), \quad (2.7\text{-}24)$$

wherein the integrand is the solution of the differential equation

$$\frac{d}{dt_2} \text{cov}\{S(t_1), X(t_2)\} = \langle S(t_1) B_1(X(t_2), t_2) \rangle - \langle S(t_1) \rangle \langle B_1(X(t_2), t_2) \rangle$$

$$(t_0 \le t_1 \le t_2) \quad (2.7\text{-}25a)$$

that satisfies the initial condition

$$\text{cov}\{S(t_1), X(t_2 = t_1)\} = \text{cov}\{S(t_1), X(t_1)\}, \tag{2.7-25b}$$

which in turn is known from the solution to Eq. (2.7-23a).

The equations (2.7-21) – (2.7-23) for the mean and variance of $S(t)$ are simple consequences of the formulas developed in the preceding subsection. Thus, Eq. (2.7-10) reads for $m = 1$ and $m = 2$,

$$\frac{d}{dt} \langle S(t) \rangle = \langle X(t) \rangle \tag{2.7-26a}$$

and

$$\frac{d}{dt}\langle S^2(t)\rangle = 2\langle S(t)\,X(t)\rangle. \qquad (2.7\text{-}26b)$$

The first equation, together with the required initial condition $S(t_0)=0$, clearly implies Eq. (2.7-21). And by using both of the above equations, we find that

$$\frac{d}{dt}\text{var}\{S(t)\} = \frac{d}{dt}\Big[\langle S^2(t)\rangle - \langle S(t)\rangle^2\Big]$$

$$= \frac{d}{dt}\langle S^2(t)\rangle - 2\langle S(t)\rangle\frac{d}{dt}\langle S(t)\rangle$$

$$= 2\langle S(t)\,X(t)\rangle - 2\langle S(t)\rangle\langle X(t)\rangle$$

$$= 2\,\text{cov}\{S(t),X(t)\}.$$

This result, together with the initial condition $S(t_0)=0$, implies Eq. (2.7-22). To establish Eqs. (2.7-23), we first note that

$$\frac{d}{dt}\text{cov}\{S(t),X(t)\} = \frac{d}{dt}\Big[\langle S(t)\,X(t)\rangle - \langle S(t)\rangle\langle X(t)\rangle\Big]$$

$$= \frac{d}{dt}\langle S(t)\,X(t)\rangle - \langle S(t)\rangle\frac{d}{dt}\langle X(t)\rangle - \langle X(t)\rangle\frac{d}{dt}\langle S(t)\rangle.$$

Then we evaluate the first term on the right from Eq. (2.7-12), the second term from Eq. (2.7-14), and the third term from Eq. (2.7-26a). We get

$$\frac{d}{dt}\text{cov}\{S(t),X(t)\} = \Big[\langle X^2(t)\rangle + \langle\,S(t)\,B_1(X(t),t)\,\rangle\Big]$$

$$- \Big[\langle S(t)\rangle\langle B_1(X(t),t)\rangle\Big] - \Big[\langle X(t)\rangle\langle X(t)\rangle\Big],$$

which is Eq. (2.7-23a). The companion initial condition (2.7-23b) is a consequence of the fact that $S(t)$ and $X(t)$ are both sharply defined at $t=t_0$.

To derive equations (2.7-24) and (2.7-25) for the covariance of $S(t)$, we start by noting that, for $t_0 \le t_1 \le t_2$,

$$\frac{d}{dt_2}\text{cov}\{S(t_1),S(t_2)\} = \frac{d}{dt_2}\Big[\langle S(t_1)\,S(t_2)\rangle - \langle S(t_1)\rangle\langle S(t_2)\rangle\Big]$$

$$= \frac{d}{dt_2}\langle S(t_1)\,S(t_2)\rangle - \langle S(t_1)\rangle\frac{d}{dt_2}\langle S(t_2)\rangle$$

$$\frac{d}{dt_2}\text{cov}\{S(t_1),S(t_2)\} = \frac{d}{dt_2}\langle S(t_1)\,S(t_2)\rangle - \langle S(t_1)\rangle\langle X(t_2)\rangle.$$

where the last step follows from Eq. (2.7-26a). To calculate the t_2-derivative on the right, we first replace t in Eq. (2.7-8) everywhere by t_2 and then multiply the resulting equation through by $S(t_1)$. This gives

$$S(t_1)\,S(t_2+dt_2) = S(t_1)\,S(t_2) + S(t_1)\,X(t_2)dt_2 + o(dt_2).$$

Averaging, transposing the first term on the right, dividing through by dt_2, and then taking the limit $dt_2\to0$, we get

$$\frac{d}{dt_2}\langle S(t_1)\,S(t_2)\rangle = \langle S(t_1)\,X(t_2)\rangle.$$

Inserting this into the above equation for the t_2-derivative of $\text{cov}\{S(t_1),S(t_2)\}$, we get

$$\frac{d}{dt_2}\text{cov}\{S(t_1),S(t_2)\} = \langle S(t_1)\,X(t_2)\rangle - \langle S(t_1)\rangle\langle X(t_2)\rangle \equiv \text{cov}\{S(t_1),X(t_2)\}.$$

Now integrating over t_2 from its lower limit t_1, we get

$$\text{cov}\{S(t_1),S(t_2)\} - \text{cov}\{S(t_1),S(t_1)\} = \int_{t_1}^{t_2}\text{cov}\{S(t_1),X(t_2')\}\,dt_2'.$$

This evidently establishes Eq. (2.7-24). Finally, to prove Eqs. (2.7-25), we start by calculating, for $t_0\le t_1\le t_2$,

$$\frac{d}{dt_2}\text{cov}\{S(t_1),X(t_2)\} = \frac{d}{dt_2}\left[\langle S(t_1)\,X(t_2)\rangle - \langle S(t_1)\rangle\langle X(t_2)\rangle\right]$$

$$= \frac{d}{dt_2}\langle S(t_1)\,X(t_2)\rangle - \langle S(t_1)\rangle\frac{d}{dt_2}\langle X(t_2)\rangle$$

$$\frac{d}{dt_2}\text{cov}\{S(t_1),X(t_2)\} = \frac{d}{dt_2}\langle S(t_1)\,X(t_2)\rangle - \langle S(t_1)\rangle\langle B_1(X(t_2),t_2)\rangle,$$

where the last step follows from Eq. (2.7-14). To calculate the derivative on the right, we first replace t in Eq. (2.7-1) everywhere by t_2 and then multiply the resulting equation through by $S(t_1)$. This gives

$$S(t_1) X(t_2 + dt_2) = S(t_1) X(t_2) + S(t_1) \Xi(dt_2; X(t_2), t_2).$$

Transposing the first term on the right and averaging, we get

$$\langle S(t_1) X(t_2 + dt_2) \rangle - \langle S(t_1) X(t_2) \rangle = \langle S(t_1) \Xi(dt_2; X(t_2), t_2) \rangle$$

$$= \langle S(t_1) B_1(X(t_2), t_2) \rangle dt_2 + o(dt_2),$$

where the last step follows from a straightforward generalization of Eq. (2.7-5). Dividing through by dt_2 and taking the limit $dt_2 \to 0$ then yields

$$\frac{d}{dt_2} \langle S(t_1) X(t_2) \rangle = \langle S(t_1) B_1(X(t_2), t_2) \rangle.$$

Substituting this result into the previous equation for the t_2-derivative of cov$\{S(t_1), X(t_2)\}$, we obtain at last Eq. (2.7-25a). Of course, the companion initial condition (2.7-25b) is self-evident.

In the next two subsections we shall illustrate the formulas derived in this subsection by applying them to two simple examples. The results obtained for these two examples will be invoked several times in the later chapters.

2.7.C EXAMPLE: $B_1(x,t) = b_1$ and $B_2(x,t) = b_2$

We consider here a Markov process $X(t)$ for which the first two propagator moment functions are *both constants*:

$$B_1(x,t) = b_1 \quad \text{and} \quad B_2(x,t) = b_2 \geq 0. \tag{2.7-27}$$

The nonnegativity of b_2 is of course mandated by the nonnegativity of the integrand in Eq. (2.4-3b) for $n=2$. Making no assumptions concerning the form (or even the existence) of the other propagator moment functions, we shall deduce from the equations developed in Subsection 2.7.B the following explicit formulas for the mean, variance and covariance of $X(t)$:

$$\langle X(t) \rangle = x_0 + b_1(t - t_0) \qquad (t_0 \leq t), \tag{2.7-28a}$$

$$\text{var}\{X(t)\} = b_2(t - t_0) \qquad (t_0 \leq t), \tag{2.7-28b}$$

$$\text{cov}\{X(t_1), X(t_2)\} = b_2(t_1 - t_0) \qquad (t_0 \leq t_1 \leq t_2). \tag{2.7-28c}$$

And we shall also deduce the following companion equations for the mean, variance and covariance of the integral $S(t)$ of $X(t)$:

$$\langle S(t)\rangle = x_0(t-t_0) + (b_1/2)(t-t_0)^2 \qquad (t_0 \le t), \quad (2.7\text{-}29a)$$

$$\text{var}\{S(t)\} = (b_2/3)(t-t_0)^3 \qquad (t_0 \le t), \quad (2.7\text{-}29b)$$

$$\text{cov}\{S(t_1),S(t_2)\} = b_2(t_1-t_0)^2\,[(t_1-t_0)/3 + (t_2-t_1)/2]$$
$$(t_0 \le t_1 \le t_2). \quad (2.7\text{-}29c)$$

Before proving these results, two comments should be made about them. First, if it happens that $b_2=0$, then Eqs. (2.7-28b) and (2.7-29b) imply that the variances of both $X(t)$ and $S(t)$ will vanish identically; this implies [see (1.3-10)] that $X(t)$ and $S(t)$ will be *deterministic* processes:

$$b_2 = 0 \;\Rightarrow\; \begin{cases} X(t) = \langle X(t)\rangle = x_0 + b_1(t-t_0), & (2.7\text{-}30a) \\[2mm] S(t) = \langle S(t)\rangle = x_0(t-t_0) + (b_1/2)(t-t_0)^2. & (2.7\text{-}30b) \end{cases}$$

This illustrates the validity of our earlier assertion, following Eq. (2.6-3), that a nonvanishing $B_2(x,t)$ is required in order for $X(t)$ to be "genuinely stochastic."

Secondly, for the more interesting case $b_2>0$, it is often useful to measure the *relative fluctuations* in $X(t)$ by the ratio of sdev$\{X(t)\}$ to $\langle X(t)\rangle$. In the present case, the *long-time limit* of that ratio is seen to satisfy

$$\lim_{t\to\infty} \frac{\text{sdev}\{X(t)\}}{\langle X(t)\rangle} = \lim_{t\to\infty} \frac{[b_2(t-t_0)]^{1/2}}{x_0 + b_1(t-t_0)} = \begin{cases} 0, & \text{if } b_1 \ne 0, \\ \infty, & \text{if } b_1 = 0. \end{cases} \quad (2.7\text{-}31)$$

The result for $b_1 \ne 0$ is especially interesting. In that case, although the fluctuations in $X(t)$ about its mean increase without bound, the mean itself grows even faster; as a consequence, the fluctuations ultimately become a negligibly small fraction of the mean, so the process may *appear* to eventually become deterministic in character.

To establish Eqs. (2.7-28), we proceed as follows: Eq. (2.7-28a) is a simple consequence of Eq. (2.7-14), which reads in this case

$$\frac{d}{dt}\langle X(t)\rangle = \langle b_1\rangle = b_1.$$

Integrating, and using the initial condition (2.7-17), we obtain Eq. (2.7-28a). For Eq. (2.7-28b) we apply Eq. (2.7-15), which reads in this case

$$\frac{d}{dt}\text{var}\{X(t)\} = 2[\langle X(t)b_1 \rangle - \langle X(t) \rangle \langle b_1 \rangle] + \langle b_2 \rangle = b_2.$$

Integrating, and using the initial condition (2.7-18), we obtain Eq. (2.7-28b). Finally, Eq. (2.7-28c) is most easily deduced as a consequence of theorem (2.7-20) and Eq. (2.7-28b).

To prove Eqs. (2.7-29), we proceed as follows: Eq. (2.7-29a) is obtained simply by inserting the formula for $\langle X(t) \rangle$ in Eq. (2.7-28a) into the formula (2.7-21) and integrating. To calculate the variance of $S(t)$ from Eq. (2.7-22) we must first solve Eq. (2.7-23a), which reads in this case

$$\frac{d}{dt}\text{cov}\{S(t),X(t)\} = \text{var}\{X(t)\} + \langle S(t)b_1 \rangle - \langle S(t) \rangle \langle b_1 \rangle = \text{var}\{X(t)\} = b_2(t-t_0).$$

Integrating with the initial condition (2.7-23b) gives

$$\text{cov}\{S(t),X(t)\} = (b_2/2)(t-t_0)^2. \tag{2.7-32}$$

Substituting this result into Eq. (2.7-22) and integrating, we get

$$\text{var}\{S(t)\} = 2\int_{t_0}^{t}(b_2/2)(t'-t_0)^2\,dt' = (b_2/3)(t-t_0)^3,$$

which proves Eq. (2.7-29b). And finally, to calculate the covariance of $S(t)$ from Eq. (2.7-24), we must first solve Eq. (2.7-25a), which reads in this case

$$\frac{d}{dt_2}\text{cov}\{S(t_1), X(t_2)\} = \langle S(t_1)b_1 \rangle - \langle S(t_1) \rangle \langle b_1 \rangle = 0.$$

Invoking the initial condition (2.7-25b), we deduce that

$$\text{cov}\{S(t_1), X(t_2)\} = \text{cov}\{S(t_1), X(t_1)\} = (b_2/2)(t_1-t_0)^2,$$

where the last step follows from our earlier result (2.7-32). Now substituting this into Eq. (2.7-24) and invoking Eq. (2.7-29b), we find

$$\text{cov}\{S(t_1), S(t_2)\} = \text{var}\{S(t_1)\} + \int_{t_1}^{t_2}(b_2/2)(t_1-t_0)^2\,dt_2$$

$$= (b_2/3)(t_1-t_0)^3 + (b_2/2)(t_1-t_0)^2(t_2-t_1).$$

This evidently establishes Eq. (2.7-29c).

2.7.D EXAMPLE: $B_1(x,t) = -\beta x$ and $B_2(x,t) = c$

For our second example, we consider a Markov process $X(t)$ for which the first two propagator moment functions have the forms

$$B_1(x,t) = -\beta x \quad \text{and} \quad B_2(x,t) = c \quad (\beta > 0, c \geq 0). \qquad (2.7\text{-}33)$$

Notice that the case $\beta = 0$, which is explicitly excluded here, is actually covered by our first example [see Eqs. (2.7-27) with $b_1 = 0$ and $b_2 = c$]. Using the equations developed in Subsection 2.7.B, we shall derive the following moment formulas for $X(t)$ and its integral $S(t)$:

$$\langle X(t) \rangle = x_0 \exp[-\beta(t-t_0)] \qquad (t_0 \leq t), \qquad (2.7\text{-}34\text{a})$$

$$\text{var}\{X(t)\} = (c/2\beta)(1 - \exp[-2\beta(t-t_0)]) \qquad (t_0 \leq t), \qquad (2.7\text{-}34\text{b})$$

$$\text{cov}\{X(t_1),X(t_2)\} = (c/2\beta)\exp[-\beta(t_2-t_1)](1 - \exp[-2\beta(t_1-t_0)])$$
$$(t_0 \leq t_1 \leq t_2); \qquad (2.7\text{-}34\text{c})$$

and

$$\langle S(t) \rangle = (x_0/\beta)(1 - \exp[-\beta(t-t_0)]) \qquad (t_0 \leq t), \qquad (2.7\text{-}35\text{a})$$

$$\text{var}\{S(t)\} = (c/\beta^2)\,[(t-t_0) - (2/\beta)(1 - \exp[-\beta(t-t_0)])$$
$$+ (1/2\beta)(1 - \exp[-2\beta(t-t_0)])] \qquad (t_0 \leq t), \qquad (2.7\text{-}35\text{b})$$

$$\text{cov}\{S(t_1),S(t_2)\} = \text{var}\{S(t_1)\} + (c/2\beta^3)(1 - \exp[-\beta(t_2-t_1)])$$
$$\times (1 - 2\exp[-\beta(t_1-t_0)] + \exp[-2\beta(t_1-t_0)])$$
$$(t_0 \leq t_1 \leq t_2). \qquad (2.7\text{-}35\text{c})$$

Before deriving these formulas, let us consider a few of their implications. First, if it happens that $c = 0$, then it follows from Eqs. (2.7-34b) and (2.7-35b) that the variances of both $X(t)$ and $S(t)$ will vanish identically; that of course implies that both $X(t)$ and $S(t)$ will be *deterministic* processes:

$$c = 0 \;\Rightarrow\; \begin{cases} X(t) = \langle X(t) \rangle = x_0 \exp[-\beta(t-t_0)], & (2.7\text{-}36\text{a}) \\[2mm] S(t) = \langle S(t) \rangle = (x_0/\beta)(1 - \exp[-\beta(t-t_0)]). & (2.7\text{-}36\text{b}) \end{cases}$$

This result is analogous to our finding (2.7-30) in the preceding example, and it illustrates once again that a nonvanishing $B_2(x,t)$ is essential if $X(t)$ is to be genuinely stochastic.

Secondly, we note the following behaviors of the means and variances of $X(t)$ and $S(t)$ in the limit $(t-t_0)\rightarrow\infty$:

$$(t-t_0)\rightarrow\infty \quad \Rightarrow \quad \begin{cases} \langle X(t)\rangle \rightarrow 0, & \text{(2.7-37a)} \\ \\ \text{var}\{X(t)\} \rightarrow (c/2\beta). & \text{(2.7-37b)} \end{cases}$$

and

$$(t-t_0)\rightarrow\infty \quad \Rightarrow \quad \begin{cases} \langle S(t)\rangle \rightarrow (x_0/\beta), & \text{(2.7-38a)} \\ \\ \text{var}\{S(t)\} \rightarrow (c/\beta^2)(t-t_0). & \text{(2.7-38b)} \end{cases}$$

We also note that these long-time limits are effectively reached whenever $(t-t_0)\gg\beta^{-1}$; therefore, β^{-1} represents a **relaxation time** for the process. Furthermore, since it follows from Eq. (2.7-34c) that

$$(t_2-t_1) \gg \beta^{-1} \quad \Rightarrow \quad \text{cov}\{X(t_1),X(t_2)\} \approx 0, \qquad \text{(2.7-39)}$$

then β^{-1} also represents a **decorrelation time** for the process; that is, $X(t_1)$ and $X(t_2)$ will for all practical purposes be *uncorrelated* whenever (t_2-t_1) is very large compared to β^{-1}.

Now let us derive the three equations (2.7-34). We begin by writing down Eq. (2.7-14) for the given propagator moment functions (2.7-33):

$$\frac{d}{dt}\langle X(t)\rangle = \langle -\beta X(t)\rangle = -\beta\langle X(t)\rangle.$$

The solution to this differential equation for the initial condition (2.7-17) is obviously Eq. (2.7-34a). Turning next to Eq. (2.7-15), we have

$$\frac{d}{dt}\text{var}\{X(t)\} = 2\Big(\langle X(t)[-\beta X(t)]\rangle - \langle X(t)\rangle\langle -\beta X(t)\rangle\Big) + \langle c\rangle$$

$$= -2\beta[\langle X^2(t)\rangle - \langle X(t)\rangle^2] + c$$

$$= -2\beta\,\text{var}\{X(t)\} + c.$$

This differential equation is seen to be of the form (A-11a), so using the initial condition (2.7-18) we obtain from Eq. (A-11b) the solution

$$\text{var}\{X(t)\} = e^{-2\beta(t-t_0)}\left[0 + \int_{t_0}^{t}dt'\,c\,e^{+2\beta(t'-t_0)}\right]$$

$$= c\,e^{-2\beta(t-t_0)}\int_{0}^{t-t_0}d\tau\,e^{2\beta\tau}$$

$$= c e^{-2\beta(t-t_0)} \frac{1}{2\beta} \left(e^{2\beta(t-t_0)} - 1 \right).$$

This establishes Eq. (2.7-34b). Finally, Eq. (2.7-16) takes the form

$$\frac{d}{dt_2} \mathrm{cov}\{X(t_1),X(t_2)\} = \langle X(t_1)[-\beta X(t_2)]\rangle - \langle X(t_1)\rangle \langle -\beta X(t_2)\rangle$$

$$= -\beta [\langle X(t_1) X(t_2)\rangle - \langle X(t_1)\rangle \langle X(t_2)\rangle]$$

$$= -\beta \, \mathrm{cov}\{X(t_1),X(t_2)\}.$$

The solution to this differential equation for the initial condition (2.7-19) is obviously

$$\mathrm{cov}\{X(t_1),X(t_2)\} = \mathrm{var}\{X(t_1)\} e^{-\beta(t_2-t_1)}$$

$$= \left[\frac{c}{2\beta} \left(1 - e^{-2\beta(t_1-t_0)} \right) \right] e^{-\beta(t_2-t_1)},$$

where the last step has invoked Eq. (2.7-34b). This result is precisely Eq. (2.7-34c).

Next we shall establish the moment formulas (2.7-35) for the integral process $S(t)$. We start by inserting our result (2.7-34a) for $\langle X(t)\rangle$ into Eq. (2.7-21) and then integrating:

$$\langle S(t)\rangle = \int_{t_0}^{t} x_0 e^{-\beta(t'-t_0)} \, dt' = \frac{x_0}{\beta} \left(1 - e^{-\beta(t-t_0)} \right).$$

This is evidently the result claimed in Eq. (2.7-35a). To calculate var$\{S(t)\}$ from Eq. (2.7-22), we must first solve Eq. (2.7-23a) for the propagator moment functions (2.7-33):

$$\frac{d}{dt} \mathrm{cov}\{S(t),X(t)\} = \mathrm{var}\{X(t)\} + \langle S(t)[-\beta X(t)]\rangle - \langle S(t)\rangle \langle -\beta X(t)\rangle$$

$$= -\beta [\langle S(t) X(t)\rangle - \langle S(t)\rangle \langle X(t)\rangle] + \mathrm{var}\{X(t)\}$$

$$= -\beta \, \mathrm{cov}\{S(t),X(t)\} + \frac{c}{2\beta} \left(1 - e^{-2\beta(t-t_0)} \right),$$

where in the last step we have used the result (2.7-34b). This differential equation is of the form (A-11a), so with the initial condition (2.7-23b) we have from Eq. (A-11b) the solution

$\text{cov}\{S(t),X(t)\}$

$$= e^{-\beta(t-t_0)}\left[0 + \int_{t_0}^{t}dt'\,\frac{c}{2\beta}\left(1 - e^{-2\beta(t'-t_0)}\right)e^{+\beta(t'-t_0)}\right]$$

$$= \frac{c}{2\beta}\,e^{-\beta(t-t_0)}\int_{t_0}^{t}dt'\left(e^{\beta(t'-t_0)} - e^{-\beta(t'-t_0)}\right)$$

$$= \frac{c}{2\beta}\,e^{-\beta(t-t_0)}\left[\frac{1}{\beta}\left(e^{\beta(t-t_0)} - 1\right) - \frac{1}{\beta}\left(1 - e^{-\beta(t-t_0)}\right)\right]$$

$$\text{cov}\{S(t),X(t)\} = \left(\frac{c}{2\beta^2}\right)\left[1 - 2e^{-\beta(t-t_0)} + e^{-2\beta(t-t_0)}\right]. \quad (2.7\text{-}40)$$

Inserting this auxiliary result into Eq. (2.7-22) and then integrating, we get

$$\text{var}\{S(t)\} = 2\int_{t_0}^{t}\left(\frac{c}{2\beta^2}\right)\left[1 - 2e^{-\beta(t'-t_0)} + e^{-2\beta(t'-t_0)}\right]dt'$$

$$= \left(\frac{c}{\beta^2}\right)\left[(t-t_0) - \frac{2}{\beta}\left(1 - e^{-\beta(t-t_0)}\right)\right.$$

$$\left. + \frac{1}{2\beta}\left(1 - e^{-2\beta(t-t_0)}\right)\right],$$

which is precisely Eq. (2.7-35b)

Finally, to calculate $\text{cov}\{S(t_1),S(t_2)\}$ from Eq. (2.7-24), we must first solve Eq. (2.7-25a). The latter equation reads in this case

$$\frac{d}{dt_2}\text{cov}\{S(t_1),X(t_2)\} = \langle S(t_1)[-\beta X(t_2)]\rangle - \langle S(t_1)\rangle\langle-\beta X(t_2)\rangle$$

$$= -\beta[\langle S(t_1)X(t_2)\rangle - \langle S(t_1)\rangle\langle X(t_2)\rangle]$$

$$= -\beta\,\text{cov}\{S(t_1),X(t_2)\}.$$

The solution to this differential equation for the initial condition (2.7-25b) is obviously

$$\text{cov}\{S(t_1),X(t_2)\} = \text{cov}\{S(t_1),X(t_1)\}e^{-\beta(t_2-t_1)}.$$

Inserting this result into Eq. (2.7-24) and then integrating, we get

$$\text{cov}\{S(t_1),S(t_2)\} = \text{var}\{S(t_1)\} + \int_{t_1}^{t_2} \text{cov}\{S(t_1),X(t_1)\} e^{-\beta(t_2'-t_1)} dt_2'$$

$$= \text{var}\{S(t_1)\} + \text{cov}\{S(t_1),X(t_1)\} \int_0^{t_2-t_1} e^{-\beta\tau} d\tau$$

$$= \text{var}\{S(t_1)\} + \text{cov}\{S(t_1),X(t_1)\} \frac{1}{\beta}\left(1 - e^{-\beta(t_2-t_1')}\right)$$

$$= \text{var}\{S(t_1)\} + \left(\frac{c}{2\beta^3}\right)\left(1 - e^{-\beta(t_2-t_1')}\right)$$

$$\times \left(1 - 2e^{-\beta(t_1-t_0)} + e^{-2\beta(t_1-t_0)}\right),$$

where in the last step we have invoked the auxiliary result (2.7-40). This establishes Eq. (2.7-35c), which of course can be made still more explicit by substituting for var$\{S(t_1)\}$ from Eq. (2.7-35b).

2.8 HOMOGENEITY

In this section we shall focus on some properties shared by many physically interesting Markov processes that can greatly simplify their analyses. Recalling that any Markov process $X(t)$ is completely characterized by its propagator density function $\Pi(\xi \mid dt; x,t)$, we make the following three *definitions*:

$X(t)$ is **temporally homogeneous**

$\Leftrightarrow$ $\Pi(\xi \mid dt; x,t)$ is explicitly independent of t; (2.8-1a)

$X(t)$ is **spatially homogeneous**

$\Leftrightarrow$ $\Pi(\xi \mid dt; x,t)$ is explicitly independent of x; (2.8-1b)

$X(t)$ is **completely homogeneous**

$\Leftrightarrow$ $\Pi(\xi \mid dt; x,t)$ is explicitly independent of both x and t. (2.8-1c)

Of these three categories of Markov processes, only the first and last seem to be encountered with any great regularity in physical applications; i.e., we seldom encounter a spatially homogeneous Markov process that is not also temporally homogeneous. We shall therefore focus here only on temporally homogeneous and completely homogeneous Markov processes, considering them in turn with a view to seeing how the general theory simplifies in their cases.

2.8.A TEMPORALLY HOMOGENEOUS MARKOV PROCESSES

Most Markov processes $X(t)$ encountered in practice are *temporally* homogeneous, in that the propagator density function $\Pi(\xi \mid dt; x,t)$, and hence also the propagator $\Xi(dt; x,t)$, do not depend explicitly on t. It must be emphasized that a lack of explicit t-dependence in $\Xi(dt; x,t)$ and its density function $\Pi(\xi \mid dt; x,t)$ does *not* necessarily imply that $X(t)$ and its density function $P(x,t \mid x_0,t_0)$ are likewise independent of t. The situation is analogous to that in particle dynamics [see Eq. (2.4-13)], where a lack of explicit time dependence in the force function $F(x,p;t)$ does not necessarily result in x_t and p_t being constant in time.

But if temporal homogeneity does not remove the t-dependence from the function $P(x,t \mid x_0,t_0)$, it does simplify that dependence somewhat. To see specifically how, we may first observe that if $\Pi(\xi \mid dt; x,t)$ is explicitly independent of t then the integral equation (2.4-12) and its auxiliary equations (2.4-11) simplify to

$$P(x,t \mid x_0,t_0) = \int_{-\infty}^{\infty} d\xi_1 \cdots \int_{-\infty}^{\infty} d\xi_{n-1} \prod_{i=1}^{n} \Pi(\xi_i \mid dt; x_{i-1}), \quad (2.8\text{-}2)$$

wherein

$$x_i = x_0 + \xi_1 + \dots + \xi_i \quad (i=1,\dots,n-1), \qquad (2.8\text{-}3a)$$

$$\xi_n \equiv x - x_0 - \xi_1 - \dots - \xi_{n-1}. \qquad (2.8\text{-}3b)$$

$$(t-t_0)/n = dt, \text{ an infinitesimal.} \qquad (2.8\text{-}3c)$$

Unlike Eqs. (2.4-11) and (2.4-12) [see especially Eq. (2.4-11a) with $i=1$], these equations involve the two variables t and t_0 only through their *difference* $(t-t_0)$. So we may immediately conclude that, when $X(t)$ is temporally homogeneous, then $P(x,t \mid x_0,t_0)$ has the property that

$$P(x,t \mid x_0,t_0) = P(x,t-t_0 \mid x_0,0) \quad (t_0 \le t). \qquad (2.8\text{-}4)$$

One noteworthy consequence of the property (2.8-4) is that

$$\frac{\partial}{\partial t} P(x,t \,|\, x_0,t_0) = -\frac{\partial}{\partial t_0} P(x,t \,|\, x_0,t_0). \qquad (2.8\text{-}5)$$

To prove this, simply set $s = t - t_0$, so that Eq. (2.8-4) becomes

$$P(x,t \,|\, x_0,t_0) = P(x,s \,|\, x_0,0),$$

and then use the chain rule for derivatives to get

$$\frac{\partial}{\partial t} P(x,t \,|\, x_0,t_0) = \frac{\partial}{\partial t} P(x,s \,|\, x_0,0) = \frac{\partial}{\partial s} P(x,s \,|\, x_0,0) \frac{\partial s}{\partial t} = \frac{\partial}{\partial s} P(x,s \,|\, x_0,0),$$

$$\frac{\partial}{\partial t_0} P(x,t \,|\, x_0,t_0) = \frac{\partial}{\partial t_0} P(x,s \,|\, x_0,0) = \frac{\partial}{\partial s} P(x,s \,|\, x_0,0) \frac{\partial s}{\partial t_0} = -\frac{\partial}{\partial s} P(x,s \,|\, x_0,0).$$

Combining these two formulas obviously gives Eq. (2.8-5).

Another consequence of no explicit t-dependence in the propagator density function Π is that the propagator moment functions B_n will likewise be explicitly independent of t; i.e., Eq. (2.4-3b) becomes:

$$B_n(x) \equiv (dt)^{-1} \int_{-\infty}^{\infty} d\xi \, \xi^n \, \Pi(\xi \,|\, dt; x) \qquad (n = 1,2,\ldots). \qquad (2.8\text{-}6)$$

In that case, the *forward* Kramers-Moyal equation (2.5-1) evidently simplifies to

$$\frac{\partial}{\partial t} P(x,t \,|\, x_0,t_0) = \sum_{n=1}^{\infty} \frac{(-1)^n}{n!} \frac{\partial^n}{\partial x^n} \Big[B_n(x) \, P(x,t \,|\, x_0,t_0) \Big], \qquad (2.8\text{-}7)$$

while the *backward* Kramers-Moyal equation (2.5-2) simplifies to

$$-\frac{\partial}{\partial t_0} P(x,t \,|\, x_0,t_0) = \sum_{n=1}^{\infty} \frac{1}{n!} B_n(x_0) \frac{\partial^n}{\partial x_0^n} P(x,t \,|\, x_0,t_0). \qquad (2.8\text{-}8)$$

Furthermore, owing to Eq. (2.8-5), *the left hand sides* of the forward and backward Kramers-Moyal equations are now *interchangeable*.

The time-evolution equations for the moments of $X(t)$ and its integral $S(t)$ that were derived in Section 2.7 also become somewhat simpler when the propagator moment functions in those equations lose their explicit dependence on t. Since the forms of those simplified equations are easy to read off, we shall not bother to write them down here. Of course, we may expect on the basis of property (2.8-4) that the solutions to the temporally homogeneous moment evolution equations will depend on the times t and t_0 only through the difference $(t - t_0)$.

For *some* temporally homogeneous Markov processes, but certainly not all, it happens that

$$\lim_{(t-t_0)\to\infty} P(x,t\,|\,x_0,t_0) = P_{\mathrm{s}}(x), \tag{2.8-9}$$

where $P_{\mathrm{s}}(x)$ is a well defined function of x that is independent of both t and x_0. In such a circumstance we say that $X(t)$ is a **stable process**, and we call $P_{\mathrm{s}}(x)$ the **stationary Markov state density function**. Since Eq. (2.8-9) would imply that $\partial P/\partial t$ approaches zero as $t\to\infty$, then we may expect $P_{\mathrm{s}}(x)$ to be a "stationary solution" of the forward Kramers-Moyal equation (2.8-7); i.e., $P_{\mathrm{s}}(x)$, if it exists at all, will be the solution of the infinite order ordinary differential equation

$$\sum_{n=1}^{\infty} \frac{(-1)^n}{n!} \frac{d^n}{dx^n}\left[B_n(x)\,P_{\mathrm{s}}(x)\right] = 0. \tag{2.8-10}$$

That $P_{\mathrm{s}}(x)$ does not exist for all temporally homogeneous Markov processes is illustrated by the fact, proved in the next subsection, that a stationary Markov state density function *never* exists for a *completely* homogeneous Markov process.

If a stationary Markov state density function $P_{\mathrm{s}}(x)$ does exist, then by definition the Markov state density function $P(x,t\,|\,x_0,t_0)$ eventually becomes time-independent. It is important to understand that this does *not* mean that the process $X(t)$ itself eventually becomes static, but only that all *statistical properties* of $X(t)$ eventually become static. For example, taking the $(t-t_0)\to\infty$ limit of the average of $g(X(t))$ in Eq. (2.3-3) and then invoking Eq. (2.8-9), we get

$$\lim_{(t-t_0)\to\infty} \langle g(X(t))\rangle = \int_{-\infty}^{\infty} dx\, g(x)\, P_{\mathrm{s}}(x) \equiv \langle g(X(t))\rangle_{\mathrm{s}}. \tag{2.8-11}$$

The middle expression shows that this quantity is independent of t. In particular, for the mean and variance of $X(t)$ we would have

$$\lim_{(t-t_0)\to\infty} \langle X(t)\rangle = \langle X(t)\rangle_{\mathrm{s}}, \tag{2.8-12}$$

and

$$\lim_{(t-t_0)\to\infty} \mathrm{var}\{X(t)\} = \mathrm{var}_{\mathrm{s}}\{X(t)\} = \langle X^2(t)\rangle_{\mathrm{s}} - \langle X(t)\rangle_{\mathrm{s}}^2, \tag{2.8-13}$$

both of which may be calculated using the integral form in Eq. (2.8-11), and both of which are therefore independent of t. We call $\langle X(t)\rangle_{\mathrm{s}}$ the **stationary mean** of $X(t)$, and $\mathrm{var}_{\mathrm{s}}\{X(t)\}$ the **stationary variance** of $X(t)$.

On the other hand, if we take the $(t_1-t_0)\to\infty$ limit of the average of $g(X(t_1),X(t_2))$ in Eq. (2.3-4), then we find using Eqs. (2.8-4) and (2.8-9), that

$$\lim_{(t_1-t_0)\to\infty} \langle g(X(t_1),X(t_2))\rangle$$

$$= \int_{-\infty}^{\infty} dx_1 \int_{-\infty}^{\infty} dx_2\, g(x_1,x_2)\, P(x_2,t_2-t_1\,|\,x_1,0)\, P_s(x_1)$$

$$\equiv \langle g(X(t_1),X(t_2))\rangle_s. \qquad (2.8\text{-}14)$$

It is obvious from the second line that this quantity depends on the times t_1 and t_2 only through their difference (t_2-t_1). In particular, the **stationary covariance** of $X(t)$ is defined by

$$\lim_{(t_1-t_0)\to\infty} \mathrm{cov}\{X(t_1),X(t_2)\} \equiv \mathrm{cov}_s\{X(t_1),X(t_2)\} = \langle X(t_1)\,X(t_2)\rangle_s - \langle X(t)\rangle_s^2.$$

$$(2.8\text{-}15)$$

Here, the *first* average on the right is evaluated using the integral formula (2.8-14), and thus depends only on (t_2-t_1), while the *second* average on the right is evaluated using the integral formula (2.8-11), and thus has no time dependence at all. So $\mathrm{cov}_s\{X(t_1),X(t_2)\}$ depends solely on the difference (t_2-t_1).

It is worth emphasizing again that the stationary mean, variance and covariance of $X(t)$ exist *only* when condition (2.8-9) is satisfied — i.e., only when $X(t)$ does in fact have a stationary Markov state density function.

2.8.B COMPLETELY HOMOGENEOUS MARKOV PROCESSES

Completely homogeneous Markov processes, as defined in (2.8-1c), can evidently be viewed as a subclass of temporally homogeneous Markov processes in which the propagator density function Π is explicitly independent of x as well as t. Although this circumstance does *not* imply that the Markov state density function $P(x,t\,|\,x_0,t_0)$ becomes independent of either x or t, it does engender a number of simplifications.

With spatial homogeneity compounding temporal homogeneity, the integral equation (2.8-2) for $P(x,t\,|\,x_0,t_0)$ now takes the simpler form

$$P(x,t\,|\,x_0,t_0) = \int_{-\infty}^{\infty} d\xi_1 \cdots \int_{-\infty}^{\infty} d\xi_{n-1} \prod_{j=1}^{n} \Pi(\xi_j\,|\,dt), \qquad (2.8\text{-}16)$$

with the three auxiliary equations (2.8-3) reducing to the two auxiliary equations

$$\xi_n \equiv x - x_0 - \sum_{i=1}^{n-1} \xi_i , \qquad (2.8\text{-}17a)$$

$$(t-t_0)/n = dt, \text{ an infinitesimal.} \qquad (2.8\text{-}17b)$$

We can simplify Eq. (2.8-16) still further by proceeding as follows. First, we incorporate the definition (2.8-17a) of ξ_n into Eq. (2.8-16) by the trick of integrating Eq. (2.8-16) over ξ_n with an appropriate Dirac delta function in the integrand:

$$P(x,t|x_0,t_0) = \int_{-\infty}^{\infty} d\xi_1 \cdots \int_{-\infty}^{\infty} d\xi_{n-1}$$

$$\times \int_{-\infty}^{\infty} d\xi_n \prod_{j=1}^{n} \Pi(\xi_j|dt)\, \delta\!\left(\xi_n - \left[x - x_0 - \sum_{i=1}^{n-1} \xi_i\right]\right)$$

$$= \prod_{j=1}^{n} \int_{-\infty}^{\infty} d\xi_j\, \Pi(\xi_j|dt)\, \delta\!\left(x - x_0 - \sum_{i=1}^{n} \xi_i\right).$$

Now using the integral representation (B-7) of the Dirac delta function, we get

$$\delta\!\left(x - x_0 - \sum_{i=1}^{n} \xi_i\right) = (2\pi)^{-1} \oint_{-\infty}^{\infty} du \exp\!\left[iu\!\left(x - x_0 - \sum_{i=1}^{n} \xi_i\right)\right]$$

$$= (2\pi)^{-1} \oint_{-\infty}^{\infty} du \exp[iu(x - x_0)] \prod_{j=1}^{n} \exp(-iu\xi_j).$$

Substituting this into the preceding equation, we get

$$P(x,t|x_0,t_0)$$

$$= (2\pi)^{-1} \oint_{-\infty}^{\infty} du \exp[iu(x - x_0)] \prod_{j=1}^{n} \int_{-\infty}^{\infty} d\xi_j\, \Pi(\xi_j|dt)\exp(-iu\xi_j)$$

$$= (2\pi)^{-1} \oint_{-\infty}^{\infty} du \exp[iu(x - x_0)] \left[\int_{-\infty}^{\infty} d\xi\, \Pi(\xi|dt)\exp(-iu\xi)\right]^n .$$

And so we conclude that, for a *completely homogeneous* Markov process with propagator density function $\Pi(\xi|dt)$, the Markov state density function is given by the *explicit integral formula*

$$P(x,t \mid x_0, t_0) = (2\pi)^{-1} \oint_{-\infty}^{\infty} du \, \exp[iu(x - x_0)] \left[\Pi^{\star}(u \mid dt) \right]^n, \quad (2.8\text{-}18)$$

where $\Pi^{\star}(u \mid dt)$ is essentially the Fourier transform of the propagator density function,

$$\Pi^{\star}(u \mid dt) \equiv \int_{-\infty}^{\infty} d\xi \, \Pi(\xi \mid dt) \exp(-iu\xi), \quad (2.8\text{-}19a)$$

and where n is *infinitely large* and dt is *infinitesimally small* in such a way that

$$n \, dt = t - t_0. \quad (2.8\text{-}19b)$$

The integral formula (2.8-18) for the Markov state density function of a completely homogeneous Markov process has several noteworthy aspects. First, as we already expect of any *temporally* homogeneous Markov process [see Eq. (2.8-4)], t and t_0 enter Eqs. (2.8-18) and (2.8-19) only through their difference $(t - t_0)$. But notice that x and x_0 also enter Eqs. (2.8-18) and (2.8-19) only through their difference $(x - x_0)$. Therefore, for a *completely* homogeneous Markov process the Markov state density function has the property that

$$P(x,t \mid x_0, t_0) = P(x - x_0, t - t_0 \mid 0, 0) \quad (t_0 \leq t). \quad (2.8\text{-}20)$$

And just as the temporally homogeneous symmetry (2.8-4) had the consequence (2.8-5), so does the completely homogeneous symmetry (2.8-20) have the consequence

$$\frac{\partial P}{\partial t} = -\frac{\partial P}{\partial t_0} \quad \text{and} \quad \frac{\partial P}{\partial x} = -\frac{\partial P}{\partial x_0}. \quad (2.8\text{-}21)$$

Another implication of Eqs. (2.8-18) and (2.8-19) is that for any completely homogeneous Markov process $X(t)$ we are only two direct integrations away from a fully explicit formula for the Markov state density function $P(x,t \mid x_0, t_0)$. The first of those two integrations, the ξ-integration in Eq. (2.8-19a), basically computes the Fourier transform of the propagator density function $\Pi(\xi \mid dt)$. As we shall see in the following chapters, this integration is easily performed for most completely homogeneous Markov processes of practical interest. [And once this integral has been evaluated and the result raised to the power n, it will become clear that the two variables n and dt appear in the dual limit $n \to \infty$ and $dt \to 0$ *only* in the combination (2.8-19b).] As for the second integration, the one over u in Eq. (2.8-18), it turns out to be very straightforward in the *continuous* case, but not in the *jump* case. In the jump case, the u-integration is made difficult by the fact that convergence usually requires one to take explicit account of the special way in which that integral is defined [see Appendix B, especially Eq. (B-6)].

It follows from Eqs. (2.4-3) that, when the propagator density function Π is independent of both x and t, then the propagator moment functions B_n will likewise be independent of both x and t:

$$B_n(x,t) = (dt)^{-1} \int_{-\infty}^{\infty} d\xi\, \xi^n\, \Pi(\xi\,|\,dt) \equiv b_n \quad (n=1,2,\dots). \quad (2.8\text{-}22)$$

Provided the ξ-dependence of Π is such that all these constants exist, then the forward Kramers-Moyal equation, which simplified in the temporally homogeneous case to Eq. (2.8-7), now assumes the even simpler form

$$\frac{\partial}{\partial t} P(x,t\,|\,x_0,t_0) = \sum_{n=1}^{\infty} \frac{(-1)^n}{n!}\, b_n\, \frac{\partial^n}{\partial x^n} P(x,t\,|\,x_0,t_0). \quad (2.8\text{-}23)$$

And the backward Kramers-Moyal equation (2.8-8) becomes

$$-\frac{\partial}{\partial t_0} P(x,t\,|\,x_0,t_0) = \sum_{n=1}^{\infty} \frac{1}{n!}\, b_n\, \frac{\partial^n}{\partial x_0^n} P(x,t\,|\,x_0,t_0).$$

But because of relations (2.8-21), this last equation is seen to be mathematically identical to the preceding equation; thus, for a *completely* homogeneous Markov process, the two Kramers-Moyal equations reduce to the *single* equation (2.8-23). But it seems likely that Eq. (2.8-23) will be much harder to solve for $P(x,t\,|\,x_0,t_0)$ than will the integral equation (2.8-18); indeed, it could be argued that Eq. (2.8-18), as supplemented by Eqs. (2.8-19), is "the solution" of the infinite order partial differential equation (2.8-23) for the particular initial condition (2.5-3).

Regardless of whether or not $P(x,t\,|\,x_0,t_0)$ can be calculated explicitly for a given completely homogeneous Markov process, the constancy of the propagator moment functions (2.8-22), assuming that they all exist, makes possible an explicit calculation of all the moments of $X(t)$ and its integral $S(t)$. In fact, in the example treated in Subsection 2.7.C, we have already calculated the mean, variance and covariance of a completely homogeneous Markov process $X(t)$ [see Eqs. (2.7-28)], as well as the mean, variance and covariance of its integral $S(t)$ [see Eqs. (2.7-29)]. We note that all those formulas involve only the first two propagator moment constants b_1 and b_2. If we require a knowledge of moments of $X(t)$ higher than the second, then we need only turn to the moment evolution equation (2.7-6), which in this case has the relatively simple form

$$\frac{d}{dt}\langle X^n(t)\rangle = \sum_{k=1}^{n} \binom{n}{k} b_k \langle X^{n-k}(t)\rangle \quad (t_0 \le t;\, n=1,2,...).\quad (2.8\text{-}24)$$

It is clearly possible to solve this equation under the initial condition (2.7-7) for the successive n-values 1, 2, 3, etc.; indeed, we already have in Eq. (2.7-28a) the solution for $n=1$. Likewise, the moment equation (2.7-10) for $S(t)$ can be recursively solved in tandem with Eqs. (2.7-12), the latter now having the somewhat simpler form

$$\frac{d}{dt}\langle S^m(t)\,X^n(t)\rangle = m\langle S^{m-1}(t)\,X^{n+1}(t)\rangle + \sum_{k=1}^{n} \binom{n}{k} b_k \langle S^m(t)\,X^{n-k}(t)\rangle$$

$$(t_0 \le t;\, m=1,2,...;\, n=1,2,...).\quad (2.8\text{-}25)$$

We shall not undertake here the straightforward but somewhat tedious solution of Eqs. (2.8-24) and (2.8-25) for the third and higher order moments, because we shall not require those moments in the sequel.

A final important fact about completely homogeneous Markov processes can be deduced from the formulas for $\langle X(t)\rangle$ and $\mathrm{var}\{X(t)\}$ in Eqs. (2.7-28): Excepting the uninteresting case $b_1=b_2=0$ [for which $X(t)=x_0$ for all $t \ge t_0$], either or both of $\langle X(t)\rangle$ and $\mathrm{var}\{X(t)\}$ will vary linearly with t. This implies that the mean and variance of $X(t)$ cannot *both* approach stationary values, as contemplated in Eqs. (2.8-12) and (2.8-13), and therefore that $P(x,t\,|\,x_0,t_0)$ cannot approach a stationary form $P_s(x)$, as contemplated in Eq. (2.8-9). So *a completely homogeneous Markov process cannot be a stable process.* Apparently, the temporally homogeneous propagator density function $\Pi(\xi\,|\,dt;\,x)$ must have a special kind of x-dependence in order for $P(x,t\,|\,x_0,t_0)$ to exhibit the asymptotic behavior (2.8-9). We shall address this point more fully in later chapters.

2.9 THE MONTE CARLO APPROACH

It should be apparent from our discussion thus far that the time evolution of a Markov process is often not easy to describe analytically. Tractable solutions to the time-evolution equations for the density function of $X(t)$ can only rarely be found, and closure problems often hamper the solving of the time-evolution equations for the moments of

$X(t)$. When analytical approaches fail, numerical approaches must be considered. One type of numerical approach that often succeeds where others do not is a strategy called *Monte Carlo simulation*. Certain features of this approach have already been touched on in Section 1.8.

The first (and usually hardest) step in a Monte Carlo simulation is to generate a set of very many, say N, **realizations** $x^{(1)}(t)$, $x^{(2)}(t)$, ..., $x^{(N)}(t)$ of the Markov process $X(t)$. For a given value of t, these realizations are just "sample values" of the random variable $X(t)$. Of course, since we normally will not know the density function $P(x,t \mid x_0,t_0)$ of $X(t)$, then we cannot expect to generate these sample values by subjecting $P(x,t \mid x_0,t_0)$ to any of the conventional random number generating methods described in Section 1.8. Practical Monte Carlo simulation schemes produce sample values of $X(t)$ by more pedestrian procedures that essentially *mimic* the actual time evolution of the process. Specifically, the ith realization of the process $X(t)$ generally consists of a set of sample values $x^{(i)}(t_0)$, $x^{(i)}(t_1)$, $x^{(i)}(t_2)$, ... of the process at successive instants t_0, t_1, t_2, ... We shall defer to later chapters a detailed description of the exact procedures used to construct these realizations, with the simulation procedure for *continuous* Markov processes being described in Section 3.9, and the simulation procedures for *jump* Markov processes being described in Sections 4.6, 5.1 and 6.1.

Once a set of realizations $\{x^{(i)}(t)\}$ has been generated, then virtually any dynamical quantity can be numerically estimated. The analyst is then in the position of an "experimentalist" with unlimited measuring capabilities. For example, the one-time and two-time averages in Eqs. (2.3-3) and (2.3-4) can be estimated respectively as

$$\langle g(X(t)) \rangle \approx \frac{1}{N} \sum_{i=1}^{N} g(x^{(i)}(t)) \qquad (2.9\text{-}1)$$

and

$$\langle g(X(t_1), X(t_2)) \rangle \approx \frac{1}{N} \sum_{i=1}^{N} g(x^{(i)}(t_1), x^{(i)}(t_2)). \qquad (2.9\text{-}2)$$

where, to repeat, $x^{(i)}(t)$ is the value of the ith realization of the process at time t. From such formulas we can easily obtain estimates of the mean, variance and covariance of the process $X(t)$. Or, we can estimate the curve $P(x,t \mid x_0,t_0)$-versus-x for any fixed $t > t_0$ as a *normalized frequency histogram* of the values $x^{(1)}(t)$, $x^{(2)}(t)$, ..., $x^{(N)}(t)$; this is simply a "bar-graph" whose height over the x-interval $[x, x + \Delta x)$ is equal to the number of the sample values falling inside that interval divided by $N\Delta x$.

The estimates (2.9-1) and (2.9-2) will become exact in the limit $N\to\infty$, and are evidently just applications of the general rule (1.8-1). In practice of course, N will be *finite*, a circumstance that gives rise to the uncertainties hinted at in Eqs. (2.9-1) and (2.9-2). But if N is large enough, we can get a quantitative handle on those uncertainties by appealing to the central limit theorem result in Eqs. (1.8-2) and (1.8-3). Thus, for example, we may refine the estimate (2.9-1) by simply calculating the *two* averages

$$\langle g(X(t))\rangle_N \equiv \frac{1}{N}\sum_{i=1}^{N} g(x^{(i)}(t)) \quad \text{and} \quad \langle g^2(X(t))\rangle_N \equiv \frac{1}{N}\sum_{i=1}^{N} g^2(x^{(i)}(t)),$$

$$(2.9\text{-}3)$$

and then asserting that

$$\langle g(X(t))\rangle \approx \langle g(X(t))\rangle_N \pm \gamma N^{-1/2}\left[\langle g^2(X(t))\rangle_N - \langle g(X(t))\rangle_N^2\right]^{1/2}. \quad (2.9\text{-}4)$$

Taking $\gamma = 1$, 2 or 3 in Eq. (2.9-4) gives *confidence intervals* of 68%, 95% or 99.7%, respectively. And of course, analogous formulas can be written down to refine the two-time estimate (2.9-2). It must be emphasized that the confidence-interval interpretation of Eq. (2.9-4) has validity *only* if N is "sufficiently large" — a condition that the central limit theorem unfortunately does not render more specifically.

As mentioned earlier, the construction of a specific realization $x(t)$ of a particular Markov process $X(t)$ generally consists of generating successive sample values $x(t_0)$, $x(t_1)$, $x(t_2)$, ... of the process at successive instants t_0, t_1, t_2, ... If each time-sampling interval $[t_{i-1}, t_i)$ is made *small enough* that $x(t)$ can be reckoned to be approximately constant over that time interval, then we can also construct an approximate companion realization $s(t)$ of the integral process $S(t)$ through the recursive formulas

$$\begin{cases} s(t_0) = 0, & (2.9\text{-}5a) \\ s(t_i) \approx s(t_{i-1}) + x(t_{i-1})\,(t_i - t_{i-1}) & (i=1,2,...). \quad (2.9\text{-}5b) \end{cases}$$

The first equation here follows from the assumed initial condition (2.6-7), while the second equation is an approximation of the defining equation (2.6-6) of $S(t)$, in which the infinitesimal dt has been replaced by the finite time increment $(t_i - t_{i-1})$. To the extent that such a non-infinitesimal approximation is accurate, this procedure for constructing realizations of the integral process $S(t)$ allows us to numerically estimate virtually any statistical property of $S(t)$ by using formulas entirely analogous to Eqs. (2.9-1) – (2.9-4).

The major strength of the Monte Carlo method is its ability to *straightforwardly* compute numerical estimates of virtually *any* dynamical property of a specified Markov process. The major weakness of the method lies in its potentially time consuming nature, which is a consequence of the relatively slow rate of decrease of the uncertainty in Eq. (2.9-4) with increasing N; e.g., to *halve* the uncertainty we must evidently generate *four times* as many realizations. Because of this weakness, the Monte Carlo method will not be useful in all circumstances; however, it can be quite useful in some situations, as will be demonstrated in later chapters of this book, so it should always be kept in mind. The Monte Carlo simulation method also has a considerable heuristic value, in that a detailed knowledge of how a Monte Carlo simulation can be carried out, and an inspection of a few simulation results, can greatly enhance one's intuitive sense of how a particular Markov process "really works."

- 3 -

CONTINUOUS MARKOV PROCESSES

In this chapter we shall enlarge upon the results of Chapter 2 for a special class of Markov processes called "continuous" Markov processes. It would be possible to simply *define* a continuous Markov process as any Markov process whose propagator density function $\Pi(\xi \mid dt; x,t)$ has the special functional form (3.1-11). However, we are going to take in Section 3.1 a more heuristic approach which reveals why that functional form for $\Pi(\xi \mid dt; x,t)$, and *only* that functional form, gives rise to a Markov process that can aptly be termed "continuous." In Section 3.2 we shall see how the various time-evolution equations derived in Chapter 2 present themselves for the continuous process propagator. In Section 3.3 we shall analyze in detail the three simplest kinds of continuous Markov process. In the remaining sections we shall examine various other aspects of continuous Markov process theory, concluding in Section 3.9 with a detailed discussion of Monte Carlo simulation.

3.1 THE CONTINUOUS PROPAGATOR AND ITS CHARACTERIZING FUNCTIONS

Roughly speaking, a continuous Markov process is one for which the propagator $\Xi(dt; x,t)$ defined in (2.4-1) behaves as much like a "smooth infinitesimal" as possible. We shall in fact simply *define* a continuous Markov process $X(t)$ as one whose propagator density function $\Pi(\xi \mid dt; x,t)$ satisfies the following two conditions:

- $\Pi(\xi \mid dt; x,t)$ varies smoothly with each of its three
  parameters dt, x and t;   (3.1-1a)

- $\Pi(\xi \mid dt; x,t)$ is practically zero everywhere outside of an
  infinitesimally small neighborhood of $\xi = 0$.   (3.1-1b)

The apparent mathematical vagueness in these two conditions will gradually abate as we develop their full implications below. Notice that condition (3.1-1b) implies, because of the definitions (2.4-1) and (2.4-2), that $X(t+dt)-X(t)$ will *almost always be infinitesimally small*; this property is what prompts us to call $X(t)$ a "continuous" Markov process. What is remarkable is that the two seemingly vague requirements (3.1-1) ultimately require $\Pi(\xi \mid dt; x,t)$ to have the specific functional form given in Eq. (3.1-11). We now proceed to show how this comes about.

First of all, on the basis of the stipulation in (3.1-1b) that $\Pi(\xi \mid dt; x,t)$ is concentrated in an infinitesimally small neighborhood of $\xi = 0$, we shall assume that the first and second moments of that density function are finite; i.e., we shall *postulate* that the random variable $\Xi(dt; x,t)$ for a continuous Markov process has a well defined mean and variance. [We thereby rule out the possibility that $\Xi(dt; x,t)$ is a Cauchy random variable, whose density function (1.4-13) is not sufficiently confined to finite values to allow existence of the first and second moments.] And on the basis of condition (3.1-1a), we shall further postulate that the mean and variance of $\Xi(dt; x,t)$ for a continuous Markov process are both smooth functions of the three parameters dt, x and t; later in our analysis we shall discover just how smooth $\langle \Xi(dt; x,t)\rangle$ and var$\{\Xi(dt; x,t)\}$ need to be in order for everything to work out satisfactorily.

Assuming now that the process has some known value x at time t, let us focus our attention on the behavior of the process during a subsequent infinitesimal time interval $[t,t+dt]$. For that purpose, we imagine the interval $[t,t+dt]$ to be divided into $n \geq 2$ subintervals of equal length dt/n by means of the division points

$$t_i = t_{i-1} + dt/n \quad (i=1,...,n), \tag{3.1-2}$$

where for the present we define $t_0 \equiv t$ and $t_n \equiv t+dt$. With the value of the process at time t specified as just mentioned according to

$$X(t) \equiv X(t_0) = x, \tag{3.1-3a}$$

then the definition (2.4-1) of the propagator allows us to write the process at the various times t_1, t_2, ..., t_n in the recursive form

$$X(t_i) = X(t_{i-1}) + \Xi_i(dt/n; X(t_{i-1}),t_{i-1}) \quad (i=1,...,n). \tag{3.1-3b}$$

Rewriting this relation as

$$X(t_i) - X(t_{i-1}) = \Xi_i(dt/n; X(t_{i-1}),t_{i-1}),$$

and observing that

$$\sum_{i=1}^{n} \left[X(t_i) - X(t_{i-1}) \right] \equiv X(t_n) - X(t_0) \equiv X(t+dt) - x \equiv \Xi(dt; x,t),$$

then we see that by simply summing the preceding equation over i we may deduce that

$$\Xi(dt; x,t) = \sum_{i=1}^{n} \Xi_i(dt/n; X(t_{i-1}),t_{i-1}), \tag{3.1-4}$$

at least to lowest order in dt. Notice that the parameter $X(t_{i-1})$ in this last equation can be regarded as being defined in terms of $X(t_0) \equiv x$ and the $i-1$ random variables $\Xi_1, ..., \Xi_{i-1}$ through the set of recursion relations (3.1-3).

Equation (3.1-4) is a mathematical statement of the simple fact that the change in the process from time $t \equiv t_0$ to time $t+dt \equiv t_n$ is equal to the sum of the changes in the process over the successive time intervals $[t_0,t_1], [t_1,t_2], ..., [t_{n-1},t_n]$. It follows that Eq. (3.1-4), together with the auxiliary relations (3.1-3), is just a "compounded" version of the Chapman-Kolmogorov condition (2.4-9) on the propagator of the process.

The n random variables $\Xi_1, ..., \Xi_n$ appearing in Eq. (3.1-4) are evidently *not* statistically independent of each other, because Ξ_i, through its dependence on $X(t_{i-1})$, depends on $\Xi_{i-1}, \Xi_{i-2}, ..., \Xi_1$. But we are now going to argue that the defining conditions (3.1-1) of a *continuous* Markov process allow us to get around that dependency by making the replacements

$$t_{i-1} \to t \quad \text{and} \quad X(t_{i-1}) \to x \tag{3.1-5}$$

in the right side of Eq. (3.1-4). We first make this argument in a heuristic, qualitative manner as follows: We take the "smooth dependence" of the density function of $\Xi(dt; x,t)$ on the parameters x and t called for in (3.1-1a) as implying that, if we displace the values of x and t by arbitrary infinitesimal amounts, then we will incur only infinitesimally small changes in the random variable $\Xi(dt; x,t)$. Therefore, in the summand $\Xi_i(dt/n; X(t_{i-1}),t_{i-1})$ in Eq. (3.1-4), we should be able to replace the last argument t_{i-1} by the infinitesimally close value t without harming the equality. Furthermore, we take the condition (3.1-1b) as implying that a *non*-infinitesimal excursion of the process during $[t,t+dt]$ is "practically impossible." Therefore, $X(t_{i-1})$ will "practically always" be infinitesimally close to $X(t_0) \equiv x$, so we should likewise be able, in the summand $\Xi_i(dt/n; X(t_{i-1}),t_{i-1})$ in Eq. (3.1-4), to replace the argument $X(t_{i-1})$ by the value x without harming the equality. In short, we take the conditions (3.1-1) that define a continuous Markov process to be our warrant for making the replacements (3.1-5) in Eq. (3.1-4), thereby giving, at least to lowest order in dt,

$$\Xi(dt; x,t) = \sum_{i=1}^{n} \Xi_i(dt/n; x,t). \tag{3.1-6}$$

Later in this section we shall deduce explicit x- and t-smoothness conditions on $\Pi(\xi \,|\, dt; x,t)$ that will ensure that Eq. (3.1-6) is indeed correct to lowest order in dt. But for now, let us examine the surprisingly far-reaching implications of this result.

To begin with, the n random variables $\Xi_1, ..., \Xi_n$ being summed in Eq. (3.1-6) are seen to be just n statistically independent copies of the random variable $\Xi(dt/n; x,t)$. Because of the statistical independence of these random variables, it follows from rule (1.6-22) that, at least to lowest order in dt,

$$\text{mean}\{\Xi(dt; x,t)\} = \sum_{i=1}^{n} \text{mean}\{\Xi_i(dt/n; x,t)\},$$

$$\text{var}\{\Xi(dt; x,t)\} = \sum_{i=1}^{n} \text{var}\{\Xi_i(dt/n; x,t)\}.$$

Furthermore, because the Ξ_i here all have the same density function as $\Xi(dt/n; x,t)$, then they all have the same mean and variance as $\Xi(dt/n; x,t)$; therefore, again to lowest order in dt,

$$\text{mean}\{\Xi(dt; x,t)\} = n \cdot \text{mean}\{\Xi(dt/n; x,t)\}, \tag{3.1-7a}$$

$$\text{var}\{\Xi(dt; x,t)\} = n \cdot \text{var}\{\Xi(dt/n; x,t)\}. \tag{3.1-7b}$$

Now these last two equations imply something rather significant about the dt-dependence of the mean and variance of $\Xi(dt; x,t)$. To extract the implication, we make use of the following lemma.

Lemma. If $h(z)$ is any smooth function of z satisfying $h(z) = nh(z/n)$ for all positive integers n, then it must be true that $h(z) = Cz$, where C is independent of z.

Proof. Differentiate $h(z) = nh(z/n)$ with respect to z to obtain $h'(z) = nh'(z/n)(1/n) = h'(z/n)$. Observe that the only way that this can be true for *all* positive integers n, given that h' is a smooth function, is to have $h'(z) = h'(z/\infty) = h'(0) \equiv C$, where C is independent of z. Also note that, since $h(z/n) = h(z)/n$, then by taking n arbitrarily large we may deduce that $h(0) = 0$. Finally recognize that the only function h satisfying both $h'(z) = C$ and $h(0) = 0$ is the function $h(z) = Cz$.   QED

Applying this lemma to Eqs. (3.1-7), with dt playing the role of z, we conclude that the mean and variance of $\Xi(\mathrm{d}t; x,t)$ must both be simply linear in dt; i.e., we must have

$$\text{mean}\{\Xi(\mathrm{d}t; x,t)\} = A(x,t)\mathrm{d}t + o(\mathrm{d}t), \tag{3.1-8a}$$

$$\text{var}\{\Xi(\mathrm{d}t; x,t)\} = D(x,t)\mathrm{d}t + o(\mathrm{d}t), \tag{3.1-8b}$$

where $A(x,t)$ and $D(x,t)$ are independent of dt. The $o(\mathrm{d}t)$ terms here have been appended to account for any errors incurred when we replaced Eq. (3.1-4) by Eq. (3.1-6). Notice that since the variance cannot be negative, then $D(x,t)$ must satisfy the condition

$$D(x,t) \geq 0. \tag{3.1-9}$$

Returning now to Eq. (3.1-6), we see that it expresses the random variable $\Xi(\mathrm{d}t; x,t)$ as the sum of n statistically independent, identically distributed random variables, each with a well defined mean, namely $A(x,t)(\mathrm{d}t/n)$, and a well defined variance, namely $D(x,t)(\mathrm{d}t/n)$. Now, the central limit theorem tells us [see Eqs. (1.6-24)] that if the number of summands n in Eq. (3.1-6) is taken sufficiently large, then the sum itself can be made arbitrarily close to a *normal* random variable. But n here can be taken to be as large as we please; consequently, the sum $\Xi(\mathrm{d}t; x,t)$ in Eq. (3.1-6) must be *exactly* a normal random variable. Combining that fact with the two results (3.1-8), we conclude that

$$\Xi(\mathrm{d}t; x,t) = N(A(x,t)\mathrm{d}t, D(x,t)\mathrm{d}t), \tag{3.1-10}$$

at least to lowest order in dt. It then follows from Eq. (1.4-9) that the density function of the random variable $\Xi(\mathrm{d}t; x,t)$ — i.e., the propagator density function of the process $X(t)$ — must have the form

$$\Pi(\xi \,|\, \mathrm{d}t; x,t) = \frac{1}{[2\pi D(x,t)\mathrm{d}t]^{1/2}} \exp\left(-\frac{(\xi - A(x,t)\mathrm{d}t)^2}{2D(x,t)\mathrm{d}t} \right), \tag{3.1-11}$$

again to lowest order in dt.

A useful alternate form of Eq. (3.1-10) can be obtained by appealing to the theorem (1.6-7). That theorem tells us that if $N = N(0,1)$, and a and d are independent of N, then

$$N d^{1/2} + a = N(d^{1/2}{\cdot}0 + a, (d^{1/2})^2{\cdot}1) = N(a,d).$$

Since the choice $a = A(x,t)dt$ and $d = D(x,t)dt$ makes the right side of this equation identical to the right side of Eq. (3.1-10), then we may conclude that, to lowest order in dt,

$$\Xi(dt; x,t) = N[D(x,t)dt]^{1/2} + A(x,t)dt. \qquad (3.1\text{-}12)$$

Equations (3.1-10), (3.1-11) and (3.1-12) are *three equivalent ways* of stating the principle result of this section. That result evidently represents quite an advance from our original premises (3.1-1a) and (3.1-1b), and we shall comment more broadly on that in a moment. But first we must tidy up an earlier point: We must investigate how much "x and t smoothness" in $\Pi(\xi \mid dt; x,t)$ is needed to guarantee that the approximation of Eq. (3.1-4) by Eq. (3.1-6) is really valid to lowest order in dt. As noted earlier, Eq. (3.1-4) is just a "compounded" version of the Chapman-Kolmogorov condition (2.4-9), which in turn is the requirement that the change in the process from time t to time $t+dt$ must, to lowest order in dt, be equal to {the change in the process from t to $t+adt$} plus {the *subsequent* change in the process from $t+adt$ to $t+dt$}, for *any* value of a between 0 and 1. We can therefore justify our heuristic assertion of Eq. (3.1-6) by proving that specific smoothness conditions on the two functions $A(x,t)$ and $D(x,t)$ will ensure that the propagator defined in Eq. (3.1-12) will satisfy Eq. (2.4-9), at least to lowest order in dt. Our proof of this point will not be an entirely academic exercise, because one of its ancillary results will be useful later for developing a condition to control the accuracy of Monte Carlo simulations of continuous Markov processes.

Let N_1 and N_2 be two independent unit normal random variables. Then using the propagator in Eq. (3.1-12), we get for the right hand side of Eq. (2.4-9),

RHS(2.4-9)

$$= \{N_1 D^{1/2}(X(t),t)(adt)^{1/2} + A(X(t),t)adt\}$$
$$+ \{N_2 D^{1/2}(X(t) + [N_1 D^{1/2}(X(t),t)(adt)^{1/2} + A(X(t),t)adt], t+adt)$$
$$\times ((1-a)dt)^{1/2}$$
$$+ A(X(t) + [N_1 D^{1/2}(X(t),t)(adt)^{1/2} + A(X(t),t)adt], t+adt) (1-a)dt\}.$$

Now let us require the function $A(x,t)$ to be "analytic" in the following sense: for any Δx and Δt sufficiently small, then

$$A(x+\Delta x, t+\Delta t) = A(x,t) + \partial_x A(x,t)\Delta x + \partial_t A(x,t)\Delta t$$
$$+ \partial_{xt} A(x,t)\Delta x \Delta t + o(\Delta x) + o(\Delta t), \qquad (3.1\text{-}13)$$

where the partial derivatives $\partial_x A(x,t)$, $\partial_t A(x,t)$ and $\partial_{xt} A(x,t)$ are all bounded functions of x and t. And let us impose the same requirement on the square root of the function $D(x,t)$. Then, abbreviating $A(X(t),t) \equiv A$, $\partial_x A(X(t),t) \equiv \partial_x A$, etc., we may expand the preceding equation as follows:

RHS(2.4-9)

$$= N_1 D^{1/2} a^{1/2} (dt)^{1/2} + A a dt$$
$$+ N_2 \{ D^{1/2} + \partial_x D^{1/2} [N_1 D^{1/2} a^{1/2} (dt)^{1/2} + A a dt] + \partial_t D^{1/2} [a dt]$$
$$+ o(dt) + o((dt)^{1/2}) + o(dt) \} (1-a)^{1/2} (dt)^{1/2}$$
$$+ \{ A + \partial_x A [N_1 D^{1/2} a^{1/2} (dt)^{1/2} + A a dt] + \partial_t A [a dt]$$
$$+ o(dt) + o((dt)^{1/2}) + o(dt) \} (1-a) dt .$$

Collecting and rearranging terms, we get

RHS(2.4-9)

$$= \{ [a^{1/2} N_1 + (1-a)^{1/2} N_2] D^{1/2} (dt)^{1/2} + A dt \}$$
$$+ (dt)^{1/2} a^{1/2} N_1 D^{1/2} \{ N_2 \partial_x D^{1/2} (1-a)^{1/2} (dt)^{1/2} + \partial_x A (1-a) dt \}$$
$$+ o(dt),$$

wherein we have retained explicitly one term of order 3/2 in dt (namely, the term containing $\partial_x A$), but have subsumed under $o(dt)$ all other terms of order $\geq 3/2$ in dt. Now, according to theorem (1.6-7),

$$a^{1/2} N_1 \equiv a^{1/2} N(0,1) = N(0,a)$$

and

$$(1-a)^{1/2} N_2 \equiv (1-a)^{1/2} N(0,1) = N(0,1-a).$$

Therefore, since N_1 and N_2 are statistically independent, it follows from the normal sum theorem [see Section 1.6] that

$$a^{1/2} N_1 + (1-a)^{1/2} N_2 = N(0,a) + N(0,1-a)$$
$$= N(0+0, a+(1-a)) = N(0,1) \equiv N.$$

Inserting this into the preceding formula for RHS(2.4-9), we conclude that our formula (3.1-12) for the continuous propagator implies

RHS(2.4-9)

$$= \{ N D^{1/2} (dt)^{1/2} + A dt \}$$
$$+ (dt)^{1/2} a^{1/2} N_1 D^{1/2} \{ N_2 \partial_x D^{1/2} (1-a)^{1/2} (dt)^{1/2} + \partial_x A (1-a) dt \}$$
$$+ o(dt). \tag{3.1-14a}$$

And of course, formula (3.1-12) gives for the left hand side of the Chapman-Kolmogorov condition (2.4-9) simply

$$\text{LHS}(2.4\text{-}9) = N D^{1/2}(dt)^{1/2} + A dt. \qquad (3.1\text{-}14\text{b})$$

Equations (3.1-14a) and (3.1-14b) will evidently be compatible with each other if and only if the second term on the right side of Eq. (3.1-14a) is of higher order in dt than the first term. That this is indeed the case follows from the fact that the two quantities *in braces* in Eq. (3.1-14a) are of the *same* order in dt, while the second quantity has $(dt)^{1/2}$ as a prefactor. We thus conclude that if the functions $A(x,t)$ and $D^{1/2}(x,t)$ are *simply analytic* in x and t, then Eq. (3.1-12), and hence also Eqs. (3.1-10) and (3.1-11), will indeed define a legitimate Markov propagator — i.e., one that satisfies the Chapman-Kolmogorov condition. The heuristic, qualitative argument that we used earlier to deduce Eq. (3.1-6) for a continuous Markov process has now been quantitatively vindicated.

Notice that a continuous Markov process $X(t)$ is *completely characterized* by the forms of the two functions $A(x,t)$ and $D(x,t)$. This follows from the evident fact that those two functions completely determine the propagator density function $\Pi(\xi \,|\, dt; x,t)$, which in turn completely determines the Markov state density function $P(x,t \,|\, x_0,t_0)$. We shall therefore call $A(x,t)$ and $D(x,t)$ the **characterizing functions** of the continuous Markov process $X(t)$. Evidently, continuous Markov processes differ from one another only insofar as they have different characterizing functions.

If a continuous Markov process is *given* to be in state x at time t, then its Markov state density function at time t is just a delta function spike at x. Eq. (3.1-11) tells us that, by the infinitesimally later time $t+dt$, that infinite spike will have relaxed to a normal or Gaussian shaped peak, which is centered on $x+A(x,t)dt$ and which has a width of $2[D(x,t)dt]^{1/2}$. The characterizing function $A(x,t)$ is often called the **drift function** of the process, because it controls the "drifting" of the peak. And the characterizing function $D(x,t)$ is often called the **diffusion function** of the process, because it controls the "diffusive spreading" of the peak.

Although $\Pi(\xi \,|\, dt; x,t)$ in Eq. (3.1-11) is positive for *all* values of ξ, it is evidently appreciably different from zero only in a ξ-interval that extends for a few multiples of $[D(x,t)dt]^{1/2}$ to either side of the ξ-value $A(x,t)dt$. Since dt is infinitesimally small, then $(dt)^{1/2}$, although also infinitesimally small, is very much larger than dt. So we can say that, for a continuous Markov process, $\Pi(\xi \,|\, dt; x,t)$ is appreciably different from zero only in a neighborhood of order $(dt)^{1/2}$ about $\xi=0$. This is a surprisingly explicit refinement of our initial requirement (3.1-1b).

Indeed, we have found that for a continuous Markov process the propagator $\Xi(dt; x,t)$ is *necessarily* a normal random variable with a mean of order dt and a standard deviation of order $(dt)^{1/2}$. In retrospect, we can see that this enormous gain in specificity over the rather mild initial requirements (3.1-1) is due to the combined effects of the Chapman-Kolmogorov condition and the central limit theorem. The functional form of the continuous propagator density function in Eq. (3.1-11) is therefore not the result of a capricious definition: If we want $\Pi(\xi \mid dt; x,t)$ to have the continuous characteristics set forth in conditions (3.1-1), then we simply have no choice but to give it the functional form (3.1-11).

The remainder of this chapter will be devoted to developing the principal consequences of Eqs. (3.1-10) – (3.1-12). There is one important consequence that may be noted immediately: Since $\Pi(\xi \mid dt; x,t)$ evidently depends on x and t *only* through the characterizing functions $A(x,t)$ and $D(x,t)$, then the conditions (2.8-1) for homogeneity apply directly to the characterizing functions. In particular, for any *continuous* Markov process $X(t)$,

$X(t)$ is **temporally homogeneous**

$$\Leftrightarrow \quad A(x,t) = A(x) \text{ and } D(x,t) = D(x), \quad (3.1\text{-}15a)$$

$X(t)$ is **completely homogeneous**

$$\Leftrightarrow \quad A(x,t) = A \text{ and } D(x,t) = D. \quad (3.1\text{-}15b)$$

3.2 TIME-EVOLUTION EQUATIONS

In Chapter 2 we found that the differential equations that govern the time evolutions of $P(x,t \mid x_0,t_0)$, $\langle X^n(t) \rangle$ and $\langle S^n(t) \rangle$ are all expressed in terms of the propagator moment functions, $B_n(x,t)$. The latter were defined in Eqs. (2.4-3), but with the important proviso that $\Pi(\xi \mid dt; x,t)$ be such that those definitions actually make sense. With the result (3.1-11), we are now in a position to show that the propagator moment functions are indeed well defined for any continuous Markov process. And once those propagator moment functions are explicitly in hand, we can then proceed to write down more explicit forms of the various time-evolution equations developed in Chapter 2.

According to the definition (2.4-3a), a determination of $B_n(x,t)$ requires that we calculate the nth moment of the propagator $\Xi(dt; x,t)$.

We showed in the last section that, for a continuous Markov process $X(t)$ with characterizing functions $A(x,t)$ and $D(x,t)$, the propagator $\Xi(dt; x,t)$ is a *normal* random variable with mean $A(x,t)dt$ and variance $D(x,t)dt$. Therefore, it follows from the general formula (1.4-10) that the nth moment of this normal random variable is

$$\langle \Xi^n(dt; x,t) \rangle = n! \sum_{\substack{k=0 \\ (k\,\text{even})}}^{n} \frac{[A(x,t)dt]^{n-k}\,[D(x,t)dt]^{k/2}}{(n-k)!\,(k/2)!\,2^{k/2}} + o(dt)$$

$$= n! \sum_{\substack{k=0 \\ (k\,\text{even})}}^{n} \frac{[A(x,t)]^{n-k}\,[D(x,t)]^{k/2}\,(dt)^{n-k/2}}{(n-k)!\,(k/2)!\,2^{k/2}} + o(dt).$$

Evaluating this sum to first order in dt for explicit values of n is straightforward, and yields

$$\langle \Xi^n(dt; x,t) \rangle = \begin{cases} A(x,t)dt + o(dt), & \text{for } n=1, \\ D(x,t)dt + o(dt), & \text{for } n=2, \\ o(dt), & \text{for } n \geq 3. \end{cases}$$

Comparing this with the definition (2.4-3a), we conclude that for a continuous Markov process $X(t)$ with drift function $A(x,t)$ and diffusion function $D(x,t)$, the propagator moment functions are

$$B_n(x,t) = \begin{cases} A(x,t), & \text{for } n=1, \\ D(x,t), & \text{for } n=2, \\ 0, & \text{for } n \geq 3. \end{cases} \qquad (3.2\text{-}1)$$

Therefore, for a continuous Markov process, the first two propagator moment functions are the respective characterizing functions of the process, and all other propagator moment functions are zero.

3.2.A THE FOKKER-PLANCK EQUATIONS

The result (3.2-1) implies that the Kramers-Moyal equations derived in Section 2.5 simplify dramatically for continuous Markov processes. Specifically, for the continuous propagator moment functions $B_n(x,t)$ in Eq. (3.2-1), the forward Kramers-Moyal equation (2.5-1) reduces to the second order partial differential equation

$$\frac{\partial}{\partial t} P(x,t\,|\,x_0,t_0) = -\frac{\partial}{\partial x}[A(x,t)\,P(x,t\,|\,x_0,t_0)] + \frac{1}{2}\frac{\partial^2}{\partial x^2}[D(x,t)\,P(x,t\,|\,x_0,t_0)],$$

(3.2-2)

which is called the **forward Fokker-Planck equation**. And the backward Kramers-Moyal equation (2.5-2) now becomes the second order partial differential equation

$$-\frac{\partial}{\partial t_0} P(x,t\,|\,x_0,t_0) = A(x_0,t_0)\frac{\partial}{\partial x_0} P(x,t\,|\,x_0,t_0) + \frac{1}{2} D(x_0,t_0)\frac{\partial^2}{\partial x_0^2} P(x,t\,|\,x_0,t_0),$$

(3.2-3)

which is called the **backward Fokker-Planck equation**. Given the initial condition (2.5-3), the forward Fokker-Planck equation constitutes a t-evolution equation for $P(x,t\,|\,x_0,t_0)$, for fixed x_0 and t_0, for the continuous Markov process with characterizing functions $A(x,t)$ and $D(x,t)$. Similarly, given the final condition (2.5-4), the backward Fokker-Planck equation constitutes a t_0-evolution equation for $P(x,t\,|\,x_0,t_0)$, for fixed x and t, for the same process.

It is sometimes useful to define for a continuous Markov process the function $J(x,t;x_0,t_0)$ by

$$J(x,t;x_0,t_0) \equiv A(x,t)\,P(x,t\,|\,x_0,t_0) - \frac{1}{2}\frac{\partial}{\partial x}[D(x,t)\,P(x,t\,|\,x_0,t_0)]. \quad (3.2\text{-}4)$$

Using this function, the forward Fokker-Planck equation (3.2-2) can be written somewhat more compactly as

$$\frac{\partial}{\partial t} P(x,t\,|\,x_0,t_0) = -\frac{\partial}{\partial x} J(x,t;x_0,t_0). \quad (3.2\text{-}5)$$

This has the form of a *continuity equation* for the probability density function $P(x,t\,|\,x_0,t_0)$, with the function $J(x,t;x_0,t_0)$ playing the role of a **probability current**. Integrating this equation over the x-interval $[a,b]$ gives

$$\frac{\partial}{\partial t}\left[\int_a^b dx\,P(x,t\,|\,x_0,t_0)\right] = J(a,t;x_0,t_0) - J(b,t;x_0,t_0). \quad (3.2\text{-}6)$$

This says that the time-rate of change of $\mathrm{Prob}\{X(t)\in[a,b)\}$ is equal to the rate at which "occupation probability" flows into the interval $[a,b)$ at $x=a$, minus the rate at which it flows out of the interval $[a,b)$ at $x=b$.

Most continuous Markov processes of practical interest are *temporally homogeneous*, in that, as noted in (3.1-15a),

$$A(x,t) = A(x) \quad \text{and} \quad D(x,t) = D(x). \tag{3.2-7a}$$

In such a case, the forward Fokker-Planck equation (3.2-2) becomes

$$\frac{\partial}{\partial t} P(x,t \,|\, x_0,t_0) = -\frac{\partial}{\partial x}[A(x)\, P(x,t \,|\, x_0,t_0)] + \frac{1}{2}\frac{\partial^2}{\partial x^2}[D(x)\, P(x,t \,|\, x_0,t_0)],$$

$$\tag{3.2-7b}$$

and the backward Fokker-Planck equation (3.2-3) becomes

$$-\frac{\partial}{\partial t_0} P(x,t \,|\, x_0,t_0) = A(x_0)\,\frac{\partial}{\partial x_0} P(x,t \,|\, x_0,t_0) + \frac{1}{2} D(x_0)\,\frac{\partial^2}{\partial x_0^2} P(x,t \,|\, x_0,t_0).$$

$$\tag{3.2-7c}$$

Furthermore, because of the temporally homogeneous property (2.8-5), *the left hand sides of these two equations are interchangeable.*

For temporally homogeneous Markov processes it sometimes happens that $P(x,t \,|\, x_0,t_0)$ approaches, as $(t-t_0)\to\infty$, a stationary density function $P_s(x)$, as was discussed more generally in connection with Eq. (2.8-9). We shall defer to Section 3.5 a consideration of the existence of $P_s(x)$, and of the interesting consequences thereof.

If the characterizing functions $A(x,t)$ and $D(x,t)$ of a continuous Markov process are independent of *both* x and t, then the process is *completely homogeneous*, and things simplify considerably. We shall give a detailed analysis of completely homogeneous continuous Markov processes in Subsection 3.3.B.

3.2.B THE MOMENT EVOLUTION EQUATIONS

The natural truncation (3.2-1) of the sequence of propagator moment functions for a continuous Markov process $X(t)$ simplifies not only the time-evolution equations derived in Section 2.5 for the density function of $X(t)$, but also the time-evolution equations derived in Section 2.7 for the moments of $X(t)$ and its integral $S(t)$. With Eqs. (3.2-1), the general moment evolution equation (2.7-6) for $X(t)$ becomes

$$\frac{d}{dt}\langle X^n(t)\rangle = n\langle X^{n-1}(t)\, A(X(t),t)\rangle + \frac{n(n-1)}{2}\langle X^{n-2}(t)\, D(X(t),t)\rangle$$

$$(t_0 \le t;\ n = 1,2,\ldots), \quad (3.2-8)$$

and is to be solved subject to the initial condition (2.7-7):

$$\langle X^n(t_0) \rangle = x_0^{\ n} \qquad (n=1,2,...). \tag{3.2-9}$$

We have deduced Eq. (3.2-8) as a special case of Eq. (2.7-6), which in turn was derived from the general propagator relation (2.7-1). It is instructive to see how Eq. (3.2-8) can also be deduced from the forward Fokker-Planck equation (3.2-2): Abbreviating $P(x,t \mid x_0,t_0) \equiv P(x,t)$, we have for any positive integer n,

$$\frac{d}{dt} \langle X^n(t) \rangle = \frac{d}{dt} \int_{-\infty}^{\infty} dx\, x^n\, P(x,t) = \int_{-\infty}^{\infty} dx\, x^n\, \frac{\partial}{\partial t} P(x,t)$$

$$= -\int_{-\infty}^{\infty} dx\, x^n\, \frac{\partial}{\partial x}[A(x,t)\,P(x,t)] + \frac{1}{2}\int_{-\infty}^{\infty} dx\, x^n\, \frac{\partial^2}{\partial x^2}[D(x,t)\,P(x,t)],$$

where the last step has invoked the forward Fokker-Planck equation (3.2-2). If we now integrate by parts the first term *once* and the second term *twice*, we get

$$\int_{-\infty}^{\infty} dx\, x^n\, \frac{\partial}{\partial x}[A(x,t)\,P(x,t)]$$

$$= x^n\,A(x,t)\,P(x,t)\Big|_{-\infty}^{\infty} - \int_{-\infty}^{\infty} [A(x,t)\,P(x,t)]\,[n\,x^{n-1}dx],$$

and

$$\int_{-\infty}^{\infty} dx\, x^n\, \frac{\partial^2}{\partial x^2}[D(x,t)\,P(x,t)]$$

$$= x^n\,\frac{\partial}{\partial x}[D(x,t)\,P(x,t)]\Big|_{-\infty}^{\infty} - \int_{-\infty}^{\infty} \frac{\partial}{\partial x}[D(x,t)\,P(x,t)]\,[n\,x^{n-1}dx]$$

$$= x^n\,\frac{\partial}{\partial x}[D(x,t)\,P(x,t)]\Big|_{-\infty}^{\infty} - n\,x^{n-1}\,D(x,t)\,P(x,t)\Big|_{-\infty}^{\infty}$$

$$+ \int_{-\infty}^{\infty} [D(x,t)\,P(x,t)]\,[n(n-1)\,x^{n-2}dx].$$

Since the assumed existence of $\langle X^n(t) \rangle = \int x^n P(x,t)dx$ implies that $P(x,t) \to 0$ very strongly as $x \to \pm\infty$, then all the integrated or boundary terms in the above equations will normally vanish. From the surviving terms we get, invoking the averaging notation of Eq. (2.3-3),

$$\int_{-\infty}^{\infty} dx\, x^n\, \frac{\partial}{\partial x}[A(x,t)\,P(x,t)] = -\langle n\,X^{n-1}(t)\,A(X(t),t) \rangle,$$

and

$$\int_{-\infty}^{\infty} dx\, x^n\, \frac{\partial^2}{\partial x^2} [D(x,t)\, P(x,t)] = \langle\, n(n-1)\, X^{n-2}(t)\, D(X(t),t)\,\rangle.$$

Substituting these two results into our last equation for the time derivative of $\langle X^n(t)\rangle$, we obtain the expected result (3.2-8).

For the integral process $S(t)$, the moment evolution equations (2.7-10) and (2.7-11) apply as written; however, the auxiliary cross-moment equations (2.7-12) now simplify to

$$\frac{d}{dt} \langle\, S^m(t)\, X^n(t)\,\rangle = m\, \langle\, S^{m-1}(t)\, X^{n+1}(t)\,\rangle$$

$$+ n\, \langle\, S^m(t)\, X^{n-1}(t)\, A(X(t),t)\,\rangle$$

$$+ \frac{n(n-1)}{2}\, \langle\, S^m(t)\, X^{n-2}(t)\, D(X(t),t)\,\rangle$$

$$(t_0 \le t;\ n=1,2,...;\ m=1,2,...), \qquad (3.2\text{-}10)$$

for which we must use the initial condition (2.7-13),

$$\langle\, S^m(t_0)\, X^n(t_0)\,\rangle = 0 \quad (n=1,2,...;\ m=1,2,...). \qquad (3.2\text{-}11)$$

Closure of the continuous moment evolution equations (3.2-8) and (3.2-10) evidently requires that $A(x,t)$ be a polynomial in x of degree ≤ 1, and $D(x,t)$ be a polynomial in x of degree ≤ 2. If these rather stringent conditions are not met, then the best we can hope for are *approximate* solutions, obtained either along the lines of the procedure outlined in Appendix C, or else by some problem-specific closure approximation.

In Section 2.7 we derived explicit time-evolution equations for the mean, variance and covariance of any Markov process $X(t)$, and also of its integral $S(t)$. For later reference, we now rewrite those equations for a continuous Markov process with characterizing functions $A(x,t)$ and $D(x,t)$. Equations (2.7-14) – (2.7-16) for the time-derivatives of the mean, variance and covariance of X(t) now take the forms

$$\frac{d}{dt}\langle X(t)\rangle = \langle A(X(t),t)\rangle \qquad (t_0 \le t), \qquad (3.2\text{-}12)$$

$$\frac{d}{dt}\text{var}\{X(t)\} = 2\Big(\langle\, X(t)\, A(X(t),t)\,\rangle - \langle X(t)\rangle \langle A(X(t),t)\rangle\Big) + \langle D(X(t),t)\rangle$$

$$(t_0 \le t), \qquad (3.2\text{-}13)$$

$$\frac{d}{dt_2}\text{cov}\{X(t_1),X(t_2)\} = \langle X(t_1) A(X(t_2),t_2)\rangle - \langle X(t_1)\rangle\langle A(X(t_2),t_2)\rangle$$

$$(t_0\le t_1\le t_2). \qquad (3.2\text{-}14)$$

These equations have the respective initial conditions (2.7-17) – (2.7-19),

$$\langle X(t=t_0)\rangle = x_0, \qquad (3.2\text{-}15)$$

$$\text{var}\{X(t=t_0)\} = 0, \qquad (3.2\text{-}16)$$

$$\text{cov}\{X(t_1),X(t_2=t_1)\} = \text{var}\{X(t_1)\}. \qquad (3.2\text{-}17)$$

The aforementioned closure conditions apply here as well. Notice from Eq. (3.2-17) that the solution of the variance evolution equation (3.2-13) provides the initial condition for the solution of the covariance evolution equation (3.2-14).

For the integral process $S(t)$, the mean, variance and covariance are to be obtained through the respective equations (2.7-21), (2.7-22) and (2.7-24). However, the auxiliary equation (2.7-23a) for the variance equation (2.7-22) now reads

$$\frac{d}{dt}\text{cov}\{S(t),X(t)\} = \text{var}\{X(t)\} + \langle S(t) A(X(t),t)\rangle - \langle S(t)\rangle\langle A(X(t),t)\rangle$$

$$(t_0\le t); \qquad (3.2\text{-}18a)$$

it is to be solved subject to the initial condition (2.7-23b),

$$\text{cov}\{S(t=t_0),X(t=t_0)\} = 0. \qquad (3.2\text{-}18b)$$

And the auxiliary equation (2.7-25a) for the covariance equation (2.7-24) now reads

$$\frac{d}{dt_2}\text{cov}\{S(t_1),X(t_2)\} = \langle S(t_1) A(X(t_2),t_2)\rangle - \langle S(t_1)\rangle\langle A(X(t_2),t_2)\rangle$$

$$(t_0\le t_1\le t_2); \qquad (3.2\text{-}19a)$$

it is to be solved subject to the initial condition (2.7-25b),

$$\text{cov}\{S(t_1),X(t_2=t_1)\} = \text{cov}\{S(t_1),X(t_1)\}. \qquad (3.2\text{-}19b)$$

Again the aforementioned closure conditions apply, in part now because the formulas (2.7-21) and (2.7-22) for $\langle S(t)\rangle$ and $\text{var}\{S(t)\}$ evidently require a prior determination of $\langle X(t)\rangle$ and $\text{var}\{X(t)\}$. Notice that the solution to the differential equation (3.2-18a) provides the initial condition for the differential equation (3.2-19a).

3.3 THREE IMPORTANT CONTINUOUS MARKOV PROCESSES

In this section we shall take note of three simple but nonetheless important examples of continuous Markov processes, each corresponding to a particular choice for the pair of characterizing functions $A(x,t)$ and $D(x,t)$. For each of these three examples we shall derive the "statistics" of $X(t)$ and its integral $S(t)$. We shall also briefly discuss the so-called "thermodynamic limit" of each process, wherein $X(t)$ is regarded as an *extensive* state variable of a dynamical system with "volume" Ω, and attention is focused on the behavior of the system's *intensive* state variable,

$$Z(t) \equiv X(t)/\Omega, \qquad (3.3\text{-}1)$$

in the limit that Ω becomes very large.

3.3.A LIOUVILLE PROCESSES

Any continuous Markov process whose diffusion function vanishes identically,

$$D(x,t) \equiv 0, \qquad (3.3\text{-}2)$$

is called a **Liouville process**. Recalling the result (3.1-10), we have for the propagator of a Liouville process,

$$\Xi(dt; x,t) = N(A(x,t)dt, 0).$$

Thus, by (1.4-12), $\Xi(dt; x,t)$ is the *sure* variable $A(x,t)dt$. This in turn implies, recalling the propagator definition (2.4-1), that

$$X(t+dt) - X(t) = A(X(t),t)dt,$$

which is equivalent to saying that $X(t)$ is the *deterministic process* that is defined by the ordinary differential equation

$$\frac{dX(t)}{dt} = A(X(t),t), \qquad (3.3\text{-}3a)$$

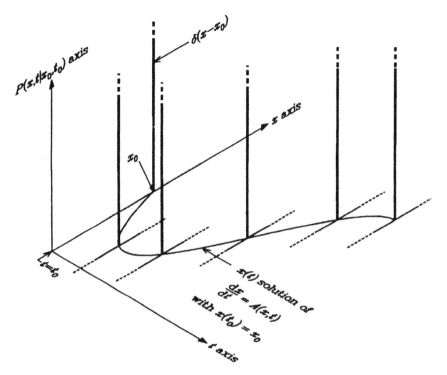

Figure 3-1. Time-evolution of the Markov state density function for a typical *Liouville process.* Because $D(x,t) \equiv 0$, the initial delta function shape of $P(x,t \mid x_0,t_0)$ never spreads out. For any $t > t_0$ we have $P(x,t \mid x_0,t_0) = \delta(x - x(t))$, where $x(t)$ is the function satisfying Eqs. (3.3-3). The process $X(t)$ is accordingly *deterministic*, and in fact is precisely the function $x(t)$.

together of course with the initial condition

$$X(t_0) = x_0. \qquad (3.3\text{-}3b)$$

Since $X(t)$ in this case is a *sure* variable, then it follows from (1.2-7) that its density function $P(x,t \mid x_0,t_0)$ is always a Dirac delta function. The time evolution of $P(x,t \mid x_0,t_0)$ is therefore as illustrated in Fig. 3-1.

We found in Eqs. (3.2-1) that all the propagator moment functions $B_n(x,t)$ for $n \geq 3$ vanish identically for any continuous Markov process. We now see that if $B_2(x,t) \equiv D(x,t)$ also vanishes identically, then the continuous Markov process becomes a *deterministic* process. That this deterministic process satisfies the differential equation (3.3-3a) is seen to be entirely consistent with the implications of Eq. (2.6-3) regarding the existence of $dX(t)/dt$. In fact, since continuity is a prerequisite for differentiability, then Liouville processes not only are the only *continuous deterministic* Markov processes, they are also the only Markov processes of *any* kind that have a well defined time derivative.

With $X(t)$ a sure function, we naturally expect that its integral $S(t)$ will also be a sure function, namely the function

$$S(t) = \int_{t_0}^{t} X(t')\, dt'. \tag{3.3-4}$$

That this is indeed the case follows from the definition (2.6-6) of $S(t)$, which says that if $S(t)$ and $X(t)$ are both given then $S(t+dt)$ is obtained as

$$S(t+dt) = S(t) + X(t)dt + o(dt).$$

This relation obviously implies that the derivative of $S(t)$ is $X(t)$, so that, in light of the initial condition $S(t_0)=0$, $S(t)$ is indeed the sure function given by Eq. (3.3-4).

A simple example of a Liouville process is provided by the characterizing functions

$$A(x,t) = k, \quad D(x,t) = 0, \tag{3.3-5a}$$

where k is any constant. Substituting into Eq. (3.3-3a) and solving subject to the initial condition (3.3-3b), we find that

$$X(t) = x_0 + k(t-t_0). \tag{3.3-5b}$$

Another example of a Liouville process is provided by the characterizing functions

$$A(x,t) = -kx, \quad D(x,t) = 0, \tag{3.3-6a}$$

where k again is some constant. In this case we find from Eqs. (3.3-3) that

$$X(t) = x_0 \exp[-k(t-t_0)]. \tag{3.3-6b}$$

If $Z(t)$ is defined by Eq. (3.3-1), with Ω independent of t, then it follows upon dividing the differential equation (3.3-3a) through by Ω that $Z(t)$ for a Liouville process is the solution of

$$\frac{dZ(t)}{dt} = \Omega^{-1}A(\Omega Z(t),t) \tag{3.3-7a}$$

for the initial condition

$$Z(t_0) = x_0/\Omega \equiv z_0. \tag{3.3-7b}$$

Clearly, the behavior of $Z(t)$ as $\Omega \to \infty$ will depend crucially on the functional form of $A(x,t)$. Thus, for the example (3.3-5), we find by dividing Eq. (3.3-5b) through by Ω that

$$Z(t) = z_0 + \Omega^{-1}k(t-t_0).$$

This will be independent of Ω in the limit $\Omega\to\infty$ if and only if k is directly proportional to Ω in that limit — i.e., if and only if k is an "extensive" parameter of the system. On the other hand, for the example (3.3-6), we find by dividing Eq. (3.3-6b) through by Ω that

$$Z(t) = z_0 \exp[-k(t-t_0)].$$

This will be independent of Ω in the limit $\Omega\to\infty$ if and only if k is independent of Ω in that limit — i.e., if and only if k is an "intensive" parameter of the system.

Since Liouville processes are purely deterministic processes, there is little reason to study them in a stochastic context. Their chief interest to us here is simply that they allow us to see how continuous, memoryless, *deterministic* processes, long familiar to us from ordinary calculus, can actually be viewed as a *special class* of continuous, memoryless, *stochastic* processes — namely, the class for which the diffusion function $D(x,t)$ is identically zero. Reexamining the result (3.1-10) at this point, we can now appreciate that *the size of $D(x,t)$ relative to $|A(x,t)|$ is a direct measure of the "inherent stochasticity" of a continuous Markov process.*

3.3.B WIENER PROCESSES

A *completely homogeneous* continuous Markov process will, according to (3.1-15b), have characterizing functions that are *constants*:

$$A(x,t) \equiv A, \quad D(x,t) \equiv D \geq 0. \tag{3.3-8}$$

Such a process is commonly called a **Wiener process**.

The Wiener process with $A=0$ and $D=1$ is called the **special Wiener process**. The special Wiener process is of sufficient theoretical interest that it is usually denoted by a special symbol, namely $W(t)$, instead of $X(t)$:

$$A(x,t) \equiv 0, \quad D(x,t) \equiv 1 \quad \Leftrightarrow \quad X(t) \equiv W(t). \tag{3.3-9}$$

Since the first two propagator moment functions for a Wiener process are $B_1(x,t) \equiv A$ and $B_2(x,t) \equiv D$, then the results of our analysis in Subsection 2.7.C are directly applicable here; we need only replace in those results the constant b_1 by A and the constant b_2 by D. In particular, Eqs. (2.7-28) give for the mean, variance and covariance of the general Wiener process,

$$\langle X(t) \rangle = x_0 + A(t - t_0) \qquad (t_0 \leq t), \qquad \text{(3.3-10a)}$$

$$\text{var}\{X(t)\} = D(t - t_0) \qquad (t_0 \leq t), \qquad \text{(3.3-10b)}$$

$$\text{cov}\{X(t_1), X(t_2)\} = D(t_1 - t_0) \qquad (t_0 \leq t_1 \leq t_2). \qquad \text{(3.3-10c)}$$

And Eqs. (2.7-29) give for the mean, variance and covariance of the *integral* of the general Wiener process,

$$\langle S(t) \rangle = x_0(t - t_0) + (A/2)(t - t_0)^2 \qquad (t_0 \leq t), \qquad \text{(3.3-11a)}$$

$$\text{var}\{S(t)\} = (D/3)(t - t_0)^3 \qquad (t_0 \leq t), \qquad \text{(3.3-11b)}$$

$$\text{cov}\{S(t_1), S(t_2)\} = D(t_1 - t_0)^2 [(t_1 - t_0)/3 + (t_2 - t_1)/2]$$
$$(t_0 \leq t_1 \leq t_2). \qquad \text{(3.3-11c)}$$

We can also derive for the general Wiener process an explicit expression for its Markov state density function $P(x,t \mid x_0,t_0)$. One way to do that would be to solve the forward Fokker-Planck equation (3.2-2) for the characterizing functions (3.3-8),

$$\frac{\partial}{\partial t} P(x,t \mid x_0,t_0) = -A \frac{\partial}{\partial x} P(x,t \mid x_0,t_0) + \frac{D}{2} \frac{\partial^2}{\partial x^2} P(x,t \mid x_0,t_0), \qquad \text{(3.3-12)}$$

subject to the initial condition $P(x,t_0 \mid x_0,t_0) = \delta(x - x_0)$. [Notice that since we are dealing here with a *completely homogeneous* process, then by our discussion of Eq. (2.8-23) the above equation is *also* the *backward* Fokker-Planck equation.] But we shall instead derive $P(x,t \mid x_0,t_0)$ for the Wiener process from the *integral equation* for P, which in this completely homogeneous case takes the form of Eq. (2.8-18).

We start by noting from Eq. (3.1-11) that the propagator density function for the general Wiener process is given by

$$\Pi(\xi \mid dt) = (2\pi D dt)^{-1/2} \exp\left(-\frac{(\xi - A dt)^2}{2 D dt} \right).$$

The function $\Pi^\star(u \mid dt)$ defined in Eq. (2.8-19a) is therefore

$$\Pi^\star(u \mid dt) = \int_{-\infty}^{\infty} d\xi \, e^{-iu\xi} (2\pi D dt)^{-1/2} \exp\left(-\frac{(\xi - A dt)^2}{2 D dt} \right)$$

$$= e^{-iuA dt} \exp\left(-u^2 D dt/2 \right),$$

where the second equality follows from the integral identity (A-7). Thus,

$$\left[\Pi^{\star}(u\,|\,dt)\right]^n = \left[e^{-iuAdt}\exp\left(-u^2 Ddt/2\right)\right]^n$$

$$= e^{-iuAndt}\exp\left(-u^2 Dndt/2\right)$$

$$\left[\Pi^{\star}(u\,|\,dt)\right]^n = \exp\left(-iuA(t-t_0)\right)\exp\left(-u^2 D(t-t_0)/2\right),$$

where in the last step we have invoked the auxiliary condition (2.8-19b) that $ndt = (t-t_0)$. Substituting this result into the completely homogeneous integral equation (2.8-18) gives

$$P(x,t\,|\,x_0,t_0) = (2\pi)^{-1}\oint_{-\infty}^{\infty} du\,\exp\left(iu(x-x_0)\right)$$

$$\times \exp\left(-iuA(t-t_0)\right)\exp\left(-u^2 D(t-t_0)/2\right)$$

$$= (2\pi)^{-1}\oint_{-\infty}^{\infty} du\,\exp\left(iu[x-x_0-A(t-t_0)]\right)$$

$$\times \exp\left(-u^2 D(t-t_0)/2\right).$$

The u-integrand in the last integral is seen to have the form of the x-integrand in Eq. (A-6) with

$$b = x - x_0 - A(t-t_0), \qquad a^2 = D(t-t_0)/2.$$

Hence, this integral converges as an ordinary infinite integral; indeed, Eq. (A-6) gives at once the result

$$P(x,t\,|\,x_0,t_0) = \frac{1}{[2\pi D(t-t_0)]^{1/2}}\exp\left(-\frac{|x-x_0-A(t-t_0)|^2}{2D(t-t_0)}\right). \qquad (3.3\text{-}13)$$

As a check on this formula for $P(x,t\,|\,x_0,t_0)$ for the general Wiener process, one can straightforwardly verify by direct differentiation that it does indeed satisfy the Fokker-Planck equation (3.3-12).

Comparing Eq. (3.3-13) with the generic normal density function (1.4-9), we see that the general Wiener process $X(t)$ is just the *normal* random variable with mean $x_0 + A(t-t_0)$ and variance $D(t-t_0)$:

$$X(t) = N(x_0 + A(t-t_0), D(t-t_0)). \qquad (3.3\text{-}14)$$

The mean and variance of $X(t)$ implied by this result quite obviously agree with Eqs. (3.3-10a) and (3.3-10b). Also, the covariance formula

(3.3-10c) can now be straightforwardly verified by starting with

$$\text{cov}\{X(t_1), X(t_2)\} \equiv \langle X(t_1) X(t_2) \rangle - \langle X(t_1) \rangle \langle X(t_2) \rangle$$

$$= \int_{-\infty}^{\infty} dx_1 \int_{-\infty}^{\infty} dx_2 \, x_1 x_2 \, P(x_2, t_2 | x_1, t_1) P(x_1, t_1 | x_0, t_0)$$

$$- [x_0 + A(t_1 - t_0)][x_0 + A(t_2 - t_0)],$$

inserting the formula (3.3-13) for P, and then integrating over x_2 and x_1 in that order.

We recall that the random variable $N(m,0)$ is in fact the sure variable m [see Eq. (1.4-12], and thus has the density function $\delta(x - m)$ [see (1.2-7)]. Therefore, the result (3.3-14) implies that $X(t_0) = N(x_0, 0)$, and $P(x, t_0 | x_0, t_0) = \delta(x - x_0)$, just as we should expect. Similarly, in the case $D = 0$ we evidently get $P(x, t | x_0, t_0) = \delta(x - [x_0 + A(t - t_0)])$, which is the expected Liouville process result [see Eqs. (3.3-5)].

A schematic plot of the time evolution of $P(x, t | x_0, t_0)$ for the Wiener process is shown in Fig. 3-2. Notice that $P(x, t | x_0, t_0)$ does *not* approach a stationary form $P_s(x)$ as $t \to \infty$, a fact that we had deduced earlier in Subsection 2.8.B for *any* completely homogeneous Markov process. A Wiener process is not a stable Markov process.

For the special Wiener process (3.3-9), we have from Eq. (3.3-14) that

$$W(t) = N(x_0, t - t_0). \tag{3.3-15}$$

Eqs. (3.3-10) in this case reduce to

$$\langle W(t) \rangle = x_0 \qquad (t_0 \le t), \tag{3.3-16a}$$

$$\text{var}\{W(t)\} = t - t_0 \qquad (t_0 \le t), \tag{3.3-16b}$$

$$\text{cov}\{W(t_1), W(t_2)\} = t_1 - t_0 \qquad (t_0 \le t_1 \le t_2). \tag{3.3-16c}$$

Formulas for the mean, variance and covariance of the *integral* of $W(t)$ similarly follow by putting $A = 0$ and $D = 1$ in Eqs. (3.3-11); however, we shall not bother to write those formulas out explicitly here.

The statistics of the intensive state variable $Z(t)$ defined in Eq. (3.3-1) can be deduced by noting that, since $X(t)$ is given by Eq. (3.3-14), then we have by theorem (1.6-7) that

$$\Omega^{-1} X(t) = N(\Omega^{-1}[x_0 + A(t - t_0)], \Omega^{-2} D(t - t_0)).$$

Thus, letting $z_0 = x_0/\Omega$, we conclude that for the general Wiener process the intensive state variable is

$$Z(t) = N(z_0 + (A/\Omega)(t - t_0)], (D/\Omega^2)(t - t_0)). \tag{3.3-17}$$

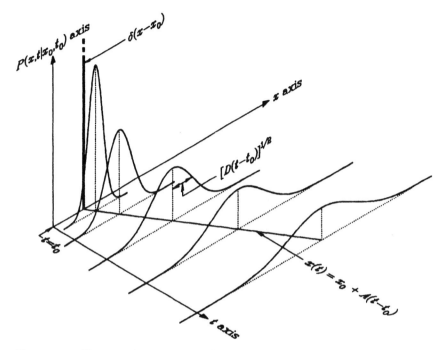

Figure 3-2. Time-evolution of the Markov state density function for a typical *Wiener process.* This is the *completely homogeneous* continuous Markov process, since its characterizing functions are both constants: $A(x,t)=A$ and $D(x,t)=D$. For any $t>t_0$, $P(x,t\,|\,x_0,t_0)$ is the normal density function with mean $x_0+A(t-t_0)$ and standard deviation $[D(t-t_0)]^{1/2}$. As $t\to\infty$, $P(x,t\,|\,x_0,t_0)$ just keeps flattening out, and does not approach a stationary form; so the Wiener process is *not* a stable process. For the *special* Wiener process $W(t)$, we have $A=0$ and $D=1$, and the center of the spreading curve remains at x_0.

Various possibilities thus arise for the thermodynamic limit of large Ω, depending upon precisely how A and D vary with Ω. To give but one example, suppose that A and D are both *extensive* system parameters; this means that, for all Ω sufficiently large,

$$A = A_1\Omega \quad \text{and} \quad D = D_1\Omega, \tag{3.3-18a}$$

where A_1 and D_1 are independent of Ω. Then Eq. (3.3-17) would give

$$Z(t) = \mathbf{N}(z_0+A_1(t-t_0)|,(D_1/\Omega)(t-t_0)). \tag{3.3-18b}$$

In this case, $Z(t)$ would exhibit normal fluctuations of approximate size $[D_1(t-t_0)/\Omega]^{1/2}$ about an Ω-independent mean, $z_0+A_1(t-t_0)$. As $\Omega\to\infty$, the fluctuations in $Z(t)$ would then go to zero like $\Omega^{-1/2}$.

3.3.C ORNSTEIN-UHLENBECK PROCESSES

Any continuous Markov process with characterizing functions of the form

$$A(x,t) = -kx, \quad D(x,t) = D \quad (k>0, D\geq 0), \qquad (3.3\text{-}19)$$

where k and D are constants, is called an **Ornstein-Uhlenbeck process**. Notice that we require k here to be *strictly* positive, the circumstance $k=0$ being more appropriately treated as a Wiener process [see the preceding subsection].

Since the first two propagator moment functions for this process are $B_1(x,t) = -kx$ and $B_2(x,t) = D$, then the results of our analysis in Subsection 2.7.D are directly applicable here; we need only replace in those results β by k and c by D. In particular, Eqs. (2.7-34) give for the mean, variance and covariance of the general Ornstein-Uhlenbeck process,

$$\langle X(t) \rangle = x_0 \exp[-k(t-t_0)] \qquad (t_0 \leq t), \quad (3.3\text{-}20a)$$

$$\text{var}\{X(t)\} = (D/2k)(1 - \exp[-2k(t-t_0)]) \qquad (t_0 \leq t), \quad (3.3\text{-}20b)$$

$$\text{cov}\{X(t_1), X(t_2)\} = (D/2k)\exp[-k(t_2-t_1)](1 - \exp[-2k(t_1-t_0)])$$
$$(t_0 \leq t_1 \leq t_2). \quad (3.3\text{-}20c)$$

And Eqs. (2.7-35) give for the mean, variance and covariance of the *integral* of the general Ornstein-Uhlenbeck process,

$$\langle S(t) \rangle = (x_0/k)(1 - \exp[-k(t-t_0)]) \qquad (t_0 \leq t), \quad (3.3\text{-}21a)$$

$$\text{var}\{S(t)\} = (D/k^2)\left[\, (t-t_0) - (2/k)(1 - \exp[-k(t-t_0)]) \right.$$
$$\left. + (1/2k)(1 - \exp[-2k(t-t_0)]) \,\right]$$
$$(t_0 \leq t), \quad (3.3\text{-}21b)$$

$$\text{cov}\{S(t_1), S(t_2)\} = \text{var}\{S(t_1)\} + (D/2k^3)(1 - \exp[-k(t_2-t_1)))$$
$$\times (1 - 2\exp[-k(t_1-t_0)] + \exp[-2k(t_1-t_0)])$$
$$(t_0 \leq t_1 \leq t_2). \quad (3.3\text{-}21c)$$

Notice that if $D=0$ then the variance of $X(t)$ vanishes identically, and $X(t)$ reduces to a deterministic Liouville process [see Eqs. (3.3-6)].

As with the Wiener process, it is possible to obtain for the Ornstein-Uhlenbeck process an exact formula for the Markov state density function $P(x,t \mid x_0,t_0)$. For that, we appeal to the forward Fokker-Planck equation (3.2-2), which reads in this case

$$\frac{\partial}{\partial t} P(x,t \mid x_0,t_0) = k \frac{\partial}{\partial x} [x P(x,t \mid x_0,t_0)] + \frac{D}{2} \frac{\partial^2}{\partial x^2} P(x,t \mid x_0,t_0). \quad (3.3\text{-}22)$$

As may be proved by simply computing the required partial derivatives, the solution to this partial differential equation that satisfies the requisite initial condition $P(x,t_0 \mid x_0,t_0) = \delta(x - x_0)$ is

$$P(x,t \mid x_0,t_0) = \left[2\pi(D/2k)\Big(1 - \exp[-2k(t-t_0)]\Big) \right]^{-1/2}$$

$$\times \exp\left\{ - \frac{\Big(x - x_0 \exp[-k(t-t_0)]\Big)^2}{2(D/2k)\Big(1 - \exp[-2k(t-t_0)]\Big)} \right\}. \quad (3.3\text{-}23)$$

Comparing this result with the generic normal density function (1.4-9), we conclude that the general Ornstein-Uhlenbeck process $X(t)$ is the normal random variable with mean $x_0 \exp[-k(t-t_0)]$ and variance $(D/2k)(1 - \exp[-2k(t-t_0)])$:

$$X(t) = \mathbf{N}(x_0 \exp[-k(t-t_0)], (D/2k)(1 - \exp[-2k(t-t_0)])). \quad (3.3\text{-}24)$$

The mean and variance of $X(t)$ implied by this result obviously agree with Eqs. (3.3-20a) and (3.3-20b).   Also, the covariance formula (3.3-20c) can now be straightforwardly verified by starting with

$$\text{cov}\{X(t_1),X(t_2)\} \equiv \langle X(t_1) X(t_2) \rangle - \langle X(t_1) \rangle \langle X(t_2) \rangle$$

$$= \int_{-\infty}^{\infty} dx_1 \int_{-\infty}^{\infty} dx_2\, x_1 x_2\, P(x_2,t_2 \mid x_1,t_1) P(x_1,t_1 \mid x_0,t_0)$$

$$- x_0 \exp[-k(t_1-t_0)]\, x_0 \exp[-k(t_2-t_0)],$$

inserting the formula (3.3-23) for P, and then integrating over x_2 and x_1 in that order.

A plot of the time evolution of $P(x,t \mid x_0,t_0)$ for the Ornstein-Uhlenbeck process is shown in Fig. 3-3. Notice that in this *temporally* homogeneous but *not completely* homogeneous case, $P(x,t \mid x_0,t_0)$ *does* approach a stationary form as $t \to \infty$. Indeed, Eq. (3.3-23) shows that the Ornstein-Uhlenbeck process has the stationary density function

$$P(x,\infty \mid x_0,t_0) \equiv P_s(x) = \left[2\pi(D/2k) \right]^{-1/2} \exp\left(- \frac{x^2}{2(D/2k)} \right); \quad (3.3\text{-}25a)$$

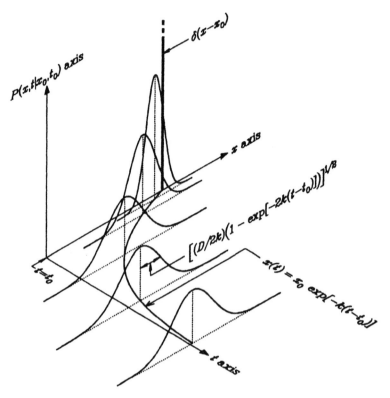

Figure 3-3. Time-evolution of the Markov state density function for a typical *Ornstein-Uhlenbeck process*, defined by the characterizing functions $A(x,t) = -kx$ and $D(x,t) = D$, with $k > 0$ and $D \geq 0$. For any $t > t_0$, $P(x,t \mid x_0, t_0)$ is the normal density function with mean $x_0 \exp[-k(t - t_0)]$ and standard deviation $[(D/2k)(1 - \exp[-2k(t - t_0)])]^{1/2}$. As $t \to \infty$, $P(x,t \mid x_0, t_0)$ approaches a stationary form $P_s(x)$, which is the normal density function with mean 0 and standard deviation $(D/2k)^{1/2}$; so the Ornstein-Uhlenbeck process is a *stable* process.

thus, as is also clear from Eq. (3.3-24),

$$X(\infty) = \mathbf{N}(0, D/2k). \qquad (3.3\text{-}25b)$$

But it is important to understand that the eventual lack of time dependence of $P(x,t \mid x_0, t_0)$ does *not* imply that the process eventually becomes quiescent and ceases to move about on the x-axis. All that is implied here is that a measurement of the process at any sufficiently large time t will be equivalent to sampling a random variable whose density function does not depend upon t -- in this case, the normal random variable with mean 0 and variance $D/2k$. Only in the Liouville case, $D=0$, would the process $X(t)$ eventually become truly quiescent [see Eq. (3.3-6b) for $t\to\infty$].

Although the mean and standard deviation of the Ornstein-Uhlenbeck process $X(t)$ become constants as $t\to\infty$, that is not quite the case with the integral $S(t)$ of the Ornstein-Uhlenbeck process. We see from Eqs. (3.3-21) that $\langle S(t)\rangle$ does eventually settle down to a constant value, namely x_0/k, but that sdev$\{S(t)\}$ increases indefinitely, eventually growing as $[D(t-t_0)]^{1/2}/k$.

We can see from Eq. (3.3-24) that k^{-1} characterizes the time scale on which the Markov state density function relaxes from its original delta function form to its final stationary form. And from Eq. (3.3-20c) we can see that k^{-1} also characterizes the time interval t_2-t_1 required for $X(t_2)$ to become effectively uncorrelated with $X(t_1)$. For these reasons, k^{-1} is often called the *relaxation time* or *decorrelation time* of the Ornstein-Uhlenbeck process.

The existence of the stationary density function $P_s(x)$ implies that cov$\{X(t_1),X(t_2)\}$ will approach, in the limit $t_0\to-\infty$, a stationary value cov$_s\{X(t_1),X(t_2)\}$, as was discussed in a more general context in Eq. (2.8-15). Indeed, it is clear from the covariance result (3.3-20c) that this stationary covariance is given by

$$\text{cov}_s\{X(t_1),X(t_2)\} = (D/2k)\exp[-k(t_2-t_1)]\quad(t_1\le t_2).\quad(3.3\text{-}26)$$

But it is instructive to derive this result directly from Eqs. (2.8-15) and (2.8-14). Thus,

$$\begin{aligned}
\text{cov}_s\{X(t_1),X(t_2)\} &= \langle X(t_1)\,X(t_2)\rangle_s - \langle X(t)\rangle_s^2 \\
&= \int_{-\infty}^{\infty}dx_1\int_{-\infty}^{\infty}dx_2\,x_1x_2\,P(x_2,t_2\,|\,x_1,t_1)\,P_s(x_1)\;-\;0 \\
&= \int_{-\infty}^{\infty}dx_1\,x_1\left[\int_{-\infty}^{\infty}dx_2\,x_2\,P(x_2,t_2\,|\,x_1,t_1)\right]P_s(x_1) \\
&= \int_{-\infty}^{\infty}dx_1\,x_1\,\langle\,X(t_2)\,|\,X(t_1)=x_1\,\rangle\,P_s(x_1) \\
&= \int_{-\infty}^{\infty}dx_1\,x_1\left(x_1\exp[-k(t_2-t_1)]\right)P_s(x_1) \\
&= \exp[-k(t_2-t_1)]\int_{-\infty}^{\infty}dx_1\,x_1^2\,P_s(x_1) \\
&= \exp[-k(t_2-t_1)](D/2k).
\end{aligned}$$

The behavior of the intensive state variable $Z(t)$, defined in Eq. (3.3-1), can be easily deduced by noting that, since $X(t)$ is the *normal* random variable in Eq. (3.3-24), then theorem (1.6-7) implies that

$$\Omega^{-1}X(t) = N(\Omega^{-1}x_0\exp[-k(t-t_0)], \Omega^{-2}(D/2k)(1-\exp[-2k(t-t_0)])).$$

Therefore, with $z_0 \equiv x_0/\Omega$, we have

$$Z(t) = N(z_0\exp[-k(t-t_0)], (D/2k\Omega^2)(1-\exp[-2k(t-t_0)])). \quad (3.3\text{-}27)$$

Evidently, the behavior of $Z(t)$ in the thermodynamic limit ($\Omega \to \infty$) will depend upon how the parameters k and D vary with Ω. As an example, suppose it happens that k is an *intensive* parameter (meaning that k is independent of the system volume Ω when that volume is large), while D is an *extensive* parameter (meaning that $D = D_1\Omega$ when Ω is large, where D_1 is independent of Ω). Then in the thermodynamic limit, $\langle Z(t)\rangle$ will be independent of Ω, while var$\{Z(t)\}$ will asymptotically approach the value $(D_1/2k\Omega)$; so the asymptotic fluctuations in the intensive state variable, as measured by sdev$\{Z(t\to\infty)\}$, will go to zero in the thermodynamic limit like $\Omega^{-1/2}$.

Since the Markov state density functions for the Liouville, the Wiener and the Ornstein-Uhlenbeck processes were all found to be *normal* (albeit a "degenerate" normal in the case of the Liouville process), one might wonder whether this is true of *all* continuous Markov processes. It is not. We will later encounter in Section 3.6 several continuous Markov processes that have density functions that are obviously far from normal. It is only the *propagator* of a continuous Markov process that is necessarily normal.

3.4 THE LANGEVIN EQUATION

We have seen that the time evolution of any continuous Markov process $X(t)$ is completely determined, in the stochastic sense of that word, by the forms of its two characterizing functions $A(x,t)$ and $D(x,t)$. One way in which that determination shows itself is through the forward Fokker-Planck equation (3.2-2). That equation is evidently completely defined by the two characterizing functions, and with appropriate initial and boundary conditions it uniquely determines the Markov state density function $P(x,t\,|\,x_0,t_0)$ of the process. The Markov state density function in turn suffices, because the process is Markovian, to calculate everything knowable about the process.

But the question arises, does there exist some sort of differential equation (presumably also involving the characterizing functions) for the continuous Markov process $X(t)$ *itself*, as opposed to its density function $P(x,t \mid x_0,t_0)$? The answer to that question is a qualified yes, the qualification being that we must be willing to deviate from the conventional notions of just what constitutes a "differential equation" and just how such an equation is to be used. The differential equation for the continuous Markov process itself is called the **Langevin equation**. In this section we shall derive and discuss three versions of that equation. As we shall discover, the Langevin equation is quite useful, and we have already reaped some of its benefits in earlier stages of our work. The role of the Langevin equation in the theory of continuous Markov processes is coequal with that of the forward Fokker-Planck equation, which it effectively complements as an analytical tool.

3.4.A TWO EQUIVALENT FORMS OF THE LANGEVIN EQUATION

We began this chapter by defining a "continuous" Markov process as a Markov process whose propagator $\Xi(dt; x,t)$ is "essentially an infinitesimal," a condition that we subsequently discovered means "usually of order $(dt)^{1/2}$." To reflect the infinitesimal nature of the continuous Markov propagator, the symbol dX is often used in place of Ξ; i.e., for a *continuous* Markov process we shall often write, instead of Eq. (2.4-1),

$$dX(dt; x,t) \equiv X(t+dt) - X(t), \text{ given } X(t)=x. \qquad (3.4\text{-}1)$$

According to our results (3.1-10) and (3.1-12), if the process has characterizing functions $A(x,t)$ and $D(x,t)$, then the random variable $dX(dt; x,t)$ can be written in either of the two equivalent forms

$$dX(dt; x,t) = N(A(x,t)dt, D(x,t)dt) \qquad (3.4\text{-}2a)$$

and

$$dX(dt; x,t) = N[D(x,t)dt]^{1/2} + A(x,t)dt, \qquad (3.4\text{-}2b)$$

where in the second equation N is the unit normal random variable, $N(0,1)$. [We may recall that the equivalence of these two equations is a simple consequence of theorem (1.6-7).] In particular, the propagator of the special Wiener process $W(t)$, for which $A(x,t)=0$ and $D(x,t)=1$, is evidently independent of x and t, and so can be written either as

$$dW(dt) = N(0, dt), \qquad (3.4\text{-}3a)$$

or as

$$dW(dt) = N(dt)^{1/2}. \qquad (3.4\text{-}3b)$$

Now if in Eq. (3.4-2b) we substitute Eq. (3.4-1) on the left side, and then take explicit note of the condition $x = X(t)$ on the right side, we obtain after a simple term transposition,

$$X(t+dt) = X(t) + N[D(X(t),t)dt]^{1/2} + A(X(t),t)dt. \qquad (3.4\text{-}4)$$

We shall call this equation the **first form of the Langevin equation.** It evidently expresses the random variable $X(t+dt)$ functionally in terms of the two random variables $X(t)$ and N. Given sample values of $X(t)$ and N, Eq. (3.4-4) thus allows us to calculate a sample value of $X(t+dt)$ for any given value of the infinitesimal variable dt.

Notice that the two random variables $X(t)$ and N on the right side of Eq. (3.4-4) are *statistically independent* of each other, with $X(t)$ being distributed according to the (often unknown) density function $P(x,t \mid x_0,t_0)$, and N being normally distributed with mean 0 and variance 1. Thus, the *joint* density function of these two random variables is just the product of their separate density functions, namely

$$P(x,t \mid x_0,t_0) \times (2\pi)^{-1/2} \exp(-n^2/2).$$

Therefore, to compute the average of any function g of $X(t)$ and $X(t+dt)$, we first use Eq. (3.4-4) to write

$$g(X(t),X(t+dt)) = g(X(t), X(t) + N[D(X(t),t)dt]^{1/2} + A(X(t),t)dt).$$

Then applying the general rule (1.5-15), we conclude that the average of g can be calculated as

$$\langle g(X(t),X(t+dt)) \rangle = \int_{-\infty}^{\infty} dx \int_{-\infty}^{\infty} dn \; g(x, x + n[D(x,t)dt]^{1/2} + A(x,t)dt)$$

$$\times P(x,t \mid x_0,t_0) (2\pi)^{-1/2} \exp(-n^2/2). \quad (3.4\text{-}5)$$

As an illustrative application of Eq. (3.4-5), let us use it to calculate the mean of the random variable $X(t+dt)$. We have

$$\langle X(t+dt) \rangle = \int_{-\infty}^{\infty} dx \int_{-\infty}^{\infty} dn \left(x + n[D(x,t)dt]^{1/2} + A(x,t)dt \right)$$

$$\times P(x,t \mid x_0,t_0) (2\pi)^{-1/2} \exp(-n^2/2)$$

$$= \int_{-\infty}^{\infty} dx \left(x + A(x,t)dt \right) P(x,t \mid x_0,t_0) (2\pi)^{-1/2} \int_{-\infty}^{\infty} dn \exp(-n^2/2)$$

$$+ \int_{-\infty}^{\infty} dx \, [D(x,t)dt]^{1/2} P(x,t \mid x_0,t_0) (2\pi)^{-1/2} \int_{-\infty}^{\infty} dn \, n \exp(-n^2/2)$$

$$= \int_{-\infty}^{\infty} dx \left(x + A(x,t)dt \right) P(x,t \mid x_0,t_0) + 0$$

$$= \langle X(t) \rangle + \langle A(X(t),t) \rangle dt.$$

If in this result we subtract $\langle X(t) \rangle$ from both sides, divide through by dt, and then take the limit $dt \to 0$, we recover the time evolution equation (3.2-12) for $\langle X(t) \rangle$. In a similar way, by using Eq. (3.4-5) to calculate the mean of the random variable $X^n(t+dt)$ for any $n \geq 1$, we can derive the general moment evolution equation (3.2-8). The derivation of such moment evolution equations constitutes one important use of the Langevin equation (3.4-4), and in that particular use it rivals the forward Fokker-Planck equation. In fact, it will be recalled that our derivation in Chapter 2 of the *general* moment evolution equation (2.7-6) was based not on the time-evolution equation for $P(x,t \mid x_0,t_0)$, but rather on the propagator equation (2.7-1). Since the Langevin equation (3.4-4) is just the "continuous version" of the propagator equation (2.7-1), then it follows that all the continuous process moment evolution equations set forth in Subsection 3.2.B were essentially derived from the Langevin equation! Two other important uses of the Langevin equation occur in the derivation of certain "weak noise" approximation formulas for continuous Markov processes (in Section 3.8), and, perhaps most importantly, in the derivation of the Monte Carlo simulation algorithm for continuous Markov processes (in Section 3.9).

In addition to these practical uses, the Langevin equation (3.4-4) also has the heuristic value that it conveys a strong sense of how a continuous Markov process $X(t)$ "really works." The equation tells us that the change in the process from time t to time $t+dt$ is the sum of two components: a *randomly fluctuating component*, $N[D(X(t),t)dt]^{1/2}$, and a *purely deterministic component*, $A(X(t),t)dt$. If the characterizing function $D(x,t)$ vanishes identically, then so also does the random component of the process change, and $X(t)$ evolves in a completely deterministic fashion. In the more interesting case in which the characterizing function $D(x,t)$ is positive, the process change has a random component that is proportional to $D(x,t)^{1/2}$. And being also proportional to $(dt)^{1/2}$, this random component is, though infinitesimally small, very much *larger* than the deterministic component, which is proportional to dt. In fact, we might very well wonder why the deterministic component makes any difference at all, since terms of order dt would seem to be negligibly small compared to terms of order $(dt)^{1/2}$. The answer lies in the random variable N that multiplies the $(dt)^{1/2}$ term. Since N is equally likely to be positive or negative, then the $(dt)^{1/2}$-contributions of the fluctuating term over a *succession* of dt-intervals will tend to cancel each other out. In contrast, the dt-contributions of the deterministic term, although comparatively small, are single-mindedly consistent over a succession of dt-intervals. In effect, the term $A(X(t),t)dt$ behaves like the proverbial *tortoise*, being slow but steady, while the term $N[D(X(t),t)dt]^{1/2}$ behaves like the proverbial *hare*, being fast but inconsistent. The *net* contributions of these two terms over a *finite* time interval turn out to be of comparable importance.

We can also understand from the Langevin equation (3.4-4) why the Markov process that it defines is *continuous* but usually *not differentiable*: The continuity of the process is a simple consequence of the fact, clearly evident from Eq. (3.4-4), that

$$X(t+dt) \rightarrow X(t) \quad \text{as} \quad dt \rightarrow 0. \tag{3.4-6}$$

To investigate whether the process is differentiable, we first subtract $X(t)$ from both sides of Eq. (3.4-4) and then divide through by dt to get

$$\frac{X(t+dt) - X(t)}{dt} = \frac{N D^{1/2}(X(t),t)}{(dt)^{1/2}} + A(X(t),t). \tag{3.4-7}$$

Clearly, if $D(x,t)$ is *not* identically zero, then the first term on the right will be undefined in the limit d$t\rightarrow0$, and "dX/dt" will not exist. On the other hand, if $D(x,t)$ *is* identically zero, then the limit as d$t\rightarrow0$ will exist, and in fact gives the ordinary differential equation dX/d$t=A(X,t)$ that governs the resulting deterministic Liouville process.

A form of the Langevin equation that differs slightly from the first form (3.4-4) can be obtained by substituting Eq. (3.4-3b) into the right side of Eq. (3.4-4). This evidently results in

$$X(t+dt) = X(t) + D^{1/2}(X(t),t) \, dW(dt) + A(X(t),t)dt, \tag{3.4-8}$$

which we shall call the **second form of the Langevin equation.** One interesting way of looking at the second form of the Langevin equation is to note that it expresses the propagator $X(t+dt) - X(t)$ of the continuous Markov process with characterizing functions $A(x,t)$ and $D(x,t)$ as a sum of two terms: the first term is $D^{1/2}(X(t),t)$ times the propagator of the special Wiener process $N(x_0, t - t_0)$, and the second term is $A(X(t),t)$ times the propagator of the special Liouville process $[x_0 + (t - t_0)]$. But it is generally more fruitful to view Eq. (3.4-8) in the same way that we viewed Eq. (3.4-4); that is, Eq. (3.4-8) expresses the random variable $X(t + dt)$ as a function of the two random variables $X(t)$ and $dW(dt)$, thus allowing sample values of $X(t + dt)$ to be calculated from sample values of $X(t)$ and $dW(dt)$. And since $X(t)$ and $dW(dt)$ are *statistically independent* random variables, with $X(t)$ having the density function $P(x,t \mid x_0,t_0)$ and $dW(dt)$ being normally distributed with mean 0 and variance dt, then their *joint* density function is

$$P(x,t \mid x_0, t_0) \times (2\pi dt)^{-1/2} \exp(-w^2/2dt).$$

From here we can go on to deduce the useful equation (3.4-5), but now with the simple (and wholly inconsequential) integration variable change $n \rightarrow w = n(dt)^{1/2}$.

Since the first and second forms of the Langevin equation differ only in whether we write $(dt)^{1/2} N(0,1)$ or $N(0,dt)$ in the fluctuation term, then it should be clear that the two equations are entirely equivalent. The second form of the Langevin equation seems to appear more frequently in the literature, but we shall often find the first form to be a bit more convenient.

3.4.B THE "WHITE NOISE" FORM OF THE LANGEVIN EQUATION

Although we shall not make use of it in our work here, there is a third form of the Langevin equation that is frequently encountered in the literature. To derive that third form, we must introduce a new random variable $\Gamma(dt)$, which we *define* to be the normal random variable with mean 0 and variance $1/dt$:

$$\Gamma(dt) = N(0, 1/dt). \tag{3.4-9}$$

This random variable is closely related to the propagator of the special Wiener process $dW(dt)$, which [see Eq. (3.4-3a)] is the normal random variable with mean 0 and variance dt:

$$dW(dt) = N(0, dt).$$

Indeed, since theorem (1.6-7) tells us that

$$dt\, N(0, 1/dt) = N(dt \cdot 0, (dt)^2 \cdot (1/dt)) = N(0, dt),$$

then we see that the connection between the two random variables $\Gamma(dt)$ and $dW(dt)$ is

$$dt\,\Gamma(dt) = dW(dt), \qquad (3.4\text{-}10a)$$

or equivalently,

$$\Gamma(dt) = dW(dt)/dt. \qquad (3.4\text{-}10b)$$

It is clear that the random variable $dW(dt)$ approaches, in the limit $dt\to0$, the normal random variable with mean 0 and variance 0, which of course is just the *sure* number 0. But it is not so clear how we should regard the $dt\to0$ limit of the random variable $\Gamma(dt)$; nevertheless, we *define*

$$\Gamma(0) \equiv \lim_{dt\to0} \Gamma(dt) \equiv \lim_{dt\to0} N(0,1/dt), \qquad (3.4\text{-}11)$$

and call it the **white noise random variable**. It would appear that the white noise random variable $\Gamma(0)$ is the normal random variable with mean 0 and variance ∞.

If we substitute Eq. (3.4-10a) into the second form (3.4-8) of the Langevin equation, and then make some simple algebraic rearrangements, we get

$$\frac{X(t+dt) - X(t)}{dt} = D^{1/2}(X(t),t)\,\Gamma(dt) + A(X(t),t).$$

If we next pass to the limit $dt\to0$, we obtain

$$\lim_{dt\to0} \frac{X(t+dt)-X(t)}{dt} \equiv \frac{dX(t)}{dt} = D^{1/2}(X(t),t)\,\Gamma(0) + A(X(t),t), \qquad (3.4\text{-}12)$$

which we shall refer to as the **white noise form of the Langevin equation**. But we must consider carefully whether this equation has any sensible meaning.

Evidently, Eq. (3.4-12) expresses the random variable $dX(t)/dt$ as a function of the two statistically independent random variables $X(t)$ and $\Gamma(0)$. But we know from earlier considerations that when the characterizing function $D(x,t)$ is not identically zero, then the process $X(t)$ *has no* derivative. Does Eq. (3.4-12) somehow escape the difficulties associated with the $dt\to0$ limit of Eq. (3.4-7)? We shall take the arguable position here that it does *not*, and that the flaw in Eq. (3.4-12) is simply that the white noise random variable $\Gamma(0)$ is *ill-defined*. Indeed, notice that if $A(x,t)\equiv0$ and $D(x,t)\equiv1$, so that $X(t)$ is just the special Wiener process $W(t)$, then Eq. (3.4-12) becomes

$$\frac{dW(t)}{dt} = \Gamma(0). \tag{3.4-13}$$

This equation, which was already implicit in Eq. (3.4-10b), asserts that the derivative of the special Wiener process is the white noise random variable. But since we have maintained throughout that the derivative of the special Wiener process is ill-defined, then we cannot consistently maintain that the white noise random variable is well-defined.

Another way to appreciate the severe difficulties associated with the random variable $\Gamma(0)$ is to reason as follows: Let K be an arbitrarily large number. Then the probability that a sample value of $\Gamma(dt)$ will have a magnitude *larger* than K is

$\text{Prob}\{\,|\Gamma(dt)| > K\,\}$

$\quad = \text{Prob}\{\,|dW(dt)/dt| > K\,\} \qquad\qquad$ [by (3.4-10b)]

$\quad = \text{Prob}\{\,|dW(dt)| > Kdt\,\}$

$\quad = 1 - \text{Prob}\{\,|dW(dt)| \le Kdt\,\}$

$\quad = 1 - \displaystyle\int_{-Kdt}^{+Kdt} dw\,(2\pi dt)^{-1/2}\exp(-w^2/2dt) \quad$ [by (3.4-3a)]

$\quad = 1 - \displaystyle\int_{-K\sqrt{dt}}^{+K\sqrt{dt}} dz\,(2\pi)^{-1/2}\exp(-z^2/2) \qquad [z = w(dt)^{-1/2}].$

Now, for any fixed K this last integral can evidently be made as close to zero as desired simply by taking dt sufficiently small. Hence, no matter how large K is, we can make $\text{Prob}\{|\Gamma(dt)| > K\}$ arbitrarily close to 1 simply by taking dt small enough. This implies that any sampling of the white noise random variable $\Gamma(0)$ is *virtually certain* to yield a value that is *infinitely large*.

This pathological characteristic of the white noise random variable prompts us to conclude that the white noise form (3.4-12) of the Langevin equation is simply a relation between two ill-defined quantities, namely $dX(t)/dt$ and $\Gamma(0)$. Although the white noise form of the Langevin equation appears quite frequently in the literature, a close examination usually reveals that the only purpose it serves is as a mnemonic for writing down either the first or second forms of the Langevin equation. Our position here will be that whatever may be found to be useful in the white noise form of the Langevin equation may also be found, but with fewer liabilities, in either the first or second forms of that equation. So we shall not have occasion to invoke Eq. (3.4-12) in the sequel.

Having thus disparaged the white noise random variable, we cannot dismiss it entirely because it is so pervasive in the working literature of stochastic process theory. Actually, what one usually encounters in the literature is not the white noise random *variable*, but the closely related white noise random *process*. The **white noise process** $\Gamma_0(t)$ can be defined as the stochastic process for which

$$\Gamma_0(t) = \Gamma(0) \quad \text{for any } t. \tag{3.4-14}$$

In words, the value of the stochastic process $\Gamma_0(t)$ at any time t is obtained by sampling the random variable $\Gamma(0)$, this sampling being made without regard for the values assumed by the process at any time prior to t. Notice that Eq. (3.4-14) allows one to write Eq. (3.4-12) with the white noise random variable $\Gamma(0)$ replaced by the white noise process $\Gamma_0(t)$ (indeed, that is the version of the Langevin equation most frequently encountered in the literature). A like replacement of $\Gamma(0)$ by $\Gamma_0(t)$ in Eq. (3.4-13) reveals that the white noise process $\Gamma_0(t)$ can be regarded as the "derivative" of the special Wiener process $W(t)$.

The white noise process $\Gamma_0(t)$ has two often quoted properties, namely

$$\langle \Gamma_0(t) \rangle = 0, \tag{3.4-15a}$$

and

$$\langle \Gamma_0(t_1) \Gamma_0(t_2) \rangle = \delta(t_1 - t_2). \tag{3.4-15b}$$

Equation (3.4-15a) follows of course from the fact that the random variable $\Gamma(0)$ has, by definition, zero mean. To establish Eq. (3.4-15b), we start by noting that

$$\text{cov}\{\Gamma_0(t_1), \Gamma_0(t_2)\} \equiv \langle \Gamma_0(t_1) \Gamma_0(t_2) \rangle - \langle \Gamma_0(t_1) \rangle \langle \Gamma_0(t_2) \rangle = \langle \Gamma_0(t_1) \Gamma_0(t_2) \rangle.$$

Since the values of the white noise process $\Gamma_0(t)$ at any two different times t_1 and t_2 are by definition statistically independent, then this covariance must vanish whenever $t_1 \neq t_2$; this proves Eq. (3.4-15b) when $t_1 \neq t_2$. For $t_1 = t_2$, Eq. (3.4-15b) evidently amounts to the assertion that

$$\langle \Gamma_0^2(t) \rangle = \delta(0).$$

To see that this assertion is consistent with the definition (3.4-14), we can reason heuristically as follows: Since the random variable $\Gamma_0(t) = \Gamma(0)$ has zero mean, then its second moment is equal to its variance; so, recalling the definition (3.4-11) of $\Gamma(0)$, we have

$$\langle \Gamma_0^2(t) \rangle = \text{var}\{\Gamma(0)\} = \text{var}\left\{ \lim_{dt \to 0} \text{N}(0,1/dt) \right\}$$

$$= \lim_{dt \to 0} \text{var}\{\text{N}(0,1/dt)\} = \lim_{dt \to 0} \frac{1}{dt} \, .$$

The last limit, if viewed charitably from the perspective of the Dirac delta function definition (1.2-6), can be regarded as the "value" $\delta(0)$.

The white noise process $\Gamma_0(t)$, as defined through either of Eqs. (3.4-14) or (3.4-15), is clearly a pathological stochastic process. We shall encounter it again in Subsection 3.5.C, where we shall explain the rationale for its name and also prove the surprising fact that it can be regarded as a special limiting case of the Ornstein-Uhlenbeck process. However, the white noise process itself will *not* form a building block for our theory of Markov processes.

3.4.C LANGEVIN'S ANALYSIS OF BROWNIAN MOTION

If a sufficiently small macroscopic particle is suspended in a fluid that is in thermal equilibrium, the particle will move about erratically in response to natural collisional bombardments by the individual molecules of the fluid. This erratic motion is called "Brownian motion" after the nineteenth century scientist Robert Brown who first observed it (in legend if not in fact). The first successful mathematical descriptions of Brownian motion were given by Albert Einstein in 1905 and Max von Smoluchowski in 1906; their independent but essentially equivalent analyses resulted in an explicit formula for the root-mean-square displacement of the particle over a macroscopic time t. The subsequent experimental confirmation of that formula constituted a crucial verification of the picture of a fluid as "molecules-in-motion."

An improved treatment of Brownian motion, which however gave the same root-mean-square displacement formula, was presented in 1908 by Paul Langevin.† Langevin's analysis of Brownian motion has come to be regarded as the premier application of continuous Markov process theory. In this subsection we shall present Langevin's analysis of Brownian motion, not in the precise form that he originally gave it, but rather in the context of the more fully developed theory that is now at our disposal. We shall find that Langevin's analysis relied heavily on several ad hoc

† Chapter 1 of Gardiner's *Handbook of Stochastic Methods* (see Bibliography) has concise summaries of the original Brownian motion papers of both Langevin and Einstein.

phenomenological assumptions; so too did the analyses of Einstein and Smoluchowski. An analysis of Brownian motion that proceeds in a more deductive way from the microphysics of particle-molecule collisions will be presented in Section 4.5; however, in that more rigorous analysis, the bath for the Brownian particle is not taken to be a real liquid, but rather a highly idealized fluid known as a "one-dimensional hard-sphere gas."

Langevin's analysis of Brownian motion begins by considering a *macroscopic* sphere of mass M moving with velocity v through a fluid that is in thermal equilibrium at absolute temperature T. The sphere experiences a retarding force that can usually be written $-\gamma v$, where the positive constant γ is called the *drag coefficient*. If the fluid happens to be a liquid with viscosity η, then γ will be equal to $6\pi\eta a$, where a is the diameter of the sphere. In any event, in the absence of other forces Newton's second law for the sphere reads

$$M\,(dv/dt) = -\gamma v, \qquad (3.4\text{-}16a)$$

or equivalently,

$$v(t+dt) = v(t) - (\gamma/M)\,v(t)\,dt. \qquad (3.4\text{-}16b)$$

The "drag impulse" $-\gamma v dt$ in the last equation is the *average* net impulse imparted to the sphere by the fluid molecules that collide with it in time dt. Owing to the inherent randomness of molecular motion, there will always be *fluctuations* in that net impulse about its average. For a sufficiently large sphere these fluctuations will not be noticeable, and the velocity of the sphere will behave like a *sure* variable $v(t)$, as implied by Eqs. (3.4-16). However, if the sphere happens to be very small, then the effects of the impulse fluctuations cannot be ignored; in that case the velocity of our sphere or "particle" must be regarded as a *stochastic process*, which we shall denote by $V(t)$.

Our first assumption is that the stochastic process $V(t)$ will be a *continuous Markov* process. Our second assumption is that the Langevin equation defining that process can be written as the following generalization of the deterministic equation (3.4-16b):

$$V(t+dt) = V(t) - (\gamma/M)\,V(t)\,dt + N\,[cdt]^{1/2}. \qquad (3.4\text{-}17)$$

Here, N is the unit normal random variable $N(0,1)$, and c is a positive constant whose value remains to be determined. The last term on the right side of Eq. (3.4-17) is supposed to represent the effects of random impulse fluctuations in time dt. But we must frankly admit that the only reason we have for describing the fluctuation effects by a term of this particular form is that this is the *simplest* form that is allowed by the theory of continuous Markov processes.

A comparison of the Langevin equation (3.4-17) to the canonical Langevin form (3.4-4) shows that the characterizing functions of $V(t)$ are $A(v,t) = -(\gamma/M)v$ and $D(v,t) = c$. Therefore, we may infer from our discussion in Subsection 3.3.C that $V(t)$ is an *Ornstein-Uhlenbeck process* with $k = \gamma/M$ and $D = c$. Given the initial condition $V(0) = v_0$, it follows from the Ornstein-Uhlenbeck equations (3.3-20) that the mean and variance of $V(t)$ are

$$\langle V(t) \rangle = v_0 e^{-(\gamma/M)t}, \tag{3.4-18a}$$

$$\text{var}\{V(t)\} = (cM/2\gamma)(1 - e^{-2(\gamma/M)t}). \tag{3.4-18b}$$

Furthermore, since the Brownian particle's *position* $X(t)$ is just the time-integral of its velocity $V(t)$, then we have from Eqs. (3.3-21) the following expressions for the mean and variance of $X(t)$:

$$\langle X(t) \rangle = (v_0 M/\gamma)(1 - e^{-(\gamma/M)t}), \tag{3.4-19a}$$

$$\text{var}\{X(t)\} = (cM^2/\gamma^2)\,[\,t - (2M/\gamma)(1 - e^{-(\gamma/M)t})$$
$$+ (M/2\gamma)(1 - e^{-2(\gamma/M)t})\,]. \tag{3.4-19b}$$

To fix the constant c, we argue as follows. We first observe from Eqs. (3.4-18) that

$$\langle V(t \to \infty) \rangle = 0 \quad \text{and} \quad \text{var}\{V(t \to \infty)\} = cM/2\gamma. \tag{3.4-20}$$

Now on thermodynamic grounds we expect that, after a very long time, our particle should behave like a molecule of mass M that is "in thermal equilibrium" at the fluid temperature T; in other words, we expect $V(t \to \infty)$ to be a normal random variable with mean zero and variance $k_B T/M$, where k_B is Boltzmann's constant. Equations (3.4-20) show that the mean of $V(t \to \infty)$ is indeed zero, but that in order for its variance to equal the anticipated value we would have to have $cM/2\gamma = k_B T/M$. We can ensure this condition by simply requiring c to have the value

$$c = 2\gamma k_B T/M^2. \tag{3.4-21}$$

Substituting Eq. (3.4-21) into Eq. (3.4-19b) and then passing to the long-time limit, we get

$$\text{var}\{X(t)\} = (2k_B T/\gamma)\,t \quad (t \gg M/\gamma). \tag{3.4-22a}$$

Thus we arrive at the now famous formula for the root-mean-square displacement of the Brownian particle after a sufficiently long time t:

$$\text{sdev}\{X(t)\} = (2k_B T/\gamma)^{1/2}\,t^{1/2} \quad (t \gg M/\gamma). \tag{3.4-22b}$$

We note that the root-mean-square displacement of the Brownian particle increases with t like the *square root* of t, and also that it depends on the fluid temperature T and the drag coefficient γ, but *not* on the particle's mass M.

We shall present a more thorough and rigorous analysis of the phenomenon of Brownian motion in Section 4.5.

3.5 STABLE PROCESSES

As discussed in Section 2.8, *some* temporally homogeneous Markov processes possess a stationary Markov state density function $P_s(x)$, as defined in Eq. (2.8-9). Such processes are said to be *stable*. In this section we shall deduce the conditions that the characterizing functions $A(x,t)$ and $D(x,t)$ must satisfy in order for a continuous Markov process $X(t)$ to be stable. We shall derive under those conditions an explicit formula for $P_s(x)$ in terms of the characterizing functions, and we shall deduce some of the interesting features that attend processes that are stable.

3.5.A THE STATIONARY DENSITY AND POTENTIAL FUNCTIONS

The first condition to be met in order for a process $X(t)$ to be stable is that it be *temporally homogeneous*. For a continuous Markov process, temporal homogeneity means that the characterizing functions $A(x,t)$ and $D(x,t)$ do not depend explicitly on t, and the forward Fokker-Planck equation takes the form (3.2-7b). So one way to calculate $P_s(x)$, if it exists, is to first solve Eq. (3.2-7b) explicitly for $P(x,t \mid x_0,t_0)$, and then compute the limit

$$\lim_{t \to \infty} P(x,t \mid x_0,t_0) = P_s(x). \qquad (3.5\text{-}1a)$$

Indeed, this is the procedure that we followed in Section 3.3.C when we calculated the stationary density function (3.3-25a) for the Ornstein-Uhlenbeck process. Another approach to calculating $P_s(x)$ is to recognize that, if Eq. (3.5-1a) is true, then it must also be true that

$$\lim_{t \to \infty} \frac{\partial}{\partial t} P(x,t \mid x_0,t_0) = 0. \qquad (3.5\text{-}1b)$$

Therefore, taking the $t \to \infty$ limit of the temporally homogeneous Fokker-Planck equation (3.2-7b), and invoking both of Eqs. (3.5-1), we get

$$0 = -\frac{d}{dx}[A(x) P_s(x)] + \frac{1}{2}\frac{d^2}{dx^2}[D(x) P_s(x)]. \qquad (3.5\text{-}2)$$

This is evidently an *ordinary* differential equation for $P_s(x)$, and it usually affords a much easier way of calculating $P_s(x)$ than by first solving the partial differential equation (3.2-7b) for $P(x,t \mid x_0,t_0)$.

We can in fact obtain a formal solution to Eq. (3.5-2) as follows: Writing that equation as

$$\frac{d}{dx}\left\{ -A(x) P_s(x) + \frac{1}{2}\frac{d}{dx}[D(x) P_s(x)] \right\} = 0,$$

we see that the quantity in braces must be constant with respect to the variable x. If we assume that the normalization condition on the density function $P_s(x)$ implies a sufficiently strong convergence $P_s(x) \to 0$ as $|x| \to \infty$, then the constant in question must be zero; hence,

$$\frac{d}{dx}[D(x) P_s(x)] = 2A(x) P_s(x).$$

Dividing both sides by $D(x)P_s(x)$, and thereby *assuming* that both $D(x)$ and $P_s(x)$ never vanish, we get

$$\frac{d[D(x) P_s(x)]}{[D(x) P_s(x)]} = \frac{2A(x)}{D(x)}.$$

Integrating this gives

$$\ln[D(x) P_s(x)] = \int^x \frac{2A(x')}{D(x')}\,dx' + \text{constant},$$

or

$$D(x) P_s(x) = \text{constant} \times \exp\left(\int^x \frac{2A(x')}{D(x')}\,dx' \right).$$

Solving for $P_s(x)$, and choosing the integration constant so that this density function is properly normalized, we conclude that $P_s(x)$ is given in terms of the characterizing functions by the explicit formula

$$P_s(x) = \frac{2K}{D(x)}\exp(-\phi(x)), \qquad (3.5\text{-}3)$$

where we have defined the **potential function** $\phi(x)$ by

$$\phi(x) \equiv - \int^x \frac{2A(x')}{D(x')}\, dx', \tag{3.5-4}$$

and the **normalization constant** K by

$$\frac{1}{K} \equiv \int_{-\infty}^{\infty} \frac{2}{D(x)} \exp(-\phi(x))\, dx. \tag{3.5-5}$$

The reason for calling the function ϕ a "potential" will be explained shortly.

The prime requirement for the *existence* of $P_s(x)$ is that the normalization constant K, as defined in Eq. (3.5-5), must exist as a *finite, nonzero* number:

$$0 < K < \infty. \tag{3.5-6}$$

It should be clear from the formula (3.5-3) that when this condition is satisfied then $P_s(x)$ will not only exist but will also be everywhere nonzero:

$$P_s(x) > 0 \quad \text{(strictly)}. \tag{3.5-7}$$

Indeed, it will be recalled that the nonvanishing of both $D(x)$ and $P_s(x)$ was specifically *assumed* in our derivation of Eq. (3.5-3) from Eq. (3.5-2).

To illustrate the existence condition (3.5-6), consider first the Wiener process, for which $A(x) \equiv A$ and $D(x) \equiv D$. The formula (3.5-4) defining the potential function gives in this case

$$\phi(x) = - \int^x [2A/D]\, dx' = -2Ax/D,$$

and the formula (3.5-5) then gives for $1/K$

$$\frac{1}{K} = \int_{-\infty}^{\infty} \frac{2}{D} \exp(+2Ax/D)\, dx = \infty.$$

So condition (3.5-6) is *not* satisfied, and we recover our earlier conclusion that the Wiener process does not have a stationary Markov state density function. On the other hand, for the Ornstein-Uhlenbeck process with $A(x) = -kx$ and $D(x) \equiv D$, the formula for the potential function gives

$$\phi(x) = - \int^x [-2kx'/D]\, dx' = kx^2/D.$$

The formula for $1/K$ then gives, using the integral identity (A-3),

$$\frac{1}{K} = \int_{-\infty}^{\infty} \frac{2}{D} \exp(-kx^2/D)\,dx = \frac{2}{D}\left(\frac{\pi}{k/D}\right)^{1/2}.$$

Now condition (3.5-6) *is* satisfied, so $P_s(x)$ does exist. And upon substituting these expressions for $\phi(x)$ and K into Eq. (3.5-3), we straightforwardly obtain

$$P_s(x) = \left(\frac{k}{\pi D}\right)^{1/2} \exp(-kx^2/D),$$

which agrees exactly with our earlier result (3.3-25a) for the Ornstein-Uhlenbeck process.

To understand why we call $\phi(x)$ the "potential" function, observe from its definition (3.5-4) that its slope at x is given by

$$\phi'(x) = -2A(x)/D(x). \tag{3.5-8}$$

Since $D(x)$ is always positive, then ϕ slopes downward or upward at x according to whether $A(x)$ is positive or negative. Now, if the process is in state x at time t, then it is clear from the Langevin equation (3.4-4) that, in the next dt, the process will be *more likely* to move to the *right* if $A(x)>0$ and ϕ is sloping *downward*, or to the *left* if $A(x)<0$ and ϕ is sloping *upward*. Furthermore, this movement bias will in either case be more pronounced the larger $|A(x)|$ is relative to $D(x)$, and hence the larger the magnitude of $\phi'(x)$ is. So we see that the process, in state x at time t, has a *probabilistic bias* to move in the next dt in the direction that *decreases* the function $\phi(x)$; moreover, the steeper the local slope of $\phi(x)$ is, the greater will be this probabilistic movement bias. We conclude that $\phi(x)$ is somewhat analogous to the potential energy function in classical mechanics, whose slope at x is the negative of the x-component of the force on a particle at x. Of course, this analogy cannot be pushed too far: Our Markov process lacks "inertia," and the classical particle lacks "stochasticity."

3.5.B STABLE STATES

We have just seen that a continuous Markov process $X(t)$ has a stochastic tendency to move toward the nearest local minimum of the potential function $\phi(x)$. The consequent statistical favoring of regions of the x-axis where $\phi(x)$ is relatively small is in the main consistent with the formula (3.5-3) for $P_s(x)$, which shows that such regions are usually where $P_s(x)$ is relatively large; i.e., those are the regions where we would be most likely to find the process $X(t)$ a sufficiently long time after it is initialized.

Because $X(t)$ tends to move toward states that are local minimums of $\phi(x)$, and tends to be found near states that are local maximums of $P_s(x)$, we call such states the **stable states** of the process. If $D(x)$ happens to be constant with respect to x, then it is clear from Eq. (3.5-3) that every local minimum of $\phi(x)$ will be a local maximum of $P_s(x)$, and vice versa. But if $D(x)$ is not a constant, then the local minimums of $\phi(x)$ and the local maximums of $P_s(x)$ will usually not coincide, and our definition of stable state becomes ambiguous. However, in most cases where the notion of stable state is of *practical use*, the distance between the center of a valley in $\phi(x)$ and the center of the nearest peak in $P_s(x)$ will be small compared to the widths of those two structures, and the ambiguity in our definition of stable state will be of no consequence. In those cases, each valley of $\phi(x)$ overlies one and only one peak of $P_s(x)$, and $X(t)$ is said to be "in" the corresponding stable state whenever it is *anywhere* inside that $\phi(x)$-valley or that $P_s(x)$-peak.

The relative minimums of $\phi(x)$ are of course defined by the two conditions $\phi'(x)=0$ and $\phi''(x)>0$. Using Eq. (3.5-8), and remembering that $D(x)$ is always positive here, it is easy to express these two conditions in terms of the characterizing functions as follows:

x_i is a relative minimum of $\phi(x)$

$$\Leftrightarrow \quad A(x_i) = 0 \text{ and } A'(x_i) < 0. \qquad (3.5\text{-}9)$$

In other words, the relative minimums of the potential function $\phi(x)$ are the down-going roots of the drift function $A(x)$. Now, a case could be made for letting condition (3.5-9) define the "nominal values" of the stable states of the process. Certainly that criterion is simple, and it might apply even in circumstances where $P_s(x)$ does not exist -- i.e., when condition (3.5-6) is not satisfied. However, when $P_s(x)$ *does* exist, we shall prefer to define the nominal values of the stable states to be the relative maximums of $P_s(x)$.

The relative maximums of $P_s(x)$ are of course defined by the two conditions $P_s'(x)=0$ and $P_s''(x)<0$. To express these two conditions more explicitly in terms of the characterizing functions, we first introduce the function

$$\alpha(x) \equiv A(x) - \tfrac{1}{2}D'(x). \qquad (3.5\text{-}10)$$

Now invoking Eqs. (3.5-3) and (3.5-8), we calculate

$$P_s{}'(x) = 2\,\frac{P_s(x)}{D(x)}\,\alpha(x). \qquad (3.5\text{-}11a)$$

A second differentiation then yields

$$P_s{}''(x) = 2\,\frac{P_s(x)}{D(x)}\,\alpha'(x) + 2\,\alpha(x)\,\frac{\mathrm{d}}{\mathrm{d}x}\!\left(\frac{P_s(x)}{D(x)}\right). \qquad (3.5\text{-}11b)$$

Because $P_s(x)$ and $D(x)$ are both strictly positive, Eq. (3.5-11a) shows that $P_s{}'(x)$ vanishes if and only if $\alpha(x)$ vanishes. And if $\alpha(x)$ does vanish, then the second equation shows that $P_s{}''(x)$ will be negative if and only if $\alpha'(x)$ is negative. Thus we conclude that [compare condition (3.5-9)]

x_i is a relative maximum of $P_s(x)$

$$\Leftrightarrow \quad \alpha(x_i) = 0 \ \text{ and } \ \alpha'(x_i) < 0. \qquad (3.5\text{-}12)$$

So the relative maximums of $P_s(x)$ are the down-going roots of the function $\alpha(x)$. Notice from the definition (3.5-10) that when $D(x)$ is independent of x then $\alpha(x) = A(x)$, and condition (3.5-12) coincides with condition (3.5-9), just as we should expect.

In addition to assigning to each stable state a *nominal position* x_i, which we usually take to be the associated relative maximum of $P_s(x)$, it is also convenient to assign to each stable state a *nominal width* σ_i. We shall define σ_i to be the so-called "effective width" of the normal or Gaussian-shaped curve that best fits $P_s(x)$ in the immediate neighborhood of the peak point x_i. The exact procedure for calculating σ_i is detailed in Appendix D, and requires that we evaluate the second derivative of $P_s(x)$ at $x = x_i$. Recalling Eq. (3.5-11b), and taking note of conditions (3.5-12), we see that

$$P_s{}''(x_i) = -\,P_s(x_i)\left|\frac{2\alpha'(x_i)}{D(x_i)}\right|. \qquad (3.5\text{-}13)$$

This is of the form of the last of Eqs. (D-1) provided we identify the constant c there with $|2\alpha'(x_i)/D(x_i)|$. Thus it follows from Eq. (D-7) that the nominal width of the peak in $P_s(x)$ at $x = x_i$ is given by

$$\sigma_i = \left|\frac{\pi D(x_i)}{\alpha'(x_i)}\right|^{1/2}. \qquad (3.5\text{-}14)$$

To the extent that the peak in the function $P_s(x)$ at $x = x_i$ is roughly Gaussian and does not significantly overlap any adjacent peak, we shall say that

"$X(t)$ is in the stable state x_i"

$$\Leftrightarrow \quad X(t) \in [x_i - \sigma_i/2, x_i + \sigma_i/2]. \qquad (3.5\text{-}15)$$

Whenever $P_s(x)$ exists, we can of course just visually read off the positions x_i and widths σ_i of the stable states from a graph of that function, as plotted from Eqs. (3.5-3) – (3.5-5). However, with equations (3.5-10), (3.5-12) and (3.5-14), we can now *compute* x_i and σ_i directly from the characterizing functions. For example, in the case of the Ornstein-Uhlenbeck process with $A(x) = -kx$ and $D(x) = D$, we easily calculate from Eq. (3.5-10) that

$$\alpha(x) = -kx \quad \text{and} \quad \alpha'(x) = -k.$$

Clearly, $x_i = 0$ is the only state that satisfies the two conditions (3.5-12) in this case, so we may conclude that it is the lone stable state of this process. And applying Eq. (3.5-14), we find that the width of that stable state is

$$\sigma_i = (\pi D/k)^{1/2}.$$

This is about 25% larger than twice the standard deviation $(D/2k)^{1/2}$ of the normal density function $P_s(x)$ in Eq. (3.3-25a), and so is quite reasonable. Further applications of the concepts and formulas developed in this and the preceding subsection will be given in Section 3.6.

3.5.C THE SPECTRAL DENSITY FUNCTION

In this subsection we shall consider an aspect of stable Markov processes which, although only marginally related to the rest of our exposition here, receives considerable attention in many practical applications of stochastic process theory. Our focus here will be on a continuous Markov process $X(t)$ that satisfies the following three conditions:

- $X(t)$ is a *stable* process; $\qquad\qquad\qquad\qquad$ (3.5-16a)

- $\langle X(t) \rangle = 0$ in the limit $t_0 \to -\infty$; $\qquad\qquad$ (3.5-16b)

- $X^2(t) = E(t)$, the *energy* of the process at time t. $\quad$ (3.5-16c)

The first condition above means of course that the stationary limit of the Markov state density function, namely

$$\lim_{t_0 \to -\infty} P(x,t \mid x_0, t_0) \equiv P_{\mathrm{s}}(x), \tag{3.5-17}$$

must exist as a well-defined function of x, independently of x_0 and t; thus, the characterizing functions of the process must be such that the inequality (3.5-6) is satisfied. The second condition (3.5-16b) essentially says that the mean of the stationary density function $P_{\mathrm{s}}(x)$ must be zero; this is actually a rather mild requirement, since it can usually be assured by a simple translational shift of the x-axis. The third condition (3.5-16c) attributes a distinctly physical character to the process $X(t)$. Some particular instances in which this condition would be satisfied are where $X(t)$ measures the instantaneous stretch distance of an ideal spring, in which case $X^2(t)$ would be proportional to the potential energy of the spring; or where $X(t)$ measures the instantaneous velocity of a one-dimensional particle, in which case $X^2(t)$ would be proportional to the kinetic energy of the particle; or where $X(t)$ measures the instantaneous field strength of an electromagnetic wave, in which case $X^2(t)$ would be proportional to the energy resident in the wave.

We consider, for any continuous Markov process $X(t)$ that satisfies the three conditions (3.5-16), its *stationary covariance*. According to Eq. (2.8-15), that quantity is defined for $t_1 \le t_2$ by

$$
\begin{aligned}
\mathrm{cov}_{\mathrm{s}}&\{X(t_1),X(t_2)\} \\
&\equiv \lim_{t_0 \to -\infty} [\,\langle X(t_1)\,X(t_2) \rangle - \langle X(t_1) \rangle \langle X(t_2) \rangle\,] \\
&= \lim_{t_0 \to -\infty} \langle X(t_1)\,X(t_2) \rangle - 0 \\
&= \lim_{t_0 \to -\infty} \int_{-\infty}^{\infty} dx_1 \int_{-\infty}^{\infty} dx_2 \; x_1 x_2\, P(x_2,t_2 \mid x_1,t_1)\, P(x_1,t_1 \mid x_0,t_0),
\end{aligned}
$$

where the second equality follows from condition (3.5-16b), and the last equality follows from the definition (2.3-4). Now invoking the definition (3.5-17) of the stationary density function, and also the property (2.8-4) which holds for *any* temporally homogeneous Markov process, we obtain

$$\mathrm{cov}_{\mathrm{s}}\{X(t_1),X(t_2)\} = \int_{-\infty}^{\infty} dx_1 \int_{-\infty}^{\infty} dx_2 \; x_1 x_2\, P(x_2,t_2 - t_1 \mid x_1,0)\, P_{\mathrm{s}}(x),$$

for any $t_1 \le t_2$. Just two features of this result have a direct bearing upon our analysis here: First, the right side depends on t_1 and t_2 only through their *difference* $t_2 - t_1$. And second, if we would allow t_2 to be less than t_1

on the *left* side, then the only change that would occur on the *right* side would be a replacement of $t_2 - t_1$ by $t_1 - t_2$ and an inconsequential relabeling of the integration variables. Therefore, for the class of processes under consideration here we have

$$\text{cov}_s\{X(t), X(t+\tau)\} = \lim_{t_0 \to -\infty} \langle X(t) X(t+\tau) \rangle \equiv C(\tau), \qquad (3.5\text{-}18a)$$

where the function $C(\tau)$ is independent of t and has the symmetry

$$C(-\tau) = C(\tau). \qquad (3.5\text{-}18b)$$

Next we introduce the *Fourier transform* $S(\omega)$ of the stationary covariance function $C(\tau)$:

$$S(\omega) \equiv (2\pi)^{-1} \int_{-\infty}^{\infty} d\tau \, e^{-i\omega\tau} C(\tau). \qquad (3.5\text{-}19a)$$

Since $C(\tau)$ is an even function of τ then the imaginary part of the integrand here, namely $-\sin\omega\tau \, C(\tau)$, is an odd function of τ, and so contributes nothing to the symmetric-limit integral. Therefore, $S(\omega)$ is the real-valued function

$$S(\omega) \equiv (2\pi)^{-1} \int_{-\infty}^{\infty} d\tau \cos\omega\tau \, C(\tau). \qquad (3.5\text{-}19b)$$

Also, by substituting $\tau' = -\tau$ into either of Eqs. (3.5-19) and then invoking the evenness of $C(\tau)$, it is straightforward to show that $S(\omega) = S(-\omega)$. In short, the fact that $C(\tau)$ is an even, real function of τ implies that its Fourier transform $S(\omega)$ is an even, real function of ω. The inverse Fourier transform formula reads

$$C(\tau) = \int_{-\infty}^{\infty} d\omega \, e^{i\omega\tau} S(\omega), \qquad (3.5\text{-}20a)$$

and since $S(-\omega) = S(\omega)$ this can likewise be simplified to

$$C(\tau) = \int_{-\infty}^{\infty} d\omega \cos\omega\tau \, S(\omega). \qquad (3.5\text{-}20b)$$

The reason why we have concerned ourselves with the stationary covariance function $C(\tau)$ and its Fourier transform $S(\omega)$ becomes clear upon making two simple observations. First, observe from Eq. (3.5-18a) that

$$C(0) = \lim_{t_0 \to -\infty} \langle X(t) X(t+0) \rangle = \lim_{t_0 \to -\infty} \langle X^2(t) \rangle;$$

therefore, invoking the condition (3.5-16c), we have

$$C(0) = \lim_{t_0 \to -\infty} \langle E(t) \rangle \equiv E_s,$$
(3.5-21)

where E_s is the *stationary energy* of the process. Second, observe from Eq. (3.5-20b) that $C(0)$ can also be written as

$$C(0) = \int_{-\infty}^{\infty} d\omega\, S(\omega).$$
(3.5-22)

Combining the last two equations evidently yields the formula

$$E_s = \int_{-\infty}^{\infty} S(\omega)\, d\omega.$$
(3.5-23)

We now *interpret* this formula as implying that

$S(\omega)d\omega =$ the stationary energy of the process associated
with frequencies between ω and $\omega + d\omega$.   (3.5-24)

Thus we have established the following result:

Wiener-Khintchine Theorem. For a continuous Markov process $X(t)$ that satisfies conditions (3.5-16), the Fourier transform $S(\omega)$ of the stationary covariance function $C(\tau)$ measures the *frequency density* of the stationary energy of $X(t)$.

Because the function $S(\omega)$ has this special property, it is called the **spectral density function** of the process $X(t)$.

As an illustration of these concepts, consider the Ornstein-Uhlenbeck process. It clearly satisfies conditions (3.5-16a) and (3.5-16b), and we shall assume for the sake of argument that the physical context of the process is such that condition (3.5-16c) is also satisfied. Now, we found in Eq. (3.3-26) that the stationary covariance of the Ornstein-Uhlenbeck process is

$$C(\tau) = \left(\frac{D}{2k} \right) \exp(-k|\tau|),$$
(3.5-25a)

where the absolute value operation here follows from the property (3.5-18b) and allows the variable τ to assume negative values. To calculate the spectral density function of the Ornstein-Uhlenbeck process, we need only insert this formula for $C(\tau)$ into Eq. (3.5-19b):

$$S(\omega) = (2\pi)^{-1} \int_{-\infty}^{\infty} d\tau \cos\omega\tau \left(\frac{D}{2k} \right) \exp(-k|\tau|)$$

$$= (2\pi)^{-1} \left(\frac{D}{2k} \right) 2 \int_0^\infty d\tau \cos\omega\tau \exp(-k\tau).$$

Evaluating this integral with the help of the integral identity (A-9), we get

$$S(\omega) = \left(\frac{D}{2k} \right) \frac{k/\pi}{\omega^2 + k^2}. \qquad (3.5\text{-}25b)$$

We see from the Cauchy-like form of this result that the stationary energy of the Ornstein-Uhlenbeck process is associated *mainly* with frequencies ω between $-k$ and $+k$; however, *all* frequencies contribute to the stationary energy, with the relative contribution of very large frequencies (namely those with $|\omega| \gg k$) being proportional to $1/\omega^2$.

Two interesting *limiting cases* of the Ornstein-Uhlenbeck process are suggested by the forms of Eqs. (3.5-25). In the first limiting case, we let the two parameters k and D both approach *zero*, but in such a way that $D/2k$ approaches unity. Since [cf. Eqs. (1.4-13) and (1.4-14)]

$$\lim_{k\to 0} \frac{k/\pi}{\omega^2 + k^2} = \delta(\omega),$$

then we see from Eqs. (3.5-25) that

$$k \to 0 \text{ and } D \to 0 \text{ with } D/2k \to 1$$

$$\Rightarrow \quad C(\tau) \to 1 \text{ and } S(\omega) \to \delta(\omega). \qquad (3.5\text{-}26)$$

In this limit, $E_s \to 1$, as is easily seen by substituting $S(\omega) = \delta(\omega)$ into Eq. (3.5-23). Evidently, all the stationary energy of the process in this limit is associated with the single frequency $\omega = 0$; therefore, we call this limit the *DC limit* of the Ornstein-Uhlenbeck process, in analogy with direct current (zero-frequency) electricity. The constant stationary covariance function in this limit tells us that the stationary process becomes "uniformly correlated" with itself. Since $D = 2k$ in this limit, then the Langevin equation (3.4-4) reads

$$X(t+dt) = X(t) + ND^{1/2}(dt)^{1/2} - kX(t)dt$$

$$= X(t) + N(2k)^{1/2}(dt)^{1/2} - kX(t)dt,$$

or

$$X(t+dt) = X(t) + k^{1/2}[N2^{1/2}(dt)^{1/2} - k^{1/2}X(t)dt].$$

From this relation we can see that in the hypothesized limit $k \to 0$ we have $X(t+dt) - X(t) \to 0$ for any fixed $dt > 0$. Therefore, the DC limit (3.5-26) of the Ornstein-Uhlenbeck process is a profoundly unexciting process that moves with "infinite slowness."

In the second limiting case, we let the two parameters k and D both approach *infinity*, but in such a way that D/k^2 approaches unity. To see what happens in this case, we first rewrite Eqs. (3.5-25) as

$$C(\tau) = \left(\frac{D}{k^2}\right)\frac{k}{2}\exp(-k|\tau|), \quad S(\omega) = \left(\frac{D}{k^2}\right)\frac{1/2\pi}{(\omega/k)^2 + 1},$$

and then note that [cf. Eqs. (1.4-5) and (1.4-8)]

$$\lim_{k\to\infty}\left(\frac{k}{2}\exp(-k|\tau|)\right) = \delta(\tau).$$

Thus we see that

$$k\to\infty \text{ and } D\to\infty \text{ with } D/k^2\to 1$$

$$\Rightarrow \quad C(\tau)\to\delta(\tau) \text{ and } S(\omega)\to 1/2\pi. \quad (3.5\text{-}27)$$

In this limit, $E_\mathrm{s}\to\infty$, as is easily seen by substituting $S(\omega) = 1/2\pi$ into Eq. (3.5-23). Evidently, all frequencies in this limit contribute *in equal measure* to the stationary energy. This is why the stationary energy of $X(t)$ is infinitely large, and also why we call this the *white noise limit* of the Ornstein-Uhlenbeck process (in analogy with the flat frequency spectrum of white light). The stationary covariance $C(\tau)$ in this limit is seen to be zero if $\tau\neq 0$, implying that $X(t_1)$ and $X(t_2)$ are completely uncorrelated if $t_1\neq t_2$. The prediction $C(0) = \infty$ implies that $X(t)$ has an infinite second moment, which is consistent with the process having an infinite energy. Since $D = k^2$ in this limit, then the Langevin equation (3.4-4) reads

$$X(t+dt) = X(t) + ND^{1/2}(dt)^{1/2} - kX(t)dt$$

$$= X(t) + N(k^2)^{1/2}(dt)^{1/2} - kX(t)dt,$$

or

$$X(t+dt) = X(t) + k[N(dt)^{1/2} - X(t)dt].$$

From this relation we can see that in the hypothesized limit $k\to\infty$ we have $X(t+dt) - X(t)\to\pm\infty$ for any fixed $dt > 0$. Therefore, the white noise limit (3.5-27) of the Ornstein-Uhlenbeck process is a process that moves with "infinite rapidity." Indeed, it is appropriate at this point to recall the white noise *process* $\Gamma_0(t)$, as defined in Eq. (3.4-14). From the properties (3.4-15) of that process, we see that it has a stationary mean of zero and a stationary covariance function $C(\tau) = \delta(\tau)$. Therefore, Eq. (3.5-19b) predicts that the spectral density function of the white noise process should be

$$S(\omega) = (2\pi)^{-1} \int_{-\infty}^{\infty} d\tau \cos\omega\tau \, \delta(\tau) = 1/2\pi.$$

The fact that the frequency spectrum of the stationary energy of the process $\Gamma_0(t)$ is *flat* is of course why $\Gamma_0(t)$ is called the "white noise" process. Comparing these properties of $\Gamma_0(t)$ with Eqs. (3.5-27), we conclude that the Ornstein-Uhlenbeck process $X(t)$ becomes, in the white noise *limit*, the white noise *process* $\Gamma_0(t)$. This rather surprising result in some sense parallels the result in Eq. (3.4-11), which says that the normal random variable $N(0,1/a)$ becomes, in the limit $a\to0$, the white noise random variable $\Gamma(0)$. Both results demonstrate that otherwise well-behaved entities can become very ill-behaved in certain limits.

3.6 SOME EXAMPLES OF STABLE PROCESSES

Perhaps the simplest and best known example of a *stable* continuous Markov process, i.e., one for which $P(x,t\,|\,x_0,t_0)\to P_s(x)$ as $(t-t_0)\to\infty$, is the Ornstein-Uhlenbeck process, discussed in Subsection 3.3.C. Its characterizing functions are

$$A(x,t) = -kx, \quad D(x,t) = D \quad (k>0, D>0), \qquad (3.6\text{-}1a)$$

and as we found in Subsection 3.5.A its potential function and stationary density function are

$$\phi(x) = (k/D)\,x^2, \qquad (3.6\text{-}1b)$$

and

$$P_s(x) = [2\pi(D/2k)]^{-1/2} \exp[-x^2/2(D/2k)], \qquad (3.6\text{-}1b)$$

the latter being the normal density function with mean zero and variance $(D/2k)$. Since $x=0$ is at once the single relative minimum of $\phi(x)$ and the single relative maximum of $P_s(x)$, then $x=0$ is the sole stable state of this process. And we can intuitively understand how that stable state arises: the drift term in the Langevin equation (3.4-4) persistently urges the process toward the origin with a "force" $-\phi'(x) = -(k/D)x$ that increases with the distance from the origin. In this section we shall consider some other stable processes that will give us further insight into the causes and consequences of stability.

3.6.A QUADRATIC-NOISE ORNSTEIN-UHLENBECK PROCESSES

The continuous Markov process $X(t)$ defined by the characterizing functions

$$A(x,t) = -kx, \quad D(x,t) = D + \gamma x^2 \quad (k>0, D>0, \gamma>0), \quad (3.6\text{-}2)$$

is called a **quadratic-noise Ornstein-Uhlenbeck process**, for reasons that are obvious on comparing with Eqs. (3.6-1a). Since $A(x,t)$ is linear in x and $D(x,t)$ is quadratic in x, then the moment evolution equations (3.2-8) for this process are closed, and it is possible to solve those equations exactly for $\langle X^n(t) \rangle$ for all $n \geq 1$. But we shall not do that here; instead, we shall merely focus on the question of whether or not this process has any stable states.

The quickest way to answer this question is to proceed as follows: By applying criteria (3.5-9), it is easy to see that the state $x=0$ should be the one and only relative minimum of the potential function $\phi(x)$. Furthermore, since the definition (3.5-10) implies that

$$a(x) = -(k+\gamma)x, \quad (3.6\text{-}3)$$

then it follows from criteria (3.5-12) that the state $x=0$ should also be the one and only relative maximum of the stationary density function $P_s(x)$. Clearly then, this process should have a single stable state at $x=0$. And by applying Eq. (3.5-14), it is a simple matter to deduce that the width of that stable state should be

$$\sigma = \left(\frac{nD}{k+\gamma} \right)^{1/2}. \quad (3.6\text{-}4)$$

Let us check the validity of these predictions by explicitly calculating $\phi(x)$ and $P_s(x)$ for this process.

Using the definition (3.5-4) of the potential function, we find

$$\phi(x) = -\int \frac{2(-kx)}{D+\gamma x^2} \, dx = \frac{k}{\gamma} \int \frac{d(\gamma x^2)}{D+\gamma x^2}.$$

The integration here is easily carried out, and we obtain

$$\phi(x) = (k/\gamma) \ln(D + \gamma x^2). \quad (3.6\text{-}5)$$

As expected, this function has a single relative minimum at $x=0$.

The existence of a stationary Markov state density function hinges on whether the normalization constant K, defined in Eq. (3.5-5), exists as a *finite, nonzero* number. In this case we have

$$K^{-1} = \int_{-\infty}^{\infty} \frac{2}{D + \gamma x^2} \exp\left(-(k/\gamma) \ln(D + \gamma x^2) \right) dx$$

$$= 2 \int_{-\infty}^{\infty} \frac{(D + \gamma x^2)^{-(k/\gamma)}}{D + \gamma x^2} dx$$

or

$$K^{-1} = 2 \int_{-\infty}^{\infty} \frac{dx}{(D + \gamma x^2)^{1 + (k/\gamma)}} . \qquad (3.6\text{-}6a)$$

This integral converges for all positive, finite values of k, D and γ, so a stationary density function indeed exists. And using Eq. (3.5-3), we get

$$P_s(x) = \frac{2K}{D + \gamma x^2} \exp\left(-(k/\gamma) \ln(D + \gamma x^2) \right)$$

$$= \frac{2K}{D + \gamma x^2} (D + \gamma x^2)^{-(k/\gamma)}$$

or

$$P_s(x) = \frac{2K}{(D + \gamma x^2)^{1 + (k/\gamma)}} . \qquad (3.6\text{-}6b)$$

As expected, this function has a single relative maximum at $x = 0$.

The stationary density function $P_s(x)$ is obviously well defined for this process, but some of its moments *may not* exist. The nth stationary moment is calculated as

$$\langle X^n \rangle_s \equiv \int_{-\infty}^{\infty} x^n P_s(x) dx = 2K \int_{-\infty}^{\infty} \frac{x^n dx}{(D + \gamma x^2)^{1 + (k/\gamma)}} ,$$

and we note that for $x \to \infty$ this integrand behaves like x raised to the power $[n - 2 - 2(k/\gamma)]$. Since convergence obtains only if the asymptotic exponent of x is less than -1, then we conclude that

$$\langle X^n \rangle_s \text{ exists} \quad \Leftrightarrow \quad n < 1 + 2k/\gamma. \qquad (3.6\text{-}7)$$

Now we observe that this existence condition for the stationary moments appears to be somewhat at odds with the formula for the stable state width σ in Eq. (3.6-4): If we let $\gamma \to 0$, so that we approach the regular Ornstein-Uhlenbeck process, Eq. (3.6-7) tells us that higher and higher order stationary moments $\langle X^n \rangle_s$ become finite, and that in turn suggests that $P_s(x)$ becomes more closely confined to the origin. But Eq. (3.6-4) tells us that as γ gets smaller, the width σ of the stable state at $x = 0$ gets *larger*, and that suggests that $P_s(x)$ becomes *less* closely confined to the origin. To investigate this seeming inconsistency, let us take $D = k = 1$, and then calculate $\phi(x)$ and $P_s(x)$ explicitly in the two subcases $\gamma = 1$ and $\gamma = 0$. The latter subcase of course is just a regular Ornstein-Uhlenbeck process, for which we find from Eqs. (3.6-1) that

$$\phi(x) = x^2, \quad P_s(x) = \pi^{-1/2} \exp(-x^2) \quad [D = k = 1; \gamma = 0]. \quad (3.6\text{-}8)$$

For the former case, we find from Eqs. (3.6-5) and (3.6-6) that

$$\phi(x) = \ln(1 + x^2), \quad P_s(x) = \frac{2\pi^{-1}}{(1 + x^2)^2} \quad [D = k = 1; \gamma = 1], \quad (3.6\text{-}9)$$

where the normalization constant K has been explicitly evaluated from Eq. (3.6-6a) using a common table of integrals. We have plotted the four functions (3.6-8) and (3.6-9) in Fig. 3-4, using lighter lines for the $\gamma = 0$ case and darker lines for the $\gamma = 1$ case. The shallower stable state valley of $\phi(x)$ and the heavier tails of $P_s(x)$ when $\gamma = 1$ indeed suggest that the confinement of the process is less in the $\gamma = 1$ case than in the $\gamma = 0$ case; however, the stable state peak in $P_s(x)$ is unmistakably sharper in the $\gamma = 1$ case than in the $\gamma = 0$ case!

Condition (3.6-7) with $k = 1$ implies that $\langle X^n \rangle_s$ exists (is finite) for *all* n in the $\gamma = 0$ case, but only for $n = 1$ and 2 in the $\gamma = 1$ case. Since $\langle X \rangle_s = 0$ in both cases, then $\langle X^2 \rangle_s$ is in both cases the stationary variance. Calculating the second moments in the usual way, and also calculating the stable state widths from Eq. (3.6-4), it is not hard to show that the standard deviations and widths of the $P_s(x)$ curves in the two cases are as follows:

$$\text{sdev}_s\{X\} = 0.707, \quad \sigma = 1.773 \quad [D = k = 1; \gamma = 0]; \quad (3.6\text{-}10)$$

$$\text{sdev}_s\{X\} = 1.000, \quad \sigma = 1.253 \quad [D = k = 1; \gamma = 1]. \quad (3.6\text{-}11)$$

Evidently, as γ is increased from 0 to 1, $\text{sdev}_s\{X\}$ increases while σ decreases! These effects are readily apparent in Fig. 3-4, but their joint occurrence is nevertheless puzzling: How can the additional process noise that is introduced by the larger diffusion function in the $\gamma = 1$ case cause the stable state to have a *smaller* width? We shall elucidate this

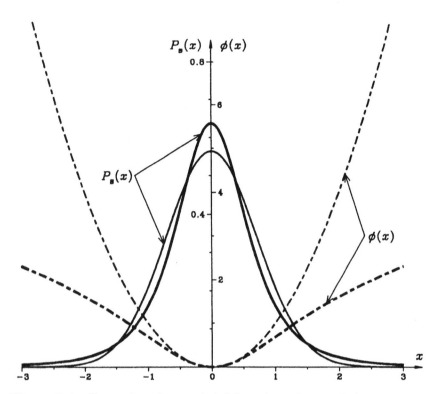

Figure 3-4. Comparing the potential functions $\phi(x)$ and the stationary Markov state density functions $P_s(x)$ of a *regular* Ornstein-Uhlenbeck process and a *quadratic-noise* Ornstein-Uhlenbeck process. The lighter curves are for the regular Ornstein-Uhlenbeck process with characterizing functions $A(x) = -x$ and $D(x) = 1$, while the darker curves are for the quadratic-noise Ornstein-Uhlenbeck process with characterizing functions $A(x) = -x$ and $D(x) = 1 + x^2$. Both processes have a single stable state at $x = 0$. The tails of the stationary density function are heavier in the quadratic-noise case, as we might expect. But curiously, the stable state is *narrower* in the quadratic-noise case: the probability of finding $X(t \to \infty)$ between $x = -0.3$ and $x = +0.3$ is evidently greater for the quadratic-noise process than for the regular process. This effect is called *noise-enhanced stability*, and its underlying cause is heuristically explained in Subsection 3.6.B.

curious phenomenon, which we call **noise-enhanced stability**, in Subsection 3.6.B; there we shall also discover that, not only can process noise occasionally *enhance* stability, it can sometimes actually *cause* stability.

3.6.B NOISE-INDUCED STABILITY

The Wiener process with $A(x)=0$ and $D(x)=1$ has a flat potential function, no stationary Markov state density function, and certainly no stable state. If we were to change that process by adding $-kx$ to the drift function, then we would obtain a regular Ornstein-Uhlenbeck process. For that process we know that the potential function has a valley at $x=0$ and the stationary Markov state density function exists with a peak at $x=0$, so $x=0$ is clearly a stable state; moreover, we can see how that stable state arises as a consequence of the persistent, origin-directed urgings of the drift term. But what if, instead of adding $-kx$ to the drift function, we add γx^2 to the diffusion function:

$$A(x,t) = 0, \quad D(x,t) = D + \gamma x^2 \quad (D>0, \gamma>0). \tag{3.6-12}$$

This process can evidently be looked upon *either* as an extension of the driftless $(A=0)$ Wiener process to include a quadratic noise term, *or* as a degenerate case of the quadratic-noise Ornstein-Uhlenbeck process of Eqs. (3.6-2) in which $k=0$.

Now, one might reasonably suppose that the addition of a quadratic term to the diffusion constant of a driftless Wiener process would surely not confer stability on that distinctly unstable process. That supposition is given support by the fact that the potential function for the process, as determined from Eqs. (3.5-4) and (3.6-12), is perfectly flat,

$$\phi(x) = - \int \frac{2\times0}{D+\gamma x^2}\, dx = 0, \tag{3.6-13}$$

just as it is for the driftless Wiener process. Since the drift term does not urge the process toward any particular state, then there is no state x_i that satisfies condition (3.5-9). However, it follows from Eq. (3.5-10) that, for the process defined by Eqs. (3.6-12),

$$a(x) = -\gamma x. \tag{3.6-14}$$

Using this in criteria (3.5-12), we are let to conclude that the state $x=0$ should be a relative maximum of $P_s(x)$. Indeed, the formula (3.5-5) for K gives in this case

$$K^{-1} = \int_{-\infty}^{\infty} \frac{2}{D+\gamma x^2}\, e^0\, dx = 2\int_{-\infty}^{\infty} \frac{dx}{D+\gamma x^2} = \frac{2\pi}{(D\gamma)^{1/2}},$$

which shows that condition (3.5-6) is satisfied, and hence that a stationary Markov state density function *does exist* for this process. And using Eq. (3.5-3) we easily deduce that stationary density function is

$$P_s(x) = \frac{2(D\gamma)^{1/2}/(2\pi)}{D+\gamma x^2} \, e^{-0},$$

which can be algebraically simplified to

$$P_s(x) = \frac{(D/\gamma)^{1/2}/\pi}{x^2 + (D/\gamma)}. \tag{3.6-15}$$

Comparing this result with Eq. (1.4-13), we see that this stationary density function is just a *Cauchy* density function that is centered at $x = 0$ with half-width $(D/\gamma)^{1/2}$. That half-width value is consistent with (although slightly more than half of) the nominal stable state width σ predicted by formula (3.5-14), which turns out to be

$$\sigma = (\pi D/\gamma)^{1/2}. \tag{3.6-16}$$

In particular, we see from both Eqs. (3.6-15) and (3.6-16) that, if the coefficient γ of the quadratic noise term is made *larger*, then the peak in $P_s(x)$ at $x = 0$ actually becomes *narrower* and *taller*, corresponding to a more sharply defined stable state!

Since the characterizing functions (3.6-12) are simply the $k=0$ cases of the characterizing functions (3.6-2), which define the quadratic-noise Ornstein-Uhlenbeck process, then all the foregoing results could also have been obtained by simply putting $k=0$ in the formulas of the preceding subsection. So the peculiar behavior of the process just considered is intimately related to the peculiar behavior noted earlier for the quadratic-noise Ornstein-Uhlenbeck process, where, according to Eq. (3.6-4), we can just as easily sharpen the stable state by increasing the coefficient γ of the quadratic noise term as by increasing the coefficient k of the attractive drift term.

A heuristic explanation for these curious behaviors is this: For $\gamma > 0$, states with larger $|x|$ obviously have more "intrinsic noise" than states with smaller $|x|$. And although that noise is nondirectional, it does encourage the process to *leave* states with larger $|x|$ *sooner* than states with smaller $|x|$. Thus, as the parameter γ is taken larger and larger, the diffusion function $D+\gamma x^2$ encourages the process to spend less and less time in states with larger $|x|$; as a consequence, the process *must* spend more and more time in states with smaller $|x|$. So we see that the observed favoring of the state $x = 0$ by the process when γ is positive arises, not because of any "overt attraction" by that state, but rather because that state makes the process feel "less unwelcomed" than does any other state.

The Markov process defined by the characterizing functions (3.6-12) is an example of a process that exhibits **noise-induced stability**. Noise-induced stability will occur with any temporally homogeneous continuous Markov process for which $A(x)$ vanishes identically and $D(x)$ tends to ∞ sufficiently strongly as $|x| \to \infty$. For, in such a case, it follows from Eq. (3.5-4) that $\phi(x) \equiv 0$, and hence from Eq. (3.5-3) that $P_s(x) = c/D(x)$ where c is a normalizing constant; therefore, every relative minimum of $D(x)$ is necessarily a relative maximum of $P_s(x)$, and hence a stable state of the process. For example, suppose that

$$A(x,t) = 0 \quad \text{and} \quad D(x,t) = \exp(x^2). \tag{3.6-17a}$$

Then we easily find from the standard formulas (3.5-3), (3.5-4) and (3.5-5) that

$$P_s(x) = \pi^{-1/2} \exp(-x^2), \tag{3.6-17b}$$

which obviously has a stable state at $x = 0$. Notice, by the way, that this function $P_s(x)$ is *also* the stationary Markov state density function of the regular Ornstein-Uhlenbeck process with $k = D = 1$. So we see that a continuous Markov process is *not* completely defined by its stationary Markov state density function. The process with characterizing functions $A(x,t) = -x$ and $D(x,t) = 1$ behaves in *most* respects quite differently from the process with characterizing functions (3.6-17a), despite the fact that they both have the stationary Markov state density function (3.6-17b).

3.6.C BISTABLE PROCESSES

If it should happen that $\phi(x)$ has more than one relative minimum, or $P_s(x)$ has more than one relative maximum, then we say that $X(t)$ is a **multistable process**. The most commonly encountered multistable processes are those that have exactly *two* stable states, and they are referred to as **bistable processes**. A simple example of a bistable continuous Markov process is the process $X(t)$ that is defined by the characterizing functions

$$A(x) = -x(x+1.5)(x-1.7), \quad D(x) = 1.0. \tag{3.6-18a}$$

Inasmuch as $D(x)$ is constant, it follows that conditions (3.5-9) determine not only the relative minimums of $\phi(x)$ but also the relative maximums of

$P_s(x)$. Because $A(x)$ is an asymptotically *down*-going cubic with roots at -1.5, 0 and 1.7, then it is clear that only the outside roots can satisfy the negative-slope condition in (3.5-9); hence, we expect exactly two stable states for this process, one at $x_1 = -1.5$ and the other at $x_2 = 1.7$.

Substituting Eqs. (3.6-18a) into the definition (3.5-4) of the potential function and integrating, we find that $\phi(x)$ is in fact given by

$$\phi(x) = (0.5)\, x^4 - (0.1333...)\, x^3 - (2.55)\, x^2. \qquad (3.6\text{-}18\text{b})$$

This function is plotted in Fig. 3-5, and we see that it indeed has two relative minima $x_1 = -1.5$ and $x_2 = 1.7$, these being separated by a

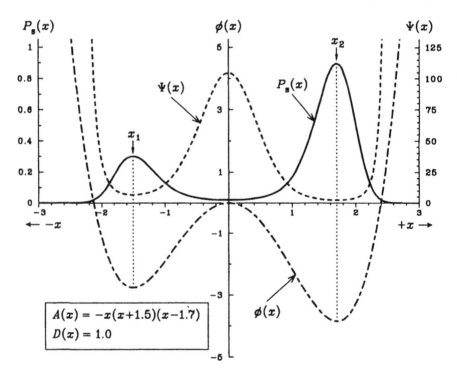

Figure 3-5. The potential function $\phi(x)$ and stationary Markov state density function $P_s(x)$ for the continuous Markov process defined by Eqs. (3.6-18a). This is a *bistable* process with stable states at $x_1 = -1.5$ and $x_2 = 1.7$, those values being both relative maximums of $P_s(x)$ and relative minimums of $\phi(x)$. Also plotted here is the *barrier function*, $\Psi(x) \equiv K^{-1}\exp[\phi(x)]$, whose significance is discussed in Section 3.7. The two stable states x_1 and x_2 are evidently separated by a *barrier state* at $x_b = 0$, which is defined as a relative maximum of the functions $\phi(x)$ and $\Psi(x)$.

relative maximum $x_b = 0$. A numerical evaluation of the integral in Eq. (3.5-5) yields in this case the result $K = 9.5564 \times 10^{-3}$. Since $0 < K < \infty$, then a stationary density function $P_s(x)$ exists for this process, and of course it is given by formula (3.5-3). That function is also plotted in Fig. 3-5, and it indeed exhibits two well defined stable state peaks at $x_1 = -1.5$ and $x_2 = 1.7$. The nominal widths of those two peaks can be straightforwardly calculated from Eqs. (3.5-10) and (3.5-14), and they turn out to be

$$\sigma_1 = 0.8090 \quad \text{and} \quad \sigma_2 = 0.7599. \tag{3.6-18c}$$

We notice in Fig. 3-5 that the stable state x_2 corresponds to the higher peak in $P_s(x)$ and the deeper valley in $\phi(x)$, and therefore would appear to be somehow "more stable" than x_1. In Section 3.7 we shall develop a quantitative measure of the relative stability of two stable states which agrees with this intuitive assessment. In that section we shall also define and explain the significance of the third function $\Psi(x)$ that is plotted in Fig. 3-5.

Another bistable continuous Markov process is defined by the characterizing functions

$$A(x) = -x(x+1.5)(x-1.7), \quad D(x) = 1.0 + 0.25\,x^2. \tag{3.6-19a}$$

This process evidently differs from the one just discussed only in the addition of a quadratic term to the diffusion function. Using the criteria (3.5-9), we see that the relative minimums of $\phi(x)$ are at -1.5 and 1.7, just as before. However, because $D(x)$ now depends on x, then the function $a(x)$ in Eq. (3.5-10) no longer coincides with the function $A(x)$. Applying to $a(x)$ the criteria (3.5-12), we straightforwardly deduce that the relative maximums of $P_s(x)$ should now be at $x_1 = -1.420$ and $x_2 = 1.620$.

An explicit analytical calculation of $\phi(x)$ from the definition (3.5-4) is not easy for the characterizing functions (3.6-19a), but it can be accomplished. The result, as can be proved simply by showing that its derivative satisfies Eq. (3.5-8), is

$$\phi(x) = 4[\,x^2 - (6.55)\ln(1.0 + 0.25\,x^2) - 0.4\,x + (0.8)\arctan(0.5\,x)\,].$$

$$\tag{3.6-19b}$$

This function is plotted in Fig. 3-6. As compared with the potential function plotted in Fig. 3-5, the valleys in this potential function at -1.5 and 1.7 are seen to be somewhat shallower, apparently as a consequence of the additional noise in this process.

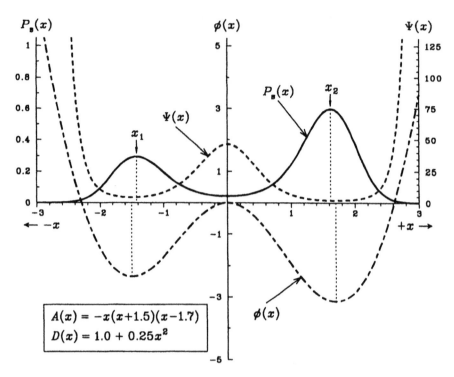

Figure 3-6. The potential function $\phi(x)$ and stationary Markov state density function $P_s(x)$ for the continuous Markov process defined by Eqs. (3.6-19a). The function $A(x)$ is the same as for the process of Fig. 3-5, but the inclusion here of an x-dependent term in the diffusion function $D(x)$ has resulted in a displacement of the relative maximums of $P_s(x)$, now at $x_1 = -1.42$ and $x_2 = 1.62$, away from the relative minimums of $\phi(x)$, which are still at -1.5 and 1.7. As compared to the process of Fig. 3-5, the stable state peaks in $P_s(x)$ here are lower and broader, and the stable state valleys in $\phi(x)$ are shallower and broader. The implications of the smaller peak in the barrier function $\Psi(x) \equiv K^{-1}\exp[\phi(x)]$ at $x_b = 0$ will be discussed in Subsection 3.7.B.

A numerical evaluation of the integral in Eq. (3.5-5) for this process gives the result $K = 2.1564 \times 10^{-2}$, thus indicating that a stationary Markov state density function $P_s(x)$ indeed exists for this process. Using the formula (3.5-3), we have calculated $P_s(x)$ and plotted it in Fig. 3-6. As expected, we find two relative maxima at $x_1 = -1.420$ and $x_2 = 1.620$, which we therefore take to be the nominal values of the stable states of this process. And using Eq. (3.5-15), we find that the nominal widths of those two stable states are

$$\sigma_1 = 1.0461 \quad \text{and} \quad \sigma_2 = 1.0278. \tag{3.6-19c}$$

Comparing these widths with the widths (3.6-18c) for the process of Fig. 3-5, we note the broadening effect of the increased noise. We shall give a quantitative assessment of the apparent "weakening" of the two stable states later in Section 3.7. The reason why the quadratic noise term in Eq. (3.6-19a) has shifted each stable state *toward the origin* is basically the same as the reason why the quadratic noise term in the dark-lined process of Fig. 3-4 squeezed that stable state peak toward the origin: the noisier large-$|x|$ states make the process feel "more unwelcomed" than the less noisy small-$|x|$ states, so the process tends to spend less time in the larger $|x|$ states than in the smaller $|x|$ states.

3.7 FIRST EXIT TIME THEORY

In this section, as in the preceding two sections, we shall be concerned exclusively with a *temporally homogeneous* continuous Markov process $X(t)$ that is characterized by a drift function $A(x)$ and a diffusion function $D(x)$. But our focus now will be on the following general question: If the process starts at time 0 in a given state x_0 inside some given x-axis interval (a,b), when will the process *first exit* that interval? To answer that question is to discover the statistics of the **first exit time**, which is the random variable

$$T(x_0; a,b) \equiv \text{time at which } X(t) \text{ first exits } (a,b),$$
$$\text{given that } X(0) = x_0 \in (a,b). \tag{3.7-1}$$

In Subsection 3.7.A we shall derive a set of differential equations for the *moments* of this random variable, namely,

$$T_n(x_0; a,b) \equiv \text{the } n\text{th moment of } T(x_0; a,b) \quad (n=1,2,...). \tag{3.7-2}$$

Of particular interest will be the **mean first exit time**, $T_1(x_0; a,b)$, which of course is the "average" or "expected" time of the first exit from (a,b).

Most applications of the theory of first exit times involve processes with one or more stable states. For example, if x_i is a stable state of the process $X(t)$, then $T_1(x_i; x_i - \varepsilon, x_i + \varepsilon)$ can evidently be regarded as the average time for $X(t)$, initially at x_i, to first experience a *spontaneous fluctuation* away from x_i of size ε. And for a bistable system with stable states x_1 and x_2, such as the systems shown in Figs. 3-5 and 3-6, $T_1(x_1; -\infty, x_2)$ can evidently be regarded as the average time for the process to make a *spontaneous transition* from stable state x_1 to stable state x_2, and $T_1(x_2; x_1, \infty)$ can be regarded as the average time for a transition from stable state x_2 to stable state x_1. In Subsection 3.7.B we shall use the formula for $T_1(x_0; a,b)$ that we derive in Subsection 3.7.A to calculate the *average stable state transition times* in a bistable system.

3.7.A EQUATIONS FOR THE MOMENTS OF THE FIRST EXIT TIME

The derivation of a set of differential equations for the moments of the first exit time $T(x_0; a,b)$ proceeds, interestingly enough, from the *backward* Fokker-Planck equation. We begin by defining the function

$$G(x_0, t; a,b) \equiv \int_a^b dx\, P(x,t \mid x_0, 0), \qquad (3.7\text{-}3a)$$

which function has the evident physical interpretation,

$$G(x_0, t; a,b) = \text{Prob}\{X(t) \in (a,b), \text{ given that } X(0) = x_0\}. \qquad (3.7\text{-}3b)$$

Now if, in the temporally homogeneous backward Fokker-Planck equation (3.2-7c), we substitute $\partial P/\partial t$ for $-\partial P/\partial t_0$, as is permitted for temporally homogeneous Markov processes by Eq. (2.8-5), and then integrate the resulting equation over the x-interval (a,b), we obtain the following partial differential equation for G:

$$\frac{\partial}{\partial t} G(x_0, t; a,b) = A(x_0) \frac{\partial}{\partial x_0} G(x_0, t; a,b) + \frac{1}{2} D(x_0) \frac{\partial^2}{\partial x_0^2} G(x_0, t; a,b).$$

$$(3.7\text{-}4)$$

A unique solution to this partial differential equation requires that we specify one initial condition and two boundary conditions. For the initial condition we shall take

$$G(x_0, 0; a,b) = 1. \qquad (3.7\text{-}5)$$

This, by Eq. (3.7-3b), is simply the statement that the initial point x_0 lies inside (a,b). For the two boundary conditions we shall impose some *one* of the following three sets:

$$G(a,t; a,b) = 0 \quad \text{and} \quad G(b,t; a,b) = 0; \qquad (3.7\text{-}6a)$$

$$G'(a,t; a,b) = 0 \quad \text{and} \quad G(b,t; a,b) = 0; \qquad (3.7\text{-}6b)$$

$$G(a,t; a,b) = 0 \quad \text{and} \quad G'(b,t; a,b) = 0. \qquad (3.7\text{-}6c)$$

The primes in the last two lines denote the partial derivative of $G(x_0,t; a,b)$ with respect to its argument x_0. To understand the implications of each of these three sets of boundary conditions, consider first the condition $G(a,t; a,b) = 0$. This says that if $X(0) = a$, then $X(t)$ will *not* be inside (a,b) for any $t > 0$; hence, because the process is temporally homogeneous, if $X(t') = a$ for any t', then $X(t)$ will not be inside (a,b) for any $t > t'$. This means that if $X(t)$ should ever wander out to the boundary point a, then $X(t)$ is immediately removed from (a,b) forever; therefore,

$$G(a,t; a,b) = 0 \; \Rightarrow \; \text{boundary point } a \text{ is } absorbing. \qquad (3.7\text{-}7a)$$

Consider next the condition $G'(a,t; a,b) = 0$. It implies that $G(a + \varepsilon,t; a,b) = G(a,t; a,b)$ for all infinitesimally small $\varepsilon > 0$; hence, if $X(0)$ is taken very close to a, then the probability of $X(t)$ being inside (a,b) for any $t > 0$ would not be reduced if $X(0)$ were taken *even closer* to a. This must mean that $X(t)$ cannot exit (a,b) at the boundary point a; therefore,

$$G'(a,t; a,b) = 0 \; \Rightarrow \; \text{boundary point } a \text{ is } reflecting. \qquad (3.7\text{-}7b)$$

So we see that the three possible sets of boundary conditions (3.7-6) correspond to at least one of the boundary points of the interval (a,b) being *absorbing*, while the other boundary point is either reflecting or absorbing. But *any one* of these three sets of boundary conditions, together with the initial condition (3.6-5), *expands* the physical interpretation (3.7-3b) of $G(x_0,t; a,b)$ so that now

$$G(x_0,t; a,b) = \text{probability that the process, at } x_0 \text{ in } (a,b) \text{ at}$$
$$\text{time 0, has } not \, yet \, exited \, (a,b) \text{ by time } t. \qquad (3.7\text{-}8)$$

Because now, for any of the three boundary conditions (3.7-6), the process can exit (a,b) only at an absorbing point, in which case it can never return to (a,b).

It follows from Eq. (3.7-8) that $1 - G(x_0,t; a,b)$ is the probability that the process exits (a,b) *before* time t, and hence by the definition (3.7-1) that

$$1 - G(x_0,t; a,b) = \text{Prob}\{ T(x_0; a,b) < t \}. \qquad (3.7\text{-}9)$$

Thus we see that if $G(x_0,t; a,b)$ is the solution to the partial differential equation (3.7-4) which satisfies the initial condition (3.7-5) and any one of the boundary conditions (3.7-6), then $1 - G(x_0,t; a,b)$ can be regarded as the *distribution function* of the random variable $T(x_0; a,b)$.

Next we make the assumption that $T(x_0; a,b)$ is a "proper" random variable, in the sense that its distribution function rises monotonically from zero at $t = 0$ to unity at $t = \infty$ [see the discussion of Eq. (1.2-8)]. That the distribution function in Eq. (3.7-9) is zero at $t=0$ is a simple consequence of the initial condition (3.7-5); however, in order for it to be unity at $t=\infty$, we evidently must *assume* that

$$G(x_0,\infty; a,b) = 0. \qquad (3.7\text{-}10)$$

If this assumption is *not* true, then it is evidently possible that $X(t)$ will *never* exit (a,b), and our analysis here must be completely revised.

Since $1 - G(x_0,t; a,b)$ is the distribution function of the random variable $T(x_0; a,b)$, then it follows from Eq. (1.2-9b) that the density function of $T(x_0; a,b)$ is

$$\frac{\partial}{\partial t}[1 - G(x_0,t; a,b)] = -\frac{\partial}{\partial t} G(x_0,t; a,b).$$

The nth moment $T_n(x_0; a,b)$ of the random variable $T(x_0; a,b)$ is therefore given by

$$T_n(x_0; a,b) = \int_0^\infty dt\, t^n\left[-\frac{\partial}{\partial t} G(x_0,t; a,b)\right] \quad (n\geq 0). \quad (3.7\text{-}11a)$$

For $n=0$ this formula gives, because of Eqs. (3.7-10) and (3.7-5),

$$T_0(x_0; a,b) = -[G(x_0,\infty; a,b) - G(x_0,0; a,b)] = 1,$$

as is required of any "proper" random variable. For $n>0$ we may obtain an alternate formula for T_n by integrating Eq. (3.7-11a) by parts. This evidently gives

$$T_n(x_0; a,b) = -t^n\, G(x_0,t; a,b)\Big|_0^\infty + \int_0^\infty G(x_0,t; a,b)\, n t^{n-1} dt \quad (n\geq 1).$$

The integrated term vanishes at both limits because of Eqs. (3.7-10) and (3.7-5) [note that if it did *not* vanish at $t=\infty$ then $T_{n+1}(x_0; a,b)$ would not exist]; hence we have the formula

$$\int_0^\infty dt\, t^{n-1} G(x_0,t; a,b) = \frac{1}{n} T_n(x_0; a,b) \quad (n\geq 1). \quad (3.7\text{-}11b)$$

With the pair of equations (3.7-11), we can now convert the partial differential equation (3.7-4) for $G(x_0,t; a,b)$ into a set of coupled, ordinary differential equations for the moments $T_n(x_0; a,b)$: Multiplying Eq. (3.7-4) through by t^{n-1}, integrating the resulting equation over all t, and then substituting from Eq. (3.7-11a) on the *left* side and from Eq. (3.7-11b) on the *right* side, we obtain the differential equation

$$-T_{n-1}(x_0; a,b) = A(x_0) \frac{d}{dx_0} \left[\frac{1}{n} T_n(x_0; a,b) \right]$$

$$+ \frac{1}{2} D(x_0) \frac{d^2}{dx_0^2} \left[\frac{1}{n} T_n(x_0; a,b) \right].$$

Multiplying this equation through by n, and recalling that $T_0(x_0; a,b) = 1$, we conclude that the moments of the first exit time $T(x_0; a,b)$ satisfy the following set of coupled, second order, ordinary differential equations (a prime here denotes differentiation with respect to the variable x_0):

$$\begin{cases} A(x_0) T_1'(x_0; a,b) + \frac{1}{2} D(x_0) T_1''(x_0; a,b) = -1, & (3.7\text{-}12a) \\[2ex] A(x_0) T_n'(x_0; a,b) + \frac{1}{2} D(x_0) T_n''(x_0; a,b) = -n T_{n-1}(x_0; a,b) \\[1ex] \hphantom{A(x_0) T_n'(x_0; a,b) + \frac{1}{2} D(x_0)} (n \geq 2). & (3.7\text{-}12b) \end{cases}$$

The *boundary conditions* for these differential equations follow simply by substituting the appropriate boundary conditions (3.7-6) into Eq. (3.7-11b):

$$T_n(a; a,b) = 0 \ \& \ T_n(b; a,b) = 0 \quad (a \text{ absorbing}, b \text{ absorbing}), \quad (3.7\text{-}13a)$$

$$T_n'(a; a,b) = 0 \ \& \ T_n(b; a,b) = 0 \quad (a \text{ reflecting}, b \text{ absorbing}), \quad (3.7\text{-}13b)$$

$$T_n(a; a,b) = 0 \ \& \ T_n'(b; a,b) = 0 \quad (a \text{ absorbing}, b \text{ reflecting}). \quad (3.7\text{-}13c)$$

Notice that the coupling in Eqs. (3.7-12) is, at least in principle, very easily managed: We first solve Eq. (3.7-12a) for $T_1(x_0; a,b)$; then we can *successively* solve Eq. (3.7-12b) for $T_2(x_0; a,b)$, $T_3(x_0; a,b)$, etc.

In practice, one is often content with knowing only $T_1(x_0; a,b)$. For that, only Eq. (3.7-12a) is needed. Probably the simplest illustration of the use of that equation is for the special Wiener process $W(t)$, for which $A(x) = 0$ and $D(x) = 1$. Eq. (3.7-12a) takes for that case the simple form

$$T_1''(x_0; a,b) = -2.$$

If we want $T(x_0; a,b)$ to be the time for $W(t)$ to first exit *either end* of the interval (a,b), then we must regard both a and b as absorbing states, and so impose the boundary conditions (3.7-13a). It is easy to see that the solution of the above differential equation which satisfies conditions (3.7-13a) is the quadratic form

$$T_1(x_0; a,b) = (b-x_0)(x_0-a); \qquad (3.7\text{-}14)$$

therefore, this is the average time for the special Wiener process $W(t)$ to first exit the interval (a,b) starting from any interior point x_0.

For arbitrary characterizing functions $A(x)$ and $D(x)$, the formal solution to Eq. (3.7-12a) that satisfies the boundary conditions (3.7-13a) is rather complicated, and we shall not write it out here. However, the formal solutions for the boundary conditions (3.7-13b) and (3.7-13c) are comparatively simple, and we shall in fact make use of them in Subsection 3.7.B. As may be verified by explicit differentiation using the familiar rule (A-12), the solutions to Eq. (3.7-12a) that satisfy the boundary conditions (3.7-13b) and (3.7-13c) are, respectively,

$$T_1(x_0; a,b) = \int_{x_0}^{b} dx \exp(\phi(x)) \int_{a}^{x} dx' \frac{2}{D(x')} \exp(-\phi(x'))$$

$$(a \text{ refl.}, b \text{ abs.}), \qquad (3.7\text{-}15a)$$

$$T_1(x_0; a,b) = \int_{a}^{x_0} dx \exp(\phi(x)) \int_{x}^{b} dx' \frac{2}{D(x')} \exp(-\phi(x'))$$

$$(a \text{ abs.}, b \text{ refl.}), \qquad (3.7\text{-}15b)$$

where $\phi(x)$ is the potential function as defined through Eq. (3.5-8).

We have thus far *not* assumed that our temporally homogeneous process $X(t)$ is *stable*. However, if it *is*, so that the constant K defined in Eq. (3.5-5) exists as a *finite, nonzero* number, then we can introduce in the last two equations a factor K just after each factor 2, and a compensating factor K^{-1} just before each positive exponential function. Then, recalling the formula (3.5-3) for the stationary density function $P_s(x)$, we may rewrite these last two equations as

$$T_1(x_0; a,b) = \int_{x_0}^{b} dx \, \Psi(x) \int_{a}^{x} dx' P_s(x') \quad (a \text{ refl.}, b \text{ abs.}), \quad (3.7\text{-}16a)$$

$$T_1(x_0; a,b) = \int_{a}^{x_0} dx \, \Psi(x) \int_{x}^{b} dx' P_s(x') \quad (a \text{ abs.}, b \text{ refl.}). \quad (3.7\text{-}16b)$$

where we have now defined a new function $\Psi(x)$ by

$$\Psi(x) \equiv K^{-1} \exp(\phi(x)) = \frac{2}{D(x) P_s(x)}. \qquad (3.7\text{-}17)$$

The second equality here follows from the fundamental formula (3.5-3). For reasons that will become clear in the next subsection, the function $\Psi(x)$ is called the **barrier function**. Notice that the extremums of $\Psi(x)$ coincide exactly with those of $\phi(x)$, and hence are approximately inversely associated with the extremums of $P_s(x)$.

3.7.B APPLICATION: MEAN TRANSITION TIMES IN A BISTABLE SYSTEM

A particularly interesting application of Eqs. (3.7-16) is the calculation of the average times for *spontaneous transitions* between the stable states of a bistable system. The functions $P_s(x)$, $\phi(x)$ and $\Psi(x)$ for two such systems are shown in Figs. 3-5 and 3-6. We have defined the stable states x_1 and x_2 ($x_1 < x_2$) in those systems to be the relative maximums of $P_s(x)$. Between x_1 and x_2 there lies a single relative maximum x_b of $\phi(x)$ and $\Psi(x)$. We shall call x_b a **barrier state** of the system. For the two processes of Figs. 3-5 and 3-6, it happens that $x_b = 0$. Now, the first exit time $T(x_1; -\infty, x_2)$, with $-\infty$ a *reflecting* boundary point and x_2 an *absorbing* boundary point, can evidently be regarded as the **transition time** $T(x_1 \rightarrow x_2)$ from stable state x_1 to stable state x_2. Similarly, $T(x_2; x_1, \infty)$, with x_1 an *absorbing* boundary point and ∞ a *reflecting* boundary point, can evidently be regarded as the transition time $T(x_2 \rightarrow x_1)$ from stable state x_2 to stable state x_1. Therefore, invoking Eqs. (3.7-16), we see that the **mean stable state transition times** for a bistable system are given by the two formulas

$$T_1(x_1 \rightarrow x_2) = \int_{x_1}^{x_2} dx\, \Psi(x) \int_{-\infty}^{x} dx'\, P_s(x'), \qquad (3.7\text{-}18a)$$

$$T_1(x_2 \rightarrow x_1) = \int_{x_1}^{x_2} dx\, \Psi(x) \int_{x}^{\infty} dx'\, P_s(x'). \qquad (3.7\text{-}18b)$$

The above two formulas can be given a simple geometric interpretation if the x_1 and x_2 peaks in $P_s(x)$ happen to be *well separated*, as they are for instance in Figs. 3-5 and 3-6. In that circumstance, the x-integrands in Eqs. (3.7-18) contribute significantly only for x in the region of the x_b-peak in $\Psi(x)$, which region we shall denote by $\{x_b\}$. And for $x \in \{x_b\}$, the x'-integrand in Eq. (3.7-18a) contributes significantly only for $x' \in \{x_1\}$, the region of the x_1-peak in $P_s(x)$, while in Eq. (3.7-18b) the x'-integrand contributes significantly only for $x' \in \{x_2\}$, the region of the x_2-

peak in $P_s(x)$. Thus we conclude that, *for well separated peaks*, the exact formulas (3.7-18) for the mean transition time from stable state x_i to stable state x_f can be *approximated* by

$$T_1(x_i \to x_f) \approx \left\{ \int_{\{x_b\}} \Psi(x)\,dx \right\} \left\{ \int_{\{x_i\}} P_s(x)\,dx \right\}. \qquad (3.7\text{-}19)$$

In words, the average waiting time for a spontaneous transition from one stable state to another is approximately equal to {the area under the barrier state peak of the $\Psi(x)$ curve} *times* {the area under the *initial* stable state peak in the $P_s(x)$ curve}.

That the mean stable state transition time should be proportional to the area under the initial stable state peak in the $P_s(x)$ curve is very reasonable; because that area is equal to the fraction of the time that the process spends in the vicinity of that stable state, "waiting" for the chance occurrence of a transition to the other stable state. That the constant of proportionality is the area under the barrier state peak in the $\Psi(x)$ curve shows that the higher and wider that peak is, the longer it will take for *either* transition to occur. The x_b-peak in $\Psi(x)$ thus represents, in a very quantitative way, an "inhibiting barrier" to transitions between the stable states x_1 and x_2.

A reasonable measure of the **relative stability** of the two stable states x_1 and x_2 should be provided by the ratio $T_1(x_1 \to x_2)/T_1(x_2 \to x_1)$: the larger that ratio is compared to unity, the more stable x_1 is relative to x_2. From Eq. (3.7-19) we see that, for well separated stable states, the stability of x_1 relative to x_2 is given approximately by

$$\frac{T_1(x_1 \to x_2)}{T_1(x_2 \to x_1)} \approx \frac{\displaystyle\int_{\{x_1\}} P_s(x)\,dx}{\displaystyle\int_{\{x_2\}} P_s(x)\,dx}, \qquad (3.7\text{-}20)$$

i.e., by the ratio of the area under the x_1-peak in $P_s(x)$ to the area under the x_2-peak in $P_s(x)$. Thus, in both Figs. 3-5 and 3-6, x_2 is "more stable" than x_1.

The result (3.7-19) shows that the mean stable state transition times in a well separated bistable system can be estimated simply by estimating the areas under three peaks in the functions $P_s(x)$ and $\Psi(x)$. We could of course just estimate those peak areas graphically from plots such as those in Figs. 3-5 and 3-6. But we can also estimate those peak areas analytically by using the Gaussian approximation procedure detailed in Appendix D. There it is shown that if a function $f(x)$ has a smooth peak at $x = x_0$ with $f''(x_0) = -f(x_0)c$, then the area under that peak is given approximately by $(2\pi/c)^{1/2} f(x_0)$. For the x_i-peak in $P_s(x)$ we have from Eq. (3.5-13) that $c = |2a'(x_i)/D(x_i)|$; hence, the area under that peak is approximately

$$\int_{\{x_i\}} P_s(x)\,dx \approx \left| \frac{2\pi D(x_i)}{2a'(x_i)} \right|^{1/2} P_s(x_i)$$

$$= \left| \frac{\pi D(x_i)}{a'(x_i)} \right|^{1/2} \frac{2K}{D(x_i)} \exp(-\phi(x_i)),$$

where the last step has invoked Eq. (3.5-3). Simplifying algebraically, we get

$$\int_{\{x_i\}} P_s(x)\,dx \approx \left| \frac{\pi}{D(x_i)a'(x_i)} \right|^{1/2} 2K\exp(-\phi(x_i)). \qquad (3.7\text{-}21a)$$

For the x_b-peak in $\Psi(x)$, we first calculate from the definition (3.7-17) the two derivatives

$$\Psi'(x) = \Psi(x)\phi'(x)$$

and

$$\Psi''(x) = \Psi(x)\phi''(x) + \Psi'(x)\phi'(x).$$

Since x_b is a relative maximum of $\Psi(x)$, then $\Psi'(x_b) = 0$, so

$$\Psi''(x_b) = \Psi(x_b)\phi''(x_b) = -\Psi(x_b)\,2A'(x_b)/D(x_b).$$

Here, the last step follows by differentiating Eq. (3.5-8) and then using the fact that x_b is a relative maximum of $\phi(x)$. So in the Gaussian peak area formula for this case, we must take $c = 2A'(x_b)/D(x_b)$. This gives

$$\int_{\{x_b\}} \Psi(x)\,dx \approx \left(\frac{\pi D(x_b)}{A'(x_b)} \right)^{1/2} \Psi(x_b),$$

or, recalling again the definition (3.7-17),

$$\int_{\{x_b\}} \Psi(x)\,dx \approx \left(\frac{\pi D(x_b)}{A'(x_b)} \right)^{1/2} K^{-1}\exp(\phi(x_b)). \qquad (3.7\text{-}21b)$$

Now substituting the two peak area estimates (3.7-21) into Eq. (3.7-19), we conclude that the mean transition time from stable state x_i to stable state x_f is *approximately* given by

$$T_1(x_i \rightarrow x_f) \approx 2\pi \left| \frac{D(x_b)}{D(x_i)\, a'(x_i)\, A'(x_b)} \right|^{1/2} \exp(\phi(x_b) - \phi(x_i)). \quad (3.7\text{-}22)$$

Notice that $T_1(x_i \rightarrow x_f)$ depends very sensitively (i.e., exponentially) on the potential valley depth $\phi(x_b) - \phi(x_i)$ [see Figs. 3-5 and 3-6]. Notice also that formula (3.7-22) does *not* require a knowledge of the normalization constant K, whose computation via Eq. (3.5-5) is often rather tedious. It follows from Eq. (3.7-22) that the "stability" of stable state x_1 relative to stable state x_2 is given *approximately* by

$$\frac{T_1(x_1 \rightarrow x_2)}{T_1(x_2 \rightarrow x_1)} \approx \left| \frac{D(x_2)\, a'(x_2)}{D(x_1)\, a'(x_1)} \right|^{1/2} \exp(\phi(x_2) - \phi(x_1)). \quad (3.7\text{-}23)$$

We note that this stability ratio depends strongly on the difference in the depths of the two stable state valleys in the potential function ϕ, with the deeper valley usually corresponding to the more stable state.

If the diffusion function happens to be independent of x — i.e., if $D(x) \equiv D$ — then Eq. (3.7-22) simplifies considerably. In that case Eq. (3.5-10) implies that the function $a(x)$ will coincide with the function $A(x)$, and hence also that the stable states x_1 and x_2 will be down-going roots of $A(x)$. Equation (3.7-22) then becomes

$$T_1(x_i \rightarrow x_f) \approx 2\pi \left| \frac{1}{A'(x_i)\, A'(x_b)} \right|^{1/2} \exp(\phi(x_b) - \phi(x_i))$$

$$= 2\pi \left| \frac{1}{A'(x_i)\, A'(x_b)} \right|^{1/2} \exp\left(-\int_{x_i}^{x_b} \frac{2A(x)}{D}\, dx \right),$$

where the last step follows from the definition (3.5-4) of the potential function. Now, the barrier state x_b, being by definition a relative maximum of $\phi(x)$, can be seen from Eq. (3.5-8) to be the *up*-going root of $A(x)$ that (necessarily) lies between the two down-going roots x_1 and x_2. Therefore, the mean transition time from stable state x_i to stable state x_f has, for $D(x) \equiv D$, the approximate value

$$T_1(x_i \rightarrow x_f) \approx 2\pi \left| \frac{1}{A'(x_i)\, A'(x_b)} \right|^{1/2} \exp\left(\frac{2}{D} \left| \int_{x_i}^{x_b} A(x) dx \right| \right). \quad (3.7\text{-}24)$$

We see from this formula that the D-dependence of $T_1(x_i \rightarrow x_f)$ is wholly contained in a factor of the form $\exp(c/D)$, where c is equal to twice the area enclosed by the function $A(x)$ and the x-axis between the two adjacent roots x_i and x_b. It follows from this result that if the diffusion constant D is allowed to approach zero, then the mean stable state transition times will approach infinity very rapidly.

For the two bistable continuous Markov processes $X(t)$ considered in Figs. 3-5 and 3-6 [see Subsection 3.6.C], calculations were made of the mean stable state transition times, both *exactly* using Eqs (3.7-18), and *approximately* using Eq. (3.7-22). Each exact calculation entails numerically evaluating a two-dimensional definite integral, whereas each approximate calculation is essentially a formula substitution. The results of these calculations are displayed in the two tables below. It is

For the Bistable Process of Fig. 3-5

	Exact Eq. (3.7-18)	Approx. Eq. (3.7-22)	Approx. Error
$T_1(x_1 \to x_2)$	33.76	28.3	-16%
$T_1(x_2 \to x_1)$	92.15	79.2	-14%
$T_1(x_1 \to x_2)/T_1(x_2 \to x_1)$	0.3664	0.357	-2%

For the Bistable Process of Fig. 3-6

	Exact Eq. (3.7-18)	Approx. Eq. (3.7-22)	Approx. Error
$T_1(x_1 \to x_2)$	19.63	15.9	-19%
$T_1(x_2 \to x_1)$	38.29	31.6	-17%
$T_1(x_1 \to x_2)/T_1(x_2 \to x_1)$	0.5127	0.503	-2%

seen that the approximation formula (3.7-22) is in error for these processes by 15 to 20 percent, with the errors being slightly larger for the process of Fig. 3-6 than for the process of Fig. 3-5. The errors in Eq. (3.7-22) arise from two approximating assumptions: first, that the peaks in the functions $P_s(x)$ and $\Psi(x)$ are *nonoverlapping* [this allows us to pass from Eq. (3.7-18) to Eq. (3.7-19)]; and second, that those peaks are *Gaussian shaped* [this allows us to pass from Eq. (3.7-19) to Eq. (3.7-22)]. The slightly larger approximation errors in the process of Fig. 3-6 is probably due to the fact that the peaks there overlap more than do the peaks in Fig. 3-5. But in any case, we shall discover in Section 3.9 that even a 20% error in the estimate of the mean stable state transition time is usually not serious, because the *standard deviation* of a stable state transition time is typically about as large as the mean; hence, actual stable state transition times that differ from the mean by factors ranging anywhere from 1/2 to 2 will not be at all uncommon.

Finally, we should mention that the *reciprocal* of a mean transition time $T_1(x_i \to x_f)$ is sometimes regarded as a *mean transition rate*:

$$R(x_i \to x_f) \approx \frac{1}{2\pi} \left| \frac{D(x_i) \, a'(x_i) \, A'(x_b)}{D(x_b)} \right|^{1/2} \exp(-[\phi(x_b) - \phi(x_i)]). \quad (3.7\text{-}25)$$

If $\phi(x)$ has the form of some potential energy divided by $k_B T$, where k_B is Boltzmann's constant and T the system "temperature," then this rate formula has the general structure of many statistical mechanical rate formulas, such as for example the Arrhenius formula for a chemical reaction rate. However, caution must be exercised in using this interpretation, because a "rate" in the physical sciences is usually tacitly assumed to be the reciprocal of the mean of an *exponential* random variable, such as a radioactive decay rate. Although it is often possible to regard the random variable $T(x_0; a,b)$ as being *approximately* exponentially distributed, it is easy to prove that this cannot be true in an exact sense: Eq. (1.4-6) shows that for any exponential random variable the second moment must be exactly twice the square of the first moment; yet, it is easy to show that if $T_1(x_0; a,b)$ is a solution of Eq. (3.7-12a), then $2T_1{}^2(x_0; a,b)$ is *not* a solution of the $n=2$ version of Eq. (3.7-12b). Therefore, $T_2(x_0; a,b) \neq 2T_1{}^2(x_0; a,b)$, implying that $T(x_0; a,b)$ cannot be an exponential random variable — at least not *exactly*. Nevertheless, in many practical situations $T(x_0; a,b)$ will be sufficiently close to exponentially distributed that the rate formula (3.7-25) can be used with caution. We shall see an illustration of this fact at the end of Section 3.9.

3.8 WEAK NOISE PROCESSES

In Section 3.4 we found that the time evolution of a continuous Markov process $X(t)$ with characterizing functions $A(x,t)$ and $D(x,t)$ can be described by the Langevin equation (3.4-8):

$$X(t+dt) = X(t) + D^{1/2}(X(t),t) \, dW(dt) + A(X(t),t) \, dt. \quad (3.8\text{-}1)$$

Since $dW(dt)$ is the random variable $N(0,dt)$, then it is clear from the form of this equation that the size of the diffusion function $D(x,t)$ controls the amount of "stochasticity" in the time evolution of the process $X(t)$. In particular, if $D(x,t)$ were identically zero, then $X(t)$ would reduce to the deterministic (Liouville) process $x^*(t)$ defined by

$$x^*(t+dt) = x^*(t) + A(x^*(t),t)\,dt \qquad (3.8\text{-}2a)$$

or equivalently

$$\frac{d}{dt}x^*(t) = A(x^*(t),t), \qquad (3.8\text{-}2b)$$

together with the initial condition

$$x^*(t_0) = x_0. \qquad (3.8\text{-}3)$$

In this section we shall consider an *approximation* procedure that can be applied when the diffusion function $D(x,t)$ is "very small," so that the process $X(t)$ is a **weak noise process**. To quantify the smallness of the diffusion function, we shall assume that

$$D(x,t) = \varepsilon\, D_1(x,t), \ \text{ with } \ 0 < \varepsilon \ll 1, \qquad (3.8\text{-}4)$$

where the function $D_1(x,t)$ is roughly of the same order of magnitude as the function $A(x,t)$. In Subsection 3.8.A we shall develop some general approximating equations for this weak noise process. Then, in Subsection 3.8.B, we shall use those equations to analyze the thermodynamic limit of the intensive state variable $Z(t) \equiv X(t)/\Omega$ when the characterizing functions of $X(t)$ "scale" in a special way with the system volume Ω.

3.8.A THE WEAK NOISE APPROXIMATION

If $X(t)$ is a weak noise process, in the sense of Eq. (3.8-4), then the Langevin equation (3.8-1) evidently reads

$$X(t+dt) = X(t) + \varepsilon^{1/2}\, D_1^{1/2}(X(t),t)\,dW(dt) + A(X(t),t)\,dt. \quad (3.8\text{-}5)$$

Observing that the $\varepsilon > 0$ version of this Langevin equation differs from the $\varepsilon = 0$ version by a stochastic term that is proportional to $\varepsilon^{1/2}$, we are led to *conjecture* that the process $X(t)$ for $\varepsilon > 0$ differs from the $\varepsilon = 0$ process $x^*(t)$ by some stochastic term that is likewise proportional to $\varepsilon^{1/2}$. Motivated by this conjecture, we shall try to find a solution to Eq. (3.8-5) of the form

$$X(t) = x^*(t) + \varepsilon^{1/2}\, Y(t), \qquad (3.8\text{-}6)$$

where $Y(t)$ is some yet-to-be-determined stochastic process. Notice that the form of this solution is a superposition of a deterministic process $x^*(t)$

and a relatively weak stochastic process $\varepsilon^{1/2}Y(t)$. We shall prove that *if ε is sufficiently small* then Eq. (3.8-6) indeed provides a viable solution to Eq. (3.8-5), with $Y(t)$ a well defined, continuous Markov process that is *independent* of ε. We reason as follows.

By substituting Eq. (3.8-6) into Eq. (3.8-5), we obtain

$$x^*(t+dt) + \varepsilon^{1/2}Y(t+dt)$$

$$= x^*(t) + \varepsilon^{1/2}Y(t) + \varepsilon^{1/2}D_1^{1/2}(x^*(t)+\varepsilon^{1/2}Y(t),t)\,dW(dt)$$

$$+ A(x^*(t)+\varepsilon^{1/2}Y(t),t)\,dt.$$

Since $A(x,t)$ and $D^{1/2}(x,t)$ are assumed to be smooth functions of x [see the remarks following Eqs. (3.1-14)], then provided ε is sufficiently close to zero we can *approximate*

$$A(x^*+\varepsilon^{1/2}Y,t) \approx A(x^*,t) + \varepsilon^{1/2}Y\,A'(x^*,t), \tag{3.8-7a}$$

$$D_1^{1/2}(x^*+\varepsilon^{1/2}Y,t) \approx D_1^{1/2}(x^*,t) + \varepsilon^{1/2}Y\,[D_1^{1/2}(x^*,t)]', \tag{3.8-7b}$$

where in each case the prime denotes partial differentiation with respect to the state variable. Substituting these two approximations into the preceding equation, we get

$$x^*(t+dt) + \varepsilon^{1/2}Y(t+dt)$$

$$\approx x^*(t) + \varepsilon^{1/2}Y(t)$$

$$+ \varepsilon^{1/2}D_1^{1/2}(x^*(t),t)\,dW(dt) + \varepsilon\,Y(t)\,[D_1^{1/2}(x^*(t),t)]'\,dW(dt)$$

$$+ A(x^*(t),t)\,dt + \varepsilon^{1/2}Y(t)\,A'(x^*(t),t)\,dt.$$

The three terms in this equation that do not involve ε evidently cancel out because of Eq. (3.8-2a). All the remaining terms are proportional to $\varepsilon^{1/2}$ except for one, and it, being proportional to ε, can be neglected relative to the $\varepsilon^{1/2}$ terms. Thus we conclude that

$$Y(t+dt) = Y(t) + D_1^{1/2}(x^*(t),t)\,dW(dt) + Y(t)\,A'(x^*(t),t)\,dt. \tag{3.8-8}$$

Comparing Eq. (3.8-8) with the canonical form of the Langevin equation (3.8-1), we deduce that Eq. (3.8-8) is in fact a Langevin equation for the process $Y(t)$. As such, it identifies the process $Y(t)$ as a continuous Markov process with drift function $yA'(x^*(t),t)$ and diffusion function $D_1(x^*(t),t)$. Eq. (3.8-8) and the initial condition

$$Y(t_0) = 0, \tag{3.8-9}$$

which follows from Eqs. (3.8-3) and (3.8-6), completely define the continuous Markov process $Y(t)$.

Another way to specify $Y(t)$ is through its Markov state density function $R(y,t \mid 0,t_0)$, which is defined so that

$$R(y,t \mid 0,t_0)dy = \text{Prob}\{ Y(t) \in [y,y+dy), \text{ given that } Y(t_0)=0 \}. \quad (3.8\text{-}10)$$

Reading off the characterizing functions of $Y(t)$ from its Langevin equation (3.8-8), we may infer from the canonical form of the Fokker-Planck equation (3.2-2) that this function R satisfies the partial differential equation

$$\frac{\partial}{\partial t}R(y,t \mid 0,t_0) = -\frac{\partial}{\partial y}\left[y\, A'(x^*(t),t)\, R(y,t \mid 0,t_0) \right]$$

$$+ \frac{1}{2}\frac{\partial^2}{\partial y^2}\left[D_1(x^*(t),t)\, R(y,t \mid 0,t_0) \right],$$

or more simply

$$\frac{\partial}{\partial t}R(y,t \mid 0,t_0) = -A'(x^*(t),t)\frac{\partial}{\partial y}[y\, R(y,t \mid 0,t_0)]$$

$$+ \frac{1}{2}D_1(x^*(t),t)\frac{\partial^2}{\partial y^2}R(y,t \mid 0,t_0). \quad (3.8\text{-}11)$$

This last equation, along with the initial condition $R(y,t_0 \mid 0,t_0)=\delta(y)$, completely determines the Markov state density function of the continuous Markov process $Y(t)$.

We may summarize our findings thus far as follows: If $X(t)$ is a continuous Markov process with drift function $A(x,t)$ and diffusion function $D(x,t) \equiv \varepsilon D_1(x,t)$, and *if the positive parameter ε is sufficiently small*, then we can write $X(t) \approx x^*(t)+\varepsilon^{1/2}Y(t)$. Here, $x^*(t)$ is the sure function that solves the ordinary differential equation $dx^*/dt=A(x^*,t)$ for the initial condition $x^*(t_0)=X(t_0)=x_0$. And $Y(t)$ is the continuous Markov process with drift function $yA'(x^*(t),t)$, diffusion function $D_1(x^*(t),t)$, and initial condition $Y(t_0)=0$; thus, the Langevin equation for $Y(t)$ is given by Eq. (3.8-8), and the associated forward Fokker-Planck equation is given by Eq. (3.8-11). Evidently, the process $X(t)$ follows the deterministic solution $x^*(t)$, but with "additive noise" that goes to zero with ε like $\varepsilon^{1/2}$.

Suppose now that our weak noise process $X(t)$ is *temporally homogeneous*, so that

$$A(x,t) = A(x) \quad \text{and} \quad D(x,t) = \varepsilon D_1(x), \quad 0 < \varepsilon \ll 1. \qquad (3.8\text{-}12)$$

And suppose further that the function $x^*(t)$ defined by Eqs. (3.8-2) asymptotically approaches some stable, stationary value x_∞:

$$x^*(t) \to x_\infty \quad \text{as} \quad t \to \infty. \qquad (3.8\text{-}13)$$

In order for Eq. (3.8-13) to be true, it is necessary that the function $A(x)$ satisfy the following two conditions [compare Eqs. (3.5-9)]:

$$A(x_\infty) = 0 \quad \text{and} \quad A'(x_\infty) < 0. \qquad (3.8\text{-}14)$$

The first condition here is obviously necessary if x_∞ is to be a *stationary* solution of the differential equation (3.8-2b). To see the necessity of the second condition, suppose that x^* is displaced from x_∞ by an infinitesimal amount h. Then according to Eq. (3.8-2a) x^* will, in the next dt, change by the infinitesimal amount

$$dx^* = A(x_\infty + h)\, dt = [A(x_\infty) + h A'(x_\infty)]\, dt = h A'(x_\infty)\, dt.$$

If x_∞ is to be a *stable* stationary solution, then a positive (negative) displacement h must result in a negative (positive) change dx^*, so that x^* will move back toward x_∞ in the next dt; hence, $A'(x_\infty)$ must be negative, which establishes the second of conditions (3.8-14).

Now we know from our preceding analysis that conditions (3.8-12) allow us to write $X(t) \approx x^*(t) + \varepsilon^{1/2} Y(t)$, where $Y(t)$ is the continuous Markov process with drift function $yA'(x^*(t))$ and diffusion function $D_1(x^*(t))$. It then follows from the properties (3.8-13) and (3.8-14) that $Y(t \to \infty)$ will have drift function $yA'(x_\infty) = -|A'(x_\infty)|y$ and diffusion function $D_1(x_\infty)$; thus, $Y(t \to \infty)$ is an Ornstein-Uhlenbeck process [cf. Eqs. (3.3-19)] with $k = |A'(x_\infty)|$ and $D = D_1(x_\infty)$. And since the *long-time* behavior of an Ornstein-Uhlenbeck process is as described by Eq. (3.3-25b), then we may infer that

$$Y(t \to \infty) = \mathbf{N}(0, |D_1(x_\infty)/2A'(x_\infty)|). \qquad (3.8\text{-}15)$$

We should note that this result could also have been deduced by solving Eq. (3.8-11) in the limit $t \to \infty$. Finally, since

$$X(t \to \infty) = x^*(t \to \infty) + \varepsilon^{1/2} Y(t \to \infty) = x_\infty + \varepsilon^{1/2}\mathbf{N}(0, |D_1(x_\infty)/2A'(x_\infty)|),$$

then we may conclude from theorem (1.6-7) that

$$X(t \to \infty) = \mathbf{N}(x_\infty, \varepsilon |D_1(x_\infty)/2A'(x_\infty)|). \qquad (3.8\text{-}16)$$

In summary: If $X(t)$ is a continuous Markov process whose characterizing functions $A(x,t)$ and $D(x,t)$ satisfy conditions (3.8-12) and (3.8-14), then after a sufficiently long period of time $X(t)$ can be approximated as a *normal* random variable with mean x_∞ and standard deviation $|\varepsilon D_1(x_\infty)/2A'(x_\infty)|^{1/2}$.

But there is a caveat to this result. The point x_∞ will clearly be a *stable state* of the process $X(t)$ [compare Eqs. (3.8-14) and Eqs. (3.5-9)]. If x_∞ is the *only* stable state — i.e., if the limit (3.8-13) is realized for all possible initial states x_0 -- then the foregoing results hold unconditionally. However, if there are, depending on the value of x_0, *two* possible values x_1 and x_2 for x_∞, then both will be stable states of $X(t)$. We know from our analysis in Section 3.7.B that $X(t)$ will not hover forever about any one stable state, but will eventually make a transition to the other stable state. The average waiting time for such a transition will be large for ε small, and rapidly becomes infinite as $\varepsilon\to 0$.[†] So we have the caveat: *Equation (3.8-16) describes $X(t\to\infty)$ only when we take x_∞ to be the particular stable state about which $X(t)$ is currently hovering.*

The breakdown in our earlier analysis that gives rise to the foregoing caveat can be traced to the Taylor series approximations (3.8-7). Since $Y(t)$, like any non-Liouville continuous Markov process, will assume arbitrarily large values if we just wait long enough, then $\varepsilon^{1/2}Y(t)$ cannot be made arbitrarily small for *all* t simply by taking ε "small enough;" hence, the Taylor series approximations (3.8-7) will eventually break down, doing so in fact when a stable state transition occurs. But if our caveat is observed, then the foregoing results can be legitimately applied. We may note in particular that, for x_∞ a stable state of $X(t)$, then formulas (3.5-14) and (3.5-10) predict that its width should be

$$
\begin{aligned}
\sigma_\infty &= \left(\frac{\pi\varepsilon D_1(x_\infty)}{|A'(x_\infty) - \tfrac{1}{2}\varepsilon D_1''(x_\infty)|} \right)^{1/2} \\
&\approx \left(\frac{\pi\varepsilon D_1(x_\infty)}{|A'(x_\infty)|} \right)^{1/2} = \left(\frac{\pi}{2} \right)^{1/2} 2 \left(\frac{\varepsilon D_1(x_\infty)}{2|A'(x_\infty)|} \right)^{1/2}.
\end{aligned}
$$

The last expression shows that σ_∞ is slightly more than twice the standard deviation predicted by Eq. (3.8-16); thus, these two independent measures of the size of the asymptotic fluctuations in $X(t)$ about the stable state x_∞ are consistent.

---

† Indeed, when $D(x) = \varepsilon D_1(x)$, the mean stable state transition times will depend upon ε through a factor of the form $\exp(c/\varepsilon)$, where c is positive and independent of ε. This (approximate) result can easily be deduced from Eq. (3.7-22) in much the same way that we deduced the result (3.7-24) for $D(x) \equiv D$.

3.8.B APPLICATION: VOLUME SCALING

As an application of weak noise approximation theory, we consider now a continuous Markov process $X(t)$ whose characterizing functions $A(x,t)$ and $D(x,t)$ (the latter now *not* assumed to be small) satisfy the **scaling conditions**

$$\left. \begin{array}{l} A(\beta x,t) \doteq \beta\, A^*(x,t) \\[2mm] D(\beta x,t) \doteq \beta\, D^*(x,t) \end{array} \right\} \quad \beta \to \infty, \qquad (3.8\text{-}17)$$

where A^* and D^* are functions that are independent of β. The dots over the equal signs in Eqs. (3.8-17) denote asymptotic equality in the limit that β becomes infinitely large; note, however, that approximate equality is *not* assumed for β small.

Suppose we regard $X(t)$ as an "extensive" state variable. Then, denoting the system "volume" by Ω, the corresponding "intensive" state variable is given by

$$Z(t) \equiv X(t)/\Omega. \qquad (3.8\text{-}18)$$

Since $Z(t)$ is a simple scalar multiple of the continuous Markov process $X(t)$, then $Z(t)$ is itself a continuous Markov process. In fact, the Langevin equation for $Z(t)$ can be obtained simply by dividing the Langevin equation (3.8-1) for $X(t)$ through by Ω; in that way we obtain

$$Z(t+dt) = Z(t) + [\Omega^{-2}D(\Omega Z(t),t)]^{1/2}\,dW(dt) + \Omega^{-1}A(\Omega Z(t),t)\,dt.$$
$$(3.8\text{-}19)$$

This Langevin equation shows that the continuous Markov process $Z(t)$ has a drift function $\Omega^{-1}A(\Omega z,t)$ and a diffusion function $\Omega^{-2}D(\Omega z,t)$. Now suppose we pass to the "thermodynamic limit" of very large Ω. Then because the functions $A(x,t)$ and $D(x,t)$ obey the scaling conditions (3.8-17), we can make the replacements

$$A(\Omega Z(t),t) \doteq \Omega\, A^*(Z(t),t) \quad \text{and} \quad D(\Omega Z(t),t) \doteq \Omega\, D^*(Z(t),t),$$

where the dots over the equal signs now denote asymptotic equality for large Ω. Therefore, the Langevin equation (3.8-19) for $Z(t)$ becomes in this large-volume limit,

$$Z(t+dt) \doteq Z(t) + [\Omega^{-1}D^*(Z(t),t)]^{1/2}\,dW(dt) + A^*(Z(t),t)\,dt. \qquad (3.8\text{-}20)$$

Equation (3.8-20) tells us that, *for Ω sufficiently large*, the process $Z(t)$ has drift function $A^*(z,t)$ and diffusion function $\Omega^{-1}D^*(z,t)$; therefore, in the thermodynamic limit $Z(t)$ is a *weak noise* process of the kind analyzed in the preceding subsection, with Ω^{-1} playing the role of the smallness parameter ε. Indeed, we see that the large-Ω Langevin equation (3.8-20) is precisely the weak-noise Langevin equation (3.8-5) with the replacements

$$X(t) \rightarrow Z(t), \quad \varepsilon \rightarrow \Omega^{-1}, \quad D_1(x,t) \rightarrow D^*(z,t), \quad A(x,t) \rightarrow A^*(z,t).$$

So the weak-noise results of the preceding subsection allow us to draw the following conclusions:

If the characterizing functions $A(x,t)$ and $D(x,t)$ of the extensive continuous Markov process $X(t)$ satisfy the scaling conditions (3.8-17), and if the system volume Ω is sufficiently large, then the intensive state variable $Z(t) \equiv X(t)/\Omega$ can be approximately written as [see Eq. (3.8-6)]

$$Z(t) = z^*(t) + \Omega^{-1/2} V(t). \tag{3.8-21}$$

Here, $z^*(t)$ is the sure function defined by the differential equation [see Eq. (3.8-2b)]

$$\frac{dz^*}{dt} = A^*(z^*,t), \tag{3.8-22a}$$

and the initial condition

$$z^*(t_0) = z_0 \equiv x_0/\Omega. \tag{3.8-22b}$$

And $V(t)$ is the continuous Markov process with drift function $vA^{*'}(z^*(t),t)$ and diffusion function $D^*(z^*(t),t)$; i.e., $V(t)$ is defined by the Langevin equation [see Eq. (3.8-8)]

$$V(t+dt) = V(t) + D^{*1/2}(z^*(t),t)\,dW(dt) + V(t)\,A^{*'}(z^*(t),t)\,dt \tag{3.8-23}$$

with initial condition $V(t_0)=0$, or equivalently by the Fokker-Planck equation [see Eq. (3.8-11)]

$$\frac{\partial}{\partial t} R(v,t\,|\,0,t_0) = -A^{*'}(z^*(t),t)\,\frac{\partial}{\partial v}[v\,R(v,t\,|\,0,t_0)]$$

$$+ \frac{1}{2}D^*(z^*(t),t)\,\frac{\partial^2}{\partial v^2}R(v,t\,|\,0,t_0) \tag{3.8-24}$$

with initial condition $R(v,t_0\,|\,0,t_0)=\delta(v)$. Furthermore, if the characterizing functions of $X(t)$ are explicitly independent of t, and if the function $A^*(z)$ is such that $z^*(t)$, as defined by Eq. (3.8-22a), approaches a constant value z_∞ as $t \rightarrow \infty$, then we will have [see Eqs. (3.8-12) – (3.8-16)]

$$Z(t\rightarrow\infty) = N(z_\infty, \Omega^{-1}|D^*(z_\infty)/2A^{*'}(z_\infty)|). \tag{3.8-25}$$

That is, $Z(t{\to}\infty)$ will be *normally* distributed about the deterministic steady state z_∞ with standard deviation $|D^*(z_\infty)/2\Omega A^{*\prime}(z_\infty)|^{1/2}$, a quantity that evidently goes to zero as $\Omega{\to}\infty$ like $\Omega^{-1/2}$. Of course, as mentioned in the discussion following Eq. (3.8-16), all this is true only to the extent that transitions among any multiple steady states can be ignored — as in fact they generally can if Ω is sufficiently large.[†]

Fluctuations in the intensive state variables of many physical systems, such as the number of molecules per unit volume in a gas in thermal equilibrium, are often predicted by statistical mechanics to be inversely proportional to the square root of the system size. The foregoing analysis may offer a way of understanding in more detail how such a volume dependence arises. However, we must emphasize again that all our results here are contingent on the satisfaction by $A(x,t)$ and $D(x,t)$ of the special scaling conditions (3.8-17).

3.9 MONTE CARLO SIMULATION OF CONTINUOUS MARKOV PROCESSES

In Section 2.9 we gave the rationale for, and a broad description of, the Monte Carlo simulation method of analyzing Markov processes. We described there how various properties of a Markov process $X(t)$ and its time-integral $S(t)$ can be quantitatively inferred from a set of Monte Carlo samples or "realizations" $x_1(t)$, $x_2(t)$, ..., $x_N(t)$ of the process [see for example Eqs. (2.9-3) and (2.9-4)]; however, we did not describe there any definite procedure for actually constructing those realizations. In this section we shall detail an algorithm for generating a realization $x(t)$ of a continuous Markov process $X(t)$ with given characterizing functions $A(x,t)$ and $D(x,t)$. We shall then illustrate that algorithm by using it to simulate several of the processes discussed earlier in this chapter.

3.9.A CONTINUOUS SIMULATION THEORY

The heart of the simulation procedure for a continuous Markov process $X(t)$ is an approximate, "non-infinitesimal" version of the Langevin equation. Specifically, we take the first form (3.4-4) of the

---

[†] Recalling the footnote on page 189, and also that $\varepsilon = 1/\Omega$ here, we see that stable state transitions in $Z(t)$ will typically take place in times of order $\exp(c\Omega)$, where $c > 0$.

Langevin equation to imply that, if the process is in state x at time t, then its state at any slightly later time $t + \Delta t$ may be approximated by

$$x(t + \Delta t) \approx x(t) + n\,[D(x(t),t)\Delta t]^{1/2} + A(x(t),t)\Delta t, \qquad (3.9\text{-}1)$$

where n is a sample value of the unit normal random variable $N \equiv N(0,1)$. The approximate equality here of course becomes exact in the limit of vanishingly small Δt. Also, as discussed in connection with Eqs. (2.9-5), if the time integral $S(t)$ of the process $X(t)$ has the value $s(t)$ at time t, then its value at any slightly later time $t + \Delta t$ may be approximated by

$$s(t + \Delta t) \approx s(t) + x(t)\Delta t, \qquad (3.9\text{-}2)$$

this approximation likewise becoming exact in the limit of vanishingly small Δt. Therefore, given that the continuous Markov process $X(t)$ is in some state x_0 at some initial time t_0, we can generate a realization $x(t)$ of $X(t)$, and also a realization $s(t)$ of the integral process $S(t)$, by carrying out the iterative procedure outlined in Fig. 3-7. That procedure is self-explanatory except for two key points: How do we decide in Step 2° what constitutes a "suitably small" value for Δt? And how do we generate in Step 3° a sample value n of the unit normal random variable? Since the second question is much the simpler of the two, let's dispose of it first.

We shall assume that the simulation calculations are to be carried out on a digital computer that affords us ready access to a *unit uniform random number generator*. As discussed in connection with Eqs. (1.8-4), such a generator is essentially a function subprogram that produces a sample value r of the unit uniform random variable $U(0,1)$. With a reliable source of such random numbers r, the production of sample values n of the unit normal random variable $N(0,1)$ is simply a matter of applying the algorithm (1.8-16). That algorithm produces two independent sample values x_1 and x_2 of $N(m,a^2)$ from two independent sample values r_1 and r_2 of $U(0,1)$. [See the derivation of Eqs. (1.8-16) for an explanation of why we are constrained to generate normal random numbers two at a time.] Therefore, if we first compute from two independent samples r_1 and r_2 of $U(0,1)$ the two values

$$s = [2\ln(1/r_1)]^{1/2} \quad \text{and} \quad \theta = 2\pi r_2, \qquad (3.9\text{-}3a)$$

then

$$n_1 = s\cos\theta \quad \text{and} \quad n_2 = s\sin\theta \qquad (3.9\text{-}3b)$$

will be two *independent* sample values of $N(0,1)$. This generating method will be exact to whatever degree the unit uniform random number generator used to get r_1 and r_2 is exact.

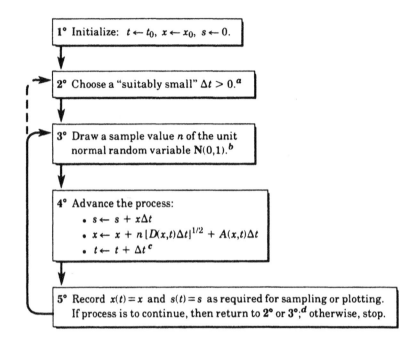

Notes:

[a] Should satisfy the *first order accuracy condition* (3.9-8) for some $\varepsilon_1 \ll 1$. Optionally, may satisfy the *plotting condition* (3.9-9) for some $\varepsilon_2 \lesssim 1$.

[b] Use the generating algorithm (3.9-3).

[c] If $(t_{max} - t_0)/\Delta t \sim 10^K$, then the sum $t + \Delta t$ should be computed with at least $K+3$ digits of precision.

[d] The value of Δt may be reset at the beginning of any cycle.

Figure 3-7. Monte Carlo simulation algorithm for the continuous Markov process with characterizing functions $A(x,t)$ and $D(x,t)$. The algorithm produces sample values $x(t)$ and $s(t)$ of the process $X(t)$ and its time-integral $S(t)$ at times $t = t_0 + \Delta t,\ t_0 + 2\Delta t,\ t_0 + 3\Delta t$, etc. The simulation algorithm is generally exact only in the limit $\Delta t \to 0$. The notation "$a \leftarrow b$" in Steps 1° and 4° means that the current value of variable a is to be replaced by the current value of the variable or expression b.

Although the procedure in Eqs. (3.9-3) for carrying out Step 3° in Fig. 3-7 is straightforward enough, the problem of quantifying the stipulation in Step 2°, that Δt be "suitably small," is decidedly not. To orient our thinking on this problem, let us begin by establishing the following result: *If both characterizing functions $A(x,t)$ and $D(x,t)$ are constants, then Eq. (3.9-1) is exact for any $\Delta t > 0$.* To prove this result, we first recall that if $A(x,t) = A$ and $D(x,t) = D$, then $X(t)$ is the Wiener process in Eq. (3.3-14). If in that equation we make the replacements $t_0 \to t$ and $t \to t + \Delta t$, so that $x_0 \to X(t)$ and $t - t_0 \to \Delta t$, then we get

$$X(t + \Delta t) = N(X(t) + A\Delta t, D\Delta t).$$

But from theorem (1.6-7) we know that if $N = N(0,1)$, then

$$(D\Delta t)^{1/2} N + [X(t) + A\Delta t] = N((D\Delta t)^{1/2} \cdot 0 + [X(t) + A\Delta t], [(D\Delta t)^{1/2}]^2)$$

$$= N(X(t) + A\Delta t, D\Delta t).$$

Comparing this result with the preceding equation, we conclude that when the characterizing functions are both constants then we have

$$X(t + \Delta t) = X(t) + N[D\Delta t]^{1/2} + A\Delta t \qquad (3.9\text{-}4a)$$

for *any* $\Delta t > 0$. In other words, if the state of the process is $x(t)$ at time t, then its state at any later time $t + \Delta t$ can be calculated *exactly* as

$$x(t + \Delta t) = x(t) + n[D\Delta t]^{1/2} + A\Delta t, \qquad (3.9\text{-}4b)$$

where n is a sample of the unit normal random variable. This proves our assertion, because Eq. (3.9-4b) is precisely Eq. (3.9-1) when the characterizing functions are constants.

The fact that Eq. (3.9-1) is exact when the functions $A(x,t)$ and $D(x,t)$ are constants should come as no surprise if we recall the details of our derivation of the form of the continuous propagator in Section 3.1. In that derivation, the only approximation we made was in writing Eq. (3.1-4) as Eq. (3.1-6). That approximation entails replacing the values $X(t')$ and t' in the random variable $\Xi(dt/n; X(t'),t')$, for any $t' \in [t, t+dt]$, by $X(t)$ and t respectively. Clearly, if $\Xi(dt; x,t)$ is independent of x and t, as Eq. (3.1-10) says it will be when the characterizing functions are constants, then those replacements can be made without incurring any errors at all. The salient implication of these considerations is that the approximate character of Eq. (3.9-1), and hence of our simulation algorithm for $X(t)$ in Fig. 3-7, is due solely to variations in the characterizing functions $A(x,t)$ and $D(x,t)$ with x and t.

If $D(x,t)$ were *identically zero*, then there would be a variety of ways to cope with the x- and t-variations in $A(x,t)$. The Langevin equation for the process in that case would read

$$X(t+dt) = X(t) + A(X(t),t)dt, \tag{3.9-5a}$$

which is equivalent to the ordinary differential equation

$$dX/dt = A(X,t). \tag{3.9-5b}$$

And the simulation algorithm (3.9-1) would read

$$x(t+\Delta t) \approx x(t) + A(x(t),t)\Delta t, \tag{3.9-6}$$

which is none other than the *simple Euler method* for integrating Eq. (3.9-5b). It is well known that this integration formula is exact when $A(x,t)$ is a constant, and our finding in Eqs. (3.9-4) has essentially generalized that result to the case in which $D(x,t)$ is a *nonzero* constant. If $A(x,t)$ is not a constant, then the simple Euler integration method (3.9-6) is not exact; moreover, it is generally not favored by numerical analysts, because, for a given time step size Δt, there are other more sophisticated algorithms (such as the Runge-Kutta and Adams-Bashforth algorithms) that will give much more accurate results for only a slight increase in computation time. It is natural to ask if any of these "super-Euler" methods can be adapted to the case in which $D(x,t)$ is not identically zero. Present indications are not encouraging. In Appendix E we sketch an attempt to improve Eq. (3.9-1) along the lines of a certain second-order Runge-Kutta procedure called the *modified Euler method*, and we find that *no improvement seems to be possible if $D(x,t)$ is not a constant*. However, we also find there that, if $D(x,t) \equiv D$ and $A(x,t) \equiv A(x)$, then Eq. (3.9-1), which now reads

$$x(t+\Delta t) \approx x(t) + n\,D^{1/2}\,(\Delta t)^{1/2} + A(x(t))\,\Delta t, \tag{3.9-7a}$$

can *probably* be improved to

$$x(t+\Delta t) \approx x(t) + [n\,D^{1/2}\,(\Delta t)^{1/2} + A(x(t))\,\Delta t]\,[1 + \tfrac{1}{2}A'(x(t))\,\Delta t], \tag{3.9-7b}$$

where $A' \equiv dA/dx$. Notice that Eq. (3.9-7b) properly reduces to Eq. (3.9-7a) if either $A(x)$ is independent of x or Δt is infinitesimally small.

The main obstacle to generalizing any of the super-Euler methods to the non-Liouville case is that most of those methods are predicated on the *assumption* that the desired solution $X(t)$ has a well-defined derivative with respect to t. That is certainly a legitimate assumption in the Liouville case (3.9-5), but not otherwise [see Section 2.6]. To improve upon Eq. (3.9-1), one must find a method that: (i) is consistent with Eq.

(3.9-1) in the limit $\Delta t \to dt$; (ii) is provably more accurate than Eq. (3.9-1) for a fixed finite Δt; and (iii) is faster to implement than simply using Eq. (3.9-1) with a suitably reduced value of Δt. The analysis in Appendix E shows how one *seemingly* plausible procedure for improving Eq. (3.9-1) when $D(x,t)$ is not a constant actually produces a formula that violates condition (i), and hence would surely give incorrect simulation results.

But for *any* finite-Δt formula, we must eventually face the following question: How large can Δt be without producing unacceptably large errors? Unfortunately, we cannot give an ironclad set of rules for answering this question in general; however, we can deduce some rough guidelines for using the approximate formula (3.9-1), upon which the simulation procedure of Fig. 3-7 is based. We reason as follows:

In our analysis in Section 3.1 we derived, as the second term on the right hand side of Eq. (3.1-14a), the lowest-order-in-dt violation of the Chapman-Kolmogorov condition (2.4-9) by the continuous Markov propagator $\Xi(dt; x,t)$. That violation term is of course negligibly small when dt is infinitesimally small. But if dt is replaced by a finite Δt, then *we should at least require that the violation term be very small in comparison with the main term.* An examination of Eq. (3.1-14a) shows that this can be ensured by picking some small number ε_1 $(0 < \varepsilon_1 \ll 1)$ and requiring that

$$\frac{(\Delta t)^{1/2} D^{1/2} \{ |\partial_x D^{1/2}| (\Delta t)^{1/2} + |\partial_x A| (\Delta t) \}}{D^{1/2} (\Delta t)^{1/2} + |A| (\Delta t)} \le 2\varepsilon_1 .$$

Since the *overall* factor multiplying the quantity in braces here evidently does not exceed unity, then we can conservatively simplify this requirement to read

$$|\partial_x D^{1/2}| (\Delta t)^{1/2} + |\partial_x A| (\Delta t) \le 2\varepsilon_1 .$$

This inequality can in turn be most easily guaranteed by simply requiring each term on the left to be less than ε_1. Thus we arrive at what we shall call the **first order accuracy condition**:

$$\Delta t \le \mathrm{Min} \left\{ \frac{\varepsilon_1}{|\partial_x A(x,t)|} , \left(\frac{\varepsilon_1}{\partial_x D^{1/2}(x,t)} \right)^2 \right\} . \qquad (3.9-8)$$

Condition (3.9-8) is "first order" in the sense that it controls the *lowest* order Δt term by which the finite Markov propagator $X(t+\Delta t) - X(t)$ violates the Chapman-Kolmogorov condition. Higher order violation terms, which will involve the derivatives $\partial_t A$, $\partial_t D$, $\partial_x^2 A$,

$\partial_x^2 D$, etc., might have to be reckoned with if the lowest order term should happen to vanish; however, we shall not try to develop any higher order accuracy conditions on Δt here. Notice that the first order accuracy condition properly allows Δt to be arbitrarily large if both characterizing functions are constants.

There is an additional consideration in choosing Δt that addresses not the accuracy of the simulation but rather the appearance of the plotted results. Usually, we shall want to simulate the process $X(t)$ on a fine enough time scale that the stochastic fluctuations in the process will be discernible over the deterministic drift of the process. For that, we should require that the magnitude $|A|\Delta t$ of the drift term in Eq. (3.9-1) be smaller than the typical magnitude $(D\Delta t)^{1/2}$ of the diffusion term. This can be ensured by picking a second control parameter ε_2 satisfying $0 < \varepsilon_2 \leq 1$ and then requiring that

$$\frac{|A|\Delta t}{(D\,\Delta t)^{1/2}} \leq \varepsilon_2.$$

Solving for Δt we obtain what we shall call the **plotting condition**:

$$\Delta t \leq \frac{\varepsilon_2^2\, D(x,t)}{A^2(x,t)}. \qquad \text{(optional)} \qquad (3.9\text{-}9)$$

If $D(x,t)$ is everywhere so small compared to $A^2(x,t)$ that this condition results in unreasonably small values for Δt, then we should probably conclude that the stochastic component of the process is so weak that we may as well simulate $X(t)$ as if it were the deterministic process of Eqs. (3.9-5). But if displaying the stochasticity of $X(t)$ is important, then condition (3.9-9) seems a reasonable requirement.

In summary, we should *always* choose Δt so as to satisfy the *first order accuracy condition* (3.9-8), using for ε_1 a value that is small compared to 1. Furthermore, if we want our simulation plots to clearly show the stochasticity of the process, then we should also require Δt to satisfy the *plotting condition* (3.9-9) for some $\varepsilon_2 \leq 1$. Notice that the smallness of ε_2 is much less critical than the smallness of ε_1.

3.9.B EXAMPLES OF CONTINUOUS SIMULATIONS

We shall now use the algorithm detailed in Fig. 3-7 to simulate some of the processes that we have analyzed earlier in this chapter. We shall find that such simulations can help us to see through much of the abstractness associated with continuous Markov processes by giving us some feeling for how these processes really evolve with time.

In Fig. 3-8a we show a simulation of the special Wiener process $X(t) = W(t)$, for which $A(x,t) = 0$ and $D(x,t) = 1$. The chosen initial condition for this simulation was $X(0) = 0$. The dots comprising the jagged curve are instantaneous samplings of the process at successive time increments of 0.0025, which was in fact the value of Δt used for this simulation. Since both characterizing functions here are constants, then the finiteness of Δt introduces no errors into the simulation; we could just as well have taken $\Delta t = 2.5$ if that courser sampling had suited our purposes. The fact that *any* positive value of Δt will produce a valid simulation for this process makes it very clear that *we should not attempt to connect successive sample points with any kind of smooth curve.* Such connections would imply that $X(t)$ is differentiable in the regions of interpolation, and we know that that is not the case for this or any other genuinely stochastic continuous Markov process. The two dashed curves in Fig. 3-8a demarcate the *one-standard deviation envelope* for the process, which is the interval

$$\langle X(t) \rangle \pm \mathrm{sdev}\{X(t)\}.$$

This envelope was calculated from Eqs. (3.3-10) by simply putting $t_0 = 0$, $x_0 = 0$, $A = 0$ and $D = 1$. Notice that the envelope widens with increasing t like $t^{1/2}$.

In Fig. 3-8b we show the time-integral $s(t)$ of the realization $x(t)$ in Fig. 3-8a. The dashed curves here demarcate the one-standard deviation envelope for the integral process $S(t)$, namely

$$\langle S(t) \rangle \pm \mathrm{sdev}\{S(t)\},$$

which was calculated from Eqs. (3.3-11) by again putting $t_0 = 0$, $x_0 = 0$, $A = 0$ and $D = 1$. The one-standard deviation envelope for $S(t)$ evidently widens with increasing t like $t^{3/2}$.

In Fig. 3-9 we show another simulation of the special Wiener process, this simulation differing from the one in Fig. 3-8 only in the choice of the "seed number" used for the unit uniform random number generator [which generates values for r_1 and r_2 in Eqs. (3.9-3a)]. This run demonstrates that the one-standard deviation envelopes are not hard boundaries for their processes; indeed, since the Markov state density function $P(x,t \mid x_0,t_0)$ for any Wiener process is a *normal* density function [see Eq. (3.3-13)], then at any instant roughly one third of any set of simulations of a Wiener process should lie outside the one-standard deviation envelope.

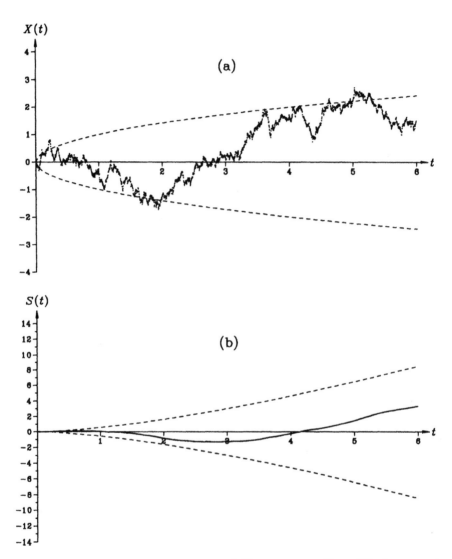

Figure 3-8. A simulation of the *special Wiener process* $X(t) = W(t)$ in (a), and its time-integral $S(t)$ in (b). The characterizing functions for the process are $A(x,t) = 0$ and $D(x,t) = 1$, and the initial condition is $X(0) = 0$. The dots comprising the jagged curves represent instantaneous samplings of the respective processes at intervals of $\Delta t = 0.0025$ time units. The dashed curves outline the respective theoretical one-standard deviation envelopes.

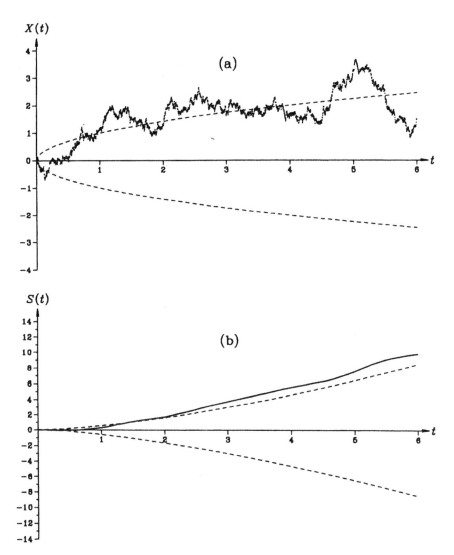

Figure 3-9. Another simulation of the special Wiener process and its integral, this one differing from the simulation in Fig. 3-8 only in the choice of the "seed number" for the unit uniform random number generator. The $X(t)$ trajectories in this and the preceding figure approximately describe the phenomenon of *self-diffusion*, as will be shown later in Subsection 4.4.C, but not, as is sometimes claimed, the phenomenon of Brownian motion.

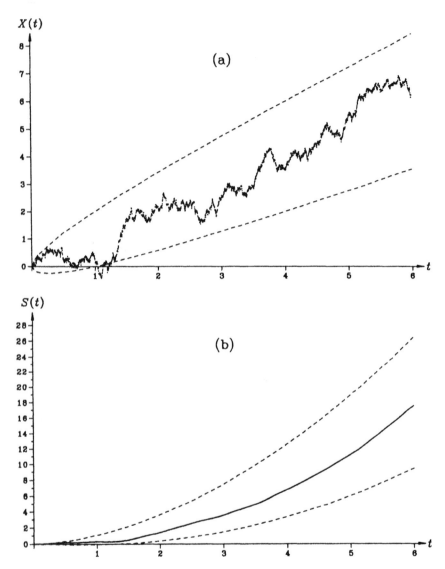

Figure 3-10. A simulation of the *Wiener process* $X(t)$, with characterizing functions $A(x,t) = 1$ and $D(x,t) = 1$, and its time-integral $S(t)$. Note the "drift" in the process induced by the nonzero drift function. We have used the same initial conditions, time-sampling interval and plot scale parameters as in the two preceding figures. The dashed curves again outline the theoretical one-standard deviation envelopes for $X(t)$ and $S(t)$.

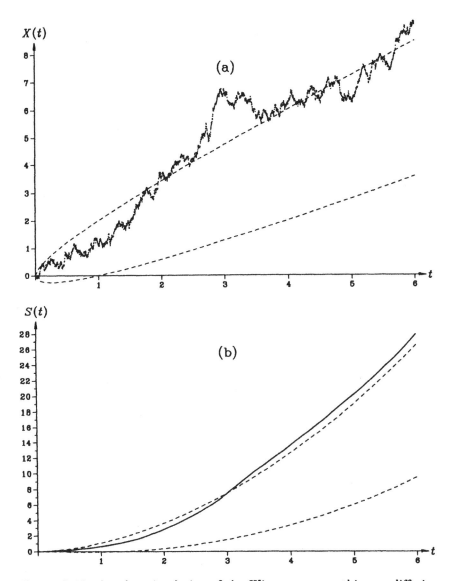

Figure 3-11. Another simulation of the Wiener process, this one differing from the simulation in Fig. 3-10 only in the seed number used for the unit uniform random number generator. Compare the $X(t)$ trajectories in Figs. 3-10a and 3-11a with the plot in Fig. 3-2 of the evolving Markov state density function $P(x,t \mid x_0,t_0)$ for a general Wiener process.

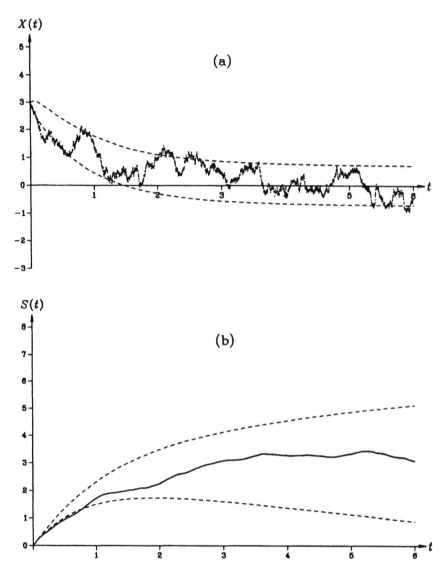

Figure 3-12. A simulation of the *Ornstein-Uhlenbeck process* with characterizing functions $A(x,t) = -x$ and $D(x,t) = 1$, and initial condition $X(0) = 3$. The process $X(t)$ is simulated in (a), and its time-integral $S(t)$ is simulated in (b). The time-step for the simulation here was $\Delta t = 0.0025$, the same as in the preceding simulations. The dashed curves are the respective one-standard deviation envelopes.

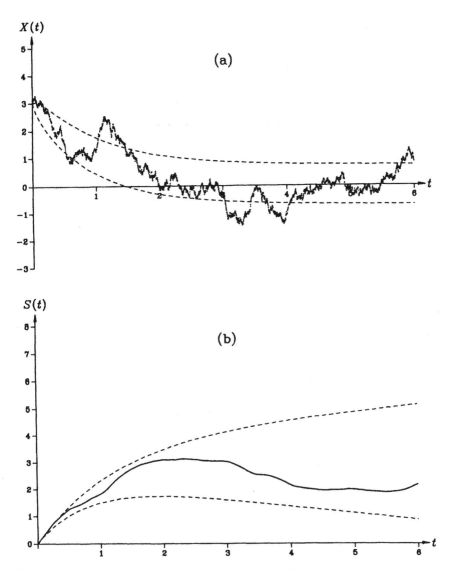

Figure 3-13. Another simulation of the Ornstein-Uhlenbeck process, this one differing from the simulation in Fig. 3-12 only in the seed number used in the random number generator. Compare the $X(t)$ trajectories in Figs. 3-12a and 3-13a with the plot of $P(x,t \mid x_0,t_0)$ in Fig. 3-3. The $S(t)$ trajectories in Figs. 3-12b and 3-13b approximately describe the phenomenon of *Brownian motion*, as shown in Subsection 3.4.C and later in Section 4.5.

It is sometimes suggested that the special Wiener process simulated in Figs. 3-8a and 3-9a mimics *Brownian motion*, the trajectory of a macroscopically small but microscopically large particle that moves about erratically in a suspending fluid because of thermal impacts from the fluid molecules. But in fact, apart from the $t^{1/2}$-spreading of its one-standard deviation envelope, the Wiener process does *not* accurately describe Brownian motion; as we found in Subsection 3.4.C, and will later confirm in Section 4.5, a much better model of Brownian motion is provided by the time-integral of the Ornstein-Uhlenbeck process. However, we shall see later in Subsection 4.4.C that the special Wiener process does gives a reasonably satisfactory description of *self-diffusion*, the trajectory of a molecule moving about in a thermally equilibrized fluid of like molecules.

In Figs. 3-10 and 3-11 we show two separate simulations of a Wiener process with D again equal to 1 but with A now equal to 1 instead of 0. The scales of the graphs here, and the size of the sampling interval Δt, are the same as for the preceding graphs. Since the characterizing functions are still constants, these simulations of $X(t)$ are still exact for any value of Δt. However, we may now apply to Δt the plotting criterion (3.9-9), which restricts Δt so that the stochastic fluctuations in $X(t)$ will be discernible over the deterministic drift. Our choice of $\Delta t = 0.0025$ in this case evidently corresponds to taking ε_2 in Eq. (3.9-9) equal to 0.05. This implies that the deterministic drift contribution to $x(t + \Delta t) - x(t)$ will typically be about 5% of the random fluctuation contribution. As may be seen from Eqs. (3.3-10), the positive value of the constant A in this case causes the center of the one-standard deviation envelope for $X(t)$ to rise linearly with t; however, the width of the envelope grows only like the square root of t. This implies that, for t sufficiently large, the width of the envelope will be very small compared to the location of the center of the envelope, and the process $X(t)$ will appear to be deterministic on the scale of its mean [recall the discussion of Eq. (2.7-31)]. The one standard deviation envelope for $X(t)$ should be visualized in the context of the plot in Fig. 3-2 of the Markov state density function of a general Wiener process. The one-standard deviation envelope for the integral process $S(t)$, again calculated from Eqs. (3.3-11), rises like t^2 and widens like $t^{3/2}$.

In Fig. 3-12a we show a simulation of the Ornstein-Uhlenbeck process with characterizing functions $A(x,t) = -x$ and $D(x,t) = 1$, and with the initial condition $X(0) = 3$. The dashed curves again demarcate the one-standard deviation envelope, this time calculated from Eqs. (3.3-20) with $t_0 = 0$, $x_0 = 3$, $k = 1$ and $D = 1$. In selecting Δt for this plot, we first noted that the first order accuracy condition (3.9-8) reads here

$$\Delta t \leq \text{Min} \left\{ \frac{\varepsilon_1}{1}, \frac{\varepsilon_1^2}{0} \right\} = \varepsilon_1.$$

The choice $\Delta t = 0.0025$ used in the preceding simulations should thus be quite acceptable from the viewpoint of the first order accuracy condition. The optional plotting condition (3.9-9) reads

$$\Delta t \leq \varepsilon_2^2 \cdot 1/x^2 = (\varepsilon_2/x)^2.$$

The value $\Delta t = 0.0025$ used in the preceding simulations would correspond to $\varepsilon_2 \leq 0.05$ for $|x| \leq 1$, and to $\varepsilon_2 \leq 0.15$ for $|x| \leq 3$; therefore, the choice $\Delta t = 0.0025$ is quite acceptable in the present case too. This simulation of $X(t)$ should be compared to the plot of the Markov state density function for the Ornstein-Uhlenbeck process shown in Fig. 3-3. A classic example of a *stable* continuous Markov process [see Eqs. (3.3-25)], $X(t)$ ultimately fluctuates about the stable state $x = 0$ with a standard deviation of $(D/2k)^{1/2} = (1/2)^{1/2} \approx 0.707$.

In Fig. 3-12b we show the time-integral of the above realization of $X(t)$. The dashed curves here represent the one-standard deviation envelope for $S(t)$ as calculated from Eq. (3.3-21) with $t_0 = 0$, $x_0 = 3$, $k = 1$ and $D = 1$. Those equations imply that the center of the envelope for $S(t)$ asymptotically approaches the constant value 3, while the envelope width asymptotically grows like $t^{1/2}$. As we found in Subsection 3.4.C [and will later elaborate on in Section 4.5], $S(t)$ here approximately describes the trajectory of a particle undergoing Brownian motion, with $X(t)$ describing the particle's velocity.

In Fig. 3-13 we show the results of another simulation of the Ornstein-Uhlenbeck process, this one differing from the preceding simulation only in the seed number used for the unit interval uniform random number generator.

In Fig. 3-14 we show two simulations of the quadratic-noise Ornstein-Uhlenbeck process with $A(x,t) = -x$ and $D(x,t) = 1 + x^2$. The initial condition and plotting frequency for these two runs are identical to those for the "simple" Ornstein-Uhlenbeck simulation runs in the preceding two figures; however, the time-step size Δt was chosen to be smaller. Since in this case

$$\partial_x A(x,t) = -1 \quad \text{and} \quad \partial_x D^{1/2}(x,t) = x(1+x^2)^{-1/2},$$

then the first order accuracy condition (3.9-8) reads

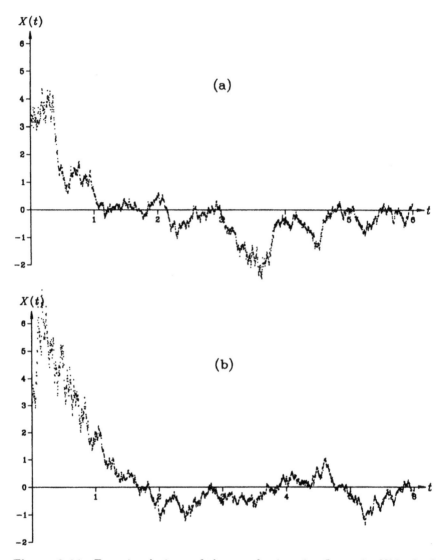

Figure 3-14. Two simulations of the *quadratic-noise Ornstein-Uhlenbeck process* with characterizing functions $A(x,t) = -x$ and $D(x,t) = 1 + x^2$, for the initial condition $X(0) = 3$. These graphs have the same scale and plotting frequency as Figs. 3-12a and 3-13a, but a smaller value of Δt was used to maintain simulation accuracy. The function $P_s(x)$ for this process is the solid *heavy* curve in Fig. 3-4; the solid *light* curve in Fig. 3-4 is the function $P_s(x)$ for the regular Ornstein-Uhlenbeck process simulated in Figs. 3-12a and 3-13a.

$$\Delta t \leq \text{Min}\left\{\frac{\varepsilon_1}{1}, \frac{\varepsilon_1^2(1+x^2)}{x^2}\right\} \leq \varepsilon_1^2,$$

where the last inequality is valid for all x. Since we don't wish to pick ε_1 larger than 0.01, we shall take $\Delta t = 0.0001$. However, to maintain for comparison purposes the same plotting frequency as used in the previous simulation graphs, we have plotted only every twenty-fifth time step. The resulting plotting interval of 0.0025 corresponds to an ε_2-value in the plotting condition (3.9-9) of

$$\varepsilon_2 = \frac{(0.0025)^{1/2}|A(x,t)|}{D^{1/2}(x,t)} = \frac{0.05\,x}{(1+x^2)^{1/2}} \leq 0.05,$$

which is quite acceptable. Notice the strong fluctuations in $X(t)$ when the process is far from zero; that of course is a consequence of the fact that the diffusion function $D(x,t)$ increases rapidly with $|x|$. Like the regular Ornstein-Uhlenbeck process in Figs 3-12a and 3-13a, the quadratic-noise Ornstein-Uhlenbeck process has $x = 0$ as a stable state. But, as we can see from Fig. 3-4, the stationary Markov state density functions of these two processes are not the same, with that for the quadratic-noise process being heavier-tailed and more sharply peaked. Despite having a larger asymptotic standard deviation than the regular process, the quadratic-noise process actually spends more time in the *immediate* vicinity of $x = 0$, a phenomenon that we have called "noise-enhanced stability." [See the discussion of Eqs. (3.6-10) and (3.6-11).]

In Fig. 3-15 we show four separate simulation runs of the continuous bistable Markov process defined by the characterizing functions in Eqs. (3.6-18a), namely

$$A(x) = -x\,(x+1.5)\,(x-1.7), \quad D(x) = 1.0.$$

We showed in Subsection 3.6.C [see Fig. 3-5] that this process has two stable states $x_1 = -1.5$ and $x_2 = 1.7$, with respective widths $\sigma_1 = 0.8090$ and $\sigma_2 = 0.7599$, these two stable states being separated by a barrier state $x_b = 0$. All four simulations shown in Fig. 3-15 used for the time-step $\Delta t = 0.00025$, but to keep the same plotting frequency as in the earlier graphs we have plotted only every tenth step. Our choice of Δt was guided by the first order accuracy condition (3.9-8), which in this case reads, since $D(x)$ is independent of x,

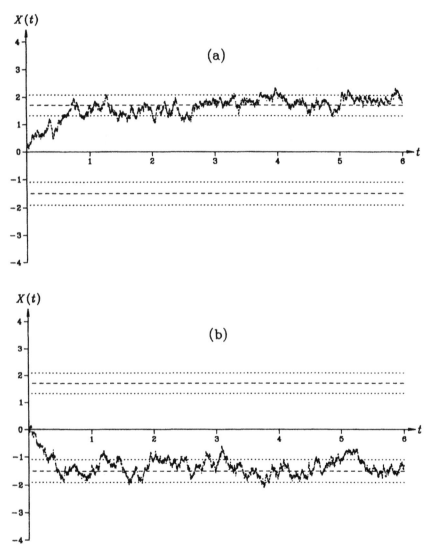

Figure 3-15. Showing the process trajectories of four separate simulation runs of the continuous bistable Markov process of Fig. 3-5, for which $A(x,t) = -x(x+1.5)(x-1.7)$ and $D(x,t) = 1.0$. Each of these four runs had $X(0) = 0$ and $\Delta t = 0.00025$, with only every tenth step point plotted in order to keep the same plotting frequency as in the preceding graphs. The dotted

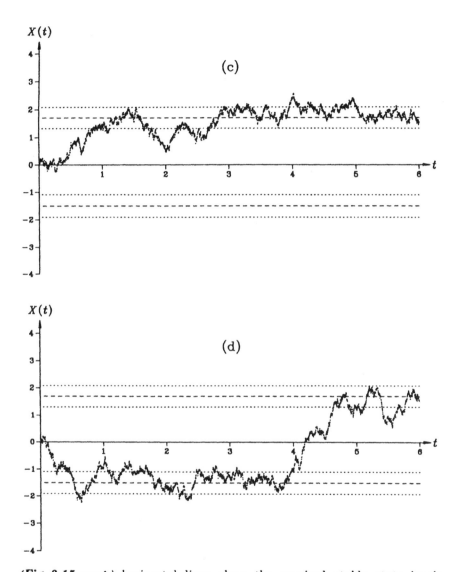

(Fig. 3-15, cont.) horizontal lines show the nominal stable state bands $[x_i - \sigma_i/2, \, x_i + \sigma_i/2]$, where x_i locates the maximum of a peak in the stationary Markov state density function $P_s(x)$, and σ_i is the Gaussian effective width of that peak. Stable state transitions are relatively rare on the time scale of these plots, but a spontaneous transition from x_1 to x_2 evidently occurs in (d).

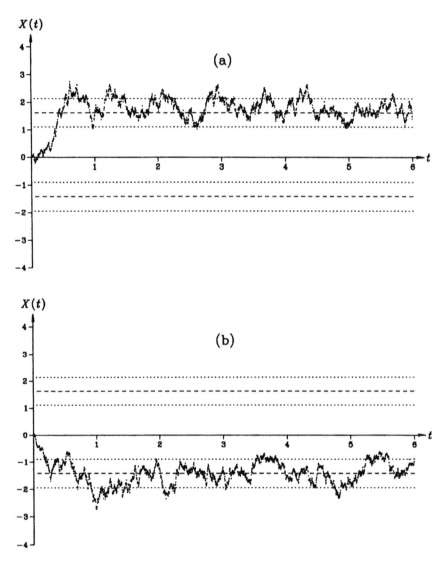

Figure 3-16. Showing the process trajectories of four separate simulation runs of the continuous bistable Markov process of Fig. 3-6, for which $A(x,t) = -x(x+1.5)(x-1.7)$ and $D(x,t) = 1.0 + 0.25x^2$. Each of these four runs was started in the barrier state $x_b = 0$, and used the same Δt and plotting frequency as did the four simulation runs in Fig. 3-15. Compared to the

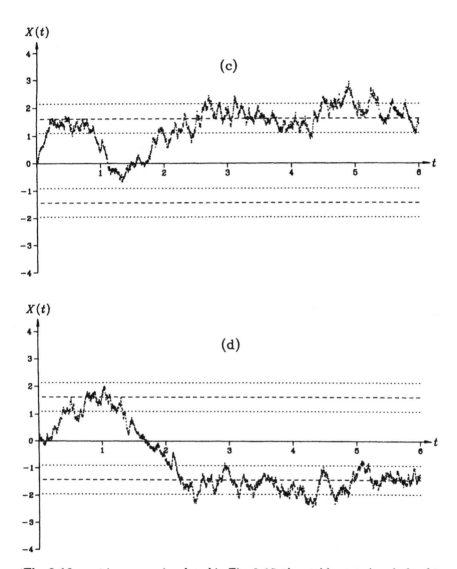

(Fig. 3-16, cont.) process simulated in Fig. 3-15, the stable state bands for this process are evidently broader and slightly shifted toward the origin [compare Figs. 3-5 and 3-6], these effects being a result of the quadratic term in the diffusion function $D(x,t)$. A spontaneous transition from stable state x_2 to stable state x_1 almost occurs in (c), and actually does occur in (d).

$$\Delta t \leq \text{Min}\left\{\frac{\varepsilon_1}{|\partial_x A(x,t)|}\right\} = \text{Min}\left\{\frac{\varepsilon_1}{|3x^2 - 0.4x - 2.55|}\right\}.$$

For $x\in[-3, 3]$ the denominator on the right has a maximum value of 25.65; thus, the choice $\Delta t = 0.00025$ means that ε_1 can be assigned the value 0.0064, which seems more prudent than the value 0.064 that would obtain if we took $\Delta t = 0.0025$.

All four runs in Fig. 3-15 had $x_0 = 0$. Since that is the barrier state x_b [see Fig. 3-5], then the process has a roughly even chance of migrating either to the upper stable state x_2, as it did in Fig. 3-15a, or to the lower stable state x_1, as it did in Fig. 3-15b. The dotted horizontal lines in these figures demarcate the intervals $[x_i - \sigma_i/2, x_i + \sigma_i/2]$, where σ_i is the aforementioned Gaussian width of the x_i-peak in the stationary Markov state density function $P_s(x)$. We see from Figs. 3-15a and 3-15b that once the process enters a particular stable state region $[x_i - \sigma_i/2, x_i + \sigma_i/2]$, it is more or less content to stay there. The simulation run in Fig. 3-15c is interesting because there the process shows considerable hesitancy at first over which stable state it wants to migrate to: initially it moves to stable state x_2, but then appears to almost have second thoughts about that decision. In fact, we know from our development in Section 3.7 that the process will *not* hover forever about any one stable state, but will make "transitions" back and forth between the two stable states. And in the simulation plotted in Fig. 3-15d, the process initially migrates to stable state x_1, but subsequently makes a spontaneous transition to stable state x_2.

In Fig. 3-16 we show four separate simulation runs of the continuous bistable Markov process defined by the characterizing functions in Eqs. (3.6-19a), namely

$$A(x) = -x(x+1.5)(x-1.7), \quad D(x) = 1.0 + 0.25\,x^2.$$

We showed in Subsection 3.6.C [see Fig. 3-6] that this process has two stable states, $x_1 = -1.42$ and $x_2 = 1.62$, with respective widths $\sigma_1 = 1.0461$ and $\sigma_2 = 1.0278$, separated by a barrier state $x_b = 0$. As in the simulation runs of Fig. 3-15, these runs here all had $x_0 = 0$ and $\Delta t = 0.00025$, and were plotted at every 0.0025 time units. To test the accuracy condition (3.9-8) for our value of Δt, we note that it reads in this case

$$\Delta t \leq \text{Min}\left\{\frac{\varepsilon_1}{|3x^2 - 0.4x - 2.55|}, \left(\frac{\varepsilon_1}{x/[2(4+x^2)^{1/2}]}\right)^2\right\}.$$

We found in the preceding example that for $x \in [-3,3]$ and $\Delta t = 0.00025$, the first quantity in the minimum braces corresponds to taking $\varepsilon_1 = 0.0064$. For the second term in the minimum braces we have

$$\varepsilon_1 = \text{Max} \left\{ \frac{x}{2(4+x^2)^{1/2}} (0.00025)^{1/2} \right\} = 0.0066,$$

where the maximization is taken over the x-interval $[-3,3]$. So the ε_1-value for these simulations becomes 0.0066, which is "suitably small."

As we should expect from a comparison of the plots of the stationary Markov state density functions $P_s(x)$ in Figs. 3-5 and 3-6, the process realizations in Fig. 3-16 have slightly "looser" stable states than do the process realizations in Fig. 3-15. In the simulation of Fig. 3-16a the process initially moves to the stable state x_2, while in the simulation of Fig. 3-16b the process initially moves to the stable state x_1. The simulation shown in Fig. 3-16c shows the process almost making a transition from stable state x_2 to stable state x_1; such a transition actually occurs in the simulation shown in Fig. 3-16d.

We shall conclude this section, and hence also our discussion of continuous Markov processes, by examining the results of some Monte Carlo studies of the transition times between the stable states of the two bistable processes of Figs. 3-15 and 3-16. We should perhaps remark at the outset that such studies are often not feasible, owing to the fact that the average time for a stable state transition to occur is typically enormously large compared to the largest acceptable simulation time step Δt. However, for the systems of Figs. 3-15 and 3-16, Monte Carlo studies are just within the realm of feasibility. These studies are worthwhile, not only because they provide concrete illustrations of the Monte Carlo method of analysis, but also because they deepen our understanding of the general nature of stable state transitions.

To study the transition time $T(x_i \rightarrow x_f)$ from stable state x_i to stable state x_f, we must make a large number N of independent simulations of the process $X(t)$, each simulation having the initial condition $X(0) = x_i$, and each terminating when the process first reaches (or first passes over) state x_f. The only datum of interest to us in the kth simulation run is the time t_k of its termination, which of course constitutes a sample value of the random variable $T(x_i \rightarrow x_f)$. From the set of numbers $t_1, t_2, ..., t_N$, we then calculate the two quantities

$$\frac{1}{N} \sum_{k=1}^{N} t_k \equiv \langle t \rangle_N, \qquad \frac{1}{N} \sum_{k=1}^{N} t_k^2 \equiv \langle t^2 \rangle_N. \qquad (3.9\text{-}10)$$

Obviously, $\langle t \rangle_N$ and $\langle t^2 \rangle_N$ approximate, respectively, the first and second moments of $T(x_i \rightarrow x_f)$, these approximations becoming exact in the limit $N \rightarrow \infty$. And

$$\sigma_N \equiv [\langle t^2 \rangle_N - \langle t \rangle_N^2]^{1/2} \qquad (3.9\text{-}11)$$

similarly approximates the standard deviation of $T(x_i \rightarrow x_f)$. So if N is "sufficiently large," we can invoke the central limit theorem corollary in Eqs. (1.8-2) and (1.8-3) to estimate the *mean* of $T(x_i \rightarrow x_f)$ as

$$T_1(x_i \rightarrow x_f) \approx \langle t \rangle_N \pm \sigma_N N^{-1/2}, \qquad (3.9\text{-}12)$$

where the $\pm$ term gives the one-standard deviation confidence interval.

The two tables below show the results of such a Monte Carlo study of the stable state transition times $T_1(x_1 \rightarrow x_2)$ and $T_1(x_2 \rightarrow x_1)$ for our two illustrative bistable systems. In these tables we show the mean transition times, first as calculated from the exact formulas (3.7-18), and then as calculated from the Monte Carlo formula (3.9-12) using $N = 1000$ samples for each transition time. Notice that in each case the Monte Carlo estimate of the mean is within one standard deviation of the exact value, although differences here of up to two standard deviations would not have been cause for alarm. If we had found differences of more than three standard deviations, then we should question whether N had been taken large enough, or whether an error had somewhere been made. But as it is, the excellent agreement shown in the tables give us strong assurance that our first-exit-time theory, and also our continuous simulation methodology, have both been correctly developed and applied.

For the Bistable Process of Figs. 3-5 and 3-15

	Exact Eq. (3.7-18)	Monte Carlo Eq. (3.9-12)	$\sigma_N / \langle t \rangle_N$
$T_1(x_1 \rightarrow x_2)$	33.76	33.1 ± 1.0	0.98
$T_1(x_2 \rightarrow x_1)$	92.15	92.3 ± 2.8	0.96

For the Bistable Process of Figs. 3-6 and 3-16

	Exact Eq. (3.7-18)	Monte Carlo Eq. (3.9-12)	$\sigma_N / \langle t \rangle_N$
$T_1(x_1 \rightarrow x_2)$	19.63	19.5 ± 0.6	0.94
$T_1(x_2 \rightarrow x_1)$	38.29	38.9 ± 1.2	0.96

To get some idea of how much computer work was involved in obtaining the Monte Carlo estimates reported above, consider the $x_2 \to x_1$ transition in the first system. This transition requires on the average 92 time units, the longest of the four transitions considered. With our time-step size $\Delta t = 0.00025$ (the same as used in generating Figs. 3-15 and 3-16), the average run for this case should take $92 \div 0.00025 = 368,000$ time steps; therefore, the $N = 1000$ simulation runs used to form our Monte Carlo estimate of $T_1(x_2 \to x_1)$ in the first system required about 368 million passes through the main loop of the simulation algorithm in Fig. 3-7. This represents by modern standards a substantial, but not Herculean, computational task.

The last column in each table above gives the Monte Carlo estimate (3.9-11) of the standard deviation of each stable state transition time, as measured in units of the corresponding mean. In each case it is seen that the standard deviation is only slightly less than the mean. This implies that actual transition times that differ from the mean by a factor of anywhere from 1/2 to 2 should be quite commonplace. And it also implies that the near 20% inaccuracy noted earlier in the approximate formula (3.7-22) is not really significant if that formula is only intended to estimate the "scale" of the transition times.

The fact that the Monte Carlo generated t_k-values allowed us to estimate not only the mean but also the standard deviation of the stable state transition times hints at the versatility of the Monte Carlo method of analysis. In fact, by simply *histogramming* the t_k-values used in computing the sums (3.9-10), we can get a fair indication of the shape of the density function of the corresponding transition time. In Fig. 3-17a we show a histogram of the 1000 t_k-values obtained in our simulations of the $x_1 \to x_2$ transition in the process of Figs. 3-6 and 3-16. And Fig. 3-17b shows the histogram of the 1000 t_k-values obtained for the $x_2 \to x_1$ transition in that same process. Although the average time for the $x_1 \to x_2$ transition is 19.6 time units, 174 of the 1000 samples fell below 5 time units, while 4 fell above 100 time units. And for the $x_2 \to x_1$ transition, with an average time of 38.3 time units, 97 of the 1000 samples fell below 5 time units and 6 fell above 200 time units. Evidently, the mean transition time $T_1(x_i \to x_f)$ provides us with little more than an order-of-magnitude estimate of how long we will have to wait before the process, initially in stable state x_i, will first reach stable state x_f.

The curves superimposed on the histograms in Fig. 3-17 are neither best fits nor expected shapes; rather, they are what the histograms *would* approach *if* the stable state transition times were *exponential* random

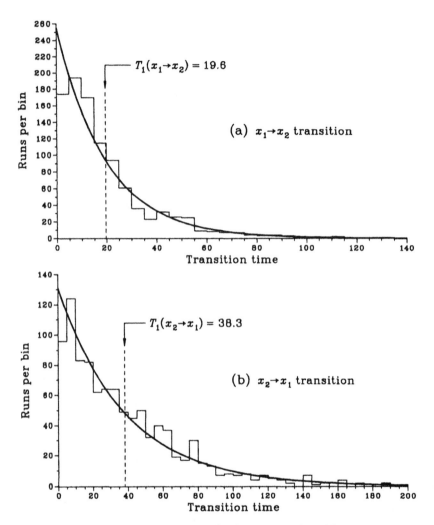

Figure 3-17. Histograms of Monte Carlo generated stable state transition times for the continuous, bistable Markov process of Figs. 3-6 and 3-16. Histogram (a) shows 1000 samples of $T(x_1{\to}x_2)$, for which the theoretically expected mean is 19.6. Histogram (b) shows 1000 samples of $T(x_2{\to}x_1)$, for which the theoretically expected mean is 38.3. The superimposed curves are what we would get asymptotically *if* the transition times were pure exponential random variables with those respective means. Although we know that the exponential forms cannot be exactly correct, they evidently provide fairly good fits in this case except for the dips in the data near zero.

variables; more specifically, setting $a_{if} \equiv 1/T_1(x_i \rightarrow x_f)$, the curves in Fig. 3-17 are plots of the function

$$[a_{if} \exp(-a_{if}t)] \, N \, \Delta_{\text{bin}} \, ,$$

where the total number of samples $N = 1000$, and the histogram bin width $\Delta_{\text{bin}} = 5$. As we proved at the end of Section 3.7, stable state transition times for a continuous Markov process are *not* purely exponentially distributed. Since our Monte Carlo data indicates that the density function of $T(x_i \rightarrow x_f)$ in this case peaks at some *strictly positive* value of time, the non-exponential character of $T(x_i \rightarrow x_f)$ is substantiated; nevertheless, we see that the fit to an exponential density function for $t \geq T_1(x_i \rightarrow x_f)$ is remarkably good.

- 4 -

JUMP MARKOV PROCESSES WITH
CONTINUUM STATES

Having concluded our discussion of "continuous" Markov processes, we now turn to consider the class of so-called "jump" Markov processes. We begin by anticipating the fact that the possible states of a jump Markov process may consist of either a continuum of real numbers or a discrete subset of the real numbers (such as the integers). This is in contrast to the situation for continuous Markov processes, where the possible states are of necessity a continuum of real numbers. Although discrete state jump Markov processes can be regarded as a subclass of continuum state jump Markov processes, we shall find it to be pedagogically convenient to consider them separately. In this chapter we shall develop the theory of jump Markov processes for continuum states, and in Chapter 5 we shall develop the theory for the discrete state case. Since jump Markov processes on discrete states are mathematically simpler than jump Markov processes on continuum states, we shall be able to develop the theory more extensively in the discrete state case. Perhaps for that reason, most jump Markov processes encountered in scientific applications turn out to be of the discrete state variety. But the two most historically significant Markov processes, namely Brownian motion and self-diffusion, are both most accurately viewed as jump Markov processes on the continuum, and thus fall within the purview of the present chapter.

Just as we could have logically defined a *continuous* Markov process as one whose propagator density function has the form given in Eq. (3.1-11), so could we now logically define a *jump* Markov process as one whose propagator density function has the form given in Eq. (4.1-13). However, we shall take in Section 4.1 a more heuristic approach which reveals how the functional form (4.1-13) arises as a consequence of a few conditions that should logically attend any Markov process that moves about by "jumping." In Section 4.2 we shall see how the various time-evolution equations derived in Chapter 2 present themselves for jump Markov processes. And we shall also deduce a new pair of time-evolution equations for $P(x,t \mid x_0,t_0)$, called the "master equations," that apply *only* to jump Markov processes. In Section 4.3 we shall define and discuss an entity called the "next-jump density function." We shall later show that this function allows us to make numerical simulations of jump Markov processes that are *exact* in the sense that we never have to approximate a differential time increment dt by a finite time increment Δt (as we do when simulating continuous Markov processes). In Section 4.4 we shall discuss completely homogeneous jump Markov processes, and in so doing discover that they are a bit more complicated and diverse than their continuous counterpart, the Wiener processes. In Section 4.5 we shall move into the realm of elementary gas kinetic theory to develop a moderately rigorous Markovian description of Brownian motion and molecular diffusion. And finally, in Section 4.6, we shall see how jump Markov processes with continuum states can be numerically simulated.

4.1 THE JUMP PROPAGATOR AND ITS CHARACTERIZING FUNCTIONS

Roughly speaking, a jump Markov process is one for which the propagator $\Xi(dt; x,t)$ defined in Eq. (2.4-1) is usually exactly zero but occasionally finitely different from zero. Thus, a jump Markov process does not evolve in a continuous, gradual manner, but rather moves in sudden "jumps." More specifically, if the process is in state x at time t, then it will remain there until some time $t' > t$ when it instantaneously jumps to a new state $x' \neq x$. The times $t^{(1)}$, $t^{(2)}$, ... at which successive jumps occur, and the states $x^{(1)}$, $x^{(2)}$, ... reached in these jumps, both form countable sets. This is *not* the case with a *continuous* Markov process, where the fundamental properties of the continuum prevent us from asking, given $X(t) = x$, when the system will "leave" state x or what will be the "next" state after x.

Since for a jump Markov process we *can* ask when the system will leave its current state and what that next state will be, then it makes sense to define for a jump Markov process the following two probabilities:

$$q(x,t; \tau) \equiv \text{probability, given } X(t) = x, \text{ that the process} \atop \text{will jump away from state } x \text{ at some instant} \atop \text{between } t \text{ and } t + \tau; \qquad (4.1\text{-}1)$$

$w(\xi \mid x,t) \, d\xi \equiv$ probability that the process, upon jumping away from state x at time t, will land in some state lying in the infinitesimal interval $[x+\xi, x+\xi+d\xi)$. (4.1-2)

It is more or less implicit in these definitions that the process $X(t)$ is Markovian, because the fate of the process in state x at time t as dictated by these definitions evidently depends only on x and t, and not on the history of the process prior to time t. We now propose to simply *define* a jump Markov process as any process $X(t)$ for which the two functions q and w exist as defined above, *and* have the following properties:

- $q(x,t; \tau)$ is a smooth function of t and τ,
  and satisfies $q(x,t; 0)=0$; (4.1-3a)

- $w(\xi \mid x,t)$ is a smooth function of t. (4.1-3b)

Later in this section we shall show explicitly that reasonable smoothness conditions on the functions q and w will guarantee that the associated process $X(t)$ is Markovian. Notice that we have *not* required any smoothness with respect to the state variable x. Other important properties of q and w, which we need not postulate since they are implicit in the definitions (4.1-1) and (4.1-2), are that

$$0 \le q(x,t; \tau) \le 1, \tag{4.1-4}$$

and

$$w(\xi \mid x,t) \ge 0, \tag{4.1-5a}$$

$$\int_{-\infty}^{\infty} d\xi \, w(\xi \mid x,t) = 1. \tag{4.1-5b}$$

A natural question at this point is, how is the propagator density function $\Pi(\xi \mid dt; x,t)$ of the process related to the two functions $q(x,t; \tau)$ and $w(\xi \mid x,t)$? To answer this question, we must first establish one additional important property of the function $q(x,t; \tau)$: If τ is an *infinitesimal* dt, then $q(x,t; \tau)$ has the functional form

$$q(x,t; dt) = a(x,t) \, dt, \tag{4.1-6}$$

where $a(x,t)$ is some nonnegative, smooth function of t. This fundamental property of q is essentially a consequence of the requirement that $X(t)$ be Markovian, and it can be proved as follows.

Imagine the infinitesimal time interval $[t,t+dt)$ to be divided into $n \ge 2$ subintervals of equal length dt/n by means of the $n-1$ division points

$$t_i = t_{i-1} + dt/n \quad (i=1, \ldots ,n-1),$$

where for the present we identify $t_0 = t$. Clearly, the probability $[1 - q(x,t; dt)]$ that the system in state x at time t does *not* jump away from that state in $[t,t+dt)$ must be equal to the probability that the system does not jump away from x in any of the successive intervals $[t,t_1)$, $[t_1,t_2)$, ..., $[t_{n-1},t+dt)$. Since the probability for not jumping away from state x in $[t_{i-1},t_i)$ is $[1 - q(x,t_{i-1}; dt/n)]$, independently of the history of the process prior to time t_{i-1} (Markovian assumption), then we have by the multiplication law of probability that

$$[1 - q(x,t; dt)] = \prod_{i=1}^{n} [1 - q(x,t_{i-1}; dt/n)]. \tag{4.1-7}$$

Now we invoke the property in (4.1-3a) that $q(x,t; \tau)$ be a "smooth function of t." This property allows us to replace t_{i-1} on the right side of Eq. (4.1-7) by the infinitesimally close value t without spoiling the equality. Upon doing that, we get

$$[1 - q(x,t; dt)] = [1 - q(x,t; dt/n)]^{n}. \tag{4.1-8}$$

Next we invoke the property in (4.1-3a) that $q(x,t; \tau)$ goes *smoothly to zero* as $\tau \to 0$. This implies, since dt/n is infinitesimally small, that $q(x,t; dt/n)$ is infinitesimally close to zero; hence, to first order in infinitesimals we can write the right side of the preceding equation as

$$[1 - q(x,t; dt/n)]^{n} = 1 - n q(x,t; dt/n). \tag{4.1-9}$$

Combining Eqs. (4.1-8) and (4.1-9) evidently gives, to first order in dt,

$$q(x,t; dt) = n q(x,t; dt/n) \quad (n \geq 2). \tag{4.1-10}$$

So $q(x,t; dt)$, by condition (4.1-3a) a smooth function of dt, satisfies Eq. (4.1-10). Now we apply to this finding the lemma that appears just after Eqs. (3.1-7), with z there replaced by dt and f replaced by q. The result is evidently Eq. (4.1-6), which is what we set out to establish.

Combining Eq. (4.1-6) with the definition (4.1-1), we arrive at the conclusion that for any jump Markov process there exists a function $a(x,t)$, which depends smoothly on t [by condition (4.1-3a)], such that

$a(x,t)dt \equiv$ probability, given $X(t)=x$, that the process will jump away from state x in the next infinitesimal time interval $[t,t+dt)$. $\quad$ (4.1-11)

It follows from this result that the probability for the process to jump once in $[t, t+adt)$ and then jump once again in $[t+adt, t+dt)$, for any a between

0 and 1, will be proportional to $(dt)^2$. We thus conclude that, to first order in dt, the process will either jump *once* or else *not at all* in the infinitesimal time interval $[t, t+dt)$.

Now we are in a position to deduce the form of the propagator density function $\Pi(\xi \mid dt; x, t)$ in terms of the two functions $a(x,t)$ and $w(\xi \mid x,t)$. Given $X(t) = x$, then by time $t+dt$ the process *either* will have jumped once, with probability $a(x,t)dt$, *or* it will not have jumped at all, with probability $1 - a(x,t)dt$. If a jump *does* occur, then by Eq. (4.1-2) the probability that the state change vector $X(t+dt) - x$ will lie in $[\xi, \xi+d\xi)$ will be $w(\xi \mid x, t')d\xi$, where t' is the precise instant in $[t, t+dt)$ when the jump occurred. If a jump does *not* occur, then the probability that the state change vector $X(t+dt) - x$ will lie in $[\xi, \xi+d\xi)$ will be $\delta(\xi)d\xi$, since that quantity is equal to unity if $\xi = 0$ and zero if $\xi \neq 0$. Therefore, appealing to the propagator density function definition (2.4-2) and the addition and multiplication laws of probability, we have

$$\Pi(\xi \mid dt; x,t)d\xi = [a(x,t)dt][w(\xi \mid x, t')d\xi] + [1 - a(x,t)dt][\delta(\xi)d\xi]. \quad (4.1\text{-}12)$$

Finally, since the jump time t' lies somewhere in the interval $[t, t+dt)$, then the smooth dependence of $w(\xi \mid x,t)$ on t assumed in condition (4.1-3b) means that we can replace t' on the right side of this last equation by the infinitesimally close value t without spoiling the equality. Upon doing so and then dividing through by $d\xi$, we conclude that

$$\Pi(\xi \mid dt; x,t) = a(x,t)dt \, w(\xi \mid x,t) + [1 - a(x,t)dt]\delta(\xi). \quad (4.1\text{-}13)$$

Equation (4.1-13) gives the general form of the propagator density function for a jump Markov process, and is the principle result of this section. The remainder of this chapter will be devoted to exploring the consequences of this result. However, we have at this point two loose ends that need to be tied up: First, we can see from Eq. (4.1-6) that the functions $q(x,t; \tau)$ and $a(x,t)$ are related according to $q(x,t; \tau) = a(x,t)\tau$ when τ is infinitesimally small. But what is the relation between these two functions when τ is *finite*? Second, in the course of our derivation of Eq. (4.1-13), we took a couple of steps [namely from Eq. (4.1-7) to Eq. (4.1-8), and from Eq. (4.1-12) to Eq. (4.1-13)] that we justified rather loosely on grounds of "t-smoothness." Can we find some explicit t-smoothness conditions for $a(x,t)$ and $w(\xi \mid x,t)$ which will guarantee that the propagator density function (4.1-13) satisfies the Chapman-Kolmogorov condition (2.4-8), and thus defines a truly Markovian process? We shall now address these two questions in turn.

To deduce a general formula for $q(x,t; \tau)$ in terms of $a(x,t)$, it is convenient to begin by defining the probability

$$q^*(x,t;\tau) \equiv 1 - q(x,t;\tau) \tag{4.1-14a}$$

$$= \text{ probability, given } X(t)=x, \text{ that the process}$$
$$\text{will } not \text{ jump away from state } x \text{ in } [t,t+\tau). \tag{4.1-14b}$$

Since the process is past-forgetting (Markovian), we have

$$q^*(x,t;\tau+d\tau) = q^*(x,t;\tau) \times q^*(x,t+\tau;d\tau) \quad \text{[by the multiplication law]},$$

$$= q^*(x,t;\tau)[1 - q(x,t+\tau;d\tau)] \qquad \text{[by (4.1-14a)]},$$

$$= q^*(x,t;\tau)[1 - a(x,t+\tau)\,d\tau] \qquad \text{[by (4.1-6)]}.$$

Writing $q^*(x,t;\tau+d\tau) - q^*(x,t;\tau)$ as $dq^*(x,t;\tau)$, we thus have

$$\frac{dq^*(x,t;\tau)}{q^*(x,t;\tau)} = -a(x,t+\tau)\,d\tau.$$

Integrating from $\tau=0$, and noting that condition (4.1-3a) and the definition (4.1-14a) together imply that $q^*(x,t;0)=1$, we obtain

$$q^*(x,t;\tau) = \exp\left(-\int_0^\tau a(x,t+\tau')\,d\tau'\right). \tag{4.1-15a}$$

So, again appealing to the definition (4.1-14a), we conclude that

$$q(x,t;\tau) = 1 - \exp\left(-\int_0^\tau a(x,t+\tau')\,d\tau'\right). \tag{4.1-15b}$$

Equation (4.1-15b) shows precisely how the function $a(x,t)$ determines the function $q(x,t;\tau)$ for all $\tau>0$.

Next we turn to the problem of proving that reasonable t-smoothness conditions on the two functions $a(x,t)$ and $w(\xi \mid x,t)$ will guarantee that the jump propagator density function (4.1-13) satisfies the Chapman-Kolmogorov condition. That condition requires Eq. (2.4-8) to hold, at least to first order in dt, for all α between 0 and 1. Substituting Eq. (4.1-13) into the right hand side of Eq. (2.4-8) gives

$$\text{RHS(2.4-8)} = \int_{-\infty}^{\infty} d\xi_1 \Big\{ a(x+\xi_1,t+\alpha dt)\,(1-\alpha)dt\, w(\xi-\xi_1 \mid x+\xi_1,t+\alpha dt)$$

$$+ [1-a(x+\xi_1,t+\alpha dt)\,(1-\alpha)dt]\,\delta(\xi-\xi_1)\Big\}$$

$$\times \Big\{ a(x,t)\,\alpha dt\, w(\xi_1 \mid x,t) + [1-a(x,t)\,\alpha dt]\,\delta(\xi_1) \Big\}.$$

$$\tag{4.1-16a}$$

Expanding the product in the integrand and retaining explicitly only terms up to first order in dt, we get

RHS(2.4–8)

$$= \int_{-\infty}^{\infty} d\xi_1 \Big\{ a(x+\xi_1, t+adt)(1-a)dt\, w(\xi-\xi_1 \,|\, x+\xi_1, t+adt)\,\delta(\xi_1) \Big\}$$

$$+ \int_{-\infty}^{\infty} d\xi_1 \Big\{ \delta(\xi-\xi_1)\,\delta(\xi_1)[1 - a(x+\xi_1, t+adt)(1-a)dt - a(x,t)\,adt] \Big\}$$

$$+ \int_{-\infty}^{\infty} d\xi_1 \Big\{ \delta(\xi-\xi_1)\,a(x,t)\,adt\, w(\xi_1 \,|\, x,t) \Big\} + o(dt). \qquad (4.1\text{-}16b)$$

The three ξ_1-integrals here are easy to evaluate because of the Dirac delta functions in their integrands. The result is evidently

$$\text{RHS}(2.4\text{-}8) = a(x,t+adt)(1-a)dt\, w(\xi \,|\, x, t+adt)$$
$$+ \delta(\xi)\{1 - a(x,t+adt)(1-a)dt - a(x,t)\,adt\}$$
$$+ a(x,t)\,adt\, w(\xi \,|\, x,t) + o(dt).$$

Now let us *require* the functions $a(x,t)$ and $w(\xi \,|\, x,t)$ to be "analytic" in t in the following sense: for all Δt sufficiently small,

$$a(x,t+\Delta t) = a(x,t) + \partial_t a(x,t)\,\Delta t + o(\Delta t), \qquad (4.1\text{-}17a)$$

$$w(\xi \,|\, x, t+\Delta t) = w(\xi \,|\, x,t) + \partial_t w(\xi \,|\, x,t)\,\Delta t + o(\Delta t), \qquad (4.1\text{-}17b)$$

where the t-derivatives $\partial_t a$ and $\partial_t w$ exist as bounded functions. Then taking $\Delta t = adt$ and substituting into the preceding equation, we get

RHS(2.4–8)

$$= [a(x,t)+\partial_t a(x,t)adt+o(dt)]\,(1-a)dt\,[w(\xi \,|\, x,t)+\partial_t w(\xi \,|\, x,t)adt+o(dt)]$$
$$+ \delta(\xi)\{1 - [a(x,t)+\partial_t a(x,t)adt+o(dt)]\,(1-a)dt - a(x,t)adt\}$$
$$+ a(x,t)\,adt\, w(\xi \,|\, x,t) + o(dt).$$

Expanding the products, and retaining explicitly only terms up to first order in dt, we find upon collecting terms,

$$\text{RHS}(2.4\text{-}8) = a(x,t)dt\, w(\xi \,|\, x,t) + [1 - a(x,t)dt]\,\delta(\xi) + o(dt). \quad (4.1\text{-}18a)$$

Of course, if we substitute the form (4.1-13) into the *left* hand side of Eq. (2.4-8), we get simply

$$\text{LHS}(2.4\text{-}8) = a(x,t)dt\, w(\xi \,|\, x,t) + [1 - a(x,t)dt]\,\delta(\xi). \qquad (4.1\text{-}18b)$$

Obviously, Eqs. (4.1-18) agree to first order in dt for all a between 0 and 1, so we conclude that the jump propagator (4.1-13) indeed satisfies the Chapman-Kolmogorov condition. Notice that the main requirement we imposed in order to ensure this result is that the functions $a(x,t)$ and

$w(\xi \,|\, x,t)$ be analytic in t in the sense of Eqs. (4.1-17).† The foregoing arguments make more explicit the smoothness requirements set forth in conditions (4.1-3), and they also serve to vindicate our somewhat heuristic derivation of Eq. (4.1-13).

Since the two functions $a(x,t)$ and $w(\xi \,|\, x,t)$ completely determine the propagator density function (4.1-13), then it follows that those two functions also completely determine the Markov state density function $P(x,t \,|\, x_0,t_0)$. Hence, a jump Markov process $X(t)$ can be uniquely specified by specifying the forms of the two functions $a(x,t)$ and $w(\xi \,|\, x,t)$. We shall therefore call $a(x,t)$ and $w(\xi \,|\, x,t)$ the **characterizing functions** of the jump Markov process $X(t)$. They are *defined* by Eqs. (4.1-11) and (4.1-2) respectively, and they can be any nonnegative, reasonably well behaved functions that are smooth in t, with w satisfying the usual normalization condition, $\int w(\xi \,|\, x,t)d\xi = 1$. Notice that the characterizing functions of a *continuous* Markov process, $A(x,t)$ and $D(x,t)$, involve only *two* independent variables, whereas the characterizing functions of a *jump* Markov process, $a(x,t)$ and $w(\xi \,|\, x,t)$, involve *three* independent variables. We may therefore anticipate that jump Markov processes will be a more "richly varied" class of processes than continuous Markov processes.

Since the propagator density function $\Pi(\xi \,|\, dt; x,t)$ in Eq. (4.1-13) depends on x and t *only* through the characterizing functions $a(x,t)$ and $w(\xi \,|\, x,t)$, then the homogeneity conditions (2.8-1) apply directly to the characterizing functions. In particular, for any jump Markov process $X(t)$ we have [cf. Eqs. (3.1-15)]

$X(t)$ is **temporally homogeneous**

$$\Leftrightarrow \quad a(x,t) = a(x) \text{ and } w(\xi \,|\, x,t) = w(\xi \,|\, x), \quad (4.1\text{-}19a)$$

$X(t)$ is **completely homogeneous**

$$\Leftrightarrow \quad a(x,t) = a \text{ and } w(\xi \,|\, x,t) = w(\xi). \quad (4.1\text{-}19b)$$

Although the product of the two characterizing functions of a *continuous* Markov process has no known use or significance, that is not the case for a *jump* Markov process. It will prove convenient in our later work to define the function

---

† Actually, a little more than Eqs. (4.1-17) is required. The "$o(dt)$" in Eq. (4.1-16b) represents a term proportional to $(dt)^2$ that would be present even if the functions a and w were independent of t. Our argument has implicitly assumed that the coefficient of $(dt)^2$ in that term, which is messy to write out explicitly, is *bounded*. It will be bounded for all "reasonable" choices of the functions a and w. But if one could find two t-analytic functions a and w that that did *not* result in the aforementioned coefficient being bounded, then those functions might not define a legitimate jump Markov process.

$$W(\xi \mid x,t) \equiv a(x,t)\, w(\xi \mid x,t), \qquad (4.1\text{-}20)$$

and call it the **consolidated characterizing function** of the jump Markov process $X(t)$. The physical meaning of this function can be inferred by multiplying Eq. (4.1-20) through by $dt\,d\xi$, invoking the definitions of $a(x,t)dt$ and $w(\xi \mid x,t)d\xi$ in Eqs. (4.1-11) and (4.1-2), and recalling that $w(\xi \mid x,t)$ is a smooth function of t; in this way we may deduce that

$W(\xi \mid x,t)\, dt\, d\xi \;\equiv\;$ probability, given $X(t)=x$, that the process
will in the time interval $[t,t+dt)$ jump from
state x to some state in $[x+\xi,\, x+\xi+d\xi)$. (4.1-21)

By integrating Eq. (4.1-20) over ξ using Eq. (4.1-5b), and then substituting the result back into Eq. (4.1-20), we may easily deduce the relations

$$a(x,t) = \int_{-\infty}^{\infty} d\xi\, W(\xi \mid x,t), \qquad (4.1\text{-}22a)$$

$$w(\xi \mid x,t) = \frac{W(\xi \mid x,t)}{\displaystyle\int_{-\infty}^{\infty} d\xi'\, W(\xi' \mid x,t)}. \qquad (4.1\text{-}22b)$$

These equations show that, had we chosen to do so, we could have defined the characterizing functions a and w in terms of the consolidated characterizing function W, instead of the other way around. So if we regard (4.1-21) as the *definition* of $W(\xi \mid x,t)$, then the specification of the form of that function will uniquely define a jump Markov process $X(t)$. We shall find in the next section that many time-evolution equations for jump Markov processes can be conveniently written in terms of the consolidated characterizing function $W(\xi \mid x,t)$.

By substituting Eqs. (4.1-22) into Eq. (4.1-13), we find that the jump propagator density function can be written in terms of the *consolidated* characterizing function as

$$\Pi(\xi \mid dt; x,t) = W(\xi \mid x,t)\, dt + \left[1 - \int_{-\infty}^{\infty} d\xi'\, W(\xi' \mid x,t)\, dt\right] \delta(\xi). \quad (4.1\text{-}23)$$

When $\xi \neq 0$ the second term in this formula vanishes, and the relation between the functions Π and W becomes quite simple; however, for *arbitrary* ξ the relation between these two functions is evidently rather complicated. Although (4.1-13) and (4.1-23) are obviously completely equivalent formulas for the jump propagator density function, we shall usually find it more convenient to work with (4.1-13) than with (4.1-23).

4.2 TIME-EVOLUTION EQUATIONS

In Chapter 2 we found that the differential equations that govern the time evolutions of $P(x,t \mid x_0,t_0)$, $\langle X^n(t) \rangle$ and $\langle S^n(t) \rangle$ are all expressed in terms of the so-called propagator moment functions, $B_n(x,t)$. The latter were defined in Eqs. (2.4-3), but with the important proviso that $\Pi(\xi \mid dt; x,t)$ be such that those definitions truly make sense. With the result (4.1-13), we are now in a position to show that the propagator moment functions are well defined for many, but not all, jump Markov processes. In this section we shall calculate the jump propagator moment functions, when they exist, in terms of the jump characterizing functions, and we shall then write down the jump versions of the various time-evolution equations developed in Chapter 2. Also in this section, we shall derive a set of time-evolution equations for $P(x,t \mid x_0,t_0)$ that are valid *only* for jump Markov processes; these are the so-called "master equations," and they, unlike the Kramers-Moyal equations, do not require the existence of the propagator moment functions.

To calculate the propagator moment function $B_n(x,t)$ for a jump Markov process $X(t)$ with characterizing functions $a(x,t)$ and $w(\xi \mid x,t)$, we simply substitute Eq. (4.1-13) into Eq. (2.4-3b):

$$B_n(x,t) = \frac{1}{dt} \int_{-\infty}^{\infty} d\xi\, \xi^n \left[a(x,t)dt\, w(\xi \mid x,t) + [1 - a(x,t)dt]\, \delta(\xi) \right]$$

$$= \frac{a(x,t)dt}{dt} \int_{-\infty}^{\infty} d\xi\, \xi^n w(\xi \mid x,t) + \frac{1 - a(x,t)dt}{dt} \int_{-\infty}^{\infty} d\xi\, \xi^n\, \delta(\xi) .$$

For any $n \geq 1$ the integral in the second term evidently vanishes; thus we conclude that

$$B_n(x,t) = a(x,t)\, w_n(x,t) = W_n(x,t) \qquad (n = 1,2,...) \qquad (4.2\text{-}1)$$

where we have defined

$$w_n(x,t) \equiv \int_{-\infty}^{\infty} \xi^n\, w(\xi \mid x,t)\, d\xi , \qquad (4.2\text{-}2a)$$

and

$$W_n(x,t) \equiv \int_{-\infty}^{\infty} \xi^n\, W(\xi \mid x,t)\, d\xi . \qquad (4.2\text{-}2b)$$

Since $w_n(x,t)$ is just the nth moment of the conditional density function $w(\xi \mid x,t)$, then it follows from Eq. (4.2-1) that

$$B_n(x,t) \text{ exists} \quad \Leftrightarrow \quad \text{the } n\text{th moment of } w(\xi \mid x,t) \text{ exists.} \qquad (4.2\text{-}3)$$

4.2.A THE JUMP KRAMERS-MOYAL EQUATIONS AND THE TRUNCATION ISSUE

With the result (4.2-1), it follows immediately from Eqs. (2.5-1) and (2.5-2) that for a jump Markov process with characterizing functions $a(x,t)$ and $w(\xi \mid x,t)$ the **forward Kramers-Moyal equation** is

$$\frac{\partial}{\partial t} P(x,t \mid x_0,t_0) = \sum_{n=1}^{\infty} \frac{(-1)^n}{n!} \frac{\partial^n}{\partial x^n} \left[a(x,t)\, w_n(x,t)\, P(x,t \mid x_0,t_0) \right] \quad (4.2\text{-}4\text{a})$$

$$= \sum_{n=1}^{\infty} \frac{(-1)^n}{n!} \frac{\partial^n}{\partial x^n} \left[W_n(x,t)\, P(x,t \mid x_0,t_0) \right], \quad (4.2\text{-}4\text{b})$$

and the **backward Kramers-Moyal equation** is

$$-\frac{\partial}{\partial t_0} P(x,t \mid x_0,t_0) = \sum_{n=1}^{\infty} \frac{1}{n!}\, a(x_0,t_0)\, w_n(x_0,t_0)\, \frac{\partial^n}{\partial x_0^n} P(x,t \mid x_0,t_0) \quad (4.2\text{-}5\text{a})$$

$$= \sum_{n=1}^{\infty} \frac{1}{n!}\, W_n(x_0,t_0)\, \frac{\partial^n}{\partial x_0^n} P(x,t \mid x_0,t_0), \quad (4.2\text{-}5\text{b})$$

where the functions w_n and W_n are as defined in Eqs. (4.2-2). Given the initial condition (2.5-3), Eq. (4.2-4) constitutes a t-evolution equation for $P(x,t \mid x_0,t_0)$ for fixed x_0 and t_0. And given the final condition (2.5-4), Eq. (4.2-5) constitutes a t_0-evolution equation for $P(x,t \mid x_0,t_0)$ for fixed x and t. If the characterizing functions $a(x,t)$ and $w(\xi \mid x,t)$ are both explicitly independent of t, then the process $X(t)$ will be temporally homogeneous; in that case, as we proved more generally in Subsection 2.8.A, the function $P(x,t \mid x_0,t_0)$ will depend on t and t_0 only through their difference $t - t_0$, and the left hand sides of Eqs. (4.2-4) and (4.2-5) will be interchangeable. Further simplifications will arise if the characterizing functions are also independent of x, but we shall defer a discussion of the completely homogeneous case to Section 4.4.

Because of the result (4.2-3), the jump Kramers-Moyal equations will make sense only if all the moments of the characterizing function $w(\xi \mid x,t)$ exist. Assuming for now that all those moments *do* exist, and remembering that for *continuous* Markov processes the Kramers-Moyal equations always truncate at $n = 2$ terms, it is natural to wonder if there are any circumstances in which the Kramers-Moyal equations for *jump*

Markov processes will truncate. Alas, as we shall prove shortly, *the jump Kramers-Moyal equations never truncate in an exact sense*; however, as we shall discover on several occasions in the sequel, it may happen that by taking certain limits of certain characterizing function parameters, we can induce viable *approximate truncations* of the jump Kramers-Moyal equations.

To prove that the jump Kramers-Moyal equations never truncate in an exact sense, we begin by observing that the function x^n for any $n \geq 1$ is "convex" on the positive x-axis. This simply means that everywhere on the positive x-axis the graph of $y = x^n$ lies on or above all of its tangent lines. Since the equation of the tangent line to the function x^n at $x = a$ is $y = a^n + [na^{n-1}](x-a)$, then we can express the convexity of x^n analytically by saying that, for any given $a \geq 0$ and all $x \geq 0$,

$$x^n \geq a^n + [n\,a^{n-1}]\,(x-a).$$

Now let X be any random variable whose sample values are always nonnegative, so that the density function $P(x)$ of X is zero for all $x < 0$. Multiplying the above inequality through by $P(x)$ (which of course is always nonnegative) and then integrating over all x, we get

$$\langle X^n \rangle \geq a^n + [n\,a^{n-1}]\,(\langle X \rangle - a).$$

Taking $a = \langle X \rangle$, this becomes

$$\langle X^n \rangle \geq \langle X \rangle^n,$$

valid for those random variables X that have strictly nonnegative sample values. Now, for any random variable X, the random variable X^2 will obviously have strictly nonnegative sample values; therefore, replacing X by X^2 in the preceding equation, we conclude that *for any random variable X,*

$$\langle X^{2n} \rangle \geq \langle X^2 \rangle^n \quad (n = 1, 2, ...). \tag{4.2-6}$$

This result is evidently a generalization of the fundamental inequality (1.3-9) to values of n greater than unity.[†]

Applying the general result (4.2-6) to the random variable whose density function is $w(\xi \mid x,t)$, we deduce that

$$w_{2n}(x,t) \geq [w_2(x,t)]^n \quad (n = 1, 2, ...). \tag{4.2-7a}$$

And since $w_2(x,t)$ is strictly positive unless $w(\xi \mid x,t) = \delta(\xi)$, then we

---

[†] The inequality (4.2-6) is in turn a special case of a more general result, known as *Jensen's inequality*, which states that $\langle f(X) \rangle \geq f(\langle X \rangle)$ for any convex function f.

conclude that, except for the very uninteresting case in which all jumps are of exactly zero length, we will have

$$w_{2n}(x,t) > 0 \quad (n=1, 2, ...). \tag{4.2-7b}$$

Clearly, this result means that there is no value of the summation index n in the jump Kramers-Moyal equations beyond which all terms vanish identically.†

So, whereas for *continuous* Markov processes the Kramers-Moyal equations are always *second order* partial differential equations, for *jump* Markov processes the Kramers-Moyal equations are always *infinite order* partial differential equations. One may reasonably surmise that the jump Kramers-Moyal equations will rarely be of much use for obtaining *exact* solutions $P(x,t \mid x_0,t_0)$. But it sometimes happens that the jump Kramers-Moyal equations can be *approximately* truncated. This possibility arises from the fact that *if $w_2(x,t)$ is very small compared to unity*, then $w_{2n}(x,t)$ could approach zero fairly rapidly with increasing n without violating the crucial inequality (4.2-7a). We shall illustrate this point with an example: We shall show that if the consolidated characterizing function $W(\xi \mid x,t)$ of an "extensive" jump Markov process $X(t)$ scales in a certain special way with x, then the forward Kramers-Moyal equation for the corresponding "intensive" process $Z(t) \equiv X(t)/\Omega$ can be approximately truncated at second order if the system volume Ω is taken sufficiently large.

It should be obvious that if $X(t)$ is a jump Markov process with initial condition x_0, then

$$Z(t) \equiv \Omega^{-1} X(t) \tag{4.2-8}$$

will be a jump Markov process with initial condition $z_0 = \Omega^{-1}x_0$. By the RVT corollary (1.6-5), the density function $Q(z,t \mid z_0,t_0)$ of $Z(t)$ will be related to the density function $P(x,t \mid x_0,t_0)$ of $X(t)$ according to

$$P(x,t \mid x_0,t_0) = Q(z,t \mid z_0,t_0) \left| \frac{dz}{dx} \right| = \Omega^{-1} Q(z,t \mid z_0,t_0),$$

where the second equality follows from the relation $z = \Omega^{-1}x$, as prescribed by the RVT corollary. This relation between z and x also implies that $\partial/\partial x = \partial/\partial(\Omega z) = \Omega^{-1}(\partial/\partial z)$, and hence that $(\partial/\partial x)^n = \Omega^{-n}(\partial/\partial z)^n$. Substituting these various relations into the forward Kramers-Moyal equation (4.2-4b) yields

---

† But the *odd* moments $w_{2n+1}(x,t)$ need *not* be strictly positive; indeed, if the density function $w(\xi \mid x,t)$ happens to be an even function of ξ, then all the odd moments will vanish.

$$\frac{\partial}{\partial t}\Omega^{-1}Q(z,t\,|\,z_0,t_0) = \sum_{n=1}^{\infty}\frac{(-1)^n}{n!}\Omega^{-n}\frac{\partial^n}{\partial z^n}\left[W_n(\Omega z,t)\Omega^{-1}Q(z,t\,|\,z_0,t_0)\right],$$

or

$$\frac{\partial}{\partial t}Q(z,t\,|\,z_0,t_0) = \sum_{n=1}^{\infty}\frac{(-1)^n}{n!}\frac{\partial^n}{\partial z^n}\left[\Omega^{-n}W_n(\Omega z,t)\,Q(z,t\,|\,z_0,t_0)\right]. \quad (4.2\text{-}9)$$

Equation (4.2-9) is evidently the forward Kramers-Moyal equation for the jump Markov process $Z(t)$, and we see upon comparing with the canonical form (2.5-1) that the nth propagator moment function for that process is $\Omega^{-n}W_n(\Omega z,t)$. The presence of the factor Ω^{-n} in the nth term of Eq. (4.2-9) suggests that in the limit of very large Ω we *might* be able to truncate the infinite sum without making any serious errors. But we cannot say anything definite in that regard unless we know how $W_n(\Omega z,t)$ behaves as a function of Ω.

As an example, let us suppose that the consolidated characterizing function $W(\xi\,|\,x,t)$ of the extensive process $X(t)$ satisfies the *scaling condition*

$$W(\xi\,|\,\beta x,t) \doteq \beta\,W(\xi\,|\,x,t) \quad (\beta\to\infty). \quad (4.2\text{-}10)$$

Here the dot over the equal sign denotes asymptotic equality in the limit that β becomes infinitely large (we do *not* assume equality for β small). Multiplying this equation through by ξ^n and integrating over all ξ we get, recalling the definition (4.2-2b),

$$W_n(\beta x,t) \doteq \beta\,W_n(x,t) \quad (\beta\to\infty).$$

Therefore, the nth propagator moment function of the intensive process $Z(t)$ satisfies

$$\Omega^{-n}W_n(\Omega z,t) \doteq \Omega^{-n}\Omega\,W_n(z,t) \equiv \Omega^{-(n-1)}W_n(z,t),$$

where the dot over the equal sign now means "for Ω sufficiently large." Substituting this into Eq. (4.2-9), we see that the nth term in that equation is now of order $(n-1)$ in the small quantity Ω^{-1}. In particular, to *first order* in that small quantity Eq. (4.2-9) reads

$$\frac{\partial}{\partial t}Q(z,t\,|\,z_0,t_0) \doteq -\frac{\partial}{\partial z}\left[W_1(z,t)\,Q(z,t\,|\,z_0,t_0)\right]$$

$$+ \frac{1}{2}\frac{\partial^2}{\partial z^2}\left[\Omega^{-1}W_2(z,t)\,Q(z,t\,|\,z_0,t_0)\right], \quad (4.2\text{-}11)$$

where again the dot over the equal sign signifies asymptotic equality for infinitely large Ω. This last equation has the canonical form of the *forward Fokker-Planck equation* (3.2-2). It therefore follows that, in the "thermodynamic limit" of very large system volume Ω, the intensive state variable $Z(t)$ behaves like a *continuous* Markov process with drift function

$$A(z,t) = W_1(z,t) \equiv \int_{-\infty}^{\infty} d\xi\,\xi\,W(\xi\,|\,z,t) \qquad (4.2\text{-}12a)$$

and diffusion function

$$D(z,t) = \Omega^{-1}\,W_2(z,t) \equiv \Omega^{-1}\int_{-\infty}^{\infty} d\xi\,\xi^2\,W(\xi\,|\,z,t). \qquad (4.2\text{-}12b)$$

We must emphasize, however, that this conclusion is predicated on the assumption that the consolidated characterizing function $W(\xi\,|\,x,t)$ of the associated extensive jump Markov process $X(t)$ satisfies the scaling condition (4.2-10).

The foregoing example demonstrates that even though the jump Kramers-Moyal equations can *never* be truncated *exactly*, it *might* be possible to truncate them *approximately*. However, approximate truncations of the jump Kramers-Moyal equations should never be simply assumed; they must be justified in terms of the specifics of the problem at hand. In the sequel, we shall encounter several other instances of approximate, two-term truncations of a jump Kramers-Moyal equation in the limit that some system parameter (such as Ω^{-1} in the example just considered) becomes very small. In each such instance, the resulting "Fokker-Planck form" of the Kramers-Moyal equation implies that the jump Markov process in question is being *approximated* as a *continuous* Markov process. This seems, in fact, to be how continuous Markov processes most often arise in physical applications, namely, as approximate or limiting versions of processes that are more accurately modeled as jump Markov processes.

We shall conclude this subsection by making some further observations about the intensive process $Z(t)$ just considered, whose corresponding "extensive" process $X(t)$ has a consolidated characterizing function $W(\xi\,|\,x,t)$ that satisfies the scaling condition (4.2-10). We found in Eq. (4.2-12) that if the system volume Ω is sufficiently large, then $Z(t)$ behaves like a *continuous* Markov process with drift function $W_1(z,t)$ and diffusion function $\Omega^{-1}W_2(z,t)$. Since Ω^{-1} is by presumption "very small," then this continuous Markov process is what we called in Section 3.8 a *weak noise process* [see Eq. (3.8-4)]. So all the results developed in Subsection 3.8.A apply here if we simply make the substitutions

$$X(t) \rightarrow Z(t), \quad \varepsilon \rightarrow \Omega^{-1}, \quad A(x,t) \rightarrow W_1(z,t), \quad D_1(x,t) \rightarrow W_2(z,t). \quad (4.2\text{-}13)$$

We may summarize those weak noise results for this case as follows: When Ω^{-1} is sufficiently small, we can write the (approximately) continuous Markov process $Z(t)$ as

$$Z(t) = z^*(t) + \Omega^{-1/2} Y(t). \quad (4.2\text{-}14)$$

Here, $z^*(t)$ is the sure function defined by the differential equation

$$\frac{d}{dt} z^*(t) = W_1(z^*(t), t) \quad (4.2\text{-}15)$$

and the initial condition $z^*(t_0) = z_0$. And $Y(t)$ is the continuous Markov process with drift function $y W_1{}'(z^*(t),t)$, diffusion function $W_2(z^*(t),t)$, and initial condition $Y(t_0) = 0$; hence, the density function $R(y,t \mid 0,t_0)$ of the process $Y(t)$ satisfies the Fokker-Planck equation

$$\frac{\partial}{\partial t} R(y,t \mid 0,t_0) = -\frac{\partial}{\partial y} \left[y\, W_1{}'(z^*(t),t)\, R(y,t \mid 0,t_0) \right]$$
$$+ \frac{1}{2} \frac{\partial^2}{\partial y^2} \left[W_2(z^*(t),t)\, R(y,t \mid 0,t_0) \right]. \quad (4.2\text{-}16)$$

Equation (4.2-14) says that $Z(t)$ follows the deterministic solution $z^*(t)$, but with "additive noise" that goes to zero as $\Omega \rightarrow \infty$ like $\Omega^{-1/2}$.

Furthermore, if the process $X(t)$ is temporally homogeneous, so that $W_1(x,t) = W_1(x)$ and $W_2(x,t) = W_2(x)$, and if also the function W_1 is such that the solution $z^*(t)$ of Eq. (4.2-15) approaches a constant value z_∞ as $t \rightarrow \infty$, then we will have [cf. Eq. (3.8-16)]

$$Z(t \rightarrow \infty) = N(z_\infty, \Omega^{-1} | W_2(z_\infty)/2 W_1{}'(z_\infty) |). \quad (4.2\text{-}17)$$

In words, for t (as well as Ω) sufficiently large, $Z(t)$ will essentially be a *normal* random variable with mean z_∞ and standard deviation $| W_2(z_\infty)/2\Omega W_1{}'(z_\infty) |^{1/2}$. Notice again that the fluctuations in $Z(t \rightarrow \infty)$ go to zero as $\Omega \rightarrow \infty$ like $\Omega^{-1/2}$.

These results illustrate the insights that can be gained whenever we are able to approximate an infinite order jump Kramers-Moyal equation by a second order Fokker-Planck type equation. But it is worth repeating that an approximate truncation of a jump Kramers-Moyal equation should never be simply assumed; it must always by justified in terms of the specifics of the problem at hand.

4.2.B THE MASTER EQUATIONS

The difficulties of working with infinite order partial differential equations motivate us to look for alternatives to the jump Kramers-Moyal equations (4.2-4) and (4.2-5). We recall that the Kramers-Moyal equations were originally derived, in Section 2.5, from the Chapman-Kolmogorov equations (2.2-5a) and (2.2-5b). Rewriting the latter two equations with $\Delta t = dt$ and $\Delta t_0 = dt_0$ respectively, we get

$$P(x,t+dt|x_0,t_0) = \int_{-\infty}^{\infty} d\xi \, P(x,t+dt|x-\xi,t) P(x-\xi,t|x_0,t_0)$$

and

$$P(x,t|x_0,t_0) = \int_{-\infty}^{\infty} d\xi \, P(x,t|x_0+\xi,t_0+dt_0) P(x_0+\xi,t_0+dt_0|x_0,t_0).$$

But by the fundamental relation (2.4-4) we have

$$P(x,t+dt|x-\xi,t) \equiv P(x-\xi+\xi,t+dt|x-\xi,t) = \Pi(\xi|dt; x-\xi,t)$$

and also

$$P(x_0+\xi,t_0+dt_0|x_0,t_0) = \Pi(\xi|dt_0; x_0,t_0).$$

Therefore, the preceding two Chapman-Kolmogorov equations may respectively be written

$$P(x,t+dt|x_0,t_0) = \int_{-\infty}^{\infty} d\xi \, \Pi(\xi|dt; x-\xi,t) P(x-\xi,t|x_0,t_0), \qquad (4.2\text{-}18)$$

and

$$P(x,t|x_0,t_0) = \int_{-\infty}^{\infty} d\xi \, P(x,t|x_0+\xi,t_0+dt_0) \Pi(\xi|dt_0; x_0,t_0). \qquad (4.2\text{-}19)$$

Equations (4.2-18) and (4.2-19) are valid for *any* Markov process $X(t)$. But now suppose that $X(t)$ is a *jump* Markov process with characterizing functions $a(x,t)$ and $w(\xi|x,t)$. In that case we can write out the propagator density functions in these two equations explicitly from formula (4.1-13). Substituting Eq. (4.1-13) into Eq. (4.2-18), we get

$$P(x,t+dt|x_0,t_0) = \int_{-\infty}^{\infty} d\xi \left[a(x-\xi,t)dt \, w(\xi|x-\xi,t) \right.$$

$$\left. + [1-a(x-\xi,t)dt] \, \delta(\xi) \right] P(x-\xi,t|x_0,t_0),$$

$$= \int_{-\infty}^{\infty} d\xi \, a(x-\xi,t)dt \, w(\xi|x-\xi,t) P(x-\xi,t|x_0,t_0)$$

$$+ [1-a(x,t)dt] P(x,t|x_0,t_0).$$

Subtracting $P(x,t \mid x_0,t_0)$ from both sides, dividing through by dt, and then taking the limit $dt \to 0$, we obtain

$$\frac{\partial}{\partial t} P(x,t \mid x_0,t_0) = \int_{-\infty}^{\infty} a(x-\xi,t)\, w(\xi \mid x-\xi,t)\, P(x-\xi,t \mid x_0,t_0)\, d\xi$$
$$- a(x,t)\, P(x,t \mid x_0,t_0). \tag{4.2-20a}$$

And if we multiply the second term on the right side by unity in the form of the integral $\int w(-\xi \mid x,t)d\xi$ [Eq. (4.1-5b) with the integration variable change $\xi \to -\xi$], and then recall the definition (4.1-20) of the consolidated characterizing function, we obtain the equivalent equation

$$\frac{\partial}{\partial t} P(x,t \mid x_0,t_0) = \int_{-\infty}^{\infty} \left[W(\xi \mid x-\xi,t)\, P(x-\xi,t \mid x_0,t_0) \right.$$
$$\left. - W(-\xi \mid x,t)\, P(x,t \mid x_0,t_0) \right] d\xi. \tag{4.2-20b}$$

Equations (4.2-20) are (both) called the **forward master equation** for the jump Markov process with characterizing functions $a(x,t)$ and $w(\xi \mid x,t)$. They are evidently differential-integral equations for $P(x,t \mid x_0,t_0)$ for fixed x_0 and t_0, and are essentially equivalent to the forward Kramers-Moyal equations (4.2-4); indeed, Eqs. (4.2-4) can be derived from Eqs. (4.2-20) by simply expanding the integrands in the latter in a Taylor series in $(-\xi)$ and then performing the ξ-integration term by term.

The backward companions to Eqs. (4.2-20) are similarly obtained by inserting into the Chapman-Kolmogorov equation (4.2-19) the explicit formula (4.1-13) for the jump propagator density function. That gives

$$P(x,t \mid x_0,t_0) = \int_{-\infty}^{\infty} d\xi\, P(x,t \mid x_0+\xi, t_0+dt_0)$$
$$\times \left[a(x_0,t_0)dt_0\, w(\xi \mid x_0,t_0) + [1 - a(x_0,t_0)dt_0]\, \delta(\xi) \right]$$
$$= \int_{-\infty}^{\infty} d\xi\, P(x,t \mid x_0+\xi, t_0+dt_0)\, a(x_0,t_0)dt_0\, w(\xi \mid x_0,t_0)$$
$$+ P(x,t \mid x_0, t_0+dt_0)\, [1 - a(x_0,t_0)dt_0].$$

Subtracting $P(x,t \mid x_0, t_0+dt_0)$ from both sides, dividing through by dt_0, and then taking the limit $dt_0 \to 0$, we obtain the result

$$-\frac{\partial}{\partial t_0} P(x,t \mid x_0,t_0) = a(x_0,t_0) \int_{-\infty}^{\infty} w(\xi \mid x_0,t_0)\, P(x,t \mid x_0+\xi, t_0)\, d\xi$$
$$- a(x_0,t_0)\, P(x,t \mid x_0,t_0). \tag{4.2-21a}$$

And if we multiply the second term on the right side by unity in the form of the integral $\int w(\xi \mid x_0,t_0)d\xi$ [Eq. (4.1-5b)], and then recall the definition (4.1-20) of the consolidated characterizing function, we obtain the equivalent equation

$$-\frac{\partial}{\partial t_0} P(x,t \mid x_0,t_0) = \int_{-\infty}^{\infty} W(\xi \mid x_0,t_0) \left[P(x,t \mid x_0+\xi,t_0) - P(x,t \mid x_0,t_0) \right] d\xi.$$

(4.2-21b)

Equations (4.2-21) are (both) called the **backward master equation** for the jump Markov process with characterizing functions $a(x,t)$ and $w(\xi \mid x,t)$. They are evidently differential-integral equations for $P(x,t \mid x_0,t_0)$ for fixed x and t, and are essentially equivalent to the backward Kramers-Moyal equations (4.2-5); indeed, Eqs. (4.2-5) can be derived from Eqs. (4.2-21) by simply expanding the integrands in the latter in a Taylor series in ξ and then performing the ξ-integration term by term.

As with the jump Kramers-Moyal equations, if the characterizing functions $a(x,t)$ and $w(\xi \mid x,t)$ are explicitly independent of t, then the left hand sides of the forward and backward master equations (4.2-20) and (4.2-21) will be *interchangeable*; this of course is a consequence of Eq. (2.8-5), which holds for any temporally homogeneous Markov processes.

It is perhaps problematical whether the master equations, being integro-differential equations, are any more tractable than the jump Kramers-Moyal equations, which are infinite order partial differential equations. In truth, none of these equations seems to promise an easy route to calculating $P(x,t \mid x_0,t_0)$. From the standpoint of general applicability though, the master equations would seem to have an edge on two counts: First, unlike the jump Kramers-Moyal equations, the master equations do not require the existence (finiteness) of all the moments $w_n(x,t)$ of the characterizing function $w(\xi \mid x,t)$. And second, unlike the jump Kramers-Moyal equations, the master equations can be easily adapted to the case in which the state variable x is confined to the integers, as we shall see in detail in Chapter 5. On the other hand, the Kramers-Moyal equations can offer significant computational advantages in those instances in which they happen to admit an *approximate truncation*, as discussed in the preceding subsection. But if an approximate truncation of the forward Kramers-Moyal equation cannot be had, it seems likely that the forward master equation would be more amenable to a brute force numerical calculation of the Markov state density function $P(x,t \mid x_0,t_0)$.

4.2.C THE MOMENT EVOLUTION EQUATIONS

Because the time-evolution equations for $P(x,t \mid x_0,t_0)$ are so formidable for most jump Markov processes, it is useful to have time-evolution equations for the mean and variance of the process. In Section 2.7 we derived time-evolution equations for the various moments of *any* Markov process $X(t)$ and its integral $S(t)$ in terms of the propagator moment functions $B_n(x,t)$ of $X(t)$. At the beginning of the present section, we found that the propagator moment functions of a *jump* Markov process with consolidated characterizing function $W(\xi \mid x,t)$ are given by

$$B_n(x,t) = W_n(x,t) \equiv \int_{-\infty}^{\infty} \xi^n \, W(\xi \mid x_0,t_0) \, d\xi \quad (n=1,2,...). \quad (4.2\text{-}22)$$

Thus, the various moment evolution equations developed in Section 2.7 can be specialized to the jump case by simply replacing B_n everywhere by W_n. For later reference, we shall now write down those equations.

The evolution equations for $\langle X^n(t) \rangle$ and $\langle S^n(t) \rangle$ for arbitrary n were derived in Subsection 2.7.A. For the moments of $X(t)$ we have the equations

$$\frac{d}{dt} \langle X^n(t) \rangle = \sum_{k=1}^{n} \binom{n}{k} \langle X^{n-k}(t) \, W_k(X(t),t) \rangle$$

$$(t_0 \leq t; n=1,2,...) \quad (4.2\text{-}23)$$

with the initial conditions $\langle X^n(t_0) \rangle = x_0^n$. For the moments of $S(t)$ we have the equations

$$\frac{d}{dt} \langle S^m(t) \rangle = m \langle S^{m-1}(t) \, X(t) \rangle \quad (t_0 \leq t; m=1,2,...) \quad (4.2\text{-}24)$$

with the initial conditions $\langle S^m(t_0) \rangle = 0$, together with the auxiliary equations

$$\frac{d}{dt} \langle S^m(t) \, X^n(t) \rangle = m \langle S^{m-1}(t) \, X^{n+1}(t) \rangle$$

$$+ \sum_{k=1}^{n} \binom{n}{k} \langle S^m(t) \, X^{n-k}(t) \, W_k(X(t),t) \rangle$$

$$(t_0 \leq t; m \geq 1; n \geq 1). \quad (4.2\text{-}25)$$

and their initial conditions $\langle S^m(t_0) \, X^n(t_0) \rangle = 0$. These moment evolution equations will be closed if and only if $W_k(x,t)$ is a polynomial in x of degree $\leq k$. If $W_k(x,t)$ does not satisfy that rather stringent condition then the most we can hope for is approximate solutions, obtained either by suitably approximating $W_k(x,t)$ or else by applying the procedure outlined in Appendix C.

The moment evolution equations in Subsection 2.7.A were all derived from the general properties of the propagators of the processes $X(t)$ and $S(t)$. It is instructive to see how the time-evolution equation for $\langle X^n(t)\rangle$ in the jump case can also be derived from the forward master equation: Abbreviating $P(x,t\,|\,x_0,t_0) \equiv P(x,t)$, we have for any positive integer n,

$$\frac{d}{dt}\langle X^n(t)\rangle = \frac{d}{dt}\int_{-\infty}^{\infty} dx\, x^n\, P(x,t) = \int_{-\infty}^{\infty} dx\, x^n\, \frac{\partial}{\partial t} P(x,t)$$

$$= \int_{-\infty}^{\infty} dx \int_{-\infty}^{\infty} d\xi\, x^n\, W(\xi\,|\,x-\xi,t)\, P(x-\xi,t)$$

$$- \int_{-\infty}^{\infty} dx \int_{-\infty}^{\infty} d\xi\, x^n\, W(-\xi\,|\,x,t)\, P(x,t),$$

where the last step has invoked the forward master equation (4.2-20b). In the first integral we change integration variable x to $x-\xi$, and in the second integral we change the integration variable ξ to $-\xi$. This gives

$$\frac{d}{dt}\langle X^n(t)\rangle = \int_{-\infty}^{\infty} dx \int_{-\infty}^{\infty} d\xi\, (x+\xi)^n\, W(\xi\,|\,x,t)\, P(x,t)$$

$$- \int_{-\infty}^{\infty} dx \int_{-\infty}^{\infty} d\xi\, x^n\, W(\xi\,|\,x,t)\, P(x,t)$$

$$= \int_{-\infty}^{\infty} dx \int_{-\infty}^{\infty} d\xi\, \left[(x+\xi)^n - x^n\right] W(\xi\,|\,x,t)\, P(x,t).$$

Expanding $(x+\xi)^n$ using the binomial formula, we get

$$\frac{d}{dt}\langle X^n(t)\rangle = \int_{-\infty}^{\infty} dx \int_{-\infty}^{\infty} d\xi \left[\sum_{k=1}^{n} \binom{n}{k} x^{n-k}\xi^k\right] W(\xi\,|\,x,t)\, P(x,t)$$

$$= \sum_{k=1}^{n} \binom{n}{k} \int_{-\infty}^{\infty} dx\, x^{n-k} \left[\int_{-\infty}^{\infty} d\xi\, \xi^k\, W(\xi\,|\,x,t)\right] P(x,t)$$

$$= \sum_{k=1}^{n} \binom{n}{k} \int_{-\infty}^{\infty} dx\, x^{n-k}\, W_k(x,t)\, P(x,t) \qquad \text{[by (4.2-2b)]}$$

$$= \sum_{k=1}^{n} \binom{n}{k} \langle X^{n-k}(t)\, W_k(X(t),t)\rangle, \qquad \text{[by (2.3-3)]}$$

in agreement with Eq. (4.2-23).

The time-evolution equations for the means, variances and covariances of any Markov process $X(t)$ and its integral $S(t)$ were derived in Subsection 2.7.B. Replacing B_n by W_n in those equations, we obtain the following equations for the case in which $X(t)$ is a *jump* Markov process.

For the mean, variance and covariance of $X(t)$ we have the differential equations:

$$\frac{d}{dt}\langle X(t)\rangle = \langle W_1(X(t),t)\rangle \qquad (t_0 \leq t), \qquad (4.2\text{-}26)$$

$$\frac{d}{dt}\text{var}\{X(t)\} = 2\Big(\langle X(t)\,W_1(X(t),t)\rangle - \langle X(t)\rangle\langle W_1(X(t),t)\rangle\Big) + \langle W_2(X(t),t)\rangle$$
$$(t_0 \leq t), \qquad (4.2\text{-}27)$$

$$\frac{d}{dt_2}\text{cov}\{X(t_1),X(t_2)\} = \langle X(t_1)\,W_1(X(t_2),t_2)\rangle - \langle X(t_1)\rangle\langle W_1(X(t_2),t_2)\rangle$$
$$(t_0 \leq t_1 \leq t_2), \qquad (4.2\text{-}28)$$

with the respective initial conditions

$$\langle X(t=t_0)\rangle = x_0, \qquad (4.2\text{-}29)$$

$$\text{var}\{X(t=t_0)\} = 0, \qquad (4.2\text{-}30)$$

$$\text{cov}\{X(t_1),X(t_2=t_1)\} = \text{var}\{X(t_1)\}. \qquad (4.2\text{-}31)$$

For the mean of $S(t)$ we have

$$\langle S(t)\rangle = \int_{t_0}^{t}\langle X(t')\rangle\,dt', \qquad (4.2\text{-}32)$$

wherein the integrand is the solution to Eq. (4.2-26). For the variance of $S(t)$ we have

$$\text{var}\{S(t)\} = 2\int_{t_0}^{t}\text{cov}\{S(t'),X(t')\}\,dt', \qquad (4.2\text{-}33)$$

wherein the integrand is the solution of the differential equation

$$\frac{d}{dt}\text{cov}\{S(t),X(t)\} = \text{var}\{X(t)\} + \langle S(t)\,W_1(X(t),t)\rangle - \langle S(t)\rangle\langle W_1(X(t),t)\rangle$$
$$(t_0 \leq t) \qquad (4.2\text{-}34a)$$

that satisfies the initial condition

$$\text{cov}\{S(t=t_0),X(t=t_0)\} = 0. \qquad (4.2\text{-}34b)$$

And finally, for the covariance of $S(t)$ we have

$$\text{cov}\{S(t_1),S(t_2)\} = \text{var}\{S(t_1)\} + \int_{t_1}^{t_2} \text{cov}\{S(t_1),X(t_2')\} \, dt_2'$$

$$(t_0 \leq t_1 \leq t_2), \quad (4.2\text{-}35)$$

wherein the integrand is the solution of the differential equation

$$\frac{d}{dt_2} \text{cov}\{S(t_1),X(t_2)\} = \langle S(t_1) \, W_1(X(t_2),t_2) \rangle - \langle S(t_1) \rangle \langle W_1(X(t_2),t_2) \rangle$$

$$(t_0 \leq t_1 \leq t_2) \quad (4.2\text{-}36a)$$

that satisfies the initial condition

$$\text{cov}\{S(t_1),X(t_2=t_1)\} = \text{cov}\{S(t_1),X(t_1)\}, \quad (4.2\text{-}36b)$$

which in turn is known from the solution to Eq. (4.2-34a).

If it should happen that $W_1(x,t)=b_1$ and $W_2(x,t)=b_2$ where b_1 and b_2 are both constants, then the means, variances and covariances of $X(t)$ and $S(t)$ will be given by the explicit formulas (2.7-28) and (2.7-29). If it should happen that $W_1(x,t)=-\beta x$ and $W_2(x,t)=c$ where β and c are both constants, then the means, variances and covariances of $X(t)$ and $S(t)$ will be given by the explicit formulas (2.7-34) and (2.7-35).

4.3 THE NEXT-JUMP DENSITY FUNCTION

In Section 4.1 we discovered how the jump characterizing functions $a(x,t)$ and $w(\xi \mid x,t)$, as defined in Eqs. (4.1-11) and (4.1-2) respectively, determine the jump propagator density function $\Pi(\xi \mid dt; x,t)$. We recall from Eqs. (2.4-1) and (2.4-2) that $\Pi(\xi \mid dt; x,t)$ is essentially defined by the statement

$\Pi(\xi \mid dt; x,t) \, d\xi \equiv$ probability that, given the process is in state x at time t, it will be found at time $t+dt$ to be in some state between $x+\xi$ and $x+\xi+d\xi$.   (4.3-1)

This definition applies, of course, for *both* continuous and jump Markov processes. We now introduce a function, $p(\tau,\xi \mid x,t)$, which is deceptively similar to $\Pi(\xi \mid dt; x,t)$, but which has meaning *only* for *jump* Markov processes. The definition of the function p is as follows:

$p(\tau,\xi \mid x,t)\,d\tau\,d\xi \equiv$ probability that, given the process is in state x at time t, its *next jump* will occur between times $t+\tau$ and $t+\tau+d\tau$, and will carry the process to some state between $x+\xi$ and $x+\xi+d\xi$.   (4.3-2)

The function $p(\tau,\xi \mid x,t)$ will be called the **next-jump density function**, because it is a density function that prescribes when and to where the process will jump next. That the definition (4.3-2) cannot be applied to a continuous Markov process follows from the fact that in a continuous Markov process we cannot speak of when the process will leave state x, or what the state subsequent to x will be.

From the viewpoint of a *jump* Markov process, both definitions (4.3-1) and (4.3-2) make sense. Notice that $\Pi(\xi \mid dt; x,t)$ is the density function for the state-change vector (ξ) over the next *specified* time interval dt. In contrast, $p(\tau,\xi \mid x,t)$ is the *joint* density function for the time (τ) to the next jump and the state-change vector (ξ) in that next jump. Unlike the density function Π, the joint density function p is not parametrically dependent upon a *preselected* time interval dt; this feature, as we shall see later in Section 4.6, will make p useful for constructing Monte Carlo simulations of jump Markov processes.

It is not difficult to derive an exact expression for $p(\tau,\xi \mid x,t)$ in terms of the process characterizing functions $a(x,t)$ and $w(\xi \mid x,t)$. First, using the multiplication law of probability theory, we write the probability in Eq. (4.3-2) as

$p(\tau,\xi \mid x,t)\,d\tau\,d\xi$

$\equiv$ Prob{ given $X(t)=x$, *no* jump occurs in $[t,t+\tau)$ }

$\times$ Prob{ given $X(t+\tau)=x$, one jump occurs in $[t+\tau, t+\tau+d\tau)$ }

$\times$ Prob{ given a jump from x at some time in $[t+\tau, t+\tau+d\tau)$, the jump vector will lie in $[\xi,\xi+d\xi)$ }.   (4.3-3)

By definition (4.1-14b), the first probability on the right side of this equation is $q^*(x,t; \tau)$. By Eq. (4.1-11), the second probability on the right side of Eq. (4.3-3) is $a(x,t+\tau)d\tau$. And by definition (4.1-2), the third probability on the right side of Eq. (4.3-3) is $w(\xi \mid x,t')d\xi$, where t' is the precise moment in $[t+\tau, t+\tau+d\tau)$ when the jump occurred; however, since the function $w(\xi \mid x,t)$ is always assumed to be smooth in t, then we can replace t' in this last probability by the infinitesimally close value $t+\tau$, at least to first order in $d\tau$. Thus, Eq. (4.3-3) becomes

$$p(\tau,\xi \mid x,t)\,d\tau\,d\xi = q^*(x,t; \tau) \times a(x,t+\tau)d\tau \times w(\xi \mid x,t+\tau)d\xi.$$

Recalling the formula for $q^*(x,t; \tau)$ in Eq. (4.1-15a), we conclude that the next-jump density function is given by the formula

$$p(\tau,\xi \,|\, x,t) = a(x,t+\tau)\exp\left(-\int_0^\tau a(x,t+\tau')d\tau'\right) w(\xi \,|\, x,t+\tau), \quad (4.3\text{-}4)$$

in which it is understood that the variable τ is restricted to nonnegative values.

It will later be convenient to "condition" the joint density function $p(\tau,\xi \,|\, x,t)$ according to

$$p(\tau,\xi \,|\, x,t) = p_1(\tau \,|\, x,t)\, p_2(\xi \,|\, \tau; x,t). \quad (4.3\text{-}5)$$

Here, as discussed more generally in Section 1.5, $p_1(\tau \,|\, x,t)$ is the density function for τ irrespective of ξ, and is obtained by integrating $p(\tau,\xi \,|\, x,t)$ over all ξ:

$$p_1(\tau \,|\, x,t) = \int_{-\infty}^{\infty} d\xi'\, p(\tau,\xi' \,|\, x,t), \quad (4.3\text{-}6a)$$

And $p_2(\xi \,|\, \tau; x,t)$ is the density function for ξ conditioned on τ, and is obtained by combining Eqs. (4.3-5) and (4.3-6a):

$$p_2(\xi \,|\, \tau; x,t) = \frac{p(\tau,\xi \,|\, x,t)}{\displaystyle\int_{-\infty}^{\infty} d\xi'\, p(\tau,\xi' \,|\, x,t)}. \quad (4.3\text{-}6b)$$

Substituting Eq. (4.3-4) into the right sides of Eqs. (4.3-6), and remembering that $w(\xi \,|\, x,t)$ is a properly normalized density function with respect to ξ, we easily find for the two subordinate density functions the formulas

$$\begin{cases} p_1(\tau \,|\, x,t) = a(x,t+\tau)\exp\left(-\int_0^\tau a(x,t+\tau')d\tau'\right), & (4.3\text{-}7a) \\[2mm] p_2(\xi \,|\, \tau; x,t) = w(\xi \,|\, x,t+\tau). & (4.3\text{-}7b) \end{cases}$$

A considerable simplification in the next-jump density function formulas occurs if the process $X(t)$ in question is *temporally homogeneous*,

$$a(x,t) \equiv a(x) \quad \text{and} \quad w(\xi \,|\, x,t) \equiv w(\xi \,|\, x),$$

as in fact most jump Markov processes encountered in practice are. In this case, the τ'-integrals appearing in Eqs. (4.3-4) and (4.3-7a) become simply $a(x)\tau$; thus, the next-jump density function (4.3-4) becomes

$$p(\tau,\xi \,|\, x,t) = a(x)\exp\left(-a(x)\,\tau\right) w(\xi \,|\, x), \quad (4.3\text{-}8)$$

and the associated conditioning density functions (4.3-7) become

$$\left\{ \begin{array}{ll} p_1(\tau\,|\,x,t) \;=\; a(x)\exp\!\left(-a(x)\,\tau\right), & \text{(4.3-9a)} \\[2ex] p_2(\xi\,|\,\tau;x,t) \;=\; w(\xi\,|\,x). & \text{(4.3-9b)} \end{array} \right.$$

Two features of these temporally homogeneous results should be marked well: First, since $p_1(\tau\,|\,x,t)$ now has the form of an exponential density function with decay constant $a(x)$, then it follows that the waiting time for the next jump from state x is an exponentially distributed random variable with mean $1/a(x)$. And second, since $p_2(\xi\,|\,\tau;x,t)$ is now independent of τ, then the next-jump displacement from state x is statistically independent of the waiting time to that jump. These two features are so important that they merit repeating in a slightly different way: For any *temporally homogeneous* jump Markov process, the characterizing functions $a(x)$ and $w(\xi\,|\,x)$ have the following interpretations:

> (i) The characterizing function $a(x)$ is the reciprocal of the
> mean of the random variable "pausing time in state x,"
> which is necessarily *exponentially distributed*.   (4.3-10a)

> (ii) The characterizing function $w(\xi\,|\,x)$ is the density
> function of the random variable "jump displacement
> from state x," which is necessarily *statistically
> independent* of the pausing time in state x.   (4.3-10b)

4.4 COMPLETELY HOMOGENEOUS JUMP MARKOV PROCESSES

As we might expect, the most tractable of the jump Markov processes are those that are *completely homogeneous*. Such processes by definition have characterizing functions of the form

$$a(x,t) \equiv a \quad \text{and} \quad w(\xi\,|\,x,t) \equiv w(\xi), \qquad \text{(4.4-1)}$$

where a is any positive constant and $w(\xi)$ is any properly normalized density function. The propagator density function for such a process simplifies from Eq. (4.1-13) to the form

$$\Pi(\xi\,|\,dt;x,t) \equiv \Pi(\xi\,|\,dt) \;=\; a\,dt\,w(\xi) + [1-a\,dt]\,\delta(\xi). \qquad \text{(4.4-2)}$$

And the next-jump density function simplifies from the temporally homogeneous form (4.3-8) to the form

$$p(\tau,\xi \mid x,t) \equiv p(\tau,\xi) = a \exp(-a\tau)\, w(\xi). \tag{4.4-3}$$

It follows from this last equation [see also the temporally homogeneous results (4.3-10)] that for the completely homogeneous jump Markov process defined by Eqs. (4.4-1), the pausing time in any state x is an exponential random variable with mean $1/a$, and the jump displacement from state x is a statistically independent random variable whose density function is $w(\xi)$. Such are the processes that ·.·e shall examine in this section.

In Subsection 4.4.A we shall specialize the various time-evolution equations developed in Section 4.2 to the completely homogeneous case. We shall find that the time-evolution equations for the moments of $X(t)$ and its integral $S(t)$ are straightforward enough, but the time-evolution equations for $P(x,t \mid x_0,t_0)$ are not as tractable as their continuous process counterparts (the exactly solvable equations of the Wiener process considered in Subsection 3.3.B). Perhaps this should come as no surprise, since a completely homogeneous *continuous* Markov process is characterized by just two constants [see Eqs. (3.3-8)], whereas one constant and one *function* are required to characterize a completely homogeneous *jump* Markov process [see Eqs. (4.4-1)]. We are able to develop a quadrature formula for $P(x,t \mid x_0,t_0)$, and although it is not easy to evaluate, it is at least an explicit closed-form expression. In Subsection 4.4.B we shall consider three examples of completely homogeneous jump Markov processes, one of which is interesting because it is an example of a well-defined Markov process that has no moments. We conclude in Subsection 4.4.C by developing and then analyzing a completely homogeneous jump Markov process that provides an approximate model of "self-diffusion" — the motion of a "labeled" molecule in a thermally equilibrized gas of otherwise identical molecules. A more rigorous approach to self-diffusion, and to the closely related phenomenon of Brownian motion, will be presented in Section 4.5.

4.4.A TIME-EVOLUTION EQUATIONS

As we noted in Section 2.8 [see Eqs. (2.8-20)], the Markov state density function $P(x,t \mid x_0,t_0)$ for any completely homogeneous Markov process depends on its four arguments only through the two differences $x - x_0$ and $t - t_0$:

$$P(x,t \mid x_0,t_0) \equiv P(x-x_0,t-t_0 \mid 0,0). \qquad (4.4\text{-}4)$$

Furthermore [see Eq. (2.8-22)], the propagator moment functions $B_n(x,t)$ for such a process are all constant with respect to x and t; indeed, it follows from Eqs. (4.2-1) and (4.2-2a) that we have in this case

$$B_n(x,t) = aw_n \qquad (n=1,2,...), \qquad (4.4\text{-}5a)$$

where w_n is the nth moment of the density function $w(\xi)$:

$$w_n \equiv \int_{-\infty}^{\infty} d\xi\, \xi^n\, w(\xi) \qquad (n=1,2,...). \qquad (4.4\text{-}5b)$$

To calculate the Markov state density function, one might appeal either to the forward Kramers-Moyal equation (4.2-4a), which in this case reads

$$\frac{\partial}{\partial t} P(x,t \mid x_0,t_0) = \sum_{n=1}^{\infty} \frac{(-1)^n}{n!} aw_n \frac{\partial^n}{\partial x^n} P(x,t \mid x_0,t_0), \qquad (4.4\text{-}6)$$

or to the forward master equation (4.2-20a), which in this case reads

$$\frac{\partial}{\partial t} P(x,t \mid x_0,t_0) = a \int_{-\infty}^{\infty} w(\xi)\, P(x-\xi,t \mid x_0,t_0)\, d\xi \; - \; a\, P(x,t \mid x_0,t_0).$$
$$(4.4\text{-}7)$$

We need not concern ourselves with the backward counterparts of these two equations, because they turn out to be the same: The backward equations can be obtained from the forward equations by simply substituting into the forward equations the relations

$$-\frac{\partial P}{\partial t_0} = \frac{\partial P}{\partial t}, \quad \frac{\partial P}{\partial x_0} = -\frac{\partial P}{\partial x}, \quad \text{and} \quad P(x,t \mid x_0+\xi,t_0) = P(x-\xi,t \mid x_0,t_0),$$

which follow from the symmetry (4.4-4).

Unfortunately, neither Eq. (4.4-6) nor Eq. (4.4-7) yields easily to solution, although Eq. (4.4-7) is to some degree approachable using Laplace transform techniques. But we recall from our general discussion of completely homogeneous Markov processes in Subsection 2.8.B that there is yet another route to calculating $P(x,t \mid x_0,t_0)$ for such processes, namely the quadrature formula (2.8-18); indeed, that is the formula that we used in Subsection 3.3.B to calculate $P(x,t \mid x_0,t_0)$ for the completely homogeneous *continuous* Markov process [see Eq. (3.3-13)]. To see what the integral formula (2.8-18) implies for the *jump* case, we must first evaluate the function $\Pi^*(u \mid dt)$ defined in the auxiliary equation

(2.8-19a) for the jump propagator density function $\Pi(\xi \,|\, dt)$ in Eq.(4.4-2). We have

$$
\begin{aligned}
\Pi^\star(u \,|\, dt) &= \int_{-\infty}^{\infty} d\xi \left[a\,dt\,w(\xi) + [1 - a\,dt]\,\delta(\xi) \right] \exp(-iu\xi) \\
&= a\,dt \int_{-\infty}^{\infty} d\xi\,w(\xi) \exp(-iu\xi) \\
&\qquad\qquad + [1 - a\,dt] \int_{-\infty}^{\infty} d\xi\,\delta(\xi) \exp(-iu\xi) \\
&= a\,dt\,w^\star(u) + [1 - a\,dt]\exp(0) \\
&= 1 + a\,dt\,[w^\star(u) - 1],
\end{aligned}
$$

where we have denoted the Fourier transform of $w(\xi)$ by $w^\star(u)$. Since Eq. (2.8-18) requires the nth power of this function, where according to Eq. (2.8-19b) $n = (t - t_0)/dt$, we next calculate

$$
\begin{aligned}
\left[\Pi^\star(u \,|\, dt) \right]^n &= \left[1 + a\,dt\,[w^\star(u) - 1] \right]^n \\
&= \left[1 + \frac{a(n\,dt)\,[w^\star(u) - 1]}{n} \right]^n.
\end{aligned}
$$

But since n here is infinitely large, with $n\,dt = t - t_0$, then this is the same as

$$
\left[\Pi^\star(u \,|\, dt) \right]^n = \exp\!\left(a(t - t_0)\,[w^\star(u) - 1] \right).
$$

Substituting this result into Eq. (2.8-18), we conclude that for the completely homogeneous jump Markov process defined by Eqs. (4.4-1) the Markov state density function is given by

$$
P(x,t \,|\, x_0,t_0) = (2\pi)^{-1} \oint_{-\infty}^{\infty} du\, \exp[iu(x - x_0)] \exp\!\left(a(t - t_0)\,[w^\star(u) - 1] \right),
$$

(4.4-8)

where

$$
w^\star(u) \equiv \int_{-\infty}^{\infty} d\xi\,w(\xi) \exp(-iu\xi).
$$
(4.4-9)

Equation (4.4-8) may be regarded as a "solution-in-quadrature" to the time-evolution equations (4.4-6) and (4.4-7) for the initial condition $P(x,t_0 \,|\, x_0,t_0) = \delta(x - x_0)$. Application of this quadrature solution can be shown to be roughly equivalent to solving Eq. (4.4-7) by Laplace transform techniques. In any event, if the Fourier transform $w^\star(u)$ of the characterizing function $w(\xi)$ can be calculated analytically, as in fact it can for most commonly encountered density functions, then Eq. (4.4-8) allows $P(x,t \mid x_0,t_0)$ to be computed by performing only one integration. However, as we shall see shortly, that one integration is usually *not* easy to effect, because convergence almost always requires that the special features of the definition (B-6) be taken explicitly into account. We note in passing though that the formula (4.4-8) does make manifest the symmetry condition (4.4-4).

Now let us see what can be said about the moments of $X(t)$ and its integral $S(t)$. Their evolution equations are most easily obtained by simply replacing the propagator moment functions $W_n(x,t)$ in the formulas of Subsection 4.2.C by aw_n. Thus, from Eq. (4.2-23) we find for the moments of $X(t)$ the set of coupled differential equations

$$\frac{d}{dt}\langle X^n(t)\rangle = \sum_{k=1}^{n} \binom{n}{k} aw_k \langle X^{n-k}(t)\rangle \quad (t_0 \le t; n=1,2,\ldots), \quad (4.4\text{-}10)$$

with the initial conditions $\langle X^n(t_0)\rangle = x_0{}^n$. And from Eqs. (4.2-24) and (4.2-25) we find for the moments of $S(t)$ the equations

$$\frac{d}{dt}\langle S^m(t)\rangle = m\langle S^{m-1}(t)X(t)\rangle \quad (t_0 \le t; m=1,2,\ldots) \quad (4.4\text{-}11)$$

with the initial conditions $\langle S^m(t_0)\rangle = 0$, together with the auxiliary equations

$$\frac{d}{dt}\langle S^m(t)X^n(t)\rangle = m\langle S^{m-1}(t)X^{n+1}(t)\rangle$$

$$+ \sum_{k=1}^{n} \binom{n}{k} aw_k \langle S^m(t)X^{n-k}(t)\rangle$$

$$(t_0 \le t; m\ge 1; n\ge 1). \quad (4.4\text{-}12)$$

and their initial conditions $\langle S^m(t_0)X^n(t_0)\rangle = 0$. These moment evolution equations are all *closed*, and can be solved recursively for as many moments as cne desires. We shall content ourselves here merely with obtaining explicit formulas for the means, variances and covariances of $X(t)$ and $S(t)$. That is most easily done by appealing to the general results obtained earlier in Subsection 2.7.C: Since the first two propagator moment functions in this case are the *constants* $B_1(x,t) = aw_1$ and $B_2(x,t) = aw_2$, then Eqs. (2.7-28) and (2.7-29) apply with $b_1 = aw_1$ and $b_2 = aw_2$. Thus we find for the mean, variance and covariance of $X(t)$,

$$\langle X(t)\rangle = x_0 + aw_1(t-t_0) \qquad (t_0 \leq t), \qquad (4.4\text{-}13\text{a})$$

$$\mathrm{var}\{X(t)\} = aw_2(t-t_0) \qquad (t_0 \leq t), \qquad (4.4\text{-}13\text{b})$$

$$\mathrm{cov}\{X(t_1),X(t_2)\} = aw_2(t_1 - t_0) \qquad (t_0 \leq t_1 \leq t_2). \qquad (4.4\text{-}13\text{c})$$

And we find for the mean, variance and covariance of the integral $S(t)$ of $X(t)$,

$$\langle S(t)\rangle = x_0(t-t_0) + (aw_1/2)(t-t_0)^2 \quad (t_0 \leq t), \qquad (4.4\text{-}14\text{a})$$

$$\mathrm{var}\{S(t)\} = (aw_2/3)(t-t_0)^3 \qquad (t_0 \leq t), \qquad (4.4\text{-}14\text{b})$$

$$\mathrm{cov}\{S(t_1),S(t_2)\} = aw_2(t_1-t_0)^2\,[(t_1-t_0)/3 + (t_2-t_1)/2]$$
$$(t_0 \leq t_1 \leq t_2). \quad (4.4\text{-}14\text{c})$$

It should be noted that these last six equations are identical to those for the Wiener process [Eqs. (3.3-10) and (3.3-11)] if we identify aw_1 with A and aw_2 with D; however, the equations for the third and higher order moments for the completely homogeneous continuous and jump Markov processes will *not* be so simply related.

Since Eq. (4.1-11) implies that the characterizing function $a(x,t)$ has the units of $[\text{time}]^{-1}$, then when that function is equal to a constant a we can make $a=1$ by suitably choosing the unit of time. This time scaling property of a for completely homogeneous jump Markov processes should be readily apparent in the various formulas derived in this section; for example, the right side of Eq. (4.4-8) depends on a and t only through the combination $t' = at$.

4.4.B THREE EXAMPLES

We shall now consider three examples of a completely homogeneous jump Markov process, obtained by taking $w(\xi)$ to be the density function of first an exponential, then a normal, and finally a Cauchy random variable. In Subsection 4.4.C we shall consider a fourth choice for $w(\xi)$ that provides us with an approximate model of the physical phenomenon of self-diffusion.

For our first example, we take $w(\xi)$ to be the density function of the exponential random variable with mean and standard deviation σ:

$$w(\xi) = \begin{cases} (1/\sigma)\exp(-\xi/\sigma) & \text{if } \xi \geq 0, \\ 0, & \text{if } \xi < 0. \end{cases} \qquad (4.4\text{-}15)$$

The moments of this density function are found by substituting $a = 1/\sigma$ into Eq. (1.4-6):

$$w_n = n!\,\sigma^n \qquad (n = 1, 2, 3, \ldots). \qquad (4.4\text{-}16)$$

In particular, we have $w_1 = \sigma$ and $w_2 = 2\sigma^2$, so the mean, variance and covariance formulas (4.4-13) and (4.4-14) become

$$\langle X(t) \rangle = x_0 + a\sigma(t - t_0) \qquad (t_0 \le t), \qquad (4.4\text{-}17a)$$

$$\mathrm{var}\{X(t)\} = 2a\sigma^2(t - t_0) \qquad (t_0 \le t), \qquad (4.4\text{-}17b)$$

$$\mathrm{cov}\{X(t_1), X(t_2)\} = 2a\sigma^2(t_1 - t_0) \qquad (t_0 \le t_1 \le t_2); \qquad (4.4\text{-}17c)$$

$$\langle S(t) \rangle = x_0(t - t_0) + (a\sigma/2)(t - t_0)^2 \qquad (t_0 \le t), \qquad (4.4\text{-}18a)$$

$$\mathrm{var}\{S(t)\} = (2a\sigma^2/3)(t - t_0)^3 \qquad (t_0 \le t), \qquad (4.4\text{-}18b)$$

$$\mathrm{cov}\{S(t_1), S(t_2)\} = 2a\sigma^2(t_1 - t_0)^2 \,[(t_1 - t_0)/3 + (t_2 - t_1)/2]$$
$$(t_0 \le t_1 \le t_2). \qquad (4.4\text{-}18c)$$

Higher order moments may be found by substituting Eq. (4.4-16) into Eqs. (4.4-10) – (4.4-12) and then systematically solving the resulting differential equations.

The Kramers-Moyal equation for this process is obtained by substituting Eq. (4.4-16) into Eq. (4.4-6), and is straightforwardly found to be

$$\frac{\partial}{\partial t} P(x,t \,|\, x_0, t_0) = \sum_{n=1}^{\infty} (-1)^n\, a\sigma^n\, \frac{\partial^n}{\partial x^n} P(x,t \,|\, x_0, t_0), \qquad (4.4\text{-}19)$$

If σ happens to be very small compared to unity, then we might be justified in *approximating* this equation with a two-term truncation; this, by Eq. (3.3-12), would correspond to a Wiener process approximation with $A = a\sigma$ and $D = 2a\sigma^2$. Without some such approximate truncation, Eq. (4.4-19) would seem to be of little use for determining $P(x,t \,|\, x_0, t_0)$. The master equation for this process is obtained by substituting Eq. (4.4-15) into Eq. (4.4-7), and is found to be

$$\frac{\partial}{\partial t} P(x,t \,|\, x_0, t_0) = (a/\sigma) \int_0^{\infty} \exp(-\xi/\sigma)\, P(x - \xi, t \,|\, x_0, t_0)\, d\xi \,-\, a\, P(x,t \,|\, x_0, t_0).$$
$$(4.4\text{-}20)$$

One might try to solve this equation by using some integral transform method (e.g., the Laplace transform). Or, one might approximate the left side by a Δt-difference formula and then try to use the resulting equation to calculate $P(x,t \mid x_0,t_0)$ numerically for $t = t_0 + \Delta t$, $t_0 + 2\Delta t$, etc.

To apply the quadrature formula (4.4-8), we see that we must first calculate the function $w^\star(u)$ defined in Eq. (4.4-9):

$$w^\star(u) = \int_0^\infty d\xi \left[(1/\sigma) \exp(-\xi/\sigma) \right] e^{-iu\xi}$$

$$= \sigma^{-1} \int_0^\infty d\xi \, \exp[-\xi(\sigma^{-1} + iu)]$$

$$= \frac{\sigma^{-1}}{\sigma^{-1} + iu} = \frac{1}{1 + i\sigma u} .$$

Therefore,

$$w^\star(u) - 1 = \frac{-i\sigma u}{1 + i\sigma u} = -\frac{(i\sigma u + \sigma^2 u^2)}{1 + \sigma^2 u^2} .$$

Substituting this result into Eq. (4.4-8), and recalling the integral definition (B-6), we get

$$P(x,t \mid x_0,t_0) = (2\pi)^{-1} \lim_{\kappa \downarrow 0} \lim_{L \to \infty} \int_{-L}^{L} du \, e^{-\kappa |u|}$$

$$\times \exp[iu(x-x_0)] \exp\left(-\frac{a(t-t_0)(i\sigma u + \sigma^2 u^2)}{1 + \sigma^2 u^2} \right).$$

Taking account of the odd-even behavior of the integrand with respect to the integration variable u, this can be simplified slightly to

$$P(x,t \mid x_0,t_0) = \pi^{-1} \lim_{\kappa \downarrow 0} \int_0^\infty du \, e^{-\kappa u} \cos\left(u \left[(x-x_0) - \frac{\sigma a(t-t_0)}{1 + \sigma^2 u^2} \right] \right)$$

$$\times \exp\left(-\frac{a(t-t_0)\sigma^2 u^2}{1 + \sigma^2 u^2} \right). \tag{4.4-21}$$

It is not clear whether the u-integral here can be evaluated analytically; however, a numerical evaluation should be feasible if the integration range is first made finite by a suitable change of integration variable, say from u to $v = e^{-\kappa u}$.

From the foregoing considerations it would appear that calculating the Markov state density function for this process will not be an easy task, and that the most straightforward description of the process will probably be provided by moment equations such as Eqs. (4.4-17). This turns out to be fairly typical of completely homogeneous jump Markov processes with continuum states. Another approach to investigating these processes is Monte Carlo simulation, and that will be discussed in Section 4.6.

For our second example, we let $w(\xi)$ be the density function of the normal random variable with mean 0 and variance σ^2:

$$w(\xi) = (2\pi\sigma^2)^{-1/2} \exp(-\xi^2/2\sigma^2). \qquad (4.4\text{-}22)$$

The moments of this density function are found by substituting $m=0$ and $a^2=\sigma^2$ into the general formula (1.4-10):

$$w_n = \begin{cases} \dfrac{n! \, (\sigma^2/2)^{n/2}}{(n/2)!}, & \text{if } n=0,2,4,\dots \\[2mm] 0, & \text{if } n=1,3,5,\dots \end{cases} \qquad (4.4\text{-}23)$$

In particular, we see that $w_1=0$ and $w_2=\sigma^2$, so the mean, variance and covariance formulas (4.4-13) and (4.4-14) become

$$\langle X(t) \rangle = x_0 \qquad\qquad (t_0 \leq t), \qquad (4.4\text{-}24\text{a})$$

$$\mathrm{var}\{X(t)\} = a\sigma^2(t-t_0) \qquad\qquad (t_0 \leq t), \qquad (4.4\text{-}24\text{b})$$

$$\mathrm{cov}\{X(t_1),X(t_2)\} = a\sigma^2(t_1-t_0) \qquad (t_0 \leq t_1 \leq t_2); \qquad (4.4\text{-}24\text{c})$$

$$\langle S(t) \rangle = x_0(t-t_0) \qquad\qquad (t_0 \leq t), \qquad (4.4\text{-}25\text{a})$$

$$\mathrm{var}\{S(t)\} = (a\sigma^2/3)(t-t_0)^3 \qquad\qquad (t_0 \leq t), \qquad (4.4\text{-}25\text{b})$$

$$\mathrm{cov}\{S(t_1),S(t_2)\} = a\sigma^2(t_1-t_0)^2 \left[(t_1-t_0)/3 + (t_2-t_1)/2\right]$$
$$(t_0 \leq t_1 \leq t_2). \qquad (4.4\text{-}25\text{c})$$

Higher order moments may be found by substituting Eq. (4.4-23) into Eqs. (4.4-10) – (4.4-12) and then systematically solving the resulting differential equations.

The Kramers-Moyal equation for this process is obtained by substituting Eq. (4.4-23) into Eq. (4.4-6), and is straightforwardly found to be

$$\frac{\partial}{\partial t} P(x,t \,|\, x_0,t_0) = \sum_{n=1}^{\infty} \frac{a\sigma^{2n}}{2^n n!} \frac{\partial^{2n}}{\partial x^{2n}} P(x,t \,|\, x_0,t_0). \qquad (4.4\text{-}26)$$

If $\sigma \ll 1$ then the coefficients of the successive terms on the right side of this equation will go rapidly to zero, and one might be justified in approximating this equation by dropping all but the $n=1$ term; that would leave us with a second order Fokker-Planck equation, corresponding to a Wiener process with $A=0$ and $D=a\sigma^2$ [see Eq. (3.3-12)]. The master equation for this process is obtained by substituting Eq. (4.4-22) into Eq. (4.4-7):

$$\frac{\partial}{\partial t} P(x,t|x_0,t_0) = a(2\pi\sigma^2)^{-1/2} \int_{-\infty}^{\infty} \exp(-\xi^2/2\sigma^2) P(x-\xi,t|x_0,t_0) d\xi$$

$$- a P(x,t|x_0,t_0). \qquad (4.4\text{-}27)$$

This equation is approachable using Laplace transform techniques, and it might also serve as a basis for numerically constructing finite time-step solutions.

To apply the quadrature formula (4.4-8), we must first calculate the function $w^\star(u)$ defined in Eq. (4.4-9). With the help of the integral identity (A-7) we easily deduce that

$$w^\star(u) = \int_{-\infty}^{\infty} d\xi \left[[2\pi\sigma^2]^{-1/2} \exp(-\xi^2/2\sigma^2) \right] e^{-iu\xi} = e^{-u^2\sigma^2/2}.$$

With this, Eq. (4.4-8) becomes, recalling the integral definition (B-6),

$$P(x,t|x_0,t_0) = (2\pi)^{-1} \exp[-a(t-t_0)] \lim_{\kappa \downarrow 0} \lim_{L \to \infty} \int_{-L}^{L} du \, e^{-\kappa|u|}$$

$$\times \exp[iu(x-x_0)] \exp\left(a(t-t_0) e^{-u^2\sigma^2/2} \right).$$

Taking account of the odd-even behavior of the integrand with respect to the integration variable u, this can easily be simplified to

$$P(x,t|x_0,t_0) = \pi^{-1} \exp[-a(t-t_0)] \lim_{\kappa \downarrow 0} \int_0^{\infty} du \, e^{-\kappa u}$$

$$\times \cos[u(x-x_0)] \exp\left(a(t-t_0) e^{-u^2\sigma^2/2} \right). \qquad (4.4\text{-}28)$$

It is not clear whether the u-integral here can be evaluated analytically; however, a numerical evaluation should be feasible if the integration variable is first changed from u to $v=e^{-\kappa u}$.

For our third example of a completely homogeneous jump Markov process, we take $w(\xi)$ to be the density function of the Cauchy random variable centered on 0 with half-width σ:

$$w(\xi) = \frac{\sigma/\pi}{\xi^2 + \sigma^2}. \qquad (4.4\text{-}29)$$

As discussed in Section 1.4, the moments w_n of this density function are not defined for any $n \geq 1$. Therefore, the propagator moment functions aw_n do not exist for this process, so there are no time-evolution equations for $\langle X^n(t) \rangle$ and $\langle S^n(t) \rangle$. We have here an example of a well-defined Markov process $X(t)$ that has *no moments* $\langle X^n(t) \rangle$. The nonexistence of the numbers w_n also means that the Kramers-Moyal equation (4.4-6) has no meaning for this process. However, the master equation (4.4-7) is well-defined, being given by

$$\frac{\partial}{\partial t} P(x,t|x_0,t_0) = (a\sigma/\pi) \int_{-\infty}^{\infty} (\xi^2 + \sigma^2)^{-1} P(x - \xi, t | x_0, t_0) \, d\xi$$
$$- a P(x,t | x_0, t_0). \qquad (4.4\text{-}30)$$

Also, the quadrature formula (4.4-8) applies for this process. For it, we must first calculate $w^\star(u)$ by substituting Eq. (4.4-29) into Eq. (4.4-9); upon doing that, and then invoking the integral identity (A-8), we get

$$w^\star(u) = \int_0^{\infty} d\xi \left[(\sigma/\pi)(\xi^2 + \sigma^2)^{-1} \right] e^{-iu\xi} = e^{-\sigma|u|}.$$

Substituting this into Eq. (4.4-8) and simplifying, we conclude that

$$P(x,t|x_0,t_0) = \pi^{-1} \exp[-a(t - t_0)] \lim_{\kappa \downarrow 0} \int_0^{\infty} du \, e^{-\kappa u}$$
$$\times \cos[u(x - x_0)] \exp\left(a(t - t_0) e^{-u\sigma} \right). \qquad (4.4\text{-}31)$$

Again, it is not clear whether the u-integral here can be evaluated analytically, but a numerical evaluation should be feasible if the integration variable is first changed from u to $v = e^{-\kappa u}$.

4.4.C A CRUDE MODELING OF SELF-DIFFUSION

As our final example of a completely homogeneous jump Markov process with continuum states, we shall develop such a process that crudely models the physical phenomenon of *self-diffusion in a hard-sphere gas*. As mentioned earlier, "self-diffusion" is the motion of a labeled molecule in a thermally equilibrized gas of otherwise identical molecules. In the so-called "hard-sphere" gas, the constituent molecules are assumed to be smooth, hard spheres that interact with one another only through instantaneous, perfectly elastic collisions. Let us denote the mass and radius of a gas molecule by m and r respectively. And let us suppose that we have a very large number N of these molecules inside a cubic box that is centered on the origin with its edges, of length L, parallel to the coordinate axes. We assume this gas to be "in thermal equilibrium at absolute temperature T." This means that the gas molecules are distributed uniformly throughout the box with a Maxwell-Boltzmann velocity distribution; more precisely, the x-, y- and z-components of the *position* of a randomly chosen gas molecule are each $U(-L/2,L/2)$, and the x-, y- and z-components of the *velocity* of a randomly chosen gas molecule are each $N(0,k_B T/m)$, where k_B is Boltzmann's constant.

We begin our physical analysis of this problem by assessing the prospects for a collision involving the labeled molecule when it is moving with velocity v. To that end, consider another gas molecule whose position is unspecified, but whose velocity is infinitesimally close to u. Such a molecule will be moving relative to the labeled molecule with speed $|u-v|$, and the two molecules will collide if and when their center-to-center separation distance decreases to $2r$. Therefore, the gas molecule will, in the next infinitesimal time interval dt, sweep out relative to the labeled molecule a "collision cylinder" of height $|u-v|\,dt$ and base area $\pi(2r)^2$, in the sense that if the center of the labeled molecule happens to lie inside this cylinder at time t then the two molecules will collide in $[t,t+dt)$. Since the positions of the gas molecules are given to be uniformly random inside the enclosing box, then the probability that the center of the labeled molecule will indeed lie inside the collision cylinder is just the ratio of the collision cylinder volume to the enclosing box volume:

$$[\pi(2r)^2 \,|u-v|\, dt] \,/\, L^3.$$

This then is the probability that the labeled molecule will collide in the next dt with a *given* gas molecule whose position is unspecified and whose velocity is infinitesimally close to u. Now there are (apart from our labeled molecule) $N-1$ molecules of unspecified position in the enclosing box, and the probability that any one of them will have a velocity within d^3u of u is $f_{MB}(u)d^3u$, where the so-called Maxwell-Boltzmann density function $f_{MB}(u)$ is a product of three normal density functions for u_x, u_y and u_z, each with mean zero and variance $k_B T/m$. So, applying the laws of probability theory, we conclude that the probability that the labeled molecule will collide in the next dt with a gas molecule whose velocity is within d^3u of u is

$$(N-1) \times f_{MB}(u)d^3u \times [\pi(2r)^2 \,|u-v|\, dt] \,/\, L^3$$
$$= \pi(2r)^2 \,\rho\,[|u-v|\,f_{MB}(u)d^3u]\, dt.$$

Here we have invoked the fact that $N \gg 1$, and let $\rho \equiv N/L^3$ denote the average number-density of the gas molecules.

We can now calculate the probability that the labeled molecule will collide in the next dt with *any* gas molecule by simply "summing" the above probability over all $\mathbf{u}$. Letting

$$u_{rel}(\mathbf{v}) \equiv \int_{all\ \mathbf{u}} |\mathbf{u} - \mathbf{v}| f_{MB}(\mathbf{u})\, d^3\mathbf{u} \qquad (4.4\text{-}32)$$

denote the average speed of the gas molecules relative to the labeled molecule when the latter is moving with velocity $\mathbf{v}$, we thus conclude that

$$\pi(2r)^2 \rho u_{rel}(\mathbf{v})dt = \text{probability that the labeled molecule,}$$
$$\text{moving with velocity } \mathbf{v}, \text{ will collide with}$$
$$\text{another gas molecule in the next}$$
$$\text{infinitesimal time interval } dt. \qquad (4.4\text{-}33a)$$

Furthermore, since the labeled molecule will move a distance $ds = v\,dt$ in the next dt, then

$$\pi(2r)^2 \rho |u_{rel}(\mathbf{v})/v|ds = \text{probability that the labeled molecule,}$$
$$\text{moving with velocity } \mathbf{v}, \text{ will collide with}$$
$$\text{another gas molecule in the next}$$
$$\text{infinitesimal travel distance } ds. \qquad (4.4\text{-}33b)$$

Thus far our arguments have been essentially exact. But now we propose to make two rather drastic *simplifying approximations*. The first of these is to baldly substitute in Eqs. (4.4-33)

$$u_{rel}(\mathbf{v}) \to u^* \quad \text{and} \quad v \to u^*, \qquad (4.4\text{-}34)$$

where u^* is some *constant* that typifies the molecular speeds in our gas:

$$u^* \sim \int_{all\ \mathbf{u}} |\mathbf{u}| f_{MB}(\mathbf{u})\, d^3\mathbf{u} \sim (k_B T/m)^{1/2}. \qquad (4.4\text{-}35)$$

The last estimate follows from the fact that $(k_B T/m)^{1/2}$ is the root-mean-square value of each Cartesian component of the molecular velocity under the Maxwell-Boltzmann density function, and is accurate only up to a multiplicative constant of order unity; however, in view of the crudeness of the approximation (4.4-34), it is pointless to try to estimate u^* any more accurately than this. The approximation (4.4-34) evidently brings Eqs. (4.4-33) into the much simpler forms

$$[1/\tau^*]\, dt = \text{probability that the labeled molecule will}$$
$$\text{collide with another gas molecule in the}$$
$$\text{next infinitesimal time interval } dt, \qquad (4.4\text{-}36a)$$

and

$[1/\lambda^*|\,ds$ = probability that the labeled molecule will
collide with another gas molecule in the
next infinitesimal travel distance ds,   (4.4-36b)

where we have defined the constant parameters τ^* and λ^* by

$$1/\tau^* \equiv \pi(2r)^2 \rho u^* \quad \text{and} \quad 1/\lambda^* \equiv \pi(2r)^2 \rho.   (4.4-37)$$

Before stating our second simplifying approximation, let us examine some important implications of this first approximation. Equation (4.4-36a) implies that the time between successive collisions of the labeled molecule is an exponentially distributed random variable with mean τ^*. To prove this, let $p_0(\tau)$ denote the probability that a time τ will elapse *without* the labeled molecule making any collisions. Then by Eq. (4.4-36a) and the laws of probability, we have

$$p_0(\tau+d\tau) = p_0(\tau) \times [1 - d\tau/\tau^*].$$

This implies the differential equation $dp_0/d\tau = -(1/\tau^*)p_0$, and the solution to that differential equation which satisfies the required initial condition $p_0(0) = 1$ is obviously

$$p_0(\tau) = \exp(-\tau/\tau^*).$$

Therefore, the probability that the next collision will occur between times τ and $\tau + d\tau$ is, invoking the multiplication law of probability,

$$p_0(\tau) \times |d\tau/\tau^*| = (1/\tau^*) \exp(-\tau/\tau^*)\, d\tau.$$

Comparing this result with Eqs. (1.4-5) – (1.4-7), we conclude that the time between successive collisions of the labeled molecule is indeed an exponentially distributed random variable with mean τ^*. Analogous reasoning applied to Eq. (4.4-36b) gives the result that the distance moved by the labeled molecule between successive collisions is an exponentially distributed random variable with mean λ^*. The apparent fact that these two random variables are *statistically independent* of each other is yet another consequence of our simplifying approximation (4.4-34). In reality of course, the time τ to the next collision and the distance s moved by the labeled molecule to that next collision are *not* statistically independent of each other, but are related by $s = v\tau$, where v is the speed of the labeled molecule. The fact that we are nevertheless regarding s and τ as statistically independent emphasizes the gross nature of our approximation (4.4-34). In particular, we are *not* simply pretending that the labeled molecule always moves with some constant speed u^*, since in that case we would have $s = u^*\tau$; rather, we are largely *ignoring* the speed of the labeled molecule.

Our second simplifying approximation is to assume that immediately *after* a collision the labeled molecule will move off in any direction with equal probability. This "isotropic scattering" approximation, like our first approximation, essentially ignores the velocity **v** of the labeled molecule; because in reality, the labeled molecule would be a bit more likely to move off in the direction of **v**, especially if $|\mathbf{v}| \equiv v$ is large. In fact, this second simplifying approximation would be intolerably inaccurate if the labeled molecule were much more massive than the other gas molecules; because in that "Brownian motion" case, the direction of motion of the labeled molecule would hardly be changed at all by a collision with a gas molecule. However, our approximation is not so bad for the present *equal-mass* case, where the actual effect of a collision between two gas molecules is to simply exchange velocity components along the line of centers at contact.

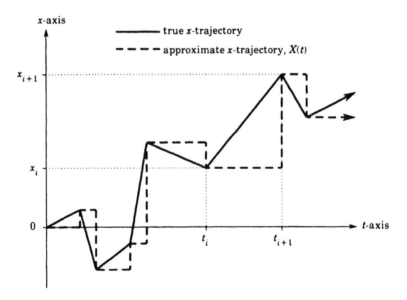

Figure 4-1. Showing the relation between the true *x*-coordinate of the labeled (self-diffusing) molecule and our approximating trajectory $X(t)$. With $t_0 = x_0 = 0$, the labeled molecule suffers its ith collision at $t = t_i$ and $x = x_i$. We can model the *approximating* trajectory $X(t)$ as a completely homogeneous jump Markov process if we make two rather heavy-handed simplifying approximations: First, that the time and distance moved by the labeled molecule between successive collisions are *statistically independent, exponentially distributed* random variables; and second, that the labeled molecule moves off *isotropically* after each scattering.

The solid graph in Fig. 4-1 shows a schematic plot of the true x-coordinate of the center of the labeled molecule versus time. We are assuming that the labeled molecule is at the origin at time $t_0 = 0$, and also that the box dimension L is so large that we need not worry about the labeled molecule striking a wall during the observation times of interest. The labeled molecule makes its first collision at time t_1 when its x-coordinate is x_1, its second collision at time t_2 when its x-coordinate is x_2, and so forth. Between times t_i and t_{i+1}, the labeled molecule moves with a constant velocity whose x-component is $(x_{i+1} - x_i)/(t_{i+1} - t_i)$; so the x-trajectory is a continuous union of linear segments. Now, we have just seen that our two simplifying approximations have the effect of essentially *ignoring* the velocity of the labeled molecule. Therefore, we should not expect our simplified treatment of self-diffusion to be able to approximate the true x-trajectory of the labeled molecule *any better* than does the dashed graph shown in Fig. 4-1, which is *defined* by

$$X(t) = x_i \quad \text{for} \quad t_i < t \le t_{i+1} \quad (i = 0, 1, \dots). \tag{4.4-38}$$

Apart from its physically unrealistic discontinuous behavior, the dashed graph $X(t)$ is seen to approximate the true x-coordinate of the labeled molecule to within an accuracy of roughly λ^*, the mean distance moved between collisions. Since λ^* is usually quite small on scales of observational interest, this level of approximation is usually acceptable.

The virtue of the approximating trajectory $X(t)$ in Fig. 4-1 is that it is potentially modelable as a *jump Markov process*. By contrast, the solid trajectory in Fig. 4-1 is obviously *not* modelable as a jump Markov process; nor for that matter is it modelable as a continuous Markov process, since it is almost everywhere differentiable. We shall now show that our two simplifying approximations imply that $X(t)$ in Fig. 4-1 is in fact the completely homogeneous jump Markov process with characterizing functions

$$a(x,t) \equiv a = 1/\tau^* \tag{4.4-39a}$$

and

$$w(\xi \mid x,t) \equiv w(\xi) = \frac{1}{2\lambda^*} \int_0^1 d\mu \, \mu^{-1} \exp(-|\xi|/\lambda^*\mu). \tag{4.4-39b}$$

The first of these two formulas follows easily from Eqs. (4.1-11) and (4.4-36a), and emphasizes the fact that τ^* is just the *mean* of the (exponentially distributed) time between collisions of the labeled molecule. But the rationale for Eq. (4.4-39b) is not at all obvious. To deduce it, we must reason as follows:

Imagine a spherical polar coordinate frame with its origin at the center of the labeled molecule at the instant of the ith collision, and with its polar axis aligned with the x-axis of the container. In that frame, let the center of the labeled molecule at the instant of its *next* collision be located by the variables (s,θ,ϕ), where s is the radius variable, θ the polar angle and ϕ the azimuthal angle. According to our first simplifying approximation, s should be exponentially distributed with mean λ^*. And according to our second simplifying approximation, θ and ϕ should be isotropically distributed. So the joint density function Q of the three variables s, $\cos\theta$ and ϕ is such that

$$Q(s,\cos\theta,\phi)\, ds\, d\cos\theta\, d\phi = [(1/\lambda^*)e^{-s/\lambda^*} ds]\,[(d\cos\theta\, d\phi)/(4\pi)],$$

where the second factor on the right is just the ratio of the differential solid angle $|\sin\theta\, d\theta\, d\phi|$ to the total solid angle 4π. Now, since our polar axis is here aligned with the container's x-axis, then the jump displacement in the process $X(t)$ between the ith and $(i+1)$th collisions is $x_{i+1} - x_i \equiv \xi = s\cos\theta$. Therefore, by the RVT theorem (1.6-4), the density function corresponding to the jump variable ξ can be calculated as

$$w(\xi) = \int_0^\infty ds \int_{-1}^1 d\cos\theta \int_0^{2\pi} d\phi \; Q(s,\cos\theta,\phi)\, \delta(\xi - s\cos\theta)$$

$$= \int_0^\infty ds \int_{-1}^1 d\mu \int_0^{2\pi} d\phi \; (4\pi\lambda^*)^{-1} e^{-s/\lambda^*} \delta(\xi - s\mu)$$

$$= (2\lambda^*)^{-1} \int_{-1}^1 d\mu \int_0^\infty ds\; e^{-s/\lambda^*} \delta(\xi - s\mu).$$

To integrate out the delta function, it is convenient to consider separately the cases $\xi<0$ and $\xi>0$. In the former case, the positivity of s implies that the argument of the delta function can vanish only if $\mu<0$; thus we have

$$w(\xi<0) = (2\lambda^*)^{-1} \int_{-1}^0 d\mu \int_0^\infty ds\; e^{-s/\lambda^*} \delta(-|\xi| - s\mu)$$

$$= (2\lambda^*)^{-1} \int_0^1 d\mu \int_0^\infty ds\; e^{-s/\lambda^*} \delta(-|\xi| + s\mu),$$

where in the last step we have changed the integration variable μ to $-\mu$. Now changing the integration variable s to $s' = s\mu$, we get

$$w(\xi<0) = (2\lambda^*)^{-1} \int_0^1 d\mu \int_0^\infty (ds'/\mu)\, \exp(-s'/\mu\lambda^*)\, \delta(-|\xi| + s')$$

$$= (2\lambda^*)^{-1} \int_0^1 d\mu\, \mu^{-1} \exp(-|\xi|/\lambda^*\mu).$$

The computation for the case $\xi > 0$ proceeds similarly and gives exactly the same result, thus establishing Eq. (4.4-39b).

At this point, our *physical* analysis is essentially completed: We have deduced that, under our two simplifying assumptions, the *approximated* diffusion process $X(t)$ is a completely homogeneous jump Markov process with characterizing functions (4.4-39). It remains now only to apply the various time-evolution equations for completely homogeneous jump Markov processes that were developed in Subsection 4.4.A. Of those equations, the only ones that will be of interest to us here are the moment evolution equations for $X(t)$ and the Kramers-Moyal equation. For those equations, we must first calculate the moments w_n of the function $w(\xi)$. Substituting the formula (4.4-39b) into the definition (4.4-5b), we get

$$
w_n = \int_{-\infty}^{\infty} d\xi \, \xi^n \left[\frac{1}{2\lambda^*} \int_0^1 d\mu \, \mu^{-1} \exp(-|\xi|/\lambda^*\mu) \right]
$$

$$
= \frac{1}{2\lambda^*} \int_0^1 d\mu \, \mu^{-1} \int_{-\infty}^{\infty} d\xi \, \xi^n \exp(-|\xi|/\lambda^*\mu).
$$

The ξ-integral here clearly vanishes if n is *odd*. If n is *even*, it gives

$$
w_n = \frac{1}{2\lambda^*} \int_0^1 d\mu \, \mu^{-1} 2 \int_0^{\infty} d\xi \, \xi^n \exp(-\xi/\lambda^*\mu)
$$

$$
= \frac{1}{\lambda^*} \int_0^1 d\mu \, \mu^{-1} n! \, (\lambda^*\mu)^{n+1} \qquad \text{[by (A-2)]}
$$

$$
= n! \, \lambda^{*n} \int_0^1 d\mu \, \mu^n = \frac{n! \, \lambda^{*n}}{n+1} .
$$

Thus, recalling Eq. (4.4-39a), we find for this process that

$$
aw_n = \begin{cases} \dfrac{n! \, \lambda^{*n}}{(n+1)\, \tau^*} & \text{if } n = 0, 2, 4, \ldots \\[2mm] 0 \, , & \text{if } n = 1, 3, 5, \ldots \end{cases} \tag{4.4-40}
$$

Remembering that we chose $t_0 = x_0 = 0$, we may now conclude from Eqs. (4.4-13) that the mean, variance and covariance of $X(t)$ are given by

$$
\langle X(t) \rangle = 0 \qquad\qquad (0 \le t), \tag{4.4-41a}
$$

$$
\mathrm{var}\{X(t)\} = (2\lambda^{*2}/3\tau^*)\, t \qquad (0 \le t), \tag{4.4-41b}
$$

$$
\mathrm{cov}\{X(t_1), X(t_2)\} = (2\lambda^{*2}/3\tau^*)\, t_1 \qquad (0 \le t_1 \le t_2). \tag{4.4-41c}
$$

Higher order moments of $X(t)$ can be found by solving the (closed) set of equations (4.4-10) for the aw_n assignment (4.4-40), but we shall not pursue those calculations here. From a physical standpoint, the most interesting aspect of these results is the implication of Eq. (4.4-41b) that

$$\text{sdev}\{X(t)\} = (2\lambda^{*2}/3\tau^*)^{1/2} t^{1/2} \equiv (2\lambda^* u^*/3)^{1/2} t^{1/2}, \qquad (4.4\text{-}42)$$

where we have invoked the fact, implicit in Eqs. (4.4-37), that $\lambda^*/\tau^* = u^*$. This formula for the root-mean-square component displacement of a labeled molecule in time t is a standard result in the elementary kinetic theory of gases.

The Kramers-Moyal equation (4.4-6) reads for the aw_n assignment (4.4-40),

$$\frac{\partial}{\partial t} P(x,t|0,0) = \sum_{\substack{n=2 \\ (n \text{ even})}}^{\infty} \frac{\lambda^{*n}}{(n+1)\,\tau^*} \frac{\partial^n}{\partial x^n} P(x,t|0,0). \qquad (4.4\text{-}43)$$

There would appear to be little hope of solving this equation as it stands. But observe that the ratio of the coefficients of successive derivative terms on the right side of this equation is

$$\frac{\lambda^{*(n+2)}}{(n+2+1)\,\tau^*} \div \frac{\lambda^{*n}}{(n+1)\,\tau^*} = \frac{n+1}{n+3} \lambda^{*2} \approx \lambda^{*2}.$$

Now it happens that λ^*, which by Eq. (4.4-37) is roughly equal to $0.1/(r^2\rho)$, is usually very small compared to 1 when expressed in "macroscopic" length units; for example, for an ordinary gas at room temperature and atmospheric pressure, $\lambda^* \approx 3 \times 10^{-4}$mm. So if x is measured in millimeters then the coefficient of each derivative term on the right side of Eq. (4.4-43) will be about 10^{-7} times its predecessor. In such a situation we can safely *approximate* the right side of Eq. (4.4-43) by discarding all but its leading ($n = 2$) term:

$$\frac{\partial}{\partial t} P(x,t|0,0) \doteq \left(\frac{\lambda^{*2}}{3\,\tau^*} \right) \frac{\partial^2}{\partial x^2} P(x,t|0,0)$$

$$[\lambda^* \ll 1 \text{ or } r^2\rho \gg 1]. \qquad (4.4\text{-}44)$$

This equation is identical in form to the so-called *diffusion equation* of elementary gas kinetic theory. From the point of view of Markov process theory, this equation is evidently a *Fokker-Planck* equation; as such, it defines a *continuous* Markov process, namely the *Wiener* process with $A = 0$ and $D = 2\lambda^{*2}/3\tau^* = 2\lambda^* u^*/3$. So we have here a physical example of a *jump* Markov process that becomes a *continuous* Markov process in a special parameter limit. This example is particularly instructive because it affords, through Fig. 4-1, a clear physical picture of what actually happens in that parameter limit: As the product $r^2 \rho$ becomes very large, the typical time τ^* between collisions and the typical distance λ^* traveled between collisions both become very small [see Eqs. (4.4-37)]; consequently, both trajectories shown in Fig. 4-1 approach, from a *macroscopic* point of view, the everywhere-continuous-but-nowhere-differentiable form of a *continuous* Markov process.

In view of the admittedly crude nature of the two simplifying approximations that we made in the preceding analysis of self-diffusion, it is worth considering what form a more rigorous analysis might take. If we stare at Fig. 4-1 long enough, we eventually come to realize that the variable in this problem that is *most naturally* modelable as a Markov process is *not* the x-position $X(t)$ of the labeled molecule, but rather its x-velocity $V(t)$. For we see from the solid curve in Fig. 4-1 that the x-velocity of the labeled molecule is constant between collisions, and makes discontinuous jumps at the instant of each collision. This is precisely the sort of behavior we expect of a jump Markov process. Furthermore, if $V(t)$ were a jump Markov process, then it would be clear why we had to employ such extreme approximation measures in order to treat $X(t)$ as a Markov process: Being the time-integral of a Markov process, $X(t)$ could not itself be Markovian [see Section 2.6].

Armed with these insights, we might at first think that a more rigorous description of self-diffusion could be had rather easily. For example, we might try to model $V(t)$ as the jump Markov process with characterizing functions

$$a(v,t) \equiv a = 1/\tau^*$$

and

$$w(\xi \mid v,t) \equiv w(\xi) = (2\pi\sigma^2)^{-1/2} \exp(-\xi^2/2\sigma^2),$$

where $\sigma^2 = k_B T/m$; in words, the jumps in $V(t)$ occur exponentially in time at the same average rate as in our preceding treatment of $X(t)$, and in each collision we augment $V(t)$ by a random sampling of the x-velocities of the other gas molecules. But this model for $V(t)$ is woefully inadequate: As may be seen from Eq. (4.4-24b), the variance of $V(t)$ in this model would be given by

$$\text{var}\{V(t)\} = (\sigma^2/\tau^*)\, t,$$

and the consequent implication that $\text{var}\{V(t \rightarrow \infty)\} = \infty$ is inconsistent with our physical expectation that $\text{var}\{V(t \rightarrow \infty)\} = \sigma^2$. Indeed, it follows from the general equations (4.4-13) that *any* modeling of $V(t)$ as a *completely homogeneous* jump Markov process will result in a divergent long-time variance. We thus conclude that $V(t)$ cannot be a *completely* homogeneous process; apparently, the v-dependence of the characterizing functions $a(v,t)$ and $w(\xi \mid v,t)$ must be taken explicitly into account.

In the next section we shall derive an exact formula for the characterizing functions of $V(t)$ when the hard-sphere gas is a so-called "one-dimensional" gas. We now conclude this section by summarizing the somewhat convoluted story that has emerged regarding a Markovian analysis of self-diffusion in a hard-sphere gas:

(i) The x-coordinate of the labeled molecule evolves with time in a manner indicated schematically by the solid curve in Fig. 4-1. Strictly speaking, that trajectory does *not* have the requisite attributes of *either* a continuous Markov process (since it is almost everywhere differentiable) *or* a jump Markov process (since it is not a discontinuous sequence of horizontal line segments).

(ii) But suppose we *approximate* the true x-trajectory of the labeled molecule by the dashed curve shown in Fig. 4-1. And suppose further that we make the rather heavy-handed assumptions that the time and distance traveled by the labeled molecule between successive collisions are *independent exponential* random variables with respective means τ^* and λ^*, and that the labeled molecule moves off *isotropically* after each collision. Then it follows that the approximating trajectory $X(t)$ will be the completely homogeneous jump Markov process with characterizing functions (4.4-39). In particular, $X(t)$ satisfies var$\{X(t)\}$ $= (2\lambda^{*2}/3\tau^*)t$, in agreement with elementary gas kinetic theory.

(iii) If we next pass to the *macroscopic limit* in which $r^2\rho$ is very large compared to unity, then τ^* and λ^* both become very small, and the Kramers-Moyal equation for $X(t)$ approaches a Fokker-Planck form. So in the macroscopic limit, our completely homogeneous *jump* Markov process becomes a completely homogeneous *continuous* Markov process; more specifically, $X(t)$ becomes the Wiener process with $A = 0$ and $D = 2\lambda^{*2}/3\tau^*$. And the Fokker-Planck equation for this Wiener process is precisely the well known "diffusion equation" of elementary gas kinetic theory.

(iv) A physically more realistic model of self-diffusion would probably start by regarding the x-velocity $V(t)$ of the labeled molecule as a jump Markov process. The labeled molecule's x-position $X(t)$, being the time-integral of $V(t)$, would then *not* be a Markov process, in agreement with our observations in (i). But any treatment of $V(t)$ as a jump Markov process will be complicated by the fact that $V(t)$ cannot be regarded as being completely homogeneous: the v-dependence of its characterizing functions cannot be blithely ignored for the sake of computational expediency.

4.5 A RIGOROUS APPROACH TO SELF-DIFFUSION AND BROWNIAN MOTION

The phenomenon of self-diffusion discussed in the preceding paragraphs has a lot in common with the phenomenon of Brownian motion, which was discussed earlier in Subsection 3.4.C. Both concern the erratic motion of a particle that is being knocked about by the molecules of a surrounding fluid in thermal equilibrium. The only difference is that in self-diffusion the particle is one of the fluid molecules, whereas in Brownian motion the particle is much larger and heavier than a fluid molecule. In view of the common ground between self-diffusion and Brownian motion, it is surprising that our analyses of these two phenomena took such different forms: In Subsection 3.4.C we approached Brownian motion in a rather heuristic and phenomenological way, treating the particle's *velocity* as a *continuous* Markov process; yet in Subsection 4.4.C we approached self-diffusion in a deductive but heavily approximated way, treating the particle's *position* as a *jump* Markov process. In this section we shall formulate a unified approach to these two phenomena based exclusively on the microphysics of particle-molecule collisions. But we shall have to pay a price for this increase in rigor: In order to keep the problem within the realm of *univariate* Markov process theory, we shall have to adopt a simplified mechanism for particle-molecule collisions.

Before we formulate this unified if somewhat idealized approach to self-diffusion and Brownian motion, let us examine briefly the strengths and weaknesses of our previous treatments of these two phenomena. As we mentioned at the conclusion of the preceding section, a more rigorous approach to self-diffusion would have tried to regard the particle's *velocity*, rather than its position, as a Markov variable. In that connection, it is interesting to note that the main reason why Langevin's analysis of *Brownian motion* constituted an improvement over the earlier analysis of Einstein is that Langevin treated the Brownian particle's *velocity* as a Markov process, whereas Einstein had tried to work directly with the Brownian particle's position. The inherently non-Markovian nature of the position of a Brownian particle is due to the particle's relatively large inertia, which causes the changes in the particle's position between successive collisions to be highly correlated.† But if Langevin was on target in recognizing that the *velocity* of a Brownian particle is the more natural Markovian variable, it would appear that he was a bit off the mark in assuming that velocity to be a *continuous* Markov process. Certainly for "hard-sphere" collisions the particle's velocity would have to be of the *jump* type rather than the continuous type. However, it turns out that this shortcoming in Langevin's analysis can be easily remedied. We now give a slightly modified version of Langevin's analysis of Brownian motion [see Subsection 3.4.C] that is completely noncommittal as to whether the Markov process $V(t)$ is a continuous type or a jump type.

Newton's second law for a macroscopic sphere of mass M moving through a fluid was written down in Eq. (3.4-16a), wherein the constant γ is the so-called drag coefficient. It is straightforward to show that the solution to that differential equation for the initial condition $v(0) = v_0$ is

$$v(t) = v_0 \exp[-(\gamma/M)t] \qquad (t \geq 0). \qquad (4.5\text{-}1)$$

Now, if the sphere is so small that its velocity must be regarded as a *stochastic* process $V(t)$ instead of a *sure* process $v(t)$, then let us assume that Eq. (4.5-1) still holds for the *average* velocity; i.e., let us *assume* that

$$\langle V(t) \rangle = v_0 \exp[-(\gamma/M)t] \qquad (t \geq 0). \qquad (4.5\text{-}2a)$$

This evidently implies that $\langle V(\infty) \rangle = 0$, so that $\mathrm{var}\{V(\infty)\} = \langle V^2(\infty) \rangle$. But the equipartition theorem of statistical mechanics suggests that $\frac{1}{2}M\langle V^2(\infty) \rangle = k_B T/2$, where k_B is Boltzmann's constant and T is the absolute temperature of the surrounding fluid. Therefore, it seems reasonable to *also assume* that

$$\mathrm{var}\{V(\infty)\} = k_B T/M. \qquad (4.5\text{-}2b)$$

Now we set our two assumptions (4.5-2) aside for the moment, and we take note of a purely mathematical fact: According to our analysis in Subsection 2.7.D, *if* $V(t)$ were a Markov process (of *any* type) whose first two propagator moment functions have the forms

$$B_1(v,t) = -\beta v \quad \text{and} \quad B_2(v,t) = c, \qquad (4.5\text{-}3)$$

---

† Einstein got around this non-Markovian behavior by the frankly awkward procedure of letting time advance in *discrete steps*, which, though macroscopically small, were nevertheless made large enough that very many particle-molecule collisions would occur during each step.

β and c being any positive constants, then under the initial condition $V(0) = v_0$ we would have [see Eqs. (2.7-34)]

$$\langle V(t) \rangle = v_0 \exp(-\beta t) \qquad (0 \le t), \qquad (4.5\text{-}4a)$$

$$\text{var}\{V(t)\} = (c/2\beta)(1 - \exp(-2\beta t)) \qquad (0 \le t), \qquad (4.5\text{-}4b)$$

$$\text{cov}\{V(t_1), V(t_2)\} = (c/2\beta) \exp[-\beta(t_2 - t_1)](1 - \exp(-2\beta t_1))$$
$$(0 \le t_1 \le t_2). \quad (4.5\text{-}4c)$$

Furthermore, since $X(t)$ is by definition the time-integral of $V(t)$, we would also have [see Eqs. (2.7-35)]

$$\langle X(t) \rangle = (v_0/\beta)[1 - \exp(-\beta t)] \qquad (0 \le t), \qquad (4.5\text{-}5a)$$

$$\text{var}\{X(t)\} = (c/\beta^2)[\, t - (2/\beta)[1 - \exp(-\beta t)]$$
$$+ (1/2\beta)[1 - \exp(-2\beta t)]\,]$$
$$(0 \le t), \qquad (4.5\text{-}5b)$$

$$\text{cov}\{X(t_1), X(t_2)\} = \text{var}\{X(t_1)\} + (c/2\beta^3)(1 - \exp[-\beta(t_2 - t_1)])$$
$$\times (1 - 2\exp(-\beta t_1) + \exp(-2\beta t_1))$$
$$(0 \le t_1 \le t_2). \quad (4.5\text{-}5c)$$

We now observe that the first two of the above six equations would be *consistent* with our hypothesized pair of equations (4.5-2) if it were true that $\beta = \gamma/M$ and $c/2\beta = k_B T/M$, or equivalently if

$$\beta = \gamma/M \quad \text{and} \quad c = 2\gamma k_B T/M^2. \qquad (4.5\text{-}6)$$

Therefore, let us simply *assume* that $V(t)$ is a Markov process whose first two propagator moment functions are given by Eqs. (4.5-3), so that β and c, by our first two assumptions, must be given by Eqs. (4.5-6). It then follows from Eq. (4.5-5b) that, when t is sufficiently large,

$$\text{var}\{X(t)\} \approx (c/\beta^2)t = (2k_B T/\gamma)t \qquad (t \gg 1/\beta), \qquad (4.5\text{-}7a)$$

and hence

$$\text{sdev}\{X(t)\} \approx (2k_B T/\gamma)^{1/2} t^{1/2} \qquad (t \gg 1/\beta). \qquad (4.5\text{-}7b)$$

This is precisely the Einstein-Langevin result (3.4-22). We have obtained it here using essentially Langevin's reasoning, but without invoking the Langevin equation (3.4-17), and hence without assuming the Markov process $V(t)$ to be of the continuous type.

But there remains the objection to Langevin's analysis that it relies heavily on phenomenologically inspired assumptions, and is not a derivation from "first principles." In defense of Langevin's analysis, it should be pointed out that early experiments on Brownian motion were all performed on a particle suspended in a *liquid*, and the enormous complexities of the microphysics of liquids would even today make an analysis from first principles extremely difficult. However, if one is willing to adopt a sufficiently idealized mechanism for the interactions between the particle and the fluid molecules, then it becomes possible to formulate a more rigorously deductive approach. That is the task to which we turn next.

4.5.A A ONE-DIMENSIONAL FORMULATION OF THE PROBLEM

A *one-dimensional gas* is an idealized gas in which the constituent molecules may move about in any direction and with any speed, but in which the forces between any two colliding molecules always act *wholly* along any one of three mutually perpendicular axes, called the "preferred axes." The molecules of a one-dimensional gas are most easily visualized as smooth cubes whose faces are constrained to be always perpendicular to some preferred axis. A collision between any two molecules can then occur only against faces that are perpendicular to the same axis, and because of the assumed smoothness of the cubes the velocity components along the other two axes will not be altered in the collision. Notice that even though collisions may occur "off-center," the colliding molecules are not allowed to rotate about their centers.

Suppose we have such a one-dimensional gas of N molecules confined to a cubic box whose edges, of length L, define the preferred axes of the gas. Let m and r denote respectively the mass and edge length of a gas molecule, with $r \ll L$. We suppose our gas to be "in thermal equilibrium at absolute temperature T;" this means that the x-, y- and z-components of the *position* of a randomly chosen gas molecule are each $U(-L/2, L/2)$, and the x-, y- and z-components of the *velocity* of a randomly chosen gas molecule are each $N(0, k_B T/m)$, k_B being Boltzmann's constant. Into this gas we place a cubic "particle" of mass M and edge length R. If we were trying to model Brownian motion we would take $M \gg m$ and $R \gg r$, whereas if we were trying to model self-diffusion we would take $M = m$ and $R = r$; however, for the present we shall make no assumptions about the values of M and R except to stipulate that R, like r, is very small compared to L. Our particle is assumed to have the same one-dimensional attributes as the gas molecules; i.e., it is a smooth cube whose faces are constrained to be always perpendicular to some preferred axis, so that a collision between the particle and any gas molecule will alter only the velocity components perpendicular to the colliding faces. We shall also assume that all particle-molecule collisions are *hard* and *elastic*, meaning that they occur instantaneously and conserve kinetic energy. We define

$V(t)$ ≡ x-component of the velocity of the particle at time t,   (4.5-8a)

$X(t)$ ≡ x-component of the position of the particle at time t,   (4.5-8b)

and we assign to these two variables the respective *initial conditions*

$$V(0) = v_0 \quad \text{and} \quad X(0) = 0. \tag{4.5-9}$$

Our assumption of a one-dimensional collision mechanism has the simplifying effect of making the time evolutions of the three velocity components of the particle completely independent of each other. In particular, $V(t)$ can change *only* when the particle experiences a collision on one of its x-faces; moreover, the prospects for and the effects of an x-face collision do not depend on the y and z velocity components of either the particle or its collision partner. So in the following analysis of the time evolution of $V(t)$, we shall for the sake of brevity refer to x-components of velocities as simply "velocities," and to x-face collisions as simply "collisions." Our aim in this subsection will be to derive an explicit formula for the consolidated characterizing function $W(\xi \mid v,t)$ for the process $V(t)$, and to thereby establish $V(t)$ as a well defined jump Markov process. According to Eq. (4.1-21), the function $W(\xi \mid v,t)$ may be defined by the statement that

$W(\xi \mid v,t)\, dt d\xi$ = probability, given $V(t) = v$, that in the time
interval $[t,t+dt)$ the particle's velocity will
increase by an amount between ξ and $\xi + d\xi$.   (4.5-10)

We assume the containing box to be so large that we need not worry about the particle colliding with the walls of the container. Our first step in calculating an expression for $W(\xi \mid v,t)$ will be to assess the probability for a collision between the particle and a gas molecule at a time t when the particle is moving with velocity v. To that end, we consider a gas molecule at time t whose position is unspecified but whose velocity is infinitesimally close to u. Such a molecule will be moving relative to the particle with speed $|u - v|$; therefore, in time $[t,t+dt)$ that molecule will sweep out relative to the particle a "collision cylinder" of height $|u - v|\, dt$ and (square) base area $(R + r)^2$, in the sense that if the center of the particle happens to lie inside that cylinder at time t then it will collide with the molecule in $[t, t+dt)$. Since the positions of the gas molecules are uniformly random inside the container, then the probability that the center of the particle will indeed lie inside that collision cylinder at time t is just the ratio of the cylinder's volume to the volume of the enclosing box:

$$[(R+r)^2 |u-v| \, dt] / L^3.$$

This then is the probability that the particle, moving with velocity v at time t, will collide in $[t, t+dt)$ with a particular gas molecule whose position is unspecified but whose velocity is infinitesimally close to u. Now, there are N molecules of unspecified position in the gas, and the probability that any one of them will have at time t a velocity between u and $u+du$ is, by the assumption of thermal equilibrium, $f(u)du$, where

$$f(u) \equiv [2\pi(k_B T/m)]^{-1/2} \exp[-u^2/2(k_B T/m)]. \qquad (4.5\text{-}11)$$

Therefore, applying the laws of probability in the usual way, we conclude that the probability that the particle, moving with velocity v at time t, will collide in $[t, t+dt)$ with a gas molecule whose velocity is between u and $u+du$ is

$$N \times f(u)du \times [(R+r)^2 |u-v| \, dt] / L^3$$

$$= \rho (R+r)^2 |u-v| f(u) \, dt \, du. \qquad (4.5\text{-}12)$$

Here, $\rho \equiv N/L^3$ is the average number density of the gas molecules. [The probability of *more than one* particle-molecule collision in $[t, t+dt)$ will be of order >1 in the infinitesimal dt, and is therefore effectively zero.]

Next we must calculate the change in the particle's velocity that occurs when, moving with velocity v, it collides with a gas molecule moving with velocity u. Letting v' and u' denote the respective velocities *after* the collision, we have by conservation of momentum and energy,[†]

$$Mv + mu = Mv' + mu',$$

$$\tfrac{1}{2}Mv^2 + \tfrac{1}{2}mu^2 = \tfrac{1}{2}Mv'^2 + \tfrac{1}{2}mu'^2.$$

Eliminating u' between these two equations, we find after a little algebra that the change in the particle's velocity is

$$v' - v \equiv \xi = \frac{2m}{M+m}(u-v). \qquad (4.5\text{-}13)$$

---

† If we had *not* assumed a one-dimensional gas, then the energy conservation equation would also involve y and z velocity components; the particle's velocity would then turn out to be a *trivariate* Markov process.

Now, for any fixed v, Eq. (4.5-13) describes a one-to-one relation between the velocity u of the struck molecule and the velocity change ξ induced in the particle. It follows that in order to experience a sudden velocity change between ξ and $\xi + d\xi$, the particle, moving with velocity v, must collide with a gas molecule whose velocity is between u and $u + du$, where by Eq. (4.5-13),

$$u = v + \frac{M+m}{2m}\, \xi , \quad du = \frac{M+m}{2m}\, d\xi .$$

(4.5-14)

Therefore, recalling the physical significance of Eq. (4.5-12), we may write the probability (4.5-10) as

$$W(\xi \,|\, v,t)\, dt\, d\xi = \rho\, (R+r)^2 |u - v| f(u)\, dt\, du$$

$$= \rho\, (R+r)^2 \left| v + \frac{M+m}{2m}\xi - v \right|$$

$$\times f\!\left(v + \frac{M+m}{2m}\xi \right) dt\, \frac{M+m}{2m}\, d\xi$$

$$= \rho\, (R+r)^2 \left(\frac{M+m}{2m} \right)^2 |\xi|\, f\!\left(v + \frac{M+m}{2m}\xi \right) dt\, d\xi .$$

So, recalling the definition (4.5-11) of the function f, and introducing for convenience the three constant parameters

$$\sigma^2 \equiv \frac{k_B T}{m}, \quad \kappa \equiv \rho\, (R+r)^2, \quad \varepsilon \equiv \frac{2m}{M+m},$$

(4.5-15)

we conclude that

$$W(\xi \,|\, v,t) \equiv W(\xi \,|\, v) = \left(2\pi\sigma^2 \right)^{-1/2} \kappa\, \varepsilon^{-2}\, |\xi|\, \exp\!\left(-\frac{(v + \xi/\varepsilon)^2}{2\sigma^2} \right).$$

(4.5-16)

Equation (4.5-16) is the desired result of our physical analysis. As an explicit formula for the function defined in Eq. (4.5-10), it establishes $V(t)$ as a *jump Markov process*. Since this formula for $W(\xi \,|\, v,t)$ is explicitly independent of t, then $V(t)$ is more precisely a *temporally homogeneous* jump Markov process; however, since this formula for $W(\xi \,|\, v,t)$ does involve v, then $V(t)$ is not *completely* homogeneous, a fact that we anticipated at the end of Section 4.4.

The characterizing functions $a(v)$ and $w(\xi \,|\, v)$ corresponding to the above consolidated characterizing function $W(\xi \,|\, v)$ can now be calculated from Eqs. (4.1-22):

$$a(v) = \int_{-\infty}^{\infty} d\xi \, W(\xi \mid v), \qquad (4.5\text{-}17a)$$

$$w(\xi \mid v) = \frac{W(\xi \mid v)}{a(v)}. \qquad (4.5\text{-}17b)$$

The integration called for in Eq. (4.5-17a) is most easily carried out after first changing the integration variable from ξ to $u \equiv v + \xi/\varepsilon$; that gives $\xi = \varepsilon(u - v)$ and $d\xi = \varepsilon du$, so

$$a(v) = \int_{-\infty}^{\infty} \varepsilon du \, (2\pi\sigma^2)^{-1/2} \kappa \, \varepsilon^{-2} \, |\varepsilon(u - v)| \exp(-u^2/2\sigma^2)$$

$$= (2\pi\sigma^2)^{-1/2} \kappa \int_{-\infty}^{\infty} du \, |u - v| \exp(-u^2/2\sigma^2)$$

$$= \frac{\kappa}{(2\pi)^{1/2}\sigma} \left\{ \int_{-\infty}^{v} du \, (v - u) \exp(-u^2/2\sigma^2) \right.$$

$$\left. + \int_{v}^{\infty} du \, (u - v) \exp(-u^2/2\sigma^2) \right\}$$

$$= \frac{\kappa}{(2\pi)^{1/2}\sigma} \left\{ v \left[\int_{-\infty}^{v} du \exp(-u^2/2\sigma^2) - \int_{v}^{\infty} du \exp(-u^2/2\sigma^2) \right] \right.$$

$$\left. + \left[-\int_{-\infty}^{v} du \, u \exp(-u^2/2\sigma^2) + \int_{v}^{\infty} du \, u \exp(-u^2/2\sigma^2) \right] \right\}$$

$$= \frac{\kappa}{(2\pi)^{1/2}\sigma} \left\{ 2v \int_{0}^{v} du \exp(-u^2/2\sigma^2) + 2 \int_{v}^{\infty} du \, u \exp(-u^2/2\sigma^2) \right\}.$$

where the last step follows from the evenness or oddness of the integrands. Changing the integration variable in the first integral from u to $z = u/\sqrt{2}\sigma$ allows that integral to be expressed in terms of the standard error function,

$$\text{erf}(x) \equiv \frac{2}{\sqrt{\pi}} \int_{0}^{x} \exp(-z^2) dz.$$

And changing the integration variable in the second integral from u to $z = u^2/2\sigma^2$ allows that integral to be explicitly evaluated in terms of the exponential function. Upon carrying out all these tasks, we find for the characterizing function $a(v)$ the explicit formula

$$a(v) = \kappa \left[v \, \mathrm{erf}(v/\sqrt{2}\sigma) + (2/\pi)^{1/2} \sigma \exp(-v^2/2\sigma^2) \right]. \qquad (4.5\text{-}18)$$

And upon substituting this along with Eq. (4.5-16) into Eq. (4.5-17b), we conclude that the characterizing function $w(\xi \mid v)$ is given by

$$w(\xi \mid v) = K(v) \, |\xi| \, \exp\left(-\frac{(\xi + \varepsilon v)^2}{2\sigma^2 \varepsilon^2} \right), \qquad (4.5\text{-}19a)$$

where

$$K(v) \equiv \left((2\pi\sigma^2)^{1/2} \varepsilon^2 \left[v \, \mathrm{erf}(v/\sqrt{2}\sigma) + (2/\pi)^{1/2} \sigma \exp(-v^2/2\sigma^2) \right] \right)^{-1}.$$

$$(4.5\text{-}19b)$$

Notice that $a(v)$ is independent of ε, while $w(\xi \mid v)$ is independent of κ.

By substituting Eqs. (4.5-18) and (4.5-19) into Eq. (4.3-8), we can obtain an explicit formula for the next-jump density function $p(\tau,\xi \mid v,t)$ for the process $V(t)$. We shall be using that next-jump density function to construct Monte Carlo simulations of $V(t)$ and its time-integral $X(t)$ in Section 4.6.

4.5.B SOLUTION IN THE BROWNIAN LIMIT

The time-evolution equations for the moments of our jump Markov process $V(t)$ and its time-integral $X(t)$ are given by Eqs. (4.2-23) – (4.2-25) with the replacements $(X,S) \rightarrow (V,X)$. Those equations evidently require that we first calculate the propagator moment functions $W_n(v,t) \equiv W_n(v)$ according to Eq. (4.2-2b). Substituting from Eq. (4.5-16), we get

$$W_n(v) \equiv \int_{-\infty}^{\infty} d\xi \, \xi^n \, W(\xi \mid v)$$

$$= \int_{-\infty}^{\infty} d\xi \, \xi^n \left(2\pi\sigma^2 \right)^{-1/2} \kappa \, \varepsilon^{-2} |\xi| \exp\left(-\frac{(v + \xi/\varepsilon)^2}{2\sigma^2} \right).$$

With the integration variable change $\xi \rightarrow u \equiv v + \xi/\varepsilon$, this becomes

$$W_n(v) = (2\pi\sigma^2)^{-1/2} \kappa \, \varepsilon^n \int_{-\infty}^{\infty} du \, (u-v)^n \, |u-v| \exp(-u^2/2\sigma^2)$$

$$(n \geq 1). \qquad (4.5\text{-}20)$$

We know that the moment evolution equations for $V(t)$ and $X(t)$ will be closed if and only if $W_n(v)$ is a polynomial in v of degree $\leq n$. Since that condition is obviously *not* satisfied by formula (4.5-20), then we should not expect to be able to get exact solutions to the moment evolution equations.

One way of proceeding here would be to apply the approximate solution procedure for open moment evolution equations outlined in Appendix C [replacing therein $X(t)$ by $V(t)$ and $B_k(x)$ by $W_k(v)$]. That procedure should in fact be quite feasible in this instance, because the key parameters

$$b_{kj} \equiv \frac{1}{j!} \frac{d^j W_k}{dv^j} \bigg|_{v=0} \qquad (k=1,2,\dots;j=0,1,\dots) \quad (4.5\text{-}21)$$

can all be calculated analytically for the function $W_k(v)$ in Eq. (4.5-20). However, these derivative calculations and the subsequent implementation of the solution-comparison procedure described in Appendix C constitute a lengthy (and ultimately numerical) computational effort. So we shall content ourselves here with a highly abbreviated analytical version of that solution procedure that works only in the "Brownian limit" $M \gg m$. This abbreviated analysis is made possible by — and so comes at the expense of — one a priori assumption: We have to assume that the typical speed of the relatively massive Brownian particle will always be very much smaller than the typical speed of a gas molecule. More precisely, recalling that the root-mean-square speed of a gas molecule is by definition $(k_B T/m)^{1/2} \equiv \sigma$, we shall assume in the following analysis that

$$M \gg m \;\Rightarrow\; \langle V^{2n}(t) \rangle \ll \sigma^{2n} \qquad (n=1,2,\dots \,;\, t \geq 0). \quad (4.5\text{-}22)$$

Naturally, we should be able to verify this assumption at the conclusion of our calculation.

The first step in our analysis is to observe that the function $W_n(v)$ in Eq. (4.5-20) satisfies

$$W_n(-v) = (2\pi\sigma^2)^{-1/2} \kappa \, \varepsilon^n \int_{-\infty}^{\infty} du \,(u+v)^n \,|u+v| \exp(-u^2/2\sigma^2)$$

$$= (2\pi\sigma^2)^{-1/2} \kappa \, \varepsilon^n \int_{+\infty}^{-\infty} d(-u) \,(-u+v)^n \,|-u+v| \exp(-(-u)^2/2\sigma^2)$$

$$= (2\pi\sigma^2)^{-1/2} \kappa \, \varepsilon^n \int_{-\infty}^{\infty} du \,(-1)^n (u-v)^n \,|u-v| \exp(-u^2/2\sigma^2),$$

or

$$W_n(-v) = (-1)^n W_n(v) \qquad (n=1,2,\dots). \quad (4.5\text{-}23)$$

Since $W_1(v)$ is thus an *odd* function of v, then its Taylor series expansion about $v = 0$ reads

$$W_1(v) = b_{1,1}v + b_{1,3}v^3 + b_{1,5}v^5 + \dots,$$

where $b_{k,j}$ is as defined in Eq. (4.5-21). Similarly, $W_2(v)$ is an *even* function of v, so its Taylor series expansion reads

$$W_2(v) = b_{2,0} + b_{2,2}v^2 + b_{2,4}v^4 + \dots.$$

Therefore, the first two propagator moment functions of $V(t)$ have the functional forms

$$W_1(v) = b_{1,1}v[1 + (b_{1,3}/b_{1,1})v^2 + (b_{1,5}/b_{1,1})v^4 + \dots], \qquad (4.5\text{-}24a)$$

$$W_2(v) = b_{2,0}[1 + (b_{2,2}/b_{2,0})v^2 + (b_{2,4}/b_{2,0})v^4 + \dots]. \qquad (4.5\text{-}24b)$$

We now argue that these last two equations must have the even more explicit forms

$$W_1(v) = b_{1,1}v[1 + c_{1,3}(v^2/\sigma^2) + c_{1,5}(v^4/\sigma^4) + \dots], \qquad (4.5\text{-}25a)$$

$$W_2(v) = b_{2,0}[1 + c_{2,2}(v^2/\sigma^2) + c_{2,4}(v^4/\sigma^4) + \dots], \qquad (4.5\text{-}25b)$$

where the $c_{i,j}$'s are pure (dimensionless) constants. We could of course prove this by directly evaluating the constants $b_{k,j}$ through their definition (4.5-21); however, we can argue more simply as follows: It is clear from the expression for $W_n(v)$ in Eq. (4.5-20) that $d^i W_n(v)/dv^i$ will depend on the two parameters κ and ε only through an overall factor of $\kappa \varepsilon^n$. Therefore, the ratio $b_{n,i}/b_{n,j}$ will be independent of both κ and ε, and hence can be a function only of the remaining parameter σ. Since σ has dimensions of velocity, then the only σ-dependence of $b_{n,i}/b_{n,j}$ that will maintain the dimensional integrity of Eqs. (4.5-24) is that which results in Eqs. (4.5-25) with the $c_{i,j}$'s dimensionless constants.

Now we observe that when the propagator moment functions W_1 and W_2 are used in the time-evolution equations for the means and variances of $V(t)$ and $X(t)$ [see Eqs. (4.2-26) – (4.2-36) with $(X,S) \to (V,X)$], they always appear in some *averaged* form. This is where we invoke our simplifying assumption (4.5-22): It implies that, when $M \gg m$, the averaged $c_{i,j}$-terms in the brackets of Eqs. (4.5-25) will all be very much less than unity. So we conclude that, at least for purposes of computing the means and variances of $V(t)$ and $X(t)$,

$$M \gg m \quad \Rightarrow \quad W_1(v) \approx b_{1,1}v \text{ and } W_2(v) \approx b_{2,0}. \qquad (4.5\text{-}26)$$

Now let us evaluate the two constants $b_{1,1}$ and $b_{2,0}$ from Eqs. (4.5-21) and (4.5-20). To calculate $b_{1,1}$, we must evidently first calculate the derivative of $W_1(v)$. Putting $n=1$ in Eq. (4.5-20), we get

$$W_1'(v) = \frac{\kappa \varepsilon}{(2\pi\sigma^2)^{1/2}} \frac{d}{dv} \int_{-\infty}^{\infty} du\,(u-v)\,|u-v|\exp(-u^2/2\sigma^2)$$

$$= \frac{\kappa \varepsilon}{(2\pi)^{1/2}\sigma} \frac{d}{dv} \left\{ \int_{-\infty}^{v} du\,(u-v)(v-u)\exp(-u^2/2\sigma^2) \right.$$

$$\left. + \int_{v}^{\infty} du\,(u-v)(u-v)\exp(-u^2/2\sigma^2) \right\}$$

$$= \frac{-2\kappa \varepsilon}{(2\pi)^{1/2}\sigma} \frac{d}{dv} \left\{ \int_{0}^{v} du\,[u^2+v^2]\exp(-u^2/2\sigma^2) \right.$$

$$\left. + \int_{v}^{\infty} du\,[2uv]\exp(-u^2/2\sigma^2) \right\}$$

$$= \frac{-2\kappa \varepsilon}{(2\pi)^{1/2}\sigma} \left\{ \int_{0}^{v} du\,[2v]\exp(-u^2/2\sigma^2) + [v^2+v^2]\exp(-v^2/2\sigma^2) \right.$$

$$\left. + \int_{v}^{\infty} du\,[2u]\exp(-u^2/2\sigma^2) - [2vv]\exp(-v^2/2\sigma^2) \right\}$$

$$W_1'(v) = \frac{-4\kappa \varepsilon}{(2\pi)^{1/2}\sigma} \left\{ v\int_{0}^{v} du\,\exp(-u^2/2\sigma^2) + \int_{v}^{\infty} du\,u\exp(-u^2/2\sigma^2) \right\},$$

where the penultimate step has invoked the rule (A-12). Therefore, using the definition (4.5-21) and the integral identity (A-4), we obtain

$$b_{1,1} \equiv W_1'(0) = -\frac{4\kappa \varepsilon}{(2\pi)^{1/2}\sigma} \int_{0}^{\infty} du\,u\exp(-u^2/2\sigma^2) = -\frac{4\kappa \varepsilon}{(2\pi)^{1/2}\sigma}(\sigma^2),$$

or

$$b_{1,1} = -(2\pi)^{-1/2}\,4\kappa \varepsilon \sigma. \qquad (4.5\text{-}27a)$$

The calculation of $b_{2,0}$ is somewhat less tedious. It proceeds from Eqs. (4.5-21) and (4.5-20) as follows:

$$b_{2,0} \equiv W_2(0) = (2\pi\sigma^2)^{-1/2}\kappa \varepsilon^2 \int_{-\infty}^{\infty} du\,u^2\,|u|\exp(-u^2/2\sigma^2)$$

$$= (2\pi\sigma^2)^{-1/2}\kappa \varepsilon^2\,2 \int_{0}^{\infty} du\,u^3\exp(-u^2/2\sigma^2)$$

$$= (2\pi\sigma^2)^{-1/2}\kappa \varepsilon^2\,2(2\sigma^4),$$

the last step following from the integral identity (A-4). Therefore,

$$b_{2,0} = (2\pi)^{-1/2} \, 4 \, \kappa \, \varepsilon^2 \, \sigma^3. \tag{4.5-27b}$$

The formulas (4.5-27) for $b_{1,1}$ and $b_{2,0}$ are *exact*. However, for the Brownian limit assumed in Eq. (4.5-26) we have $M \gg m$ and $R \gg r$, in which case the definitions (4.5-15) give $\sigma = (k_B T/m)^{1/2}$, $\kappa \approx \rho R^2$ and $\varepsilon \approx 2m/M$. So *in the Brownian limit* the formulas (4.5-27) reduce to

$$-b_{1,1} \approx 4(2/\pi)^{1/2} \, \rho \, R^2 \, (k_B T)^{1/2} \, (m^{1/2}/M) \equiv \beta \tag{4.5-28a}$$

and

$$b_{2,0} \approx 8(2/\pi)^{1/2} \, \rho \, R^2 \, (k_B T)^{3/2} \, (m^{1/2}/M^2) \equiv c. \tag{4.5-28b}$$

And Eqs. (4.5-26) then become

$$\left. \begin{array}{l} B_1(v,t) \equiv W_1(v) \approx -\beta v \\[2mm] B_2(v,t) \equiv W_2(v) \approx c \end{array} \right\} \quad \text{(Brownian limit)}. \tag{4.5-29}$$

Thus, using only the relatively mild assumption (4.5-22), we have managed to *derive* from the microphysics of our one-dimensional Brownian system equations (4.5-29) — the very equations (4.5-3) that Langevin's analysis essentially *assumed*. As a bonus, we have obtained in Eqs. (4.5-28) explicit formulas for the constants β and c in terms of the microphysical parameters of the system. Since Eqs. (4.5-29) directly imply Eqs. (4.5-4) and (4.5-5), we are now in a position to write down explicit formulas for the means, variances and covariances of $V(t)$ and $X(t)$ for Brownian motion in a one-dimensional gas. Let us examine some of the physical implications of these formulas.

First of all, Eq. (4.5-4a) shows that, from its initial value v_0, the mean velocity $\langle V(t) \rangle$ of the Brownian particle decays exponentially to zero,

$$\langle V(\infty) \rangle = 0, \tag{4.5-30}$$

the characteristic time for this exponential decay being

$$\tau_V \equiv \frac{1}{\beta} = \frac{1}{4} \left(\frac{\pi}{2} \right)^{1/2} \frac{1}{\rho \, R^2} \frac{M}{(m k_B T)^{1/2}}. \tag{4.5-31}$$

This time τ_V is also significant because Eq. (4.5-4c) implies that $V(t_1)$ and $V(t_2)$ will be effectively *uncorrelated* if $|t_2 - t_1| \gg \tau_V$.

Secondly, if we were to simply *define* the drag coefficient γ through Eq. (4.5-2a) [cf. Eqs. (3.4-16)], then it would follow from Eqs. (4.5-4a) that $\gamma = M\beta$, and hence by Eq. (4.5-28a),

$$\gamma \equiv 4(2/\pi)^{1/2} \, \rho \, R^2 \, (m k_B T)^{1/2}. \tag{4.5-32}$$

As in the case of the *liquid* drag coefficient [$\gamma = 6\pi\eta a$ with η the liquid's viscosity and a the diameter of the particle], the drag coefficient here is independent of the particle's mass M; however, γ here evidently depends quadratically, rather than linearly, on the particle's "diameter."

Thirdly, taking the $t \to \infty$ limit of Eq. (4.5-4b), we find by substituting from the β and c formulas (4.5-28) that

$$\text{var}\{V(\infty)\} = \frac{c}{2\beta} = \frac{k_B T}{M}. \tag{4.5-33}$$

This is an especially satisfying result, being precisely the formula (4.5-2b) that Langevin's analysis had to *assume*. When combined with Eq. (4.5-30), this result implies that

$$\langle V^2(\infty) \rangle = \frac{k_B T}{M} = \frac{k_B T}{m} \frac{m}{M} \equiv \sigma^2 \frac{m}{M}. \tag{4.5-34}$$

Since $m \ll M$, this relation confirms our critical assumption (4.5-22).

Turning now to the formulas (4.5-5) for the integral process $X(t)$, we first note from Eq. (4.5-5a) that $\langle X(t) \rangle$ *initially* increases like

$$\langle X(t) \rangle \approx (v_0/\beta) (1 - (1-\beta t)) = v_0 t \qquad (t \ll \tau_V), \tag{4.5-35a}$$

just as we might expect. But eventually, as $\langle V(t) \rangle$ asymptotically approaches zero, $\langle X(t) \rangle$ asymptotically approaches the constant value

$$\langle X(t) \rangle \approx (v_0/\beta) (1 - 0) = v_0 \tau_V \qquad (t \gg \tau_V). \tag{4.5-35b}$$

Superimposed on this *average* motion is a *fluctuating* motion described by var$\{X(t)\}$. It can be shown by expanding the exponentials in Eq. (4.5-5b) to third order in t that var$\{X(t)\}$ *initially* increases like

$$\text{var}\{X(t)\} \approx (c/3)t^3 \qquad (t \ll \tau_V). \tag{4.5-36a}$$

However, as t becomes large, only the linear term on the right side of Eq. (4.5-5b) survives, and we have

$$\text{var}\{X(t)\} \approx (c/\beta^2)t \qquad (t \gg \tau_V). \tag{4.5-36b}$$

Evaluating the coefficient c/β^2 from Eqs. (4.5-28) and then taking the square root, we obtain

$$\text{sdev}\{X(t)\} \approx \left[\frac{1}{2\rho R^2} \left(\frac{\pi k_B T}{2m} \right)^{1/2} \right]^{1/2} t^{1/2} \qquad (t \gg \tau_V). \tag{4.5-37}$$

Equation (4.5-37) is the one-dimensional gas version of the famous Einstein-Langevin formula (4.5-7b). It gives the average net displacement of the Brownian particle over a long time t, apart from the transient displacement (4.5-35b) that is present only when $v_0 \neq 0$. A direct comparison of the coefficients of $\sqrt{t}$ in Eqs. (4.5-37) and (4.5-7b) is difficult because of the hidden dependence of Langevin's drag coefficient on the microphysical parameters. (And of course, a real liquid medium is *not* the same as our one-dimensional gaseous medium.) But in both cases, the coefficient of $\sqrt{t}$ is independent of the particle's mass M. Apparently, the fact that a heavier particle is more difficult for the gas molecules to start moving is exactly offset by the fact that, once moving, a heavier particle is more difficult for the gas molecules to slow down. But surely the most surprising feature of formula (4.5-37) is the way in which the coefficient of $\sqrt{t}$ varies with the gas molecule mass m: Whereas common sense would suggest that *lighter* gas molecules should be *less* effective in knocking the Brownian particle about, the m-dependence of formula (4.5-37) instead implies that lighter gas molecules actually *enhance* the random movement of the Brownian particle! This seeming paradox is resolved by realizing that if m is reduced *while T is kept constant*, then the average speed $\sigma = (k_B T/m)^{1/2}$ of the gas molecules will necessarily increase. So even though the gas molecules are lighter, they will on the average be moving faster when they strike the Brownian particle, *and* they will be striking the Brownian particle at an increased average rate. The net effect is a slight enhancement of the Brownian particle's erratic motion.

The forward Kramers-Moyal equation for the Markov process $V(t)$ reads [see Eq. (4.2-4b)]

$$\frac{\partial}{\partial t} P(v,t \mid v_0, 0) = \sum_{n=1}^{\infty} \frac{(-1)^n}{n!} \frac{\partial^n}{\partial v^n} \left[W_n(v) P(v,t \mid v_0, 0) \right]. \quad (4.5\text{-}38)$$

To see if there are any circumstances in which a two-term Fokker-Planck-like truncation of this equation would be appropriate, we return to the exact formula (4.5-20) for $W_n(v)$. If we nondimensionalize the integral in that formula by writing $u' = u/\sigma$ and $v' = v/\sigma$, then the formula becomes

$$W_n(v) = (2\pi\sigma^2)^{-1/2} \kappa \, \varepsilon^n \int_{-\infty}^{\infty} \sigma du' \, \sigma^n (u' - v')^n \, \sigma \, |u' - v'| \exp(-u'^2/2),$$

or, collecting the scattered factors of σ on the right hand side,

$$W_n(v) = (2\pi)^{-1/2}(\kappa\,\sigma)(\sigma\,\varepsilon)^n \int_{-\infty}^{\infty} du'\,(u'-v')^n\,|u'-v'|\exp(-u'^2/2)$$

$$(n \geq 1). \quad (4.5\text{-}39)$$

Now, the integral over u' here is a pure number, so its n-dependence is essentially fixed. But the n-dependence of the factor $(\sigma\varepsilon)^n$ evidently depends upon our *choice of units* for measuring the velocity σ (note that ε is a pure number). If our unit of velocity is such that the numerical magnitude of σ is greater than $1/\varepsilon$, then we will have $\sigma\varepsilon > 1$, and the factor $(\sigma\varepsilon)^n$ will increase in size with increasing n. On the other hand, if we choose a unit of velocity for which

$$\sigma \ll 1/\varepsilon, \quad (4.5\text{-}40)$$

then the factor $(\sigma\varepsilon)^n$, and hence also $W_n(v)$, will decrease in size very rapidly with increasing n, and the Kramers-Moyal equation (4.5-38) will be amenable to an approximating two-term truncation. Now in the Brownian limit, we have by definition $M \gg m$, so $\varepsilon \equiv 2m/(M+m)$ will always be a "very small" number. Therefore, in the Brownian limit $1/\varepsilon$ will always be a "very large" number, and it will be easy to find a velocity unit sufficiently "macroscopic" that σ, when expressed in that unit, will satisfy the inequality (4.5-40). The resulting approximate truncation of the forward Kramers-Moyal equation to a two-term Fokker-Planck form means that the *jump* Markov process $V(t)$, *when viewed on that macroscopic velocity scale*, will look like a *continuous* Markov process.

And so we have at last arrived at a vindication of Langevin's leading assumption [see Subsection 3.4.C] that $V(t)$ can be approximately viewed as a continuous Markov process. In fact, in light of the Brownian limit formulas (4.5-29) for the first two propagator moment functions $W_1(v)$ and $W_2(v)$, we see that the continuous Markov process that approximates $V(t)$ on the macroscale is just the *Ornstein-Uhlenbeck* process [see Subsection 3.3.C] with $k = \beta$ and $D = c$. It then follows from Eq. (3.3-25b) that, in the Brownian limit, we have

$$V(\infty) \approx N(0, c/2\beta) = N(0, k_B T/M), \quad (4.5\text{-}41)$$

a result that is somewhat stronger than Eqs. (4.5-30) and (4.5-33). In words, we have proved here that a Brownian particle, *when viewed on a sufficiently macroscopic scale*, eventually comes to "thermal equilibrium" at the temperature of the gas.

In order to deduce the implications of our exact formula for $W_n(v)$ *without* the Brownian condition $M \gg m$ (such as with the self-diffusion condition $M = m$) — or even *with* the Brownian condition but *without* the *a priori* assumption (4.5-22) — we must resort to the more complicated, numerically oriented solution procedure described in Appendix C [replacing there $X(t)$ by $V(t)$ and $B_k(x)$ by $W_k(v)$]. Indeed, we would want to enlarge that solution procedure somewhat to allow a concurrent estimation of the moments of the time-integral $X(t)$ of $V(t)$. But we shall not pursue this matter any further here. Before leaving this subject, though, we should point out one very crucial assumption that has underlain our entire analysis: We have assumed throughout that the velocities of the successive gas molecules that strike the particle are *statistically independent*. In fact, that assumption is not entirely correct. After a gas molecule has collided with the particle, that gas molecule will collide with other gas molecules, and in so doing, it will communicate to them some measure of information concerning the position and velocity of the particle. The altered behavior of those molecules will, in various subtle ways, "remind the particle of its past," and will thereby spoil the purely Markovian character of the process $V(t)$. These non-Markovian influences evidently arise solely from molecule-molecule collisions, so they should be most pronounced when the gas is relatively dense and the molecules are relatively large, making molecule-molecule collisions relatively frequent. But excepting that situation, the Markovian approximation of a "perfectly forgetting gas" should be reasonably good.

4.6 MONTE CARLO SIMULATION OF CONTINUUM-STATE JUMP MARKOV PROCESSES

In Section 2.9 we described how virtually any property of a Markov process $X(t)$ can be quantitatively estimated from a set of Monte Carlo samples or "realizations" $x_1(t)$, $x_2(t)$, ..., $x_N(t)$ of the process [see for example Eqs. (2.9-3) and (2.9-4)]. A specific procedure for constructing a process realization when $X(t)$ is a *continuous* Markov process was described and illustrated in Section 3.9. In this section we shall describe and illustrate a procedure for constructing a Monte Carlo realization of $X(t)$ when $X(t)$ is a *jump* Markov process with continuum states.

4.6.A JUMP SIMULATION THEORY

The simulation of jump Markov processes is in principle easier than the simulation of continuous Markov processes, because for jump Markov processes it is possible to construct a Monte Carlo simulation algorithm that is *exact* in the sense that it never approximates an infinitesimal time increment dt by a finite time increment Δt. The jump simulation algorithm is therefore not burdened with the concern for accuracy that generally encumbers the continuous simulation algorithm outlined in Section 3.9. But in order to place this exact jump simulation algorithm in its proper pedagogical context, we must first make a brief digression.

We recall that a Markov process $X(t)$ of *any* type is defined by its propagator $\Xi(dt; x,t)$. That random variable in turn is defined by the statement that, if $X(t) = x$, then for any positive infinitesimal dt we have

$$X(t+dt) = x + \Xi(dt; x,t). \tag{4.6-1}$$

It follows that if Δt is a small but *finite* time increment, then given $X(t) = x$ we should be able to assert with some degree of accuracy that

$$X(t+\Delta t) \approx x + \Xi(\Delta t; x,t), \tag{4.6-2}$$

where $\Xi(\Delta t; x,t)$ is the random variable obtained by simply replacing dt in the process propagator by Δt. This approximate equation is in fact the basis for the continuous Markov process simulation algorithm outlined in Fig. 3-7; more specifically, since the propagator $\Xi(dt; x,t)$ of the continuous Markov process with characterizing functions $A(x,t)$ and $D(x,t)$ is the normal random variable with mean $A(x,t)dt$ and variance $D(x,t)dt$, then to advance the process in state x at time t to time $t+\Delta t$, we merely add to x a sample value ξ of the random variable $N(A(x,t)\Delta t, D(x,t)\Delta t)$, taking care of course not to choose Δt "too large."

It is quite possible to construct a simulation algorithm for *jump* Markov processes along these same lines. Since the density function of $\Xi(dt; x,t)$ for the jump Markov process with characterizing functions $a(x,t)$ and $w(\xi \mid x,t)$ has the form (4.1-13), then a sample value ξ of the random variable $\Xi(\Delta t; x,t)$ can be generated as follows: First decide with respective probabilities $a(x,t)\Delta t$ and $[1 - a(x,t)\Delta t]$ whether the process does or does not jump in $[t, t+\Delta t)$. This can be done by drawing a unit uniform random number r [a sample value of the random variable $U(0,1)$], and then saying that the process jumps if $r \leq a(x,t)\Delta t$, and does not jump otherwise.[†] Notice that by taking Δt to be "very small" in order to minimize inaccuracies, we make it very unlikely that the process will jump in $[t, t+\Delta t)$. In any case, having made a decision on whether or not to jump in $[t, t+\Delta t)$, we turn again to Eq. (4.1-13). It tells us that if the decision is to *jump*, then we must choose ξ by sampling the density function $w(\xi \mid x,t)$; however, if the decision is to *not* jump, then we must

---

† Recall that if ε is any number between 0 and 1, then the probability that a sample value r of the unit uniform random variable $U(0,1)$ will be less than or equal to ε is just ε.

choose ξ by sampling the density function $\delta(\xi)$, which of course always produces the value $\xi = 0$. Thus we have deduced the jump simulation algorithm outlined in Fig. 4-2. This simulation procedure may be regarded as the jump analog of the continuous simulation procedure outlined in Fig. 3-7.

But the jump simulation procedure of Fig. 4-2 has a serious drawback: It is very difficult to formulate a simple test condition for insuring that Δt in Step 2° has been chosen "suitably small." Recalling our logic in deriving the first order accuracy condition (3.9-8) for the continuous simulation algorithm of Fig. 3-7, it would appear that an analogous condition in this case should be obtainable by developing a $(dt)^2$-approximation to the $o(dt)$ term in Eq. (4.1-18a), which term represents the amount by which the jump propagator density function (4.1-13) fails to satisfy the fundamental Chapman-Kolmogorov condition. But the resulting formula for that $(dt)^2$-approximation turns out to be complicated and not easily dealt with [see in this connection the footnote on page 228]. Another weakness of the jump simulation method outlined in Fig. 4-2 is that *there is no circumstance in which the method is exact for all positive values of* Δt. This is in contrast to the continuous simulation algorithm of Fig. 3-7, which we found to be exact for any $\Delta t > 0$ whenever the characterizing functions are constants. That the method of Fig. 4-2 cannot be valid for arbitrary Δt even if $a(x,t) \equiv a$ and $w(\xi \mid x,t) \equiv w(\xi)$ can easily be seen by noting that the no-jump probability $[1 - a\Delta t]$ will be negative, and hence meaningless, if $\Delta t > 1/a$.

But there is really no point in developing a small-Δt criterion for the jump simulation procedure of Fig. 4-2; because, it is easy to construct an *exact* jump simulation procedure by using the next-jump density function $p(\tau,\xi \mid x,t)$. That option was of course not open to us in the continuous case [see the discussion following Eq. (4.3-2)]. By definition, the function $p(\tau,\xi \mid x,t)$ is the joint density function for the time τ to the next jump and the jump distance ξ in that next jump, given that $X(t) = x$. Therefore, if (τ,ξ) is a random sampling of the joint density function $p(\tau,\xi \mid x,t)$, then we may without further ado assert that the process, in state x at time t, remains in that state until time $t + \tau$, at which time it jumps to state $x + \xi$. So all that is needed in order to advance a jump Markov process from one jump to the next is a procedure for generating a random pair (τ,ξ) according to the joint density function $p(\tau,\xi \mid x,t)$ given in Eq. (4.3-4).

As we know from our discussion of the joint inversion generating method in Section 1.8, if the joint density function $p(\tau,\xi \mid x,t)$ can conveniently be *conditioned* in the form

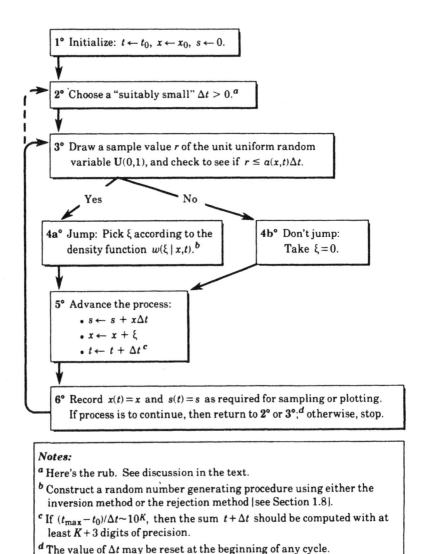

Figure 4-2. An *unrecommended* Monte Carlo simulation algorithm for the jump Markov process with characterizing functions $a(x,t)$ and $w(\xi \mid x,t)$. The procedure produces sample values $x(t)$ and $s(t)$ of the process $X(t)$ and its time-integral $S(t)$ at times $t = t_0 + \Delta t,\ t_0 + 2\Delta t,\ t_0 + 3\Delta t$, etc. This simulation algorithm becomes exact only in the limit $\Delta t \to 0$, and is generally inferior to the simulation algorithm of Fig. 4-3.

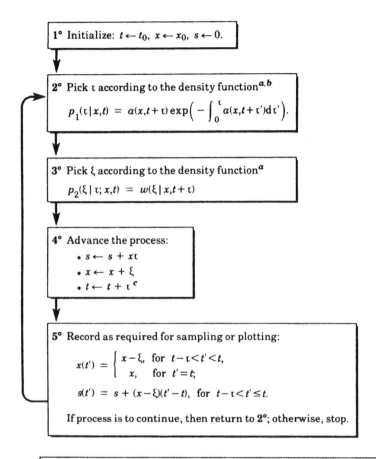

Figure 4-3. Exact Monte Carlo simulation algorithm for the jump Markov process with characterizing functions $a(x,t)$ and $w(\xi \mid x,t)$. The procedure produces exact sample values $x(t)$ and $s(t)$ of the process $X(t)$ and its time-integral $S(t)$ for all $t > t_0$.

$$p(\tau,\xi \mid x,t) = p_1(\tau \mid x,t)\, p_2(\xi \mid \tau; x,t),\qquad (4.6\text{-}3)$$

then one way to generate a random pair (τ,ξ) according to that joint density function is to first generate a random number τ according to the density function $p_1(\tau \mid x,t)$, and then generate a random number ξ according to the density function $p_2(\xi \mid \tau; x,t)$. We discussed in Section 1.8 how either the inversion generating method or the rejection generating method can be used to generate a random number x according to a given density function $P(x)$. Furthermore, we proved in Section 4.3 that the subordinate density functions $p_1(\tau \mid x,t)$ and $p_2(\xi \mid \tau; x,t)$ corresponding to the joint density function $p(\tau,\xi \mid x,t)$ are given explicitly by

$$p_1(\tau \mid x,t) = a(x,t+\tau)\exp\left(-\int_0^\tau a(x,t+\tau')d\tau'\right),\qquad (4.6\text{-}4a)$$

$$p_2(\xi \mid \tau; x,t) = w(\xi \mid x,t+\tau).\qquad (4.6\text{-}4b)$$

Thus we have established the first four steps of the procedure outlined in Fig. 4-3 for *exactly* simulating a jump Markov process with characterizing functions $a(x,t)$ and $w(\xi \mid x,t)$.

The τ-selection process of Step 2° of the procedure in Fig. 4-3 turns out to be especially simple in the case that the characterizing function $a(x,t)$ is explicitly independent of t, as in fact it is for nearly all jump Markov processes of practical interest. For, when $a(x,t) \equiv a(x)$, then the formula (4.6-4a) for $p_1(\tau \mid x,t)$ evidently simplifies to

$$p_1(\tau \mid x,t) = a(x)\exp\left(-\int_0^\tau a(x)d\tau'\right) = a(x)\exp[-a(x)\tau],$$

and this is evidently the density function of the exponential random variable with decay constant $a(x)$ [see Eq. (1.4-5)]. So it follows from our analysis in Section 1.8 [see Eq. (1.8-7)] that we can generate a random number τ according to this density function by simply drawing a random sample r of the unit uniform random variable $U(0,1)$ and then taking

$$\tau = [1/a(x)]\ln(1/r).\qquad (4.6\text{-}5)$$

A somewhat less dramatic simplification occurs in the ξ-selection process of Step 3° if $w(\xi \mid x,t)$ happens to be explicitly independent of t: In that case Eq. (4.6-4b) shows that $p_2(\xi \mid \tau; x,t)$ will be explicitly independent of both t and τ, being equal to simply $w(\xi \mid x)$, so the ξ-selection process of Step 3° will be completely independent of the result of the τ-selection process of Step 2°. Evidently, *both* of the foregoing simplifications will attend any *temporally homogeneous* jump Markov process — a fact that we could also have deduced from the properties (4.3-10).

The three advancement formulas in Step 4° of Fig. 4-3 should be obvious. Note in particular that the increase in the time-integral process $S(t)$ between times t and $t + \tau$ is *exactly* equal to $x[(t+\tau)-t] = x\tau$, because $X(t)$ had the constant value x throughout the entire time interval $[t, t+\tau)$. Of course, the s-update in Step 4° must always be done *before* the x-update. Once Step 4° has been completed, then we can assert that at the current time t the realization $x(t)$ of the process $X(t)$ will have the value x, and the realization $s(t)$ of the integral process $S(t)$ will have the value s. But notice that we can also assert precise values for those two realizations during the entire preceding time interval $(t-\tau, t)$. As just mentioned, the realization of $X(t)$ must have had the value $x - \xi$ during that interval:

$$x(t') = x - \xi \quad \text{for} \quad t' \in [t-\tau, t). \qquad (4.6\text{-}6)$$

And since, as t' increases from $t-\tau$ to t, the realization $s(t')$ increases at a constant rate $(x - \xi)$ to the final value $s(t) = s$, then we have

$$s(t') = s + (x-\xi)(t'-t) \quad \text{for} \quad t' \in [t-\tau, t]. \qquad (4.6\text{-}7)$$

Equations (4.6-6) and (4.6-7) give the realizations of $X(t)$ and $S(t)$ *exactly* during the entire time interval between the last two jumps of $X(t)$, and are the basis for Step 5° of Fig. 4-3.

When the jump Markov process $X(t)$ is *temporally homogeneous*, then the only potentially difficult step in the simulation procedure of Fig. 4-3 is Step 3°, which requires us to generate a random number ξ according to the density function $w(\xi \mid x)$. Whenever possible we should generate ξ by using the *inversion* generating method. As discussed in Section 1.8, the gist of that method is choose ξ so as to satisfy the equation

$$\int_{-\infty}^{\xi} d\xi' \, w(\xi' \mid x) = r', \qquad (4.6\text{-}8)$$

where r' is a unit uniform random number (*not* the same number as used in Step 2°). But if this equation cannot be readily solved for ξ for arbitrary values of x and r', then we should try instead the *rejection* generating method. In that method [see Section 1.8], we begin by *tentatively* choosing ξ according to

$$\xi = a(x) + (\beta(x) - a(x))r', \qquad (4.6\text{-}9a)$$

where r' is a unit uniform random number and $[a(x), \beta(x)]$ is the smallest interval that is certain, or *practically* certain, to contain any sample value of the density function $w(\xi \mid x)$. Then, letting $w^*(x)$ denote the maximum value of $w(\xi' \mid x)$ for $\xi' \in [a(x), \beta(x)]$, we draw a second unit uniform random number r'' and *accept* the tentative value ξ if and only if

$$\frac{w(\xi \mid x)}{w^*(x)} \geq r''. \tag{4.6-9b}$$

If the inequality (4.6-9b) is found *not* to be satisfied, then we must *reject* the tentative value ξ and begin again at Eq. (4.6-9a) with new unit uniform random numbers r' and r''. As was proved in Section 1.8 [see Eq. (1.8-11)], the a priori probability that the tentative ξ-value of Eq. (4.6-9a) will pass the acceptance test (4.6-9b) is

$$E = \frac{\displaystyle\int_{\alpha(x)}^{\beta(x)} w(\xi' \mid x)\, d\xi'}{w^*(x)\,(\beta(x) - \alpha(x))}. \tag{4.6-10}$$

Geometrically, this is just the ratio of the area under the function $w(\xi \mid x)$ in the interval $\alpha(x) \leq \xi \leq \beta(x)$ to the area under the rectangle of height $w^*(x)$ over that same ξ-interval.

The jump simulation examples presented in the following subsection will illustrate both of the methods described above for generating random values ξ.

4.6.B EXAMPLES OF JUMP SIMULATIONS

We shall now use the Monte Carlo simulation algorithm of Fig. 4-3 to simulate some of the jump Markov processes considered earlier in this chapter. By examining the numerical results of these simulations, we can get a good feeling for how jump Markov processes really behave. All of the jump Markov processes that we shall simulate in this subsection are *temporally homogeneous*, so the τ-selection in Step 2° of the algorithm can always be implemented by following the simple recipe in Note (b). We shall find that the ξ-selection procedure for Step 3° will sometimes be just as easy as this τ-selection recipe, but at other times it will require a bit of ingenuity.

We shall begin by simulating the three *completely* homogeneous jump Markov processes considered in Subsection 4.4.B. Each of those processes depends on only two constant parameters, a and σ: a is the constant value of the characterizing function $a(x,t)$, while σ roughly measures the "width" of the characterizing function $w(\xi \mid x,t) \equiv w(\xi)$. In each of these three processes, the τ-selection in Step 2° is most simply carried out by drawing a unit uniform random number r [a sample value of the random variable $U(0,1)$] and taking

$$\tau = (1/a)\ln(1/r).$$ (4.6-11)

In the first process considered in Subsection 4.4.B, the characterizing function $w(\xi)$ is the density function (4.4-15) for the exponential random variable with decay constant $1/\sigma$. As we proved in Section 1.8 [see Eq. (1.8-7)], we can generate a random number ξ according to this density function by drawing a unit uniform random number r' and taking

$$\xi = \sigma\ln(1/r').$$ (4.6-12)

So in using the algorithm of Fig. 4-3 to generate a Monte Carlo realization of the completely homogeneous jump Markov process defined by Eq. (4.4-15), we shall implement Steps 2° and 3° by applying formulas (4.6-11) and (4.6-12) respectively. In Fig. 4-4 we show the result of one such simulation run with $a = \sigma = 1$ and $x_0 = t_0 = 0$. Figure 4-4a shows the realization of the process $X(t)$, along with its one-standard deviation envelope $\langle X(t)\rangle \pm \mathrm{sdev}\{X(t)\}$ as calculated from Eqs. (4.4-17). Figure 4-4b shows the corresponding realization of the integral process $S(t)$, along with its one-standard deviation envelope $\langle S(t)\rangle \pm \mathrm{sdev}\{S(t)\}$ as calculated from Eqs. (4.4-18). Although we expect there to be an average of $1/a = 1$ units of time between successive jumps and $\sigma = 1$ units of $X(t)$ moved in a jump, Fig. 4-4a shows that substantial deviations from these nominally expected values are commonplace; this of course is due to the relatively large width of the exponential density function [see Fig. 1-2b]. In this particular simulation run 42 jumps occurred, which is quite compatible with the nominally expected number of 50.

In the second process considered in Subsection 4.4.B, the characterizing function $w(\xi)$ is the density function (4.4-22) for the normal random variable with mean 0 and variance σ^2. As we proved in Section 1.8 [see Eqs. (1.8-16)], we can generate a *pair* of independent random numbers ξ_1 and ξ_2 according to this density function by drawing two independent unit uniform random numbers r_1 and r_2 and taking

$$\left.\begin{aligned}\xi_1 &= \sigma\,[2\ln(1/r_1)]^{1/2}\cos(2\pi r_2)\,,\\[4pt]\xi_2 &= \sigma\,[2\ln(1/r_1)]^{1/2}\sin(2\pi r_2)\,.\end{aligned}\right\}$$ (4.6-13)

So in using the algorithm of Fig. 4-3 to generate a Monte Carlo realization of the completely homogeneous jump Markov process defined by Eq. (4.4-22), we shall implement Steps 2° and 3° by applying formulas (4.6-11) and (4.6-13) respectively. In Fig. 4-5 we show the result of one such simulation run with $a = \sigma = 1$ and $x_0 = t_0 = 0$. The realization of the process $X(t)$ is shown in Fig. 4-5a, and the corresponding realization of

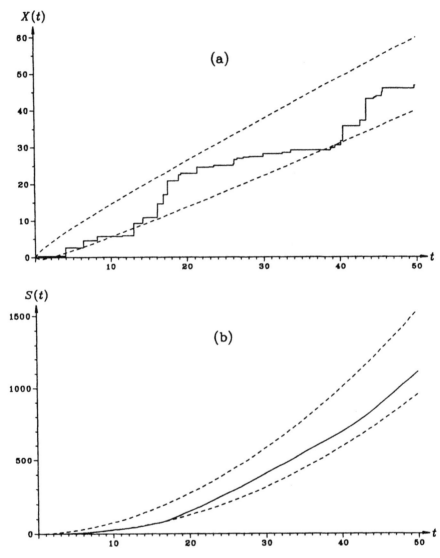

Figure 4-4. A simulation of the jump Markov process defined by the characterizing functions $a(x,t) = 1$ and $w(\xi \geq 0 \mid x,t) = \exp(-\xi)$, using the initial condition $X(0) = 0$. The realization of the process $X(t)$ is shown in (a), and the corresponding realization of its integral $S(t)$ is shown in (b). The dashed curves outline the one-standard deviation envelopes, as calculated from Eqs. (4.4-17) and (4.4-18) with $a = \sigma = 1$ and $x_0 = t_0 = 0$.

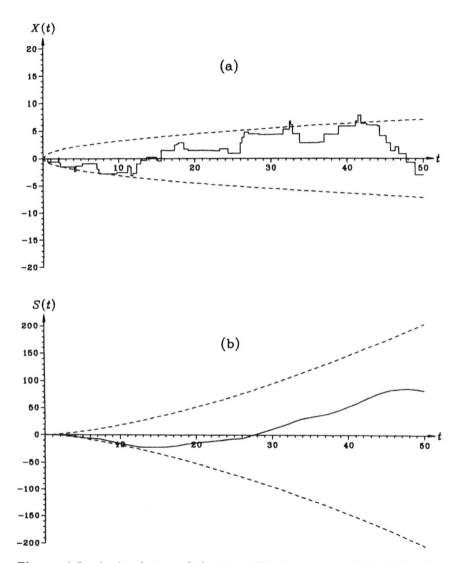

Figure 4-5. A simulation of the jump Markov process defined by the characterizing functions $a(x,t) = 1$ and $w(\xi \mid x,t) = |2\pi|^{-1/2} \exp(-\xi^2/2)$, using the initial condition $X(0) = 0$. The realization of the process $X(t)$ is shown in (a), and the corresponding realization of its integral $S(t)$ is shown in (b). The dashed curves outline the one-standard deviation envelopes, as calculated from Eqs. (4.4-24) and (4.4-25) with $a = \sigma = 1$ and $x_0 = t_0 = 0$.

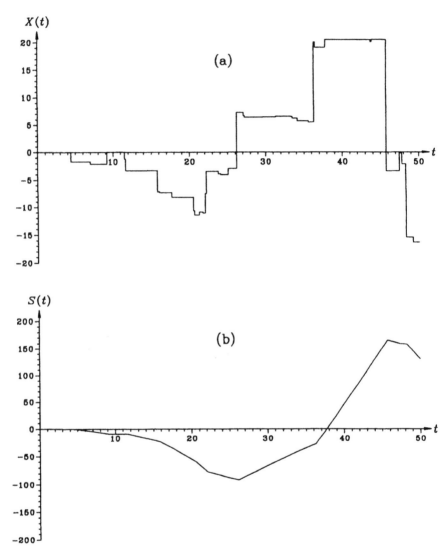

Figure 4-6. A simulation of the jump Markov process defined by the characterizing functions $a(x,t) = 1$ and $w(\xi \mid x,t) = (1/\pi)[\xi^2 + 1]^{-1}$, using the initial condition $X(0) = 0$. The realization of the process $X(t)$ is shown in (a), and the corresponding realization of its integral $S(t)$ is shown in (b). Since no moments exist for this process, there are no one-standard deviation envelopes. Note the relatively erratic behavior of this process, where ξ is sampled from $C(0,1)$, as compared to the process of Fig. 4-5, where ξ is sampled from $N(0,1)$.

the integral process $S(t)$ is shown in Fig. 4-5b. The dashed curves show the corresponding one-standard deviation envelopes as calculated from Eqs. (4.4-24) and (4.4-25). We expect there to be an average of $1/a = 1$ time units between successive jumps, and a root-mean-square distance of $\sigma = 1$ units of $X(t)$ moved in each jump. In this particular simulation run 55 jumps actually occurred, which is quite compatible with the nominally expected number of 50.

In the third process considered in Subsection 4.4.B, the characterizing function $w(\xi)$ is the density function (4.4-29) for the Cauchy random variable centered on 0 with half-width σ. As we found in Eq. (1.8-8), we can generate a random number ξ according to this density function by drawing a unit uniform random number r' and taking

$$\xi = \sigma \tan[(r' - \tfrac{1}{2})\pi]. \qquad (4.6\text{-}14)$$

So in using the algorithm of Fig. 4-3 to generate a Monte Carlo realization of the completely homogeneous jump Markov process defined by Eq. (4.4-29), we shall implement Steps 2° and 3° by using formulas (4.6-11) and (4.6-14) respectively. In Fig. 4-6 we show the result of one such simulation run with $a = \sigma = 1$ and $x_0 = t_0 = 0$. The realization of the process $X(t)$ is shown in Fig. 4-6a, and the corresponding realization of the integral process $S(t)$ is shown in Fig. 4-6b. As discussed in Subsection 4.4.B, none of the moments exist for this process, so there are no one-standard deviation envelopes. We nominally expect $1/a = 1$ time units per jump, just as in the preceding two runs; in this run 38 jumps actually occurred. Notice that the only difference between this process and the process of Fig. 4-5 is that in Fig. 4-5 we picked the jump distance ξ by sampling the random variable $N(0,1)$, whereas in Fig. 4-6 we picked the jump distance ξ by sampling the random variable $C(0,1)$. Although these two random variables are superficially similar [compare Figs 1-2c and 1-2d], the "heavy tails" of the Cauchy variable's density function make large jumps much more likely for it, which of course is why all the moments for the Cauchy process are infinite. It is interesting that this process, which is difficult to investigate analytically since we have no moment evolution equations, is nevertheless quite simple to simulate numerically.

For our next simulation example, we consider the completely homogeneous jump Markov process $X(t)$ defined by the characterizing functions in Eqs. (4.4-39). As discussed in Subsection 4.4.C, this process *approximates* the *dashed* trajectory schematized in Fig. 4-1, and that trajectory in turn crudely models the x-coordinate of a "labeled" molecule of mass m and radius r in a gas of otherwise identical molecules, the gas being in thermal equilibrium at absolute temperature T. The two parameters λ^* and τ^* appearing in the characterizing functions (4.4-39) are defined by [see Eqs. (4.4-36) and (4.4-37)]

$$\lambda^* = [\pi(2r)^2\rho]^{-1} \quad \text{and} \quad \tau^* = [\pi(2r)^2\rho u^*]^{-1}, \qquad (4.6\text{-}15)$$

where ρ is the average number of gas molecules per unit volume and $u^* = (k_B T/m)^{1/2}$, k_B being Boltzmann's constant.

Since for this completely homogeneous jump Markov process we have $a = 1/\tau^*$, then the τ-selection procedure for Step 2° of our simulation algorithm is very straightforward: We simply draw a unit uniform random number r (not to be confused with the gas radius r) and take

$$\tau = \tau^* \ln(1/r). \qquad (4.6\text{-}16)$$

But we have a slight problem with the ξ-selection procedure for Step 3°, because the density function $w(\xi)$ in Eq. (4.4-39b) cannot be calculated and inverted *analytically*, as is usually required for an easy application of the inversion generating method. There are several viable alternatives. One would be to *numerically* evaluate the distribution function

$$F(\xi) \equiv \int_{-\infty}^{\xi} w(\xi') \, d\xi'$$

for a discrete lattice of ξ-values, and then calculate $\xi = F^{-1}(r')$ for any given $r' \in (0,1)$ by using some numerical interpolation procedure. Another alternative might be to use the *rejection* generating method. However, we shall use instead a third alternative, which is basically an extension of the inversion method that is suggested by the way in which the formula (4.4-39b) for $w(\xi)$ was originally derived. Recall that the derivation of Eq. (4.4-39b) in Subsection 4.4.C proceeded from the assertion that $\xi = s\cos\theta$, where s is a sample of the random variable $E(1/\lambda^*)$ and $\cos\theta$ is a sample of the random variable $U(-1,1)$. Now, according to Eq. (1.8-7) we can generate a sample value s of $E(1/\lambda^*)$ by drawing a unit uniform random number r_1 and taking $s = \lambda^* \ln(1/r_1)$; furthermore, according to Eq. (1.8-6) we can generate a sample value $\cos\theta$ from $U(-1,1)$ by drawing a unit uniform random number r_2 and taking $\cos\theta = -1 + 2r_2$. Thus we see that a sample value ξ from the density function (4.4-39b) can be generated by simply drawing two independent unit uniform random numbers r_1 and r_2 and taking

$$\xi = \lambda^* \ln(1/r_1)(-1 + 2r_2). \qquad (4.6\text{-}17)$$

So in using the algorithm of Fig. 4-3 to generate a Monte Carlo realization of the jump Markov process defined by the characterizing functions (4.4-39), we shall implement Step 2° by using formula (4.6-16) and Step 3° by using formula (4.6-17).†

For our numerical simulation here, let us use some physically realistic values for the three parameters r, ρ and u^* in the formulas (4.6-15) for λ^* and τ^*. A reasonable value for the radius of a simple gas molecule is $r = 10^{-8}$cm ($= 1$ angstrom). The average gas density ρ can be estimated from the perfect gas equation of state, $p = \rho k_B T$; taking the pressure p to be 10^6dynes/cm^2 (one atmosphere) and the absolute temperature T to be 300° (room temperature), we find, since $k_B = 1.38 \times 10^{-16}$erg/deg, that $\rho = 2.4 \times 10^{19}$molecules/cm^3. Finally, taking $m = 4.7 \times 10^{-23}$g (28 times the proton mass, roughly the mass of the Nitrogen molecule), we get $u^* \equiv (k_B T/m)^{1/2} = 3 \times 10^4$cm/sec. For these values of r, ρ and u^*, the formulas (4.6-15) yield

$$\lambda^* = 3.3 \times 10^{-5}\text{cm} \quad \text{and} \quad \tau^* = 1.1 \times 10^{-9}\text{sec}. \qquad (4.6\text{-}18)$$

In Fig. 4-7a we show the result of a Monte Carlo simulation of this jump Markov process using the above values for λ^* and τ^* and the initial condition $X(0) = 0$. Notice that the time axis in Fig. 4-7a is measured in nanoseconds (1ns $= 10^{-9}$sec), while the distance axis is measured in micrometers (1μm $= 10^{-6}$m). We have not bothered to plot the corresponding realization of the time-integral process $S(t)$ because it has no physical significance here. The dashed curves in Fig. 4-7a show the one-standard deviation envelope $\langle X(t) \rangle \pm \text{sdev}\{X(t)\}$ as predicted by Eqs. (4.4-41). In the 50 nsec time span of this plot we nominally expect $50/1.1 \approx 45$ jumps; in the present realization there were actually 47 jumps. In Fig. 4-7b we show an *extension* of the run in Fig. 4-7a out to 5 milliseconds (1ms $= 10^{-3}$sec), a time span that encompasses roughly 4.5 million jumps. Notice that the vertical scale in Fig. 4-7b is measured in millimeters. Rather than trying to plot a connected trajectory as we did in Fig. 4-7a, we have in Fig. 4-7b simply placed a (t,x) *dot* every 0.0025 milliseconds; this gives a total of 2000 dots in the plot, with an average of 2300 jumps occurring between successive dots. The dashed curves show as before the one-standard deviation envelope predicted by Eqs. (4.4-41). We emphasize that plot (a) is essentially a magnified view of the first 0.00005 ms of plot (b), a time interval that covers only 2% of the first 0.0025 ms sampling interval in plot (b). We should also note that if we had drawn plot (a) as a series of 2000 equally spaced (t,x) dots as we did

---

† A formal proof of the generating rule (4.6-17) would evidently consist of showing that the density function of the product of the two random variables $E(1/\lambda^*)$ and $U(-1,1)$ is the function $w(\xi)$ in Eq. (4.4-39b). Such a proof may be given using the RVT theorem (1.6-4), and it is mathematically identical to our derivation of Eq. (4.4-39b) in Subsection 4.4.C.

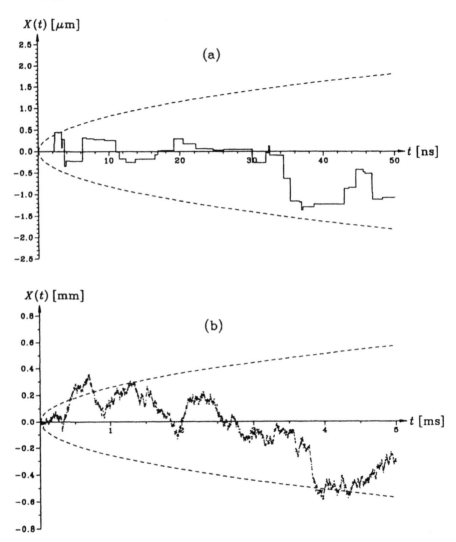

Figure 4-7. Differently scaled plots of one simulation run of the jump Markov process $X(t)$ defined by the characterizing functions (4.4-39), using the parameter values (4.6-18) and the initial condition $X(0) = 0$. Plot (a) is a magnified view of the first 0.00005 ms of Plot (b). Plot (b) has been drawn by simply placing a (t,x) dot every 0.0025 ms. This process *approximates* the *dashed* trajectory schematized in Fig. 4-1, and therefore approximately models the x-coordinate of a self-diffusing gas molecule. The dashed curves in the plots above outline the one-standard deviation envelopes predicted by Eqs. (4.4-41).

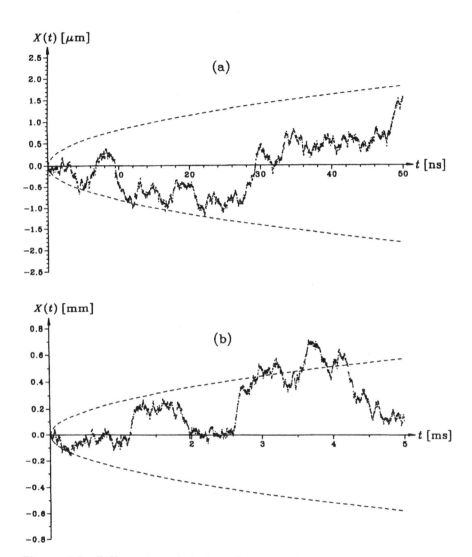

Figure 4-8. Differently scaled plots of one simulation run of the *continuous* Markov process $X(t)$ defined by the characterizing functions $A(x,t) = 0$ and $D(x,t) = 2\lambda^{*2}/3\tau^{*}$, using the parameter values (4.6-18). Plot (a) is a magnified view of the first 0.00005 ms of plot (b). Theory predicts that this continuous Markov process should provide a good *macroscale approximation* to the jump Markov process simulated in Fig. 4-7, and a comparison of the (b) plots in Figs. 4-7 and 4-8 confirms this expectation. But the two (a) plots show that the *microscale* structures of these two processes are quite different.

plot (b), then because of the finite size of those dots plot (a) would look just as it does now except that the vertical connecting lines would be missing.

We reasoned in Subsection 4.4.C [recall our derivation of Eq. (4.4-44)], that the jump Markov process $X(t)$ just simulated should look like a Wiener process with $A = 0$ and $D = 2\lambda^{*2}/3\tau^*$ *if* the process is observed on a scale where the unit of measure is "large" compared to the size of λ^*. Since $\lambda^* = 3.3 \times 10^{-4}$ mm, then the millimeter scale of Fig. 4-7b should satisfy that macroscale criterion. To test this expectation, we have used the *continuous* Markov process simulation algorithm of Fig. 3-7 to simulate the aforementioned Wiener process. Figure 4-8a shows the first 50 ns of that simulation run, sampled at 0.025 ns intervals, and Fig. 4-8b shows the first 5 ms of that same simulation run, sampled at 0.0025 ms intervals. The dashed curves show the associated one-standard deviation envelopes, now calculated from Eqs. (3.3-10). Just as our theory predicts, the two "macroscale" (b) plots in Figs. 4-7 and 4-8 look like realizations of the same continuous Markov process [compare the two realizations of the special Wiener process in Figs. 3-8a and 3-9a]. In other words, if we were to repeat each of the simulations in Figs. 4-7b and 4-8b K times, using different random number seeds each time, and then shuffle the resulting $2K$ plots, it would not be possible to sort them by inspection into K "jump" plots and K "continuous" plots. But the two "close-up" (a) plots in Figs. 4-7 and 4-8 show that these two Markov processes are easily distinguishable from each other on the *microscale*, even though their one-standard deviation envelopes are identical. In particular, the continuous process is *nowhere* differentiable, whereas the jump process is *almost everywhere* differentiable.

Finally, we note that the Wiener process simulated in Fig. 4-8 exhibits the *fractal* property of being "self-similar" over arbitrarily large scale changes: In "zooming in" on the first 0.00005 ms of the trajectory in Fig. 4-8b, we find, in Fig. 4-8a, a trajectory that looks just like a companion realization of the one in Fig. 4-8b.[†] In contrast, the jump Markov process of Fig. 4-7 clearly does not possess this unlimited scaling invariance property.

[†] The remarkable "statistical congruence" of the two continuous $X(t)$ plots in Fig. 4-8 can be understood as follows: The Langevin equation for any driftless Wiener process reads $dX = N[D dt]^{1/2}$, where N is the unit normal random variable and D the diffusion constant. If we subject X and t to the dual scaling transformation $\{X \to X' = hX, \ t \to t' = h^2 t\}$, h being any positive constant, then it is readily seen that the Langevin equation transforms to $dX' = N[D dt']^{1/2}$. Thus we see that any driftless Wiener process $X(t)$ is *dynamically invariant* to this dual scaling transformation. Now observe that, in going from Fig. 4-8b to Fig. 4-8a, we have magnified the scale of the t-axis by a factor of 10^5, and the scale of the x-axis by a factor of (very nearly) $10^{5/2}$.

The simulations of Figs. 4-7 and 4-8 concern a jump Markov process $X(t)$ which, as detailed in Subsection 4.4.C, *crudely* models a *real* physical process: the x-coordinate of a particular molecule of a gas in thermal equilibrium. We now turn to consider a jump Markov process $V(t)$ which, as detailed in Section 4.5, *rigorously* models an *idealized* physical process: the x-component of the velocity of a particle surrounded by a so-called one-dimensional gas in thermal equilibrium. The time-integral of the process $V(t)$, which we denote by $X(t)$, is of course the x-component of the position of the particle. We shall be interested here in two special cases: In the first, the particle is taken to be *much larger* than a gas molecule, and the integral process $X(t)$ is called *one-dimensional Brownian motion*. In the second case, the particle is taken to be *identical* to a gas molecule, and the integral process $X(t)$ is then called *one-dimensional self-diffusion*. We may expect the latter process to be similar to (but not necessarily identical to) the one-dimensional projection of three-dimensional self-diffusion that was crudely modeled in Figs. 4-7 and 4-8.

In Subsection 4.5.A, we proved that the process $V(t)$ is a temporally homogeneous jump Markov process with characterizing functions [see Eqs. (4.5-18) and (4.5-19)]

$$a(v) = \kappa \left[v \, \mathrm{erf}(v/\sqrt{2}\sigma) + (2/\mathrm{n})^{1/2} \sigma \exp(-v^2/2\sigma^2) \right], \quad (4.6\text{-}19a)$$

$$w(\xi \mid v) = K(v) \, |\xi| \, \exp\left(- \frac{(\xi + \varepsilon v)^2}{2\sigma^2\varepsilon^2} \right). \quad (4.6\text{-}19b)$$

The three parameters σ, κ and ε in these two functions are defined in Eqs. (4.5-15) in terms of the particle mass M and cube edge length R, the gas molecule mass m and cube edge length r, the gas density ρ, and the gas temperature T. The factor $K(v)$ is defined in Eq. (4.5-19b) in such a way that $w(\xi \mid v)$ is a properly normalized density function with respect to the variable ξ. In applying the Monte Carlo simulation algorithm of Fig. 4-3 in this instance, we must of course replace the variable pair (x,s) in that algorithm by the variable pair (v,x). We shall see shortly that we can implement this simulation algorithm for just about any set of values for the parameters σ, κ and ε; however, as we found in Subsection 4.5.B, analytical expressions for the mean and variance of $V(t)$ and of $X(t)$ can easily be obtained only if $\varepsilon \ll 1$, a condition that corresponds to the Brownian motion case, $M \gg m$.

So, how should we implement Steps 2° and 3° of the Monte Carlo simulation algorithm in Fig. 4-3 for the characterizing functions (4.6-19)? Since the first characterizing function is explicitly independent of t, then the τ-selection in Step 2° is easily accomplished: We draw a unit uniform random number r and take†

$$\tau = [1/a(v)] \ln(1/r). \qquad (4.6\text{-}20)$$

But the ξ-selection in Step 3° poses something of a problem, because the distribution function for ξ,

$$F(\xi \mid v) \equiv \int_{-\infty}^{\xi} w(\xi' \mid v) \, d\xi',$$

apparently cannot be calculated and inverted *analytically*, as is usually required for an easy application of the inversion generating method. A *numerical* inversion procedure likewise seems impractical, because the interpolating points will depend upon v and therefore will have to be recalculated after each jump. And so we turn to the *rejection* generating method.

The rejection generating method, described in conjunction with Eqs. (1.8-9) – (1.8-11), can be applied to the density function $w(\xi \mid v)$ if we make the associations

$$x \leftrightarrow \xi, \quad P(x) \leftrightarrow w(\xi \mid v), \quad C \leftrightarrow K(v), \quad h(x) \leftrightarrow h(\xi \mid v),$$

where in the last association we have defined the function

$$h(\xi \mid v) \equiv |\xi| \exp\left(-\frac{(\xi + \varepsilon v)^2}{2\sigma^2 \varepsilon^2} \right). \qquad (4.6\text{-}21)$$

Since the rejection method does not require a knowledge of C, then we need not concern ourselves with the factor $K(v)$. However, since the rejection method does require that we know the maximum value of $h(x)$, then our first step in setting up the rejection method here is to find the maximum value of $h(\xi \mid v)$. By differentiating $h(\xi \mid v)$ with respect to ξ, we discover that this function has two relative maximums at

$$\xi_{\pm} = (\varepsilon/2)[-v \pm (v^2 + 4\sigma^2)^{1/2}].$$

We note that ξ_+ is always positive while ξ_- is always negative. If $v=0$, then $h(\xi \mid v)$ is symmetric about $\xi=0$ and so assumes its absolute maximum at both $\xi_+ = \varepsilon\sigma$ and $\xi_- = -\varepsilon\sigma$. If $v>0$, then the dominating exponential factor in Eq. (4.6-21) will assume its maximum at a *negative* value of ξ, so the absolute maximum of $h(\xi \mid v>0)$ must occur at $\xi=\xi_-$.

---

† Computing $a(v)$ for Eq. (4.6-20) evidently entails evaluating the error function, erf(z). If one's computer software library does not contain that function, then one can use a rational approximation formula, such as Eq. (7.1.26) in M. Abramowitz and I. Stegun's *Handbook of Mathematical Functions*, National Bureau of Standards (1964).

But if $v < 0$, then that exponential factor will peak at a *positive* value of ξ, so the absolute maximum of $h(\xi \mid v > 0)$ must occur at $\xi = \xi_+$. Thus we conclude that the absolute maximum of $h(\xi \mid v)$ will be $h(\xi^* \mid v)$, where ξ^* is defined according to

$$\xi^* \equiv \begin{cases} \xi_+, & \text{if } v \le 0, \\ \xi_-, & \text{if } v \ge 0. \end{cases}$$

Now, the rejection generating method discussed in Section 1.8 assumes that the function $h(x)$ is zero everywhere outside some *finite* x-interval $[a,b]$, with the trial value x_t in Eq. (1.8-10) being selected as a sample of the random variable $U(a,b)$. An incentive to taking the interval $[a,b]$ to be as small as possible is provided by Eq. (1.8-11), which shows that the efficiency of this generating procedure is inversely proportional to $(b-a)$. But the function $h(\xi \mid v)$ in Eq. (4.6-21) is evidently nonzero for *all* $\xi \in (-\infty, \infty)$, and so it would appear that we ought not try to apply the rejection generating method in this instance. In actuality though, we are saved by the dominating exponential factor in $h(\xi \mid v)$. That factor is essentially the normal density function with mean $-\varepsilon v$ and standard deviation $\sigma \varepsilon$, and because of that factor the probability that a random sample from the density function $K(v) h(\xi \mid v)$ will lie outside of the interval $[-\varepsilon v - \gamma \sigma \varepsilon, -\varepsilon v + \gamma \sigma \varepsilon]$ goes to zero very rapidly as γ increases beyond 3. The point is that in any actual application of the rejection method, we shall require only a *finite* number N of random samples from the density function $w(\xi \mid v)$, and the probability that such a set of N samples will contain a ξ-value lying outside the aforementioned interval can be made arbitrarily small by taking $\gamma \equiv \gamma(N)$ large enough. A suitable value for γ can be found by empirical testing. For our purposes here, the value $\gamma = 6$ turns out to be "suitably large" from an accuracy standpoint, yet "suitably small" from an efficiency standpoint.

In light of the foregoing considerations, we see that an acceptable implementation of Step $3°$ of our jump simulation algorithm for the process $V(t)$ can be had as follows:

$3a°$ Evaluate $h(\xi^* \mid v)$ according to Eq. (4.6-21), taking

$$\xi^* \equiv \begin{cases} (\varepsilon/2)[(v^2 + 4\sigma^2)^{1/2} - v], & \text{if } v \le 0, \\ -(\varepsilon/2)[(v^2 + 4\sigma^2)^{1/2} + v], & \text{if } v > 0. \end{cases} \tag{4.6-22a}$$

$3b°$ With $\gamma = 6$, draw a unit uniform random number r' and calculate the *trial* ξ-value

$$\xi = (-\varepsilon v - \gamma \sigma \varepsilon) + 2(\gamma \sigma \varepsilon)r'. \tag{4.6-22b}$$

3c° Draw another unit uniform random number r''. If it is true that

$$h(\xi \mid v) \; \geq \; r'' \, h(\xi^* \mid v), \qquad\qquad (4.6\text{-}22c)$$

then *accept* the trial value ξ; otherwise, *reject* the trial value ξ and return to step·3b° with new samplings for both r' and r''.

We shall find in our simulations that, for $\gamma = 6$, the average ξ-acceptance ratio of this generating procedure is about 27% for the Brownian motion case and about 19% for the self-diffusion case. This inefficiency is the price we have to pay for the relative simplicity of the rejection method.

With the τ-selection procedure as in Eq. (4.6-20) and the ξ-selection procedure as in Eqs. (4.6-22), we are now ready to apply the Monte Carlo simulation algorithm of Fig. 4-3 to generate realizations of the jump Markov process $V(t)$ and its time integral $X(t)$. To maintain some semblance of continuity with the process modeled in Figs. 4-7 and 4-8, let us assume that each cube-shaped molecule of our one-dimensional gas has mass $m = 4.7 \times 10^{-23}$g and edge length $r = 2 \times 10^{-8}$cm (or 2 angstroms). And let us also assume that the gas is in thermal equilibrium at room temperature and atmospheric pressure, so that its density is $\rho = 2.4 \times 10^{19}$molecules/cm^3 and the root-mean-square speed of its constituent molecules is $u^* = 3 \times 10^4$cm/sec. We shall first consider a *Brownian motion case* in which the particle mass is $M = 10^6 m = 4.7 \times 10^{-17}$g. A cubic particle of this mass and a reasonable density could have an edge length of $R = 3 \times 10^{-6}$cm $= 300$A. With these values for the various microphysical constants, the three parameters σ, κ and ε defined in Eqs. (4.5-15) take the values

$$\sigma = 3 \times 10^4 \, \text{cm/sec}, \quad \kappa \approx 2.2 \times 10^8 \, \text{cm}^{-1}, \quad \varepsilon \approx 2 \times 10^{-6}. \quad (4.6\text{-}23)$$

Using these values for σ, κ and ε, and assuming the particle to be at rest ($v_0 = 0$) at the initial time $t_0 = 0$, we have executed our jump simulation algorithm until the time variable reached 10^{-11}sec $= 0.01$ nanosecond. Figure 4-9a shows the resulting realization of $V(t)$, measured in centimeters per second, and Fig. 4-9b shows the corresponding realization of $X(t)$ measured in milliangstroms (1 mA $= 10^{-11}$cm). A rough estimate of the average number of jumps (particle-molecule collisions) that we should expect to occur in a simulation run of this duration can be obtained from the value of $1/a(0)$, the average time between collisions when the particle velocity is zero:

$$1/a(0) = 1/[\kappa(2/\pi)^{1/2}\sigma] \approx 0.2 \times 10^{-12} \, \text{sec/jump};$$

therefore, in 10^{-11}sec we should see roughly $10^{-11}/[0.2 \times 10^{-12}] = 50$ jumps. In the realization shown in Fig. 4-9, 56 jumps actually occurred.

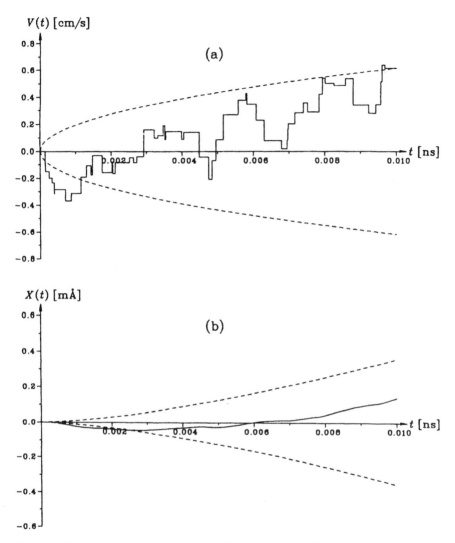

Figure 4-9. A simulation of the jump Markov process $V(t)$ and its time integral $X(t)$ as defined by the characterizing functions $a(v)$ and $w(\xi \,|\, v)$ in Eqs. (4.6-19), using the σ, κ and ε values of Eqs. (4.6-23) and the initial condition $V(0) = 0$. This is a rigorous simulation of *one-dimensional Brownian motion*, with the Brownian particle being 10^6 times more massive than a molecule of the surrounding air, which is asumed to be at standard temperature and pressure. The Brownian particle's velocity $V(t)$ is here measured in centimeters per second, while its position $X(t)$ is measured in milliangstroms. The dashed curves are the theoretically predicted one-standard deviation envelopes.

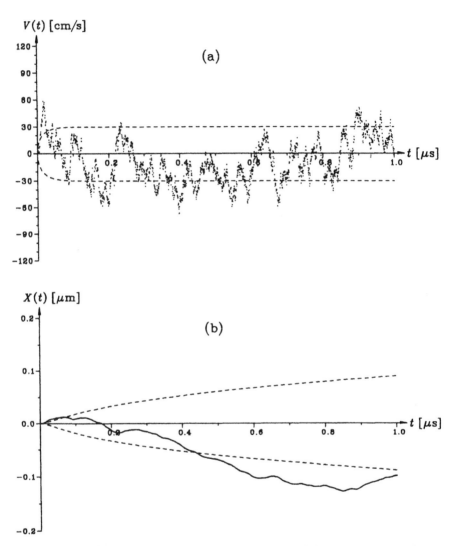

Figure 4-10. The Brownian motion simulation of Fig. 4-9 extended to 1 microsecond. The dots show the values of $V(t)$ in plot (a) and $X(t)$ in plot (b) every $5 \times 10^{-4} \mu s$. An average of 2500 particle-molecule collisions occur between successive dots, and the entire run contains about 5.27 million such collisions. Note that $X(t)$ is measured here in micrometers rather than in the milliangstrom units of Fig. 4-9 (1μm = 10^7mA). The dashed curves are as before the theoretical one-standard deviation envelopes.

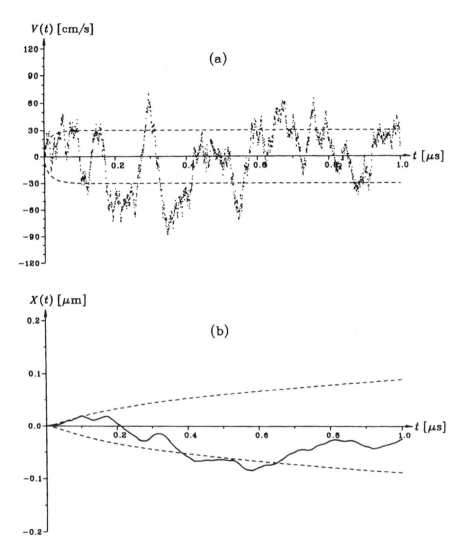

Figure 4-11. A simulation of the *continuous* Markov process $V(t)$ and its time-integral $X(t)$ as defined by the characterizing functions $A(v,t) = -\beta v$ and $D(v,t) = c$, using the β and c values of Eqs. (4.6-24) and the initial condition $V(0) = 0$. $V(t)$ here is an Ornstein-Uhlenbeck process, and theory predicts that, on the "macroscale" of this figure, it should provide a reasonably good approximation to the jump Markov process $V(t)$ simulated in Fig. 4-10. It evidently does.

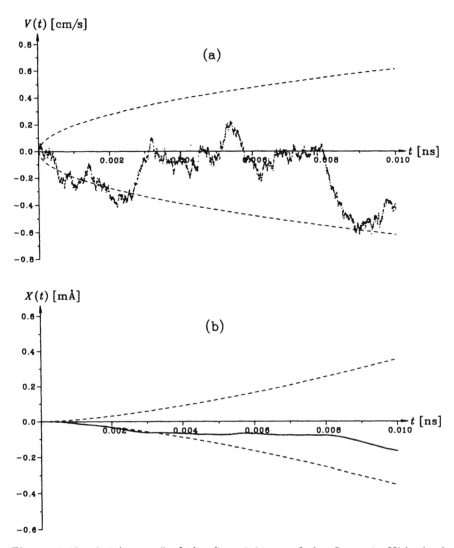

Figure 4-12. A "close-up" of the first 0.01 ns of the Ornstein-Uhlenbeck simulation run in Fig. 4-11. Comparing this *continuous* process simulation with the *jump* process simulation in Fig. 4-9, which is the analogous "close-up" of the simulation run in Fig. 4-10, we see that those two Markov processes are indeed distinguishable from each other on this more microscopic scale.

In Section 4.5 we proved that, in the Brownian limit, the first two propagator moment functions for $V(t)$ are $W_1(v) \approx -\beta v$ and $W_2(v) \approx c$, where β and c are given by Eqs. (4.5-28). As a consequence, the mean, variance and covariance of $V(t)$ are given (approximately) by Eqs. (4.5-4), while the mean, variance and covariance of $X(t)$ are given (approximately) by Eqs. (4.5-5). For the parameter values (4.6-23), we find from the formulas (4.5-28) that

$$\beta = 2.11 \times 10^7 \, \text{s}^{-1} \quad \text{and} \quad c = 3.79 \times 10^{10} \, \text{cm}^2 \text{s}^{-3}. \quad (4.6\text{-}24)$$

Substituting these values (along with $v_0 = 0$) into Eqs. (4.5-4) and (4.5-5), we can calculate explicitly the one-standard deviation envelopes $\langle V(t) \rangle \pm \text{sdev}\{V(t)\}$ and $\langle X(t) \rangle \pm \text{sdev}\{X(t)\}$. Those envelopes are plotted as the dashed curves in Fig. 4-9.

It can be seen from Eq. (4.5-4b) that after a time of the order of

$$\tau_V \equiv 1/\beta = 4.7 \times 10^{-8} \, \text{sec},$$

the standard deviation of the particle velocity should effectively reach its asymptotic value of

$$\text{sdev}\{V(\infty)\} = (c/2\beta)^{1/2} = 30 \, \text{cm/sec}.$$

Since the 0.01 nsec time interval simulated in Fig. 4-9 is only about 0.02 percent of τ_V, then clearly we are a long way from seeing the "thermal equilibration" of the particle. Figure 4-10, however, shows an *extension* of the 10^{-11}sec run in Fig. 4-9 to 10^{-6}sec = 1 microsecond. This time span contains over 5 million jumps, and, being about $20 \times \tau_V$, is easily long enough for the particle to reach thermal equilibrium. In Fig. 4-10 we have not plotted the full trajectories of $V(t)$ and $X(t)$ as we did in Fig. 4-9; instead, we have simply placed dots at the values of $V(t)$ and $X(t)$ at time intervals of 5×10^{-10}sec. That sampling interval gives a total of 2000 dots in each plot, with an average of 2500 jumps between successive dots. To repeat, the run shown in Fig. 4-9 is the initial 0.00001 μs of the run shown in Fig. 4-10, a time interval that covers only 2 percent of the first sampling interval in Fig. 4-10. The dashed one-standard deviation envelopes in Fig. 4-10 were calculated from the same formulas as used in Fig. 4-9.

We argued in Subsection 4.5.B [see the discussion of Eq. (4.5-40)] that the jump Markov process $V(t)$ just simulated should look like an Ornstein-Uhlenbeck process with $k=\beta$ and $D=c$, *if* it is observed on a scale where the natural unit of velocity is such that σ in that unit is very small compared to $1/\epsilon$. According to Eqs. (4.6-23), $\sigma = 3 \times 10^4$cm/sec and $1/\epsilon = 5 \times 10^5$. Now, the "natural velocity scale" in Fig. 4-9a is roughly 1 millimeter per second, and on that scale σ has a magnitude of 3×10^5; since that is *not* "very small" compared to $1/\varepsilon$, then we do not expect $V(t)$ in Fig. 4-9a to look like an Ornstein-Uhlenbeck process. But the natural velocity scale in Fig. 4-10 is more like 1 decimeter per second, and on that scale σ has a magnitude of 3×10^3; since that *is* "very small" compared to $1/\varepsilon$, then we do expect $V(t)$ in Fig. 4-10a to resemble the Ornstein-Uhlenbeck process with $k = \beta$ and $D = c$. To test this prediction, we have simulated that Ornstein-Uhlenbeck process using the continuous Markov process simulation algorithm of Fig. 3-7 [replacing there the variable pair (x,s) with the variable pair (v,x)]. In this continuous process simulation we have used $\Delta t = 5 \times 10^{-11}$ sec, so that the first order accuracy condition (3.9-8) is satisfied with $\varepsilon_1 = 0.001$, but we have sampled the process and its time-integral only every 5×10^{-10} sec, as was done in the simulation of Fig. 4-10. Fig. 4-11 shows the results of that continuous process simulation, and also the one-standard deviation envelopes as calculated from the Ornstein-Uhlenbeck formulas (3.3-20) and (3.3-21). As expected, the plots in Figs. 4-10 and 4-11 look pretty much like two realizations of the same Ornstein-Uhlenbeck process, even as the plots in Figs. 3-12 and 3-13 are by construction two realizations of a single Ornstein-Uhlenbeck process (but with a nonzero initial value).

To test the correlative prediction that the jump Markov process of Fig. 4-10 and the continuous Markov process of Fig. 4-11 should be *distinguishable* from each other on the finer scale of Fig. 4-9, we have simulated the Ornstein-Uhlenbeck process of Fig. 4-11 over a time interval of 10^{-11} sec in time-steps of $1/2000$th of that interval. The results of that simulation are shown in Fig. 4-12, and a comparison with Fig. 4-9 confirms our expectation: Although the one-standard deviation envelopes in Figs. 4-9 and 4-12 are identical and the $X(t)$ trajectories are very similar, the $V(t)$ trajectories are nevertheless quite different. The approximating continuous Markov process $V(t)$ does not have the piecewise-constant fine structure of the exact jump Markov process $V(t)$, and exhibits instead a "fractal" scaling character. We note, however, that the Ornstein-Uhlenbeck process does not have the same high degree of scaling invariance as the driftless Wiener process, which we observed in Fig. 4-8 [see the footnote on page 300].

It is clear from the simulation results in Fig. 4-10a that the velocity of our Brownian particle usually stays inside its asymptotic one-standard deviation envelope of ± 30 cm/sec. In Fig. 4-13 we show a simulation of the jump process $V(t)$ of Figs. 4-9 and 4-10 with the initial condition $V(0) = v_0 = 150$ cm/sec and a time duration of $0.2\mu\text{sec} \approx 4 \times \tau_V$. This run uses a dot-plot interval of 10^{-10} sec, and, being one-fifth as long as the

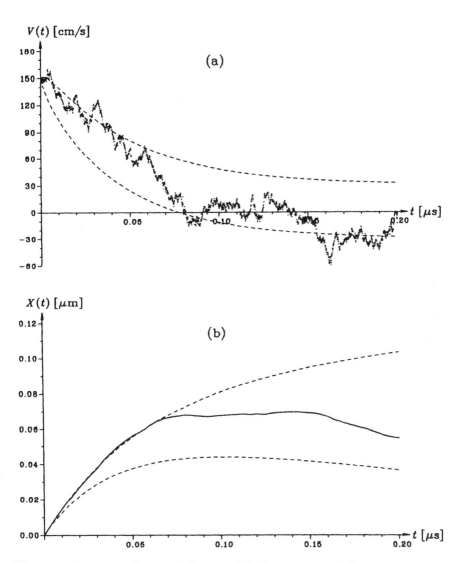

Figure 4-13. A simulation of the jump Markov process of Figs. 4-9 and 4-10 using the initial condition $V(0) = 150 \text{cm/sec}$. The indicated one-standard deviation envelopes are calculated from the same formulas as in Figs. 4-9 and 4-10, except that v_0 in those formulas is now taken to be 150 instead of 0. The statistically more frequent and harder impacts of gas molecules on the leading face of the Brownian particle create a "viscous drag" effect, which in a time of order $\tau_V = 0.047 \mu \text{sec}$ slows the particle to appropriate thermal speeds.

run in Fig. 4-10, shows the effects of about one million particle-molecule collisions. The respective one-standard envelopes are again shown as dashed curves; they were calculated from the same formulas as in Figs. 4-9 and 4-10, but taking $v_0 = 150$cm/sec. We see from Fig. 4-13a that the Brownian particle is unable to sustain its initial super-thermal velocity, and arrives in the "thermal band" of ± 30cm/sec after a time of the order of τ_V. What we are observing here is the macroscopic phenomenon of "viscous drag." We recall that in Langevin's analysis of Brownian motion, viscous drag was explicitly built in to the analysis in an overtly phenomenological way [see Eqs. (3.4-17) and (4.5-2a)]. However, in our derivation of the characterizing functions of the jump Markov process $V(t)$ simulated in Fig. 4-13 [see Subsection 4.5.A], no assumptions of viscous drag were ever made. The drag effect exhibited by the simulation in Fig. 4-13 arises solely from the propensities for and the dynamical effects of the collisions that naturally occur between the Brownian particle and the gas molecules.

As our final simulation example for jump Markov processes with continuum states, we consider again the particle velocity process $V(t)$ defined by the characterizing functions (4.6-19), but now in the *self-diffusion case* in which the particle is simply a "labeled" molecule of the one-dimensional gas just considered. In that case we have $M = m$ and $R = r$, so the three parameters σ, κ and ε defined in Eqs. (4.5-15) now take the values

$$\sigma = 3 \times 10^4 \text{ cm/sec}, \quad \kappa = 3.8 \times 10^4 \text{ cm}^{-1}, \quad \varepsilon = 1. \qquad (4.6\text{-}25)$$

Figure 4-14 shows the result of one simulation run for this process from time $t_0 = 0$, with $v_0 = 0$, to time $t = 4 \times 10^{-8}$sec = 40ns. Exactly 52 jumps (labeled molecule collisions) occurred during this simulation run. As discussed in Subsection 4.5.B, it is not possible to derive estimates of the means and variances of $V(t)$ and $X(t)$ in this self-diffusion case unless we undertake the lengthy series of numerical calculations outlined in Appendix C. Since we have not done that here, then we are unable plot the one-standard deviation envelopes for $V(t)$ and $X(t)$. However, we should expect that the labeled molecule's velocity $V(t)$ will ultimately have the same standard deviation of $\sigma = 0.3$ km/sec as the velocities of all the other gas molecules. We have indicated the $\pm\sigma$-band with dashed lines in Fig. 4-14a, and it would certainly appear that $V(t)$ does honor that one-standard deviation band. But even though it seldom strays very far from that band, $V(t)$ is clearly much more erratic than any other Markov process we have simulated thus far. As for the companion one-dimensional self-diffusion trajectory $X(t)$ in Fig. 4-14b, we note that it compares reasonably well with the projected three-dimensional self-diffusion trajectory in Fig. 4-7a, provided that we "smooth" the latter in the manner indicated in Fig. 4-1.

Figure 4-15 shows an *extension* of the simulation run in Fig. 4-14 to time $t = 4 \times 10^{-3}\text{sec} = 4\text{ms}$, and entails roughly 5.15 million labeled molecule collisions. In this figure we have dot-plotted $V(t)$ in (a) and $X(t)$ in (b) at time intervals of $2 \times 10^{-3}\text{ms}$, so roughly 2570 labeled molecule collisions occur between successive dots. Again, the trajectories in Fig. 4-14 are "close-ups" of the first 0.00004 ms of the trajectories in Fig. 4-15.

If we were to perform appropriate statistical tests on the successive $V(t)$ dots in Fig. 4-15a, we would discover that they appear to be statistically independent samplings of the random variable $N(0,\sigma^2)$. To understand why this should be so, we first note that when the labeled molecule moving with velocity v collides with another molecule moving with velocity u, the two molecules in effect just exchange velocities; this follows from Eq. (4.5-13), which for $M = m$ reduces to $v' = u$. Since our analysis is premised on the assumption that the velocity u of a randomly selected gas molecule is $N(0,\sigma^2)$, then one might be tempted to conclude from this observation that when the labeled molecule collides with another gas molecule, its new velocity can be found simply by sampling the random variable $N(0,\sigma^2)$. But that is not so, because the labeled molecule's current velocity *biases* the selection of its next collision partner; for example, if the labeled molecule has velocity v, then it cannot possibly collide with a gas molecule of velocity $u = v$ because those two molecules would be at rest relative to each other. However, the bias exerted by the labeled molecule's *current* velocity on the selection of its collision partner n collisions from now will weaken very rapidly with increasing n, and should certainly be unnoticeable for $n \approx 2570$. Therefore, the successive velocities $V(t)$ plotted in Fig. 4-15a are essentially independent samplings of $N(0,\sigma^2)$, even though the successive velocities plotted in Fig. 4-14a are not.

The plot of $X(t)$ in Fig. 4-15b is evidently much smoother than the companion plot of $V(t)$ in Fig. 4-15a; however, on this "macroscale," $X(t)$ has evidently lost the piecewise-linear flavor of its "microscale" behavior in Fig. 4-14b. In fact, we observe that the $X(t)$ plot in Fig. 4-15b bears a strong statistical resemblance to the $X(t)$ plots in Figs. 4-7b and 4-8b. This resemblance is of course to be expected from a *physical* point of view, because all three of these $X(t)$ processes purport to be macroscale models of the physical phenomenon of self-diffusion in a gas. But the fact of their resemblances is at first glance rather surprising from a *mathematical* point of view: How can we mathematically understand the resemblance

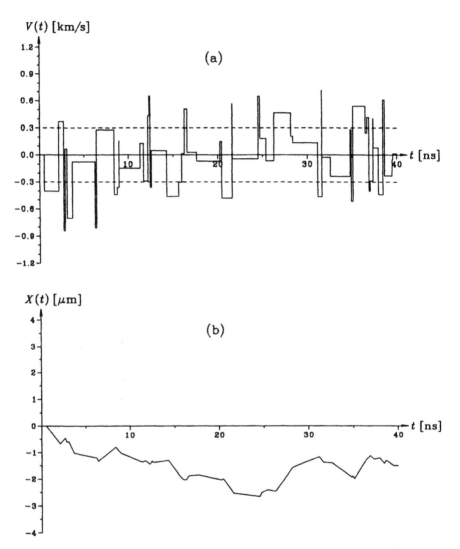

Figure 4-14. A simulation of the jump Markov process $V(t)$ and its time integral $X(t)$ as defined by the characterizing functions $a(v)$ and $w(\xi \mid v)$ in Eqs. (4.6-19), using the σ, κ and ε values of Eqs. (4.6-25) and the initial condition $V(0) = 0$. This is a rigorous modeling of the velocity $V(t)$ in (a), and the position $X(t)$ in (b), of a labeled molecule in a one-dimensional gas at standard temperature and pressure — i.e., *one-dimensional self diffusion.* The one-standard deviation envelopes have not been calculated, but the dashed lines in (a) demarcate the one-standard deviation "thermal envelope" postulated for the other gas molecules.

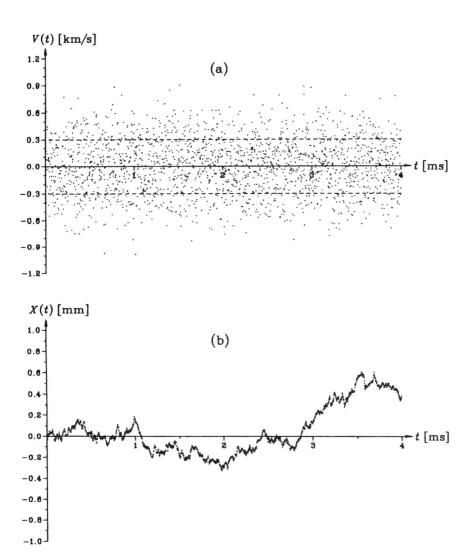

Figure 4-15. The one-dimensional self-diffusion simulation run of Fig. 4-14 extended to 4 milliseconds. The dots show the values of $V(t)$ in plot (a) and $X(t)$ in plot (b) at time intervals of 2×10^{-3} ms. About 2570 collisions involving the labeled molecule occur between successive dots, and the entire run contains about 5.15 million such collisions. Note the strong similarity between the $X(t)$ plot in (b) and the two $X(t)$ plots in Fig. 4-7b and 4-8b; the mathematical reasons underlying this physically expected similarity are discussed in the text.

of the process $X(t)$ in Fig. 4-15b to the process $X(t)$ in Fig. 4-8b in light of the fact that the former process, being the time-integral of a Markov process, is itself *not* Markovian, while the latter process, being a Wiener process, *is* Markovian? An explanation of sorts can be deduced from the simulation results for $V(t)$ in Fig. 4-15a: As we noted above, on the time scale of that figure $V(t)$ behaves like a process whose values at successive instants of time are obtained by simply sampling the random variable $N(0,\sigma^2)$. Now the velocity $\sigma = 0.3$ km/s, when measured in the mm/ms "slope unit" appropriate to the $X(t)$ trajectory in Fig. 4-15b, has the value 300. Thus, $X(t)$ in Fig. 4-15b is a process whose time-derivative $V(t)$ behaves much as if its values at successive instants of time were obtained by sampling the random variable $N(0, 9 \times 10^4) \approx N(0,\infty)$. But the process whose successive values are obtained by sampling the random variable $N(0,\infty)$ is by definition the *white noise process* $\Gamma_0(t)$. And as discussed in Subsection 3.4.B, the time-integral of the white noise process $\Gamma_0(t)$ is just the *special Weiner process* $W(t)$. Meanwhile, the process $X(t)$ in Fig. 4-8b, being the Wiener process with $A = 0$ and $D = 2\lambda^{*2}/3\tau^*$, can be viewed as the *special* Wiener process $W(t)$ by merely choosing units so that $2\lambda^{*2}/3\tau^* = 1$. Therefore, the *mathematical* reason why the $X(t)$ process in Fig. 4-15b resembles the $X(t)$ process in Fig. 4-8b is this: The derivative of the $X(t)$ process in Fig. 4-15b resembles, on the scale of that figure, the white noise process, and the antiderivative of the white noise process is, like the $X(t)$ process in Fig. 4-8b, a driftless Wiener process!

- 5 -

JUMP MARKOV PROCESSES WITH
DISCRETE STATES

In this chapter we shall continue the development of jump Markov process theory begun in Chapter 4, but now for the "discrete state" case in which the jump Markov process $X(t)$ has only *integer-valued* states. In Section 5.1 we shall obtain the discrete state versions of the fundamental concepts and equations that were developed for the continuum state case in Chapters 2 and 4. In Section 5.2 we shall discuss the completely homogeneous case. And in Section 5.3 we shall discuss the temporally homogeneous case, but only for such processes whose states are confined to the *nonnegative* integers. As an illustrative application of temporally homogeneous, nonnegative integer Markov processes, we shall show how they can be used to describe in a fairly rigorous way the time-evolution of certain kinds of chemically reacting systems. We shall continue our discussion of temporally homogeneous nonnegative integer Markov processes in Chapter 6, but there under the further restriction that only jumps of unit magnitude may occur.

5.1 FOUNDATIONAL ELEMENTS OF DISCRETE STATE MARKOV PROCESSES

The key definitions and equations for jump Markov processes with *real* variable states were developed in Chapters 2 and 4. The adaptation of those definitions and equations to the case of jump Markov processes with *integer* variable states pretty much parallels the way in which integer random variable theory follows from real random variable theory (see Section 1.7). For the most part, all we need to do is to replace the *real* variables x and ξ, which represent the values of the jump Markov process

X and its propagator Ξ, with *integer* variables n and ν respectively; of course, this will also entail replacing any integrals over x or ξ with sums over n or ν, and any Dirac delta functions of x or ξ with Kronecker delta functions of n or ν. Although it would be possible to deduce the integer variable versions of the key jump Markov process equations by routinely implementing the aforementioned replacements in Chapters 2 and 4, such an exposition would sacrifice much in clarity for only a slight gain in efficiency. So we shall instead simply begin anew in this section, and quote specific results from Chapters 2 and 4 only when the arguments leading to those results are entirely independent of whether the state variables are real or integer.

5.1.A THE CHAPMAN-KOLMOGOROV EQUATION

For any stochastic process $X(t)$ with *integer*-valued states, we define the **Markov state density function** P by

$$P(n,t \mid n_0,t_0) \equiv \text{Prob}\{X(t) = n, \text{ given } X(t_0) = n_0\} \quad (t_0 \leq t). \quad (5.1\text{-}1)$$

The probability density nature of this function requires that it satisfy the two relations

$$P(n,t \mid n_0,t_0) \geq 0 \quad (t_0 \leq t) \quad\quad (5.1\text{-}2)$$

and

$$\sum_{n=-\infty}^{\infty} P(n,t \mid n_0,t_0) = 1 \quad (t_0 \leq t). \quad\quad (5.1\text{-}3)$$

Furthermore, the conditional nature of this function requires that it satisfy the relation

$$P(n,t_0 \mid n_0,t_0) = \delta(n,n_0), \quad\quad (5.1\text{-}4)$$

where $\delta(n,n_0)$ is the Kronecker delta function (1.7-5).

The Markov state density function P is actually just one of an infinite hierarchy of state density functions of the general form

$$P_{k-j}^{(j+1)}(n_k,t_k; \ldots; n_{j+1},t_{j+1} \mid n_j,t_j; \ldots; n_0,t_0)$$
$$(0 \leq j < k; t_0 \leq t_1 \leq \ldots \leq t_k),$$

which is defined to be the probability that $X(t)$ will have the indicated values at the $k-j$ times standing to the *left* of the conditioning bar, *given* that $X(t)$ had the indicated values at the $j+1$ times standing to the *right* of the conditioning bar. The Markov state density function P defined in

Eq. (5.1-1) evidently coincides with the *first* of these hierarchical functions; i.e., $P \equiv P_1^{(1)}$.

We say that the integer-valued stochastic process $X(t)$ is *Markovian* if and only if, for all $k > 1$ and all $t_0 \le t_1 \le \ldots \le t_k$,

$$P_1^{(k)}(n_k, t_k \mid n_{k-1}, t_{k-1}; \ldots; n_0, t_0) = P(n_k, t_k \mid n_{k-1}, t_{k-1}). \quad (5.1\text{-}5)$$

This equation asserts that our ability to predict the value of $X(t_k)$ given the value of $X(t_{k-1})$ cannot be enhanced by learning the values of $X(t)$ at any times *prior* to t_{k-1}. In effect, the process has no memory of the past. The consequences of the Markov property (5.1-5) are very far-reaching. For example, since the multiplication law of probability implies quite generally that, for any three times $t_0 \le t_1 \le t_2$,

$$P_2^{(1)}(n_2, t_2; n_1, t_1 \mid n_0, t_0) = P(n_1, t_1 \mid n_0, t_0) P_1^{(2)}(n_2, t_2 \mid n_1, t_1; n_0, t_0),$$

then the Markov property (5.1-5) allows us to replace the second factor on the right by $P(n_2, t_2 \mid n_1, t_1)$, and so obtain the formula

$$P_2^{(1)}(n_2, t_2; n_1, t_1 \mid n_0, t_0) = P(n_2, t_2 \mid n_1, t_1) P(n_1, t_1 \mid n_0, t_0)$$
$$(t_0 \le t_1 \le t_2). \quad (5.1\text{-}6)$$

Thus, when the Markov condition (5.1-5) holds, then the state density function $P_2^{(1)}$ is completely determined by the Markov state density function P. Analogous arguments lead to the more general conclusion that, when the Markov condition (5.1-5) holds, then *all* the state density functions in the infinite hierarchy are determined by the Markov state density function P according to the formula

$$P_{k-j}^{(j+1)}(n_k, t_k; \ldots; n_{j+1}, t_{j+1} \mid n_j, t_j; \ldots; n_0, t_0) = \prod_{i=j+1}^{k} P(n_i, t_i \mid n_{i-1}, t_{i-1})$$
$$(0 \le j < k; t_0 \le t_1 \le \ldots \le t_k). \quad (5.1\text{-}7)$$

From the addition law of probability it follows that, for any three times $t_0 \le t_1 \le t_2$, it will *always* be true that

$$P(n_2, t_2 \mid n_0, t_0) = \sum_{n_1 = -\infty}^{\infty} P_2^{(1)}(n_2, t_2; n_1, t_1 \mid n_0, t_0).$$

But if, as we shall henceforth assume, $X(t)$ is *Markovian*, then we can substitute the relation (5.1-6) into the right side of this equation and obtain

$$P(n_2,t_2 \mid n_0,t_0) = \sum_{n_1 = -\infty}^{\infty} P(n_2,t_2 \mid n_1,t_1) \, P(n_1,t_1 \mid n_0,t_0)$$

$$(t_0 \leq t_1 \leq t_2). \quad (5.1\text{-}8)$$

This is (the discrete state version of) the **Chapman-Kolmogorov equation**. We can look upon this equation as a *condition* on the Markov state density function P, in addition to conditions (5.1-2) – (5.1-4), that arises as a consequence of $X(t)$ being Markovian. It is a straightforward matter to iterate Eq. (5.1-8) and deduce the **compounded Chapman-Kolmogorov equation**,

$$P(n_k,t_k \mid n_0,t_0) = \sum_{n_1 = -\infty}^{\infty} \cdots \sum_{n_{k-1} = -\infty}^{\infty} \prod_{i=1}^{k} P(n_i,t_i \mid n_{i-1},t_{i-1})$$

$$(k \geq 2; \, t_0 \leq t_1 \leq \ldots \leq t_k). \quad (5.1\text{-}9)$$

This formula can also be deduced by setting $j=0$ in Eq. (5.1-7) and then summing over all values of $n_1, n_2, \ldots, n_{k-1}$.

The **initially conditioned average** of any univariate function g is calculated, for any time $t \geq t_0$, as

$$\langle g(X(t)) \mid X(t_0) = n_0 \rangle \equiv \langle g(X(t)) \rangle = \sum_{n = -\infty}^{\infty} g(n) \, P(n,t \mid n_0,t_0). \quad (5.1\text{-}10)$$

Similarly, for g any *bivariate* state function, we have for any two times t_1 and t_2 satisfying $t_0 \leq t_1 \leq t_2$,

$$\langle g(X(t_1),X(t_2)) \mid X(t_0) = n_0 \rangle \equiv \langle g(X(t_1),X(t_2)) \rangle$$

$$= \sum_{n_1 = -\infty}^{\infty} \sum_{n_2 = -\infty}^{\infty} g(n_1,n_2) \, P_2^{(1)}(n_2,t_2; n_1,t_1 \mid n_0,t_0)$$

$$= \sum_{n_1 = -\infty}^{\infty} \sum_{n_2 = -\infty}^{\infty} g(n_1,n_2) \, P(n_2,t_2 \mid n_1,t_1) \, P(n_1,t_1 \mid n_0,t_0),$$

$$(5.1\text{-}11)$$

where of course the last line follows expressly from the Markov condition (5.1-6). Using the average formulas (5.1-10) and (5.1-11), we can calculate in the usual way the various moments of $X(t)$, and in particular the mean, variance, standard deviation and covariance.

5.1.B THE PROPAGATOR

The **propagator** of the discrete state Markov process $X(t)$ is defined, just as in the continuum state case, to be the random variable

$$\Xi(dt; n,t) \equiv X(t+dt) - X(t), \text{ given } X(t)=n, \qquad (5.1\text{-}12)$$

where dt is a positive infinitesimal. The density function of this random variable, namely

$$\Pi(v \mid dt; n,t) \equiv \text{Prob}\{\Xi(dt; n,t)=v\}, \qquad (5.1\text{-}13)$$

is called the **propagator density function**. Since the preceding two definitions imply that

$$\Pi(v \mid dt; n,t) = \text{Prob}\{X(t+dt)-X(t)=v, \text{ given } X(t)=n\}$$
$$= \text{Prob}\{X(t+dt)=n+v, \text{ given } X(t)=n\}.$$

then by applying the definition (5.1-1) we immediately deduce the fundamental relation

$$\Pi(v \mid dt; n,t) = P(n+v,t+dt \mid n,t). \qquad (5.1\text{-}14)$$

Equation (5.1-14) shows that the Markov state density function P uniquely determines the propagator density function Π. Less obvious, but more significant for our purposes, is the converse fact that the propagator density function Π uniquely determines the Markov state density function P. To prove this, consider the compounded Chapman-Kolmogorov equation (5.1-9) with $n_k=n$ and $t_k=t$. Let the points t_1, t_2, ..., t_{k-1} divide the interval $[t_0,t]$ into k subintervals of equal length $(t-t_0)/k$. Change the summation variables in that equation according to

$$n_i \rightarrow v_i \equiv n_i - n_{i-1} \quad (i=1,...,k-1).$$

Finally, define $v_k \equiv n - n_{k-1}$. With all these substitutions, the compounded Chapman-Kolmogorov equation (5.1-9) becomes

$$P(n,t \mid n_0,t_0)$$

$$= \sum_{v_1=-\infty}^{\infty} \cdots \sum_{v_{k-1}=-\infty}^{\infty} \prod_{i=1}^{k} P(n_{i-1}+v_i, t_{i-1}+(t-t_0)/k \mid n_{i-1},t_{i-1}),$$

wherein

$$t_i = t_{i-1} + (t-t_0)/k \quad (i=1,...,k-1), \qquad (5.1\text{-}15a)$$

$$n_i = n_0 + v_1 + ... + v_i \quad (i=1,...,k-1), \qquad (5.1\text{-}15b)$$

$$v_k \equiv n - n_0 - v_1 - ... - v_{k-1}. \qquad (5.1\text{-}15c)$$

Now choose k *so large* that

$$(t-t_0)/k = dt, \text{ an infinitesimal.} \qquad (5.1\text{-}15d)$$

Then the P-factors on the right hand side of the preceding equation for $P(n,t \,|\, n_0,t_0)$ become, by virtue of Eq. (5.1-14), Π-factors; indeed, that formula becomes

$$P(n,t \,|\, n_0,t_0) = \sum_{v_1 = -\infty}^{\infty} \cdots \sum_{v_{k-1} = -\infty}^{\infty} \prod_{i=1}^{k} \Pi(v_i \,|\, dt; n_{i-1}, t_{i-1}), \qquad (5.1\text{-}16)$$

where now *all four* of Eqs. (5.1-15) apply. This result shows that if we specify the propagator density function $\Pi(v \,|\, dt; n', t')$ as a function of v for all n', all $t' \in [t_0, t)$, and all infinitesimally small dt, then the Markov state density function $P(n,t \,|\, n_0, t_0)$ is uniquely determined for all n.

To deduce the general form of the propagator density function for a discrete state Markov process $X(t)$, we begin by recognizing that if $X(t)$ is *always* to coincide with some *integer* value, then the only way for $X(t)$ to change with time is to make *instantaneous jumps* from one integer to another. That being the case, it makes sense to define for any discrete state jump Markov process $X(t)$ the two probability functions

$$q(n,t; \tau) \equiv \text{ probability, given } X(t)=n, \text{ that the}$$
$$\text{process will jump away from state } n \text{ at}$$
$$\text{some instant between } t \text{ and } t+\tau; \qquad (5.1\text{-}17)$$

$$w(v \,|\, n,t) \equiv \text{ probability that the process, upon}$$
$$\text{jumping away from state } n \text{ at time } t,$$
$$\text{will land in state } n+v. \qquad (5.1\text{-}18)$$

In fact, we shall simply *define* a discrete state jump Markov process as any integer state process $X(t)$ for which these two functions q and w exist *and* have the following properties:

- $q(n,t; \tau)$ is a smooth function of t and τ,
  and satisfies $q(n,t; 0)=0$; $\qquad (5.1\text{-}19a)$

- $w(v \,|\, n,t)$ is a smooth function of t. $\qquad (5.1\text{-}19b)$

It is clear from the definition (5.1-18) that $w(v \,|\, n,t)$ is a density function with respect to the integer variable v; therefore, it must satisfy the two conditions

$$w(v \,|\, n,t) \geq 0 \qquad (5.1\text{-}20a)$$

and

$$\sum_{v=-\infty}^{\infty} w(v \mid n,t) = 1. \qquad (5.1\text{-}20b)$$

As for the function $q(n,t; \tau)$ defined in Eq. (5.1-17), it turns out that if τ is a positive infinitesimal dt, then the assumed Markovian nature of the process $X(t)$ demands that this function have the form

$$q(n,t; dt) = a(n,t)\, dt, \qquad (5.1\text{-}21)$$

where $a(n,t)$ is some nonnegative, smooth function of t. The proof of this fact is exactly the same as the proof of the analogous result (4.1-6) for the continuum state case, so we shall not repeat it here. If we combine Eq. (5.1-21) with the definition (5.1-17), we see that the significance of the function $a(n,t)$ is that

$$\begin{aligned} a(n,t)dt \equiv\ & \text{probability, given } X(t) = n, \text{ that the process will} \\ & \text{jump away from state } n \text{ in the next infinitesimal} \\ & \text{time interval } [t,t+dt). \end{aligned} \qquad (5.1\text{-}22)$$

It follows from this result that the probability for the system to jump once in $[t, t+\alpha dt)$ and then jump once again in $[t+\alpha dt, t+dt)$, for any α between 0 and 1, will be proportional to $(dt)^2$. We thus conclude that, to first order in dt, the system will either jump *once* or else *not at all* in the infinitesimal time interval $[t,t+dt)$.

Now we are in a position to deduce an explicit formula for the propagator density function $\Pi(v \mid dt; n,t)$ in terms of the two functions $a(n,t)$ and $w(v \mid n,t)$. Given $X(t) = n$, then by time $t+dt$ the system *either* will have jumped once, with probability $a(n,t)dt$, *or* it will not have jumped at all, with probability $1 - a(n,t)dt$. If a jump *does* occur, then by Eq. (5.1-18) the probability that the state change vector $X(t+dt) - n$ will equal v will be $w(v \mid n,t')$, where t' is the precise instant in $[t,t+dt)$ when the jump occurred. If a jump does *not* occur, then the probability that the state change vector $X(t+dt) - n$ will equal v will be $\delta(v,0)$, since that quantity is equal to unity if $v=0$ and zero if $v \neq 0$. Therefore, by the definition (5.1-13) and the multiplication and addition laws of probability, we have

$$\Pi(v \mid dt; n,t) = [a(n,t)dt]\,[w(v \mid n,t')] + [1 - a(n,t)dt]\,[\delta(v,0)].$$

Finally, since $t' \in [t,t+dt)$, then the smooth dependence of $w(v \mid n,t)$ on t assumed in condition (5.1-19b) means that we can replace t' on the right side of this last equation by the infinitesimally close value t without spoiling the equality. Thus we conclude that the propagator density function of a discrete state Markov process must be given by the formula

$$\Pi(v \mid dt; n,t) = a(n,t)dt\, w(v \mid n,t) + [1 - a(n,t)dt]\, \delta(v,0). \qquad (5.1\text{-}23)$$

This is the principle result of our analysis in this subsection.

Because the propagator density function $\Pi(v \mid dt; n,t)$ is completely determined by the two functions $a(n,t)$ and $w(v \mid n,t)$, we shall call those two functions the **characterizing functions** of the associated discrete state Markov process $X(t)$. And we shall say that

$X(t)$ is **temporally homogeneous**

$$\Leftrightarrow\quad a(n,t) = a(n) \text{ and } w(v \mid n,t) = w(v \mid n), \qquad (5.1\text{-}24a)$$

$X(t)$ is **completely homogeneous**

$$\Leftrightarrow\quad a(n,t) = a \text{ and } w(v \mid n,t) = w(v). \qquad (5.1\text{-}24b)$$

The "past forgetting" character of the definitions of $a(n,t)$ and $w(v \mid n,t)$ in Eqs. (5.1-22) and (5.1-18) should make the Markovian nature of the process $X(t)$ defined by the propagator density function (5.1-23) rather obvious. However, a formal proof of the Markov property can be obtained by showing that that propagator density function satisfies, to first order in dt and for all a between 0 and 1, the equation

$$\Pi(v \mid dt; n,t) = \sum_{v_1 = -\infty}^{\infty} \Pi(v - v_1 \mid (1-a)dt; n + v_1, t + a\,dt)\, \Pi(v_1 \mid a\,dt; n,t).$$

This condition on the discrete state propagator density function Π is called the *Chapman-Kolmogorov condition*, and it is a direct consequence of the fundamental identity (5.1-14) and the Chapman-Kolmogorov equation (5.1-8). By straightforwardly adapting the continuum state arguments leading from Eqs. (4.1-16) to Eqs. (4.1-18), one can prove explicitly that if $a(n,t)$ and $w(v \mid n,t)$ are analytic functions of t, then the propagator density function $\Pi(v \mid dt; n,t)$ in Eq. (5.1-23) does indeed satisfy the foregoing Chapman-Kolmogorov condition, and hence defines a *Markovian* process $X(t)$. We shall not exhibit the proof here because the required modifications to the continuum state proof given in Section 4.1 are so minor.

It will prove convenient for our subsequent work to define the function

$$W(v \mid n,t) \equiv a(n,t)\, w(v \mid n,t), \qquad (5.1\text{-}25)$$

and call it the **consolidated characterizing function** of the jump Markov process $X(t)$. The physical meaning of this function can straightforwardly be inferred by multiplying Eq. (5.1-25) through by dt, invoking the definitions of $a(n,t)dt$ and $w(v \mid n,t)$ in Eqs. (5.1-22) and (5.1-18) respectively, and then recalling that $w(v \mid n,t)$ is a smooth function of t; in this way we may deduce that

$W(v \mid n,t)\, dt \equiv$ probability, given $X(t) = n$, that the process
will in the time interval $[t, t+dt)$ jump from
state n to state $n+v$.   (5.1-26)

By summing Eq. (5.1-25) over v using Eq. (5.1-20b), and then substituting the result back into Eq. (5.1-25), we may easily deduce the relations

$$a(n,t) = \sum_{v=-\infty}^{\infty} W(v \mid n,t), \qquad (5.1\text{-}27a)$$

$$w(v \mid n,t) = \frac{W(v \mid n,t)}{\sum_{v'=-\infty}^{\infty} W(v' \mid n,t)}. \qquad (5.1\text{-}27b)$$

These equations show that, had we chosen to do so, we could have defined the characterizing functions a and w in terms of the consolidated characterizing function W, instead of the other way around. So if we regard (5.1-26) as the *definition* of $W(v \mid n,t)$, then it follows from Eqs. (5.1-27) that the specification of the form of that function will uniquely define a jump Markov process $X(t)$. We should note in passing the subtly close relationship between the consolidated characterizing function W and the propagator density function Π: By substituting Eqs. (5.1-27) into Eq. (5.1-23), we get

$$\Pi(v \mid dt; n,t) = W(v \mid n,t)\,dt + \left[1 - \sum_{v'=-\infty}^{\infty} W(v' \mid n,t)\,dt\right] \delta(v,0). \quad (5.1\text{-}28)$$

So the relation between Π and W is very simple, *except* when $v=0$. Because of this caveat, formula (5.1-23) is usually less confusing to work with.

The **propagator moment functions** $B_k(n,t)$ of a discrete state Markov process $X(t)$ are *defined*, when they exist, through the relation

$$\langle \Xi^k(dt;n,t)\rangle \equiv \sum_{v=-\infty}^{\infty} v^k\, \Pi(v \mid dt; n,t) \equiv B_k(n,t)\,dt + o(dt)$$

$$(k=1,2,...), \quad (5.1\text{-}29)$$

where $o(dt)/dt \to 0$ as $dt \to 0$. To deduce an explicit formula for $B_k(n,t)$ in terms of the characterizing functions of $X(t)$, we simply note from Eq.

(5.1-23) that, for any $k \geq 1$,

$$\sum_{v=-\infty}^{\infty} v^k \, \Pi(v \,|\, dt; n,t) = \sum_{v=-\infty}^{\infty} v^k \left\{ a(n,t)dt \, w(v \,|\, n,t) + [1 - a(n,t)dt] \, \delta(v,0) \right\}$$

$$= a(n,t) \left(\sum_{v=-\infty}^{\infty} v^k \, w(v \,|\, n,t) \right) dt$$

$$\equiv \left(\sum_{v=-\infty}^{\infty} v^k \, W(v \,|\, n,t) \right) dt,$$

where the last equality has invoked the definition (5.1-25). So if we *define* the quantities

$$w_k(n,t) \equiv \sum_{v=-\infty}^{\infty} v^k \, w(v \,|\, n,t), \qquad (5.1\text{-}30a)$$

$$W_k(n,t) \equiv \sum_{v=-\infty}^{\infty} v^k \, W(v \,|\, n,t), \qquad (5.1\text{-}30b)$$

then we may conclude from Eq. (5.1-29) that the propagator moment function $B_k(n,t)$ is given by the formulas

$$B_k(n,t) = a(n,t) \, w_k(n,t) = W_k(n,t) \quad (k = 1,2,\ldots). \qquad (5.1\text{-}31)$$

We see from this result that the kth propagator moment function $B_k(n,t)$ of the discrete state Markov process $X(t)$ exists *if and only if* the kth moment of the density function $w(v \,|\, n,t)$ exists.

The sense in which the propagator Ξ "propagates" the process X from time t to the infinitesimally later time $t + dt$ can be made a little more transparent by writing the propagator definition (5.1-12) in the equivalent form

$$X(t+dt) = X(t) + \Xi(dt; X(t),t). \qquad (5.1\text{-}32)$$

Now, if $\Xi(dt; X(t),t)$ in this formula were always directly proportional to dt, at least to first order in dt, then the proportionality constant could evidently be called the "time-derivative" of $X(t)$. But since $\Xi(dt; X(t),t)$ will be a *nonzero integer* for those intervals $[t,t+dt)$ that contain a jump, and since a nonzero integer certainly cannot be regarded as being proportional to an infinitesimal, then we must conclude that a discrete state Markov process $X(t)$ does *not* have a time-derivative. However, we can easily define an antiderivative or **time-integral process** $S(t)$ of $X(t)$ by simply declaring $S(t)$ to have the "propagator" $X(t)dt$:

$$S(t+dt) = S(t) + X(t)\,dt. \qquad (5.1\text{-}33)$$

This equation means simply that if the process S has the value s at time t, *and* the process X has the value n at time t, then the value of the process S at the infinitesimally later time $t+dt$ will be $s+n\,dt$; indeed, this statement evidently holds true for all $dt \in [0, \tau)$, where $t+\tau$ is the instant that the process X next jumps away from state n. We may complete this definition of $S(t)$ by adopting the convention that

$$S(t_0) = 0. \qquad (5.1\text{-}34)$$

The time-integral $S(t)$ of the discrete state Markov process $X(t)$ is itself *neither* a discrete state process (since it assumes a continuum of values), *nor* a Markov process (since it by definition has a time-derivative, which a genuinely stochastic Markov process cannot have). Nevertheless, $S(t)$ is a perfectly well defined stochastic process; we shall see shortly how it can be numerically simulated and also how its moments can be calculated analytically.

5.1.C THE NEXT-JUMP DENSITY FUNCTION AND ITS SIMULATION ALGORITHM

We define for any discrete state Markov process $X(t)$ its **next-jump density function** p by

$p(\tau,v \mid n,t)\,d\tau \equiv$ probability that, given the process is in state
n at time t, its *next jump* will occur between
times $t+\tau$ and $t+\tau+d\tau$, and will carry the
process to state $n+v$. $\qquad (5.1\text{-}35)$

Whereas the propagator density function $\Pi(v \mid dt; n,t)$ is the density function for the state-change vector (v) over the next *specified* time interval dt, the next-jump density function $p(\tau,v \mid n,t)$ is the *joint* density function for the time (τ) to the next jump and the state-change vector (v) in that next jump. Unlike the propagator density function Π, the next-jump density function p does not depend parametrically upon a preselected time interval dt. As we shall see shortly, this feature makes p useful for constructing exact Monte Carlo simulations of the discrete state Markov process $X(t)$ and its time-integral process $S(t)$.

To derive a formula for $p(\tau,v \mid n,t)$ in terms of the characterizing functions a and w, we begin by using the multiplication law to write the probability (5.1-35) as

$$p(\tau,v \mid n,t)\, d\tau \ = \ [1 - q(n,t;\tau)] \times a(n,t+\tau)d\tau \times w(v \mid n,t+\tau). \quad (5.1\text{-}36)$$

In this equation, the first factor on the right is by definition (5.1-17) the probability that the system, in state n at time t, will *not* jump away from that state in the time interval $[t,t+\tau)$; the second factor on the right is by definition (5.1-22) the probability that the system, in state n at time $t+\tau$, *will* jump away from that state in the next infinitesimal time interval $[t+\tau, t+\tau+d\tau)$; and the third factor on the right is by definition (5.1-18) the probability that the system, upon jumping away from state n at time $t+\tau$, will land in state $n+v$.† Now we have only to express $q(n,t;\tau)$, as defined in (5.1-17), explicitly in terms of the characterizing functions $a(n,t)$ and $w(v \mid n,t)$. This can be done by simply repeating the argument leading from Eqs. (4.1-14) to Eqs. (4.1-15), but replacing x there by n; the result is [see Eq. (4.1-15b)]

$$q(n,t;\tau) \ = \ 1 \ - \ \exp\left(- \int_0^\tau a(n,t+\tau')d\tau' \right). \quad (5.1\text{-}37)$$

Substituting this expression for $q(n,t;\tau)$ into Eq. (5.1-36), we conclude that the next-jump density function for $X(t)$ is given by the formula

$$p(\tau,v \mid n,t) = a(n,t+\tau)\exp\left(- \int_0^\tau a(n,t+\tau')d\tau' \right) w(v \mid n,t+\tau), \quad (5.1\text{-}38)$$

wherein it is understood that τ is a *nonnegative real* variable, and v is an *integer* variable.

It will later be convenient to "condition" the joint density function $p(\tau,v \mid n,t)$ according to

$$p(\tau,v \mid n,t) = p_1(\tau \mid n,t)\, p_2(v \mid \tau; n,t). \quad (5.1\text{-}39)$$

Here, $p_1(\tau \mid n,t)$, the density function for τ irrespective of v, is calculated by summing $p(\tau,v \mid n,t)$ in Eq. (5.1-38) over all v; this v-summation, owing to the normalization condition (5.1-20b), has the effect of simply removing the factor $w(v \mid n,t+\tau)$ from the right hand side of Eq. (5.1-38). And $p_2(v \mid \tau; n,t)$, the density function for v conditioned on τ, may then be calculated, according to Eq. (5.1-39), simply by dividing $p(\tau,v \mid n,t)$ by $p_1(\tau \mid n,t)$; that division evidently yields the result $w(v \mid n,t+\tau)$. Thus we find that the two subordinate density functions p_1 and p_2 for the next-jump-density function are given by the respective formulas

---

† The last factor in Eq. (5.1-36) should actually be $w(v \mid n,t')$, where t' is the exact instant in $[t+\tau, t+\tau+d\tau)$ at which the jump away from state n occurs. However, the t-smoothness of the function $w(v \mid n,t)$ stipulated in (5.1-19b) allows us to replace t' by the infinitesimally close value $t+\tau$ without introducing any sensible error in Eq. (5.1-36).

$$\begin{cases} p_1(\tau \mid n,t) = a(n,t+\tau)\exp\left(-\int_0^\tau a(n,t+\tau')d\tau'\right), & \text{(5.1-40a)} \\[2mm] p_2(v \mid \tau; n,t) = w(v \mid n,t+\tau). & \text{(5.1-40b)} \end{cases}$$

A considerable simplification in these next-jump density function formulas occurs if the process $X(t)$ in question is *temporally homogeneous*,

$$a(n,t) \equiv a(n) \quad \text{and} \quad w(v \mid n,t) \equiv w(v \mid n),$$

as in fact most discrete state Markov processes encountered in practice are. In that case the τ'-integrals in Eqs. (5.1-38) and (5.1-40a) become simply $a(n)\tau$; so the next-jump density function (5.1-38) becomes

$$p(\tau,v \mid n,t) = a(n)\exp\left(-a(n)\,\tau\right)w(v \mid n), \qquad \text{(5.1-41)}$$

while the associated conditioning density functions (5.1-40) become

$$\begin{cases} p_1(\tau \mid n,t) = a(n)\exp\left(-a(n)\,\tau\right), & \text{(5.1-42a)} \\[2mm] p_2(v \mid \tau; n,t) = w(v \mid n). & \text{(5.1-42b)} \end{cases}$$

Since $p_1(\tau \mid n,t)$ now has the form of an exponential density function with decay constant $a(n)$, then it follows that the waiting time to the next jump from state n is an exponential random variable with mean $1/a(n)$. And since $p_2(v \mid \tau; n,t)$ is now independent of τ, then the next-jump displacement from state n is statistically independent of the waiting time for that jump. So we see that, for any *temporally homogeneous* discrete state Markov process, the characterizing functions $a(n)$ and $w(v \mid n)$ have the following interpretations:

(i) The characterizing function $a(n)$ is the reciprocal of the mean of the random variable "pausing time in state n," which is necessarily *exponentially distributed*.   (5.1-43a)

(ii) The characterizing function $w(v \mid n)$ is the density function of the random variable "jump displacement from state n," which is necessarily *statistically independent* of the pausing time in state n.   (5.1-43b)

Returning now to the general (nonhomogeneous) case, it should be clear that if we can generate a pair of random numbers (τ,v) according to the joint density function $p(\tau,v \mid n,t)$, then we may without further ado assert that the process X, in state n at time t, will remain in that state until time $t+\tau$, at which time it will jump to state $n+\nu$. Therefore, all that is needed in order to advance a discrete-state Markov process from one jump to the next is a procedure for generating a random pair (τ,ν) according to the joint density function $p(\tau,\nu \mid n,t)$ given in Eq. (5.1-38). In most cases, the easiest way to do that is to first generate a random (real) number τ according to the density function $p_1(\tau \mid n,t)$ in Eq. (5.1-40a), and then generate a random (integer) number ν according to the density function $p_2(\nu \mid \tau; n,t)$ in Eq. (5.1-40b). Thus we have deduced the first four steps of the procedure outlined in Fig. 5-1 for *exactly* simulating a discrete state Markov process with characterizing functions $a(n,t)$ and $w(\nu \mid n,t)$. This procedure is of course the discrete state version of the continuum state jump simulation procedure in Fig. 4-3.†

The τ-selection procedure in Step 2° of Fig. 5-1 is virtually identical to the τ-selection procedure in Fig. 4-3, the only difference being the inconsequential replacement of the real state variable x with the integer state variable n. The procedure is especially simple in the *temporally homogeneous* case, when, as noted above, τ is to be selected by sampling the exponential random variable with decay constant $a(n)$. In that case, according to Eq. (1.8-7), we merely draw a unit uniform random number r and take

$$\tau = [1/a(n)]\ln(1/r).\tag{5.1-44}$$

But if $a(n,t)$ depends explicitly on t, then this generating formula is not applicable, and one will have to carefully assess whether the inversion generating method [see Eq. (1.8-5)] or the rejection generating method [see Eqs. (1.8-9) – (1.8-11)] will be easier to implement.

The ν-selection procedure in Step 3° of Fig. 5-1 requires that one implement the *integer* version of either the inversion generating method or the rejection generating method. To use the integer inversion method [see Eq. (1.8-12)], one would draw a unit uniform random number r' and then take ν to be that integer for which

$$\sum_{\nu'=-\infty}^{\nu-1} w(\nu' \mid n,t+\tau) \le r' < \sum_{\nu'=-\infty}^{\nu} w(\nu' \mid n,t+\tau).\tag{5.1-45}$$

If the ν'-sums here cannot be calculated analytically, but $w(\nu' \mid n,t+\tau)$ vanishes for all ν' less than some finite value ν_1 (which may depend on n and $t+\tau$), then the lower summation limits in Eq. (5.1-45) can be replaced by ν_1 and the sums can be computed numerically: One just cumulatively

---

† One can also formulate a discrete state version of the *approximate* continuum state jump simulation procedure in Fig. 4-2. But we shall not bother to do so here, because that procedure is almost always inferior to the exact procedure of Fig. 5-1.

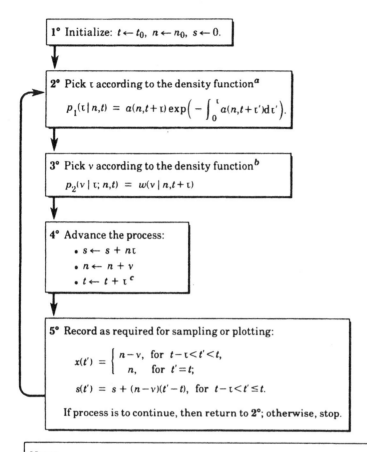

1° Initialize: $t \leftarrow t_0$, $n \leftarrow n_0$, $s \leftarrow 0$.

2° Pick τ according to the density function[a]

$$p_1(\tau \mid n,t) = a(n,t+\tau) \exp\left(-\int_0^\tau a(n,t+\tau')d\tau'\right).$$

3° Pick v according to the density function[b]

$$p_2(v \mid \tau; n,t) = w(v \mid n,t+\tau)$$

4° Advance the process:
- $s \leftarrow s + n\tau$
- $n \leftarrow n + v$
- $t \leftarrow t + \tau$ [c]

5° Record as required for sampling or plotting:

$$x(t') = \begin{cases} n-v, & \text{for } t-\tau<t'<t, \\ n, & \text{for } t'=t; \end{cases}$$

$$s(t') = s + (n-v)(t'-t), \text{ for } t-\tau<t'\leq t.$$

If process is to continue, then return to **2°**; otherwise, stop.

Notes:

[a] Use either the inversion or the rejection generating method [see Section 1.8]. If $a(n,t) \equiv a(n)$, then the inversion method is easy: Draw a unit uniform random number r, and take $\tau = [1/a(n)] \ln(1/r)$.

[b] Use either the integer inversion or the integer rejection generating method [see Section 1.8]. For the former, draw a unit uniform random number r' and take v to be the *smallest* integer for which the sum over $w(v' \mid n,t+\tau)$ from $v' = -\infty$ to $v' = v$ exceeds r'.

[c] In a simulation run containing $\sim 10^K$ jumps, the sum $t+\tau$ should be computed with at least $K+3$ digits of precision.

Figure 5-1. Exact Monte Carlo simulation algorithm for the discrete state Markov process with characterizing functions $a(n,t)$ and $w(v \mid n,t)$. The procedure produces exact sample values $x(t)$ and $s(t)$ of the process $X(t)$ and its time-integral $S(t)$ for all $t>t_0$.

adds $w(v' \mid n, t+\tau)$ for $v' = v_1, v_1+1, v_1+2, \ldots$, until that sum *first exceeds* r', and one then takes v to be the index of the last term added. In the integer version of the *rejection* generating method, one takes the interval $[a, b]$ in Eqs. (1.8-9) to be an appropriate *integer* interval $[v_1, v_2]$, and so replaces Eq. (1.8-10) with its integer counterpart [see Eq. (1.8-13)]

$$v_t = \text{greatest-integer-in}\{v_1 + (v_2 - v_1 + 1)r_1\}.$$

However, it is rare in practice that the integer rejection method will prove to be more expeditious than some form of the integer inversion method. In any case, if $w(v \mid n, t)$ is explicitly independent of t, then the v-selection process of Step 3° will be independent of the τ-value selected in Step 2° [see (5.1-43b)].

The three advancement formulas in Step 4° of Fig. 5-1 should be obvious. Note in particular that the increase in the time-integral process $S(t)$ between times t and $t+\tau$ is *exactly* equal to $n[(t+\tau) - t] = n\tau$, because the time-derivative $X(t)$ of $S(t)$ has the constant value n throughout that time interval. Note also that the s-update in Step 4° must always be done *before* the n-update. Once Step 4° has been completed, then we can assert that at the current time t the realization $x(t)$ of the process $X(t)$ will have the value n, and the realization $s(t)$ of the integral process $S(t)$ will have the value s. But notice that we can also assert precise values for those two realizations during the entire preceding time interval $(t-\tau, t)$. As just mentioned, the realization of $X(t)$ must have had the value $n-v$ during that interval:

$$x(t') = n - v \quad \text{for } t' \in [t-\tau, t). \tag{5.1-46}$$

And since, as t' increases from $t-\tau$ to t, the realization $s(t')$ increases at a constant rate $(n-v)$ to the final value $s(t) = s$, then we have

$$s(t') = s + (n-v)(t' - t) \quad \text{for } t' \in [t-\tau, t]. \tag{5.1-47}$$

Equations (5.1-46) and (5.1-47) give the realizations of $X(t)$ and $S(t)$ *exactly* during the entire time interval between the last two jumps of $X(t)$, and are the basis for Step 5° of Fig. 5-1.

5.1.D THE MASTER EQUATIONS

Since the Kramers-Moyal equations of *continuum* state Markov process theory involve *partial derivatives* with respect to the state variable x, then those equations cannot be conveniently adapted to the *discrete* state case. Consequently, in discrete state Markov process theory the description of the time-evolution of the Markov state density function $P(n,t \mid n_0,t_0)$ falls totally upon the forward and backward master equations. In this subsection we shall derive those discrete state equations.

The *formal* way of deriving the forward master equation is to start with the Chapman-Kolmogorov equation (5.1-8), written however in the form

$$P(n,t+dt \mid n_0,t_0) = \sum_{v=-\infty}^{\infty} P(n,t+dt \mid n-v,t) \, P(n-v,t \mid n_0,t_0). \quad (5.1\text{-}48)$$

Then observe, using Eqs. (5.1-14) and (5.1-23), that

$$P(n,t+dt \mid n-v,t) = P(n-v+v,t+dt \mid n-v,t)$$
$$= \Pi(v \mid dt; n-v,t)$$
$$= a(n-v,t)dt \, w(v \mid n-v,t) + [1-a(n-v,t)dt] \, \delta(v,0).$$

Substituting this last expression into the Chapman-Kolmogorov equation (5.1-48) gives

$$P(n,t+dt \mid n_0,t_0) = \sum_{v=-\infty}^{\infty} \Big(a(n-v,t)dt \, w(v \mid n-v,t) \Big) P(n-v,t \mid n_0,t_0)$$
$$+ \sum_{v=-\infty}^{\infty} \Big([1-a(n-v,t)dt] \, \delta(v,0) \Big) P(n-v,t \mid n_0,t_0),$$

or, upon carrying out the second v-summation using the Kronecker delta function,

$$P(n,t+dt \mid n_0,t_0) = \sum_{v=-\infty}^{\infty} [a(n-v,t)dt \, w(v \mid n-v,t)] \, P(n-v,t \mid n_0,t_0)$$
$$+ [1-a(n,t)dt] \, P(n,t \mid n_0,t_0). \quad (5.1\text{-}49)$$

But now observe that this last equation could actually have been written down *directly* from the definitions (5.1-22) and (5.1-18), because it merely expresses the probability of finding $X(t+dt)=n$, given $X(t_0)=n_0$, as the sum of the probabilities of all possible ways of arriving at state n at time $t+dt$ via *specified* states at time t: The vth term under the summation sign is the product of the probability that $X(t)=n-v$, given that $X(t_0)=n_0$, times the subsequent probability of a jump of size v in the next dt. And the last term is the product of the probability that $X(t)=n$, given that $X(t_0)=n_0$, times the subsequent probability of *no* jump in the next dt.

This logic ignores *multiple* jumps in $[t,t+dt)$, but that is okay because the probability for such jumps will be of order >1 in dt. Now subtracting $P(n,t \mid n_0,t_0)$ from both sides of Eq. (5.1-49), dividing through by dt, and taking the limit $dt \to 0$, we obtain

$$\frac{\partial}{\partial t} P(n,t \mid n_0,t_0) = \sum_{v=-\infty}^{\infty} [a(n-v,t)\, w(v \mid n-v,t)\, P(n-v,t \mid n_0,t_0)]$$

$$- a(n,t)\, P(n,t \mid n_0,t_0). \qquad (5.1\text{-}50a)$$

If we multiply the second term on the right by *unity* in the form of $\sum_v w(-v \mid n,t)$ [namely Eq. (5.1-20b) with the summation variable change $v \to -v$], and then recall the definition (5.1-25) of the consolidated characterizing function, we obtain the equivalent formula

$$\frac{\partial}{\partial t} P(n,t \mid n_0,t_0) = \sum_{v=-\infty}^{\infty} \left[W(v \mid n-v,t)\, P(n-v,t \mid n_0,t_0) \right.$$

$$\left. - W(-v \mid n,t)\, P(n,t \mid n_0,t_0) \right]. \qquad (5.1\text{-}50b)$$

Equations (5.1-50) are (both) called the **forward master equation** for the discrete state Markov process $X(t)$ defined by the characterizing functions a and w, or by the consolidated characterizing function W. They are evidently differential-difference equations for $P(n,t \mid n_0,t_0)$ for fixed n_0 and t_0, and they are to be solved subject to the *initial condition* $P(n,t=t_0 \mid n_0,t_0) = \delta(n,n_0)$.

To derive the backward companions to Eqs. (5.1-50), we may begin, again formally, with the Chapman-Kolmogorov equation (5.1-8), but now written in the form

$$P(n,t \mid n_0,t_0) = \sum_{v=-\infty}^{\infty} P(n,t \mid n_0+v,t_0+dt_0)\, P(n_0+v,t_0+dt_0 \mid n_0,t_0). \qquad (5.1\text{-}51)$$

Then observe, using Eqs. (5.1-14) and (5.1-23), that

$$P(n_0+v,t_0+dt_0 \mid n_0,t_0) = \Pi(v \mid dt_0; n_0,t_0)$$

$$= a(n_0,t_0)dt_0\, w(v \mid n_0,t_0) + [1 - a(n_0,t_0)dt_0]\, \delta(v,0).$$

Substituting this expression into the Chapman-Kolmogorov equation (5.1-51) gives

$$P(n,t|n_0,t_0) = \sum_{v=-\infty}^{\infty} P(n,t|n_0+v,t_0+dt_0) \left(a(n_0,t_0)dt_0\, w(v|n_0,t_0) \right)$$

$$+ \sum_{v=-\infty}^{\infty} P(n,t|n_0+v,t_0+dt_0) \left([1-a(n_0,t_0)dt_0]\, \delta(v,0) \right),$$

or, upon carrying out the second v-summation using the Kronecker delta function,

$$P(n,t|n_0,t_0) = \sum_{v=-\infty}^{\infty} P(n,t|n_0+v,t_0+dt_0) [a(n_0,t_0)dt_0\, w(v|n_0,t_0)]$$

$$+ P(n,t|n_0,t_0+dt_0)[1-a(n_0,t_0)dt_0]. \qquad (5.1\text{-}52)$$

But now observe that this last equation could actually have been written down *directly* from the definitions (5.1-22) and (5.1-18), because it merely expresses the probability of finding $X(t)=n$, given $X(t_0)=n_0$, as the sum of the probabilities of all possible ways of arriving at state n at time t via *specified* states at time t_0+dt_0: The vth term under the summation sign is the probability of jumping from n_0 at time t_0 to n_0+v by time t_0+dt_0 and then going on from there to n at time t. And the last term is the probability of staying at n_0 until time t_0+dt_0 and then going on to n by time t. This logic ignores *multiple* jumps in $[t_0,t_0+dt_0)$, but that is okay because the probability for such jumps will be of order >1 in dt_0. Now subtracting $P(n,t|n_0,t_0+dt_0)$ from both sides of Eq. (5.1-52), dividing through by dt_0, and taking the limit $dt_0\to 0$, we obtain

$$-\frac{\partial}{\partial t_0}P(n,t|n_0,t_0) = \sum_{v=-\infty}^{\infty} [a(n_0,t_0)\, w(v|n_0,t_0)\, P(n,t|n_0+v,t_0)]$$

$$- a(n_0,t_0)\, P(n,t|n_0,t_0). \qquad (5.1\text{-}53a)$$

If we multiply the second term on the right by *unity* in the form of $\sum_v w(v|n_0,t_0)$ [Eq. (5.1-20b)] and then recall the definition (5.1-25) of the consolidated characterizing function, we obtain the equivalent formula

$$-\frac{\partial}{\partial t_0}P(n,t|n_0,t_0) = \sum_{v=-\infty}^{\infty} W(v|n_0,t_0) \left[P(n,t|n_0+v,t_0) - P(n,t|n_0,t_0) \right].$$

$$(5.1\text{-}53b)$$

Equations (5.1-53) are (both) called the **backward master equation** for the discrete state Markov process $X(t)$ defined by the characterizing functions a and w, or by the consolidated characterizing function W.

They are evidently differential-difference equations for $P(n,t \mid n_0,t_0)$ for fixed n and t, and they are to be solved subject to the *final condition* $P(n,t \mid n_0,t_0 = t) = \delta(n,n_0)$.

5.1.E THE MOMENT EVOLUTION EQUATIONS

More often than not, the discrete state master equations derived in the preceding subsection cannot be directly solved for $P(n,t \mid n_0,t_0)$. It is therefore desirable to develop explicit time evolution equations for the various *moments* of the process $X(t)$ and its integral $S(t)$. In Section 2.7 we derived such equations for the *continuum* state case. Those equations are expressed in terms of the propagator moment functions B_1, B_2, ... , which are given for any *continuous* Markov process with characterizing functions $A(x,t)$ and $D(x,t)$ by [see Eqs. (3.2-1)]

$$B_k(x,t) = \begin{cases} A(x,t), & \text{for } k=1, \\ D(x,t), & \text{for } k=2, \\ 0, & \text{for } k \geq 3, \end{cases} \qquad (5.1\text{-}54)$$

and for any *continuum state jump* Markov process with consolidated characterizing function $W(\xi \mid x,t)$ by [see Eqs. (4.2-1) and (4.2-2b)]

$$B_k(x,t) = W_k(x,t) \equiv \int_{-\infty}^{\infty} d\xi\, \xi^k\, W(\xi \mid x,t), \quad \text{for } k \geq 1. \qquad (5.1\text{-}55)$$

An inspection of those moment evolution equations in Section 2.7 reveals that there is nothing about them that seems to require that the first argument of the propagator moment functions $B_k(x,t)$ be *real*-valued instead of *integer*-valued. We might therefore expect that those moment evolution equations should also be valid for a *discrete state jump* Markov process, for which the propagator moment functions are given in terms of the consolidated characterizing function $W(v \mid n,t)$ by [see Eqs. (5.1-31) and (5.1-30b)]

$$B_k(n,t) = W_k(n,t) \equiv \sum_{v=-\infty}^{\infty} v^k\, W(v \mid n,t), \quad \text{for } k \geq 1. \qquad (5.1\text{-}56)$$

In fact, as we shall prove momentarily, this expectation is entirely correct: The time-evolution equations for the moments of a discrete state Markov process $X(t)$ with propagator moment functions W_k are given precisely by Eqs. (4.2-23), and the time-evolution equations for the moments of the associated integral process $S(t)$ are given by Eqs. (4.2-24) and (4.2-25). And so it follows that the time evolution equations for the mean, variance and covariance of $X(t)$ are given by Eqs. (4.2-26) – (4.2-31), and the time evolution equations for the mean, variance and covariance of $S(t)$ are given by Eqs. (4.2-32) – (4.2-36). All of those equations for *continuum* state jump Markov processes hold equally well for *discrete* state jump Markov processes, it being immaterial whether the jump propagator moment functions W_k are given by Eqs. (5.1-55) or Eqs. (5.1-56).

To prove the foregoing statements, let us recall the specific arguments that were used in Section 2.7 to derive the various moment evolution equations. The equations derived in Subsection 2.7.A for the moments of $X(t)$ and $S(t)$ were derived wholly from *three basic relations*. The first two of those relations are the basic propagator relations (2.7-1) and (2.7-8):

$$X(t+dt) = X(t) + \Xi(dt; X(t),t), \qquad (5.1\text{-}57a)$$

$$S(t+dt) = S(t) + X(t)\,dt. \qquad (5.1\text{-}57b)$$

The third relation is the fundamental property (2.7-5):

$$\langle\, X^j(t)\,\Xi^k(dt; X(t),t)\,\rangle = \langle\, X^j(t)\, B_k(X(t),t)\,\rangle\,dt + o(dt)$$

$$(j \ge 0,\, k \ge 1). \qquad (5.1\text{-}57c)$$

Now, we have already seen in Eqs. (5.1-32) and (5.1-33) that the first two relations above are just as valid for discrete state Markov processes as for continuum state Markov processes. But it remains to be seen whether the third relation (5.1-57c), which was proved in Subsection 2.7.A for *continuum* state Markov processes, is also true for *discrete* state Markov processes. To prove that it is, we begin by noting that the *joint* density function for the two random variables $X(t)$ and $\Xi(dt; X(t),t)$ is

$$\text{Prob}\{\, X(t)=n \text{ and } \Xi(dt; X(t),t)=v \mid X(t_0)=n_0 \,\}$$

$$= P(n,t \mid n_0,t_0)\, \Pi(v \mid dt; n,t).$$

Therefore, the average on the left of Eq. (5.1-57c) is given by

$$\langle\, X^j(t)\,\Xi^k(dt; X(t),t)\,\rangle = \sum_{n=-\infty}^{\infty} \sum_{v=-\infty}^{\infty} \left[n^j v^k \right] \left[P(n,t \mid n_0,t_0)\, \Pi(v \mid dt; n,t) \right]$$

$$= \sum_{n=-\infty}^{\infty} n^j \left[\sum_{v=-\infty}^{\infty} v^k\, \Pi(v \mid dt; n,t) \right] P(n,t \mid n_0,t_0)$$

$$= \sum_{n=-\infty}^{\infty} n^j \left[B_k(n,t)\, dt + o(dt) \right] P(n,t \mid n_0,t_0),$$

where the last step follows from the definition (5.1-29). Thus we conclude that

$$\langle X^j(t)\, \Xi^k(dt;\, X(t),t) \rangle = \left(\sum_{n=-\infty}^{\infty} \left[n^j B_k(n,t) \right] P(n,t \mid n_0,t_0) \right) dt + o(dt),$$

and this, by virtue of Eq. (5.1-10), is precisely Eq. (5.1-57c). With the three relations (5.1-57) thus established, the derivation of the time-evolution equations (4.2-23) – (4.2-25) for the moments of $X(t)$ and $S(t)$ now proceeds exactly as detailed in Subsection 2.7.A.

As was shown in Subsection 2.7.B, the time-evolution equations for the means and variances of $X(t)$ and $S(t)$ are straightforward consequences of the first and second moment evolution equations. But the proof of the covariance evolution equations given in Subsection 2.7.B requires, in addition to Eqs. (5.1-57a) – (5.1-57c), the relation

$$\langle X(t_1)\, \Xi(dt_2;\, X(t_2),t_2) \rangle = \langle X(t_1)\, B_1(X(t_2),t_2) \rangle dt_2 + o(dt_2)$$

$$(t_0 \le t_1 \le t_2). \quad (5.1\text{-}57d)$$

This relation is proved by first noting that the joint density function of the three random variables $X(t_1)$, $X(t_2)$ and $\Xi(dt_2;\, X(t_2),t_2)$, for $t_0 \le t_1 \le t_2$, is

$$\text{Prob}\{ X(t_1) = n_1 \text{ and } X(t_2) = n_2 \text{ and } \Xi(dt_2;\, X(t_2),t_2) = v \mid X(t_0) = n_0 \}$$

$$= [P(n_2,t_2 \mid n_1,t_1)\, P(n_1,t_1 \mid n_0,t_0)]\, \Pi(v \mid dt_2;\, n_2,t_2).$$

Therefore, we can calculate the average on the left of Eq. (5.1-57d) as

$$\langle X(t_1)\, \Xi(dt_2;\, X(t_2),t_2) \rangle$$

$$= \sum_{n_1=-\infty}^{\infty} \sum_{n_2=-\infty}^{\infty} \sum_{v=-\infty}^{\infty} (n_1\, v)$$

$$\times P(n_2,t_2 \mid n_1,t_1)\, P(n_1,t_1 \mid n_0,t_0)\, \Pi(v \mid dt_2;\, n_2,t_2)$$

$$= \sum_{n_1=-\infty}^{\infty} \sum_{n_2=-\infty}^{\infty} n_1 \left[\sum_{v=-\infty}^{\infty} v\, \Pi(v \mid dt_2;\, n_2,t_2) \right]$$

$$\times P(n_2,t_2 \mid n_1,t_1)\, P(n_1,t_1 \mid n_0,t_0)$$

$$= \sum_{n_1 = -\infty}^{\infty} \sum_{n_2 = -\infty}^{\infty} n_1 \left[B_1(n_2,t_2) \, dt_2 + o(dt_2) \right]$$

$$\times P(n_2,t_2 \mid n_1,t_1) \, P(n_1,t_1 \mid n_0,t_0)$$

$$= \sum_{n_1 = -\infty}^{\infty} \sum_{n_2 = -\infty}^{\infty} \left[n_1 B_1(n_2,t_2) \right] P(n_2,t_2 \mid n_1,t_1) \, P(n_1,t_1 \mid n_0,t_0) \, dt_2$$

$$+ \, o(dt_2),$$

where the penultimate step follows from the definition (5.1-29). The relation (5.1-57d) now follows upon application of Eq. (5.1-11). With Eq. (5.1-57d), the derivation of the covariance evolution equations for $X(t)$ and $S(t)$ now proceeds exactly as detailed in Subsection 2.7.B.

We have now established that *the moment evolution equations for discrete state Markov processes are identical to those for continuum state Markov processes.* It follows that the same consequences and limitations of those equations noted earlier apply here as well. For example, in the special case $W_1(n,t) = b_1$ and $W_2(n,t) = b_2$, the means, variances and covariances of $X(t)$ and $S(t)$ will be given explicitly by Eqs. (2.7-28) and (2.7-29). And in the special case $W_1(n,t) = -\beta n$ and $W_2(n,t) = c$, the means, variances and covariances of $X(t)$ and $S(t)$ will be given explicitly by Eqs. (2.7-34) and (2.7-35). More generally, the hierarchy of moment evolution equations will be "closed" if and only if the function $W_k(n,t)$ is a polynomial in n of degree $\leq k$. If that rather stringent condition is not satisfied, then approximate solutions can usually be obtained, albeit with much effort, by proceeding along the lines indicated in Appendix C.

As a final note on these matters, it is instructive to see how the time evolution equations (4.2-23) for the moments of $X(t)$ can *also* be derived directly from the forward master equation (5.1-50). Abbreviating $P(n,t \mid n_0,t_0) \equiv P(n,t)$, we have for any positive integer k,

$$\frac{d}{dt} \langle X^k(t) \rangle = \frac{d}{dt} \sum_{n = -\infty}^{\infty} n^k P(n,t) = \sum_{n = -\infty}^{\infty} n^k \frac{\partial}{\partial t} P(n,t)$$

$$= \sum_{n = -\infty}^{\infty} \sum_{v = -\infty}^{\infty} n^k \, W(v \mid n-v,t) \, P(n-v,t)$$

$$- \sum_{n = -\infty}^{\infty} \sum_{v = -\infty}^{\infty} n^k \, W(-v \mid n,t) \, P(n,t)$$

where the last step has invoked the forward master equation (5.1-50b). In the first sum of this last expression we change the summation variable n to $n-v$, while in the second sum we change the summation variable v to $-v$. This gives

$$\frac{d}{dt}\langle X^k(t)\rangle = \sum_{n=-\infty}^{\infty} \sum_{v=-\infty}^{\infty} (n+v)^k \, W(v\,|\,n,t) \, P(n,t)$$

$$- \sum_{n=-\infty}^{\infty} \sum_{v=-\infty}^{\infty} n^k \, W(v\,|\,n,t) \, P(n,t).$$

$$= \sum_{n=-\infty}^{\infty} \sum_{v=-\infty}^{\infty} \left[(n+v)^k - n^k \right] W(v\,|\,n,t) \, P(n,t).$$

Expanding $(n+v)^k$ using the binomial formula, we get

$$\frac{d}{dt}\langle X^k(t)\rangle = \sum_{n=-\infty}^{\infty} \sum_{v=-\infty}^{\infty} \left[\sum_{j=1}^{k} \binom{k}{j} v^j n^{k-j} \right] W(v\,|\,n,t) \, P(n,t)$$

$$= \sum_{j=1}^{k} \binom{k}{j} \sum_{n=-\infty}^{\infty} n^{k-j} \left[\sum_{v=-\infty}^{\infty} v^j \, W(v\,|\,n,t) \right] P(n,t)$$

$$= \sum_{j=1}^{k} \binom{k}{j} \sum_{n=-\infty}^{\infty} n^{k-j} \, W_j(n,t) \, P(n,t). \qquad \text{[by (5.1-30b)]}$$

$$= \sum_{j=1}^{k} \binom{k}{j} \langle X^{k-j}(t) \, W_j(X(t),t) \rangle, \qquad \text{[by (5.1-10)]}$$

in agreement with Eq. (4.2-23).

Our earlier derivation of Eq. (4.2-23) using the process propagator has three advantages over the foregoing master equation derivation: First, the propagator derivation is slightly shorter than the master equation derivation; second, the propagator derivation applies to *all* Markov processes, not just to discrete-state jump Markov processes; and third, the propagator approach allows a concurrent derivation of the equations governing the moments of the integral process $S(t)$. We should note that it is also possible (and equally instructive) to derive the time-evolution equation (4.2-28) for $\text{cov}\{X(t_1),X(t_2)\}$ directly from the forward master equation (5.1-50b); however, that derivation likewise lacks the simplicity and generality of our earlier propagator derivation.

5.2 COMPLETELY HOMOGENEOUS DISCRETE STATE PROCESSES

For the remainder of this chapter we shall be concerned exclusively with discrete state Markov processes whose characterizing functions are explicitly independent of t. In the present section we shall focus on the subclass of those processes for which the characterizing functions are *also* explicitly independent of n, namely, the *completely homogeneous* discrete state Markov processes.

5.2.A BASIC EQUATIONS

A completely homogeneous discrete state Markov process $X(t)$ is by definition a jump Markov process whose characterizing functions have the forms

$$a(n,t) \equiv a \quad \text{and} \quad w(v \mid n,t) \equiv w(v); \qquad (5.2\text{-}1)$$

$$W(v \mid n,t) \equiv W(v) \equiv a\,w(v). \qquad (5.2\text{-}2)$$

Here, a can be any positive constant, while $w(v)$ can be any integer-variable density function. For such a process, the propagator density function (5.1-23) takes the form

$$\Pi(v \mid dt; n,t) \equiv \Pi(v \mid dt) = adt\,w(v) + [1-adt]\,\delta(v,0), \qquad (5.2\text{-}3)$$

and the next-jump density function (5.1-41) becomes

$$p(\tau,v \mid n,t) \equiv p(\tau,v) = a\,e^{-a\tau}\,w(v). \qquad (5.2\text{-}4)$$

The latter equation shows clearly that the waiting time of $X(t)$ in any state is an exponentially distributed random variable with mean $1/a$, while the state-change in the next jump is the statistically independent integer-valued random variable with density function $w(v)$ [compare statements (5.1-43)].

Monte Carlo simulations of $X(t)$ and its integral $S(t)$ can be effected by applying the algorithm of Fig. 5-1. The implementation of that algorithm's Steps 2° and 3° along the lines described in figure notes a and b will usually be fairly straightforward for processes of this type. Some illustrative examples will be given shortly [see Figs. 5-2 and 5-3].

With the characterizing functions (5.2-1), the forward master equation (5.1-50a) for $X(t)$ becomes

$$\frac{\partial}{\partial t} P(n,t \mid n_0,t_0) = a \sum_{v=-\infty}^{\infty} [w(v) P(n-v,t \mid n_0,t_0)] - a P(n,t \mid n_0,t_0).$$

(5.2-5)

And the backward master equation (5.1-53a) becomes

$$-\frac{\partial}{\partial t_0} P(n,t \mid n_0,t_0) = a \sum_{v=-\infty}^{\infty} [w(v) P(n,t \mid n_0+v,t_0)] - a P(n,t \mid n_0,t_0).$$

(5.2-6)

However, a usually more efficient way to calculate the Markov state density function $P(n,t \mid n_0,t_0)$ in the completely homogeneous case is to use formulas (5.1-16) and (5.1-15). They simplify here to

$$P(n,t \mid n_0,t_0) = \sum_{v_1=-\infty}^{\infty} \cdots \sum_{v_{k-1}=-\infty}^{\infty} \prod_{i=1}^{k} \Pi(v_i \mid dt),$$

(5.2-7)

where v_k is defined by

$$v_k = n - n_0 - \sum_{j=1}^{k-1} v_j,$$

(5.2-8a)

and where k and dt are related by

$$(t-t_0)/k = dt, \text{ an infinitesimal.}$$

(5.2-8b)

To reduce Eq. (5.2-7) to a more tractable form, we proceed as follows. We first observe that the definition (5.2-8a) of v_k can be incorporated into the formula (5.2-7) by the trick of introducing a sum over v_k with an appropriate Kronecker delta function:

$$1 = \sum_{v_k=-\infty}^{\infty} \delta\left(v_k, n - n_0 - \sum_{j=1}^{k-1} v_j\right) = \sum_{v_k=-\infty}^{\infty} \delta\left(n-n_0, \sum_{j=1}^{k} v_j\right).$$

Substituting this into the summand of Eq. (5.2-7), we get

$$P(n,t \mid n_0,t_0) = \sum_{v_1=-\infty}^{\infty} \cdots \sum_{v_k=-\infty}^{\infty} \prod_{i=1}^{k} \Pi(v_i \mid dt) \, \delta\left(n-n_0, \sum_{j=1}^{k} v_j\right).$$

Now using Eq. (B-2), we rewrite the Kronecker delta function as a definite integral

$$\delta\left(n-n_0, \sum_{j=1}^{k} v_j\right) = \frac{1}{2\pi} \int_{-\pi}^{\pi} du \, \exp\left[iu\left(n-n_0 - \sum_{j=1}^{k} v_j\right)\right]$$

$$= \frac{1}{2\pi} \int_{-\pi}^{\pi} du \, \exp \left[iu(n-n_0) \right] \prod_{j=1}^{k} \exp(-iuv_j).$$

Inserting this into the preceding equation, we get

$$P(n,t \mid n_0, t_0) = \frac{1}{2\pi} \int_{-\pi}^{\pi} du \, \exp[iu(n-n_0)]$$

$$\times \sum_{v_1 = -\infty}^{\infty} \cdots \sum_{v_k = -\infty}^{\infty} \prod_{j=1}^{k} \Pi(v_j \mid dt) \exp(-iuv_j),$$

which is evidently the same as

$$P(n,t \mid n_0, t_0) = \frac{1}{2\pi} \int_{-\pi}^{\pi} du \, \exp[iu(n-n_0)] \left(\sum_{v = -\infty}^{\infty} \Pi(v \mid dt) \exp(-iuv) \right)^{k}.$$

Now we invoke the explicit formula for $\Pi(v \mid dt)$ in Eq. (5.2-3) to calculate

$$\sum_{v = -\infty}^{\infty} \Pi(v \mid dt) e^{-iuv} = \sum_{v = -\infty}^{\infty} [a dt \, w(v)] e^{-iuv}$$

$$+ \sum_{v = -\infty}^{\infty} [1 - adt] \, \delta(v,0) \, e^{-iuv}$$

$$= a dt \, w^{\star}(u) + [1 - adt],$$

where in the last step we have defined $w^{\star}(u) \equiv \Sigma_v \, w(v) \, e^{-iuv}$. Raising this equation to the kth power, as is required for the preceding equation, we get

$$\left(\sum_{v = -\infty}^{n} \Pi(v \mid dt) e^{-iuv} \right)^{k} = \left(1 + adt[w^{\star}(u) - 1] \right)^{k}$$

$$= \left(1 + \frac{a(kdt)[w^{\star}(u) - 1]}{k} \right)^{k}.$$

Finally, invoking the dual implication of Eq. (5.2-8b) that $(kdt) = (t - t_0)$ with k being "infinitely large," we get

$$\left(\sum_{v = -\infty}^{\infty} \Pi(v \mid dt) e^{-iuv} \right)^{k} = \exp \left(a(t - t_0)[w^{\star}(u) - 1] \right).$$

Substituting this into the previous equation for $P(n,t|n_0,t_0)$ gives the final result

$$P(n,t|n_0,t_0) = \frac{1}{2\pi} \int_{-\pi}^{\pi} du \, \exp[iu(n-n_0)] \exp\left(a\,(t-t_0)[w^\star(u)-1]\right),$$

(5.2-9)

wherein $w^\star(u)$ is defined by

$$w^\star(u) \equiv \sum_{v=-\infty}^{\infty} w(v)\,e^{-iuv}.$$

(5.2-10)

Equation (5.2-9) essentially constitutes a "solution-in-quadrature" to the master equations (5.2-5) and (5.2-6) for the requisite boundary condition $P(n,t_0|n_0,t_0) = \delta(n,n_0)$. Notice that the function $w^\star(u)$ is basically just the discrete Fourier transform of the characterizing function $w(v)$. Once the function $w^\star(u)$ has been calculated, $P(n,t|n_0,t_0)$ can be found, according to Eq. (5.2-9), by evaluating a single ordinary definite integral. And if that integral cannot be evaluated analytically, it can surely be evaluated numerically.

We observe from the form of the right hand side of Eq. (5.2-9) that $P(n,t|n_0,t_0)$ depends on its four arguments only through the two differences $(n-n_0)$ and $(t-t_0)$:

$$P(n,t|n_0,t_0) = P(n-n_0,t-t_0|0,0).$$

(5.2-11)

This property of a completely homogeneous discrete state Markov process is of course analogous to the property (2.8-20) established earlier for completely homogeneous *continuum* state Markov processes. Using Eq. (5.2-11), it is easy to show that the forward and backward master equations (5.2-5) and (5.2-6) are entirely equivalent to each other.

As we discovered in Subsection 5.1.E, the time-evolution equations for the various moments of $X(t)$ and its integral $S(t)$ are the same as for the continuum jump case — namely Eqs. (4.2-23) – (4.2-36) — provided we calculate the functions W_k therein according to

$$W_k(n,t) \equiv \sum_{v=-\infty}^{\infty} v^k\,W(v|n,t) \equiv a(n,t) \sum_{v=-\infty}^{\infty} v^k\,w(v|n,t).$$

In the present completely homogeneous case, this calculation evidently gives

$$W_k(n,t) = aw_k$$

where w_k is the kth moment of the density function $w(v)$:

$$w_k \equiv \sum_{v=-\infty}^{\infty} v^k w(v). \tag{5.2-12}$$

Therefore, we need only replace $W_k(X(t),t)$ in the aforementioned time-evolution equations by aw_k. But that replacement is precisely what we did in Subsection 4.4.A to obtain the moment evolution equations (4.4-10) – (4.4-14) for a completely homogeneous continuum state jump Markov process. It follows that the moment evolution equations for completely homogeneous *discrete state* jump Markov processes are *identical* to the moment evolution equations for completely homogeneous *continuum state* jump Markov processes. Specifically, with w_k as defined in Eq. (5.2-12), the time-evolution equations for the moments of $X(t)$ are given by Eq. (4.4-10), and the time-evolution equations for the moments of $S(t)$ are given by Eqs. (4.4-11) and (4.4-12). Those time evolution equations are closed, and they can be solved exactly for all the moments. Explicit formulas for the mean, variance and covariance of $X(t)$ are given in Eqs. (4.4-13) [with x_0 replaced by n_0], and explicit formulas for the mean, variance and covariance of $S(t)$ are given in Eqs. (4.4-14).

5.2.B EXAMPLE: THE COMPLETELY HOMOGENEOUS ONE-STEP PROCESS

An example of a completely homogeneous discrete state Markov process $X(t)$ is the completely homogeneous "one-step" process, for which the function $w(v)$ is given by

$$w(v) = \begin{cases} b, & \text{for } v = +1 \\ 1-b, & \text{for } v = -1 \\ 0, & \text{for } v \neq \pm 1, \end{cases} \tag{5.2-13}$$

b being any constant that satisfies $0 \leq b \leq 1$. We see that when this process is in any state n, then its next jump will be *either* to state $n+1$, with probability b, *or* to state $n-1$, with probability $1-b$.

We showed in the preceding subsection that the moment evolution equations for this process are Eqs. (4.4-10) – (4.4-14), and those equations evidently require that we first calculate the constants aw_k. This is easily done for the characterizing function $w(v)$ specified by Eq. (5.2-13): Using the definition (5.2-12), we calculate

$$w_k = (-1)^k w(-1) + (+1)^k w(+1) = (-1)^k (1-b) + b;$$

so, upon multiplying through by a, we conclude that

$$aw_k = \begin{cases} a(2b-1), & \text{for } k=1,3,5,... \\ a, & \text{for } k=2,4,6,... \end{cases} \qquad (5.2\text{-}14)$$

Inserting this formula for aw_k into Eqs. (4.4-10) – (4.4-12), we obtain explicit differential equations for the moments of $X(t)$ and its integral $S(t)$. Those differential equations are closed, and can be solved exactly. Of most interest of course are the means, variances and covariances of $X(t)$ and $S(t)$. Explicit formulas for those quantities may be obtained by substituting Eq. (5.2-14) into Eqs. (4.4-13) and (4.4-14). In that way we find for $X(t)$ the formulas

$$\langle X(t) \rangle = n_0 + a(2b-1)(t-t_0) \quad (t_0 \le t), \qquad (5.2\text{-}15a)$$

$$\text{var}\{X(t)\} = a(t-t_0) \qquad\qquad (t_0 \le t), \qquad (5.2\text{-}15b)$$

$$\text{cov}\{X(t_1), X(t_2)\} = a(t_1-t_0) \qquad (t_0 \le t_1 \le t_2). \quad (5.2\text{-}15c)$$

and for $S(t)$ the formulas

$$\langle S(t) \rangle = n_0(t-t_0) + a(b-\tfrac{1}{2})(t-t_0)^2 \quad (t_0 \le t), \qquad (5.2\text{-}16a)$$

$$\text{var}\{S(t)\} = (a/3)(t-t_0)^3 \qquad\qquad (t_0 \le t), \qquad (5.2\text{-}16b)$$

$$\text{cov}\{S(t_1), S(t_2)\} = a(t_1-t_0)^2 [(t_1-t_0)/3 + (t_2-t_1)/2]$$
$$(t_0 \le t_1 \le t_2). \quad (5.2\text{-}16c)$$

To see what we can deduce about the form of $P(n,t \mid n_0,t_0)$ for this process, we turn to Eqs. (5.2-9) and (5.2-10). From the latter we calculate

$$w^\star(u) \equiv \sum_v w(v) e^{-iuv} = (1-b) e^{-iu(-1)} + b e^{-iu(+1)},$$

or, after expanding the complex exponentials and collecting terms,

$$w^\star(u) = \cos u - i(2b-1) \sin u. \qquad (5.2\text{-}17)$$

Inserting this into Eq. (5.2-9) gives

$$P(n,t \mid n_0,t_0) = (2\pi)^{-1} \int_{-\pi}^{\pi} du \; \exp[iu(n-n_0)]$$

$$\times \exp\Big(a(t-t_0)[\cos u - i(2b-1)\sin u - 1]\Big)$$

$$= (2\pi)^{-1} \exp[-a(t-t_0)] \int_{-\pi}^{\pi} du \; \exp[a(t-t_0)\cos u]$$

$$\times \exp[i\{u(n-n_0) - a(2b-1)(t-t_0)\sin u\}].$$

Expanding the complex exponential, and taking account of the even and odd natures of the real and imaginary parts of the integrand, we conclude that the Markov state density function for the completely homogeneous one-step Markov process defined by Eqs. (5.2-13) is

$$P(n,t\,|\,n_0,t_0) \;=\; \pi^{-1}\exp[-a(t-t_0)]\,\int_0^\pi du\;\exp[\,a(t-t_0)\cos u\,]$$

$$\times \cos[\,u(n-n_0) - a(2b-1)(t-t_0)\sin u\,]$$

$$(t_0 \le t). \qquad (5.2\text{-}18)$$

Equation (5.2-18) shows that a single ordinary integration is all that is required to evaluate the Markov state density function for a completely homogeneous one-step Markov process. And if that integral cannot be evaluated analytically, it can surely be evaluated numerically. One instance in which we *can* evaluate the u-integral in Eq. (5.2-18) analytically is for $b=1$, in which case every jump that occurs is a jump of size $v = +1$. In that case we have $(2b-1) = 1$, and the u-integral assumes the form of Eq. (A-10) with the replacements

$$a \to a(t-t_0) \quad \text{and} \quad n \to (n-n_0);$$

the integral therefore has the value $\{[a(t-t_0)]^{(n-n_0)}\pi\,/\,(n-n_0)!\}$, so Eq. (5.2-18) becomes

$$P(n,t\,|\,n_0,t_0) \;=\; \frac{e^{-a(t-t_0)}[a(t-t_0)]^{(n-n_0)}}{(n-n_0)!}$$

$$(b=1;\, t_0 \le t;\, n \ge n_0). \qquad (5.2\text{-}19)$$

Comparing with Eqs. (1.7-14), we see that $P(n,t\,|\,n_0,t_0)$, considered as a function of $(n-n_0)$, is the density function of the *Poisson* random variable with mean and variance $a(t-t_0)$. This particular process (with $b=1$) is in fact known as the *Poisson process*.

Figures 5-2 and 5-3 show simulation runs of two completely homogeneous one-step Markov processes $X(t)$ and their integrals $S(t)$. Each simulation was carried out using the Monte Carlo algorithm of Fig. 5-1 with $t_0 = n_0 = 0$ and $a = 1$. Figure 5-2 has $b = 1$, thus making $X(t)$ a *Poisson process*, whereas Fig. 5-3 has $b = \frac{1}{2}$, thus making $X(t)$ a so-called *drunkard's walk process*; both of these processes will be discussed further in Chapter 6. The dashed curves in Figs. 5-2 and 5-3 show the one-standard deviation envelopes $\langle X(t)\rangle \pm \text{sdev}\{X(t)\}$ and $\langle S(t)\rangle \pm \text{sdev}\{S(t)\}$, as calculated from Eqs. (5.2-15) and (5.2-16).

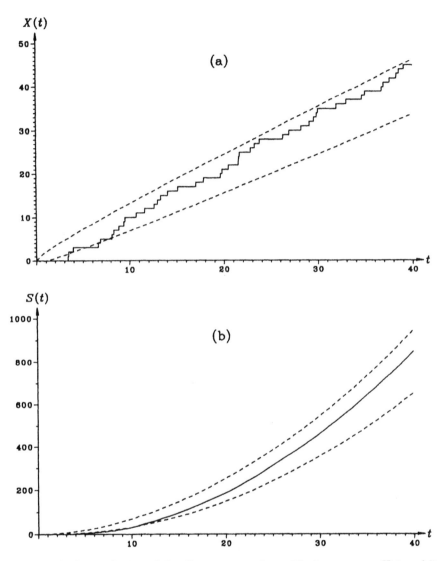

Figure 5-2. A simulation of the discrete state jump Markov process $X(t)$ and its integral $S(t)$ as defined by the characterizing functions $a(n,t) = 1$ and $w(v \mid n,t) = \delta(v, +1)$. The process $X(t)$ here is called a *Poisson process*. The simulation was carried out using the algorithm of Fig. 5-1. The dashed curves are the respective one-standard deviation envelopes as calculated from Eqs. (5.2-15) and (5.2-16) with $t_0 = n_0 = 0$ and $a = b = 1$. The function $P(n,t \mid n_0,t_0)$ for this process is given by Eq. (5.2-19) with $t_0 = n_0 = 0$ and $a = 1$.

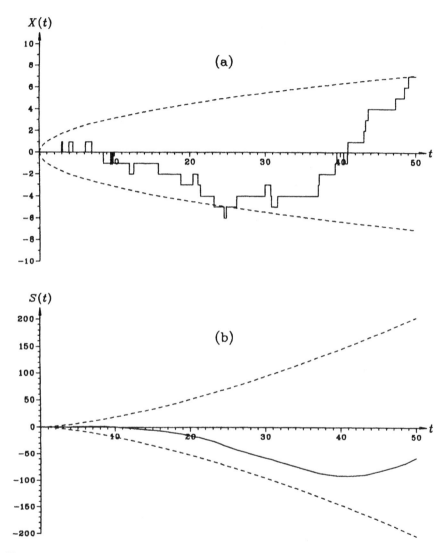

Figure 5-3. A simulation of the discrete state jump Markov process $X(t)$ and its integral $S(t)$ as defined by the characterizing functions $a(n,t) = 1$ and $w(v \mid n,t) = \frac{1}{2}\delta(v, -1) + \frac{1}{2}\delta(v, +1)$. The process $X(t)$ here is called a *drunkard's walk process*. The simulation was carried out using the algorithm of Fig. 5-1. The dashed curves are the respective one-standard deviation envelopes as calculated from Eqs. (5.2-15) and (5.2-16) with $t_0 = n_0 = 0$, $a = 1$ and $b = \frac{1}{2}$. The function $P(n,t \mid n_0, t_0)$ for this process is given in quadrature form by Eq. (5.2-18) with $n_0 = t_0 = 0$, $a = 1$ and $b = \frac{1}{2}$.

5.3 TEMPORALLY HOMOGENEOUS MARKOV PROCESSES ON THE NONNEGATIVE INTEGERS

In this section we shall discuss temporally homogeneous jump Markov processes whose states are confined to the nonnegative integers. We shall illustrate the practical applicability of such processes by showing how they can be used to mathematically model well-stirred systems of chemically reacting molecules. Our discussion of temporally homogeneous nonnegative integer Markov processes will continue in Chapter 6, but there under the additional restriction that only jumps to immediately adjacent states be allowed.

5.3.A BASIC EQUATIONS

To say that the discrete state jump Markov process $X(t)$ is *temporally homogeneous* means of course that its characterizing functions do not depend explicitly on time:

$$a(n,t) \equiv a(n) \quad \text{and} \quad w(v \mid n,t) \equiv w(v \mid n); \tag{5.3-1}$$

$$W(v \mid n,t) \equiv W(v \mid n) \equiv a(n)\,w(v \mid n). \tag{5.3-2}$$

To ensure that $X(t)$ will always be *nonnegative*, it is of course first necessary to require that the initial state be nonnegative:

$$X(t_0) = n_0 \geq 0. \tag{5.3-3}$$

It is then necessary to disallow jumps *to* negative states from nonnegative states, and in order to simplify the writing of some later formulas we shall go even further and disallow jumps *from* negative states as well. So for *nonnegative integer* processes, we shall simply assume that

$$w(v \mid n) = W(v \mid n) = 0 \quad \text{if either } n < 0 \text{ or } n+v < 0. \tag{5.3-4}$$

It will sometimes be convenient to denote the value of the consolidated characterizing function $W(v \mid n)$ for given values of n and v by

$$W(v \mid n) \equiv \omega_{n+v,n}; \tag{5.3-5a}$$

in other words, we are *defining* the quantities

$$\omega_{m,n} \equiv W(m-n \mid n) \equiv a(n)\,w(m-n \mid n). \tag{5.3-5b}$$

In view of the definition (5.1-26) of $W(v \mid n,t)dt$, it follows that

$\omega_{m,n} dt = $ probability, given $X(t) = n$, that the process will in the time interval $[t,t+dt)$ jump from state n to state m. (5.3-6)

Also, since $\Sigma_v W(v \mid n) = a(n)$, then the $\omega_{m,n}$'s satisfy

$$\sum_{m=0}^{\infty} \omega_{m,n} = a(n). \tag{5.3-7}$$

Notice that the nonnegativity restriction (5.3-4) can be written in terms of the $\omega_{m,n}$'s as simply

$$\omega_{m,n} = 0 \quad \text{if either } n < 0 \text{ or } m < 0. \tag{5.3-8}$$

The next-jump density function for temporally homogeneous $X(t)$ was written down in Eq. (5.1-41); that formula, along with its consequences (5.1-43), apply here with the constraint that n and $n+v$ both be nonnegative. Monte Carlo simulations of $X(t)$ and its integral $S(t)$ can be implemented according to the algorithm in Fig. 5-1. There, the τ-selection process in Step 2° is very straightforward because the simplification in note a applies. The v-selection process in Step 3° can be accomplished by drawing a unit uniform random number r', and then finding the value v that satisfies the double inequality

$$\sum_{v'=-n}^{v-1} w(v' \mid n) \leq r' < \sum_{v'=-n}^{v} w(v' \mid n); \tag{5.3-9a}$$

here, the lower limit on the summation index comes from the fact that $w(v' \mid n)$ vanishes for all $v' < -n$ [see Eqs. (5.3-4)]. This v-selection condition can be stated in terms of the consolidated characterizing function $W(v \mid n)$ by multiplying through by $a(n)$ and then invoking Eq. (5.3-2):

$$\sum_{v'=-n}^{v-1} W(v' \mid n) \leq a(n) r' < \sum_{v'=-n}^{v} W(v' \mid n). \tag{5.3-9b}$$

Or, it can be stated in terms of the $\omega_{m,n}$'s by first changing the summation variable v' in this last equation to $m = n+v'$, and then using Eq. (5.3-5b):

$$\sum_{m=0}^{n+v-1} \omega_{m,n} \leq a(n) r' < \sum_{m=0}^{n+v} \omega_{m,n}. \tag{5.3-9c}$$

As discussed in Subsection 5.1.E, the time-evolution equations for the various moments of $X(t)$ and its integral $S(t)$ in the *discrete state* jump case are identical to the equations for the *continuum state* jump case, namely Eqs. (4.2-23) – (4.2-36), provided that we calculate the propagator moments functions W_k therein according to the discrete state formula (5.1-56). The latter becomes, for $X(t)$ temporally homogeneous and nonnegative,

$$B_k(n) = W_k(n) \equiv \sum_{v=-n}^{\infty} v^k W(v \mid n) \quad (k \geqslant 1; n \geq 0). \quad (5.3\text{-}10)$$

So all the moment evolution equations (4.2-23) – (4.2-36) apply here, provided that we simply ignore the second argument (t) in the functions W_k appearing in those equations. For later convenience, we shall rewrite those equations here:

The time-evolution equations for the moments of $X(t)$ are

$$\frac{d}{dt}\langle X^k(t)\rangle = \sum_{l=1}^{k} \binom{k}{l} \langle X^{k-l}(t)\, W_l(X(t))\rangle$$

$$(t_0 \leq t; k=1,2,...) \quad (5.3\text{-}11)$$

with the initial conditions $\langle X^k(t_0)\rangle = n_0^k$. The time-evolution equations for the moments of $S(t)$ are

$$\frac{d}{dt}\langle S^m(t)\rangle = m\langle S^{m-1}(t)\, X(t)\rangle \quad (t_0 \leq t; m=1,2,...) \quad (5.3\text{-}12)$$

with the initial conditions $\langle S^m(t_0)\rangle = 0$, together with the auxiliary equations

$$\frac{d}{dt}\langle S^m(t)\, X^k(t)\rangle = m\langle S^{m-1}(t)\, X^{k+1}(t)\rangle$$

$$+ \sum_{l=1}^{k} \binom{k}{l} \langle S^m(t)\, X^{k-l}(t)\, W_l(X(t))\rangle$$

$$(t_0 \leq t; m \geq 1; k \geq 1), \quad (5.3\text{-}13)$$

and their initial conditions $\langle S^m(t_0)\, X^k(t_0)\rangle = 0$. These moment evolution equations will be closed if and only if $W_l(n)$ is a polynomial in n of degree $\leq l$. If $W_l(n)$ does not satisfy that rather stringent condition, then the most we can hope for is approximate solutions, obtained either by suitably approximating $W_l(n)$, or else by implementing the procedure outlined in Appendix C.

For the mean, variance and covariance of $X(t)$ we have the differential equations

$$\frac{d}{dt}\langle X(t)\rangle = \langle W_1(X(t))\rangle \qquad (t_0 \leq t), \qquad (5.3\text{-}14)$$

$$\frac{d}{dt}\text{var}\{X(t)\} = 2\Big(\langle X(t)\,W_1(X(t))\rangle - \langle X(t)\rangle\langle W_1(X(t))\rangle\Big) + \langle W_2(X(t))\rangle$$
$$(t_0 \leq t), \qquad (5.3\text{-}15)$$

$$\frac{d}{dt_2}\text{cov}\{X(t_1),X(t_2)\} = \langle X(t_1)\,W_1(X(t_2))\rangle - \langle X(t_1)\rangle\langle W_1(X(t_2))\rangle$$
$$(t_0 \leq t_1 \leq t_2), \qquad (5.3\text{-}16)$$

with the respective initial conditions

$$\langle X(t=t_0)\rangle = n_0, \qquad (5.3\text{-}17)$$

$$\text{var}\{X(t=t_0)\} = 0, \qquad (5.3\text{-}18)$$

$$\text{cov}\{X(t_1),X(t_2=t_1)\} = \text{var}\{X(t_1)\}. \qquad (5.3\text{-}19)$$

For the mean of $S(t)$ we have

$$\langle S(t)\rangle = \int_{t_0}^{t}\langle X(t')\rangle\,dt', \qquad (5.3\text{-}20)$$

wherein the integrand is the solution to Eq. (5.3-14). For the variance of $S(t)$ we have

$$\text{var}\{S(t)\} = 2\int_{t_0}^{t}\text{cov}\{S(t'),X(t')\}\,dt', \qquad (5.3\text{-}21)$$

wherein the integrand is the solution of the differential equation

$$\frac{d}{dt}\text{cov}\{S(t),X(t)\} = \text{var}\{X(t)\} + \langle S(t)\,W_1(X(t))\rangle - \langle S(t)\rangle\langle W_1(X(t))\rangle$$
$$(t_0 \leq t) \qquad (5.3\text{-}22a)$$

that satisfies the initial condition

$$\text{cov}\{S(t=t_0),X(t=t_0)\} = 0. \qquad (5.3\text{-}22b)$$

And finally, for the covariance of $S(t)$ we have

$$\text{cov}\{S(t_1),S(t_2)\} = \text{var}\{S(t_1)\} + \int_{t_1}^{t_2}\text{cov}\{S(t_1),X(t_2')\}\,dt_2'$$

$$(t_0 \leq t_1 \leq t_2), \qquad (5.3\text{-}23)$$

wherein the integrand is the solution of the differential equation

$$\frac{d}{dt_2}\text{cov}\{S(t_1),X(t_2)\} = \langle\, S(t_1)\, W_1(X(t_2))\,\rangle - \langle S(t_1)\rangle\langle W_1(X(t_2))\rangle$$

$$(t_0 \leq t_1 \leq t_2) \qquad (5.3\text{-}24a)$$

that satisfies the initial condition

$$\text{cov}\{S(t_1),X(t_2=t_1)\} = \text{cov}\{S(t_1),X(t_1)\}, \qquad (5.3\text{-}24b)$$

which in turn is known from the solution to Eq. (5.3-22a).

If it should happen that $W_1(n)=b_1$ and $W_2(n)=b_2$ where b_1 and b_2 are both constants, then the means, variances and covariances of $X(t)$ and $S(t)$ will be given by the explicit formulas (2.7-28) and (2.7-29). If it should happen that $W_1(n)=-\beta n$ and $W_2(n)=c$ where β and c are both constants, then the means, variances and covariances of $X(t)$ and $S(t)$ will be given by the explicit formulas (2.7-34) and (2.7-35).

The forward master equation (5.1-50b) reads, for $X(t)$ a temporally homogeneous nonnegative integer process,

$$\frac{\partial}{\partial t}P(n,t\,|\,n_0,t_0) = \sum_{v=-\infty}^{n}\left[\, W(v\,|\,n-v)\,P(n-v,t\,|\,n_0,t_0)\right.$$

$$\left. - W(-v\,|\,n)\,P(n,t\,|\,n_0,t_0)\right]. \qquad (5.3\text{-}25a)$$

Here the summation index restriction $v \leq n$ arises from condition (5.3-4), which implies that both $W(v\,|\,n-v)$ and $W(-v\,|\,n)$ vanish if $v>n$. It is sometimes convenient to write this equation in terms of the quantities $\omega_{m,n}$ defined in Eqs. (5.3-5). Formally this is done by first changing the summation index in Eq. (5.3-25a) from v to $m=n-v$, and then invoking the definition (5.3-5b). The result of those straightforward manipulations is

$$\frac{\partial}{\partial t}P(n,t\,|\,n_0,t_0) = \sum_{m=0}^{\infty}\left[\,\omega_{n,m}\,P(m,t\,|\,n_0,t_0) - \omega_{m,n}\,P(n,t\,|\,n_0,t_0)\right].$$

$$(5.3\text{-}25b)$$

Notice that in the two-term summands on the right hand sides of both the above forms of the forward master equation, the first term essentially describes the *increases* in $P(n,t \mid n_0,t_0)$ caused by jumps *to* state n, while the second term describes *decreases* in $P(n,t \mid n_0,t_0)$ caused jumps *away from* state n.

The backward master equation (5.1-53b) reads, for $X(t)$ a temporally homogeneous nonnegative integer process,

$$-\frac{\partial}{\partial t_0} P(n,t \mid n_0,t_0) = \sum_{v=-n_0}^{\infty} W(v \mid n_0) \left[P(n,t \mid n_0+v,t_0) - P(n,t \mid n_0,t_0) \right].$$

(5.3-26a)

Here the summation index restriction $v \geq -n_0$ arises because, according to condition (5.3-4), $W(v \mid n_0)$ vanishes whenever $v < -n_0$. By changing the summation index here from v to $m = n_0+v$, and then invoking the definition (5.3-5b), we can bring Eq. (5.3-26a) into the alternate form

$$-\frac{\partial}{\partial t_0} P(n,t \mid n_0,t_0) = \sum_{m=0}^{\infty} \omega_{m,n_0} \left[P(n,t \mid m,t_0) - P(n,t \mid n_0,t_0) \right].$$

(5.3-26b)

Neither of the above two forms of the backward master equation seems to admit a simple "rate" interpretation of the kind described above for the forward master equations (5.3-25). [Probably the closest that one can come to making the backward master equations "transparent" is the argument following Eq. (5.1-52).]

Finally, we note that when $X(t)$ is temporally homogeneous, the forward master equations (5.3-25) and the backward master equations (5.3-26) have the property that *their left hand sides are interchangeable*:

$$-\frac{\partial}{\partial t_0} P(n,t \mid n_0,t_0) = \frac{\partial}{\partial t} P(n,t \mid n_0,t_0).$$ (5.3-27)

This property is most easily proved by appealing to Eqs. (5.1-15) and (5.1-16): When the propagator density function $\Pi(v \mid dt; n,t)$ does not depend explicitly on t, then Eqs. (5.1-15a) are not required by Eq. (5.1-16), and the right hand side of Eq. (5.1-16) will depend on t and t_0 only through their difference $(t-t_0)$ in accordance with Eq. (5.1-15d); this in turn leads to the result (5.3-27) by application of the argument that we used to establish the same result (2.8-5) for the continuum state case.

5.3.B APPLICATION: CHEMICAL KINETICS

The time-evolution of the chemical species concentrations in a well-stirred, chemically reacting mixture is usually described by a set of coupled, first order, ordinary differential equations called the *reaction rate equations*. Traditional derivations of the reaction rate equations tend to ignore or gloss over the intrinsically random way in which chemical reactions actually occur on the molecular level. It turns out that a more logically satisfying derivation of the reaction rate equations can be constructed within the framework of Markov process theory. This Markovian derivation is based on a careful stochastic description of the molecular processes that are responsible for chemical reactions, and as a bonus it provides us with the means to go *beyond* the reaction rate equations when certain key assumptions that underlie them break down. In this subsection we shall outline the Markov process approach to chemical kinetics; however, our treatment will be somewhat circumscribed by the requirement that we stay within the realm of *univariate* Markov process theory.

We consider a well-stirred gaseous mixture of several chemical species, the mixture being confined to a container of volume Ω and kept in thermal equilibrium at absolute temperature T. The constituent chemical species are assumed to interact through one or more "chemical reaction channels" R_μ ($\mu = 1,2,...$). A typical *bimolecular* reaction channel R_μ has the form

$$X_1 + X_2 \rightarrow X_3 + \ldots, \tag{5.3-28}$$

and specifies the transformation of two molecules of species 1 and 2, called the "reactant" species, into molecules of one or more species 3, ... , called the "product" species. The key to treating this bimolecular gas-phase chemical reaction R_μ using the machinery of Markov process theory is the fact that there always exists a constant c_μ, called the *specific probability rate constant* of reaction R_μ, such that

$$c_\mu dt \equiv \text{probability that a randomly chosen combination}$$
of R_μ reactant molecules inside Ω at time t will
react according to R_μ in the next infinitesimal
time interval $[t, t + dt)$. $\tag{5.3-29}$

The word "rate" in the name of c_μ reflects the fact that $c_\mu dt$, and not c_μ, is a probability; the word "specific" reflects the fact that this probability is *per* randomly chosen combination of R_μ reactant molecules. We shall show later that the specific probability rate constant c_μ is closely related to, but not quite the same as, a parameter that chemists call the *reaction rate constant* k_μ.

To demonstrate that c_μ as defined in Eq. (5.3-29) should always exist for reaction (5.3-28), and further that the value of c_μ should be completely determined by T, Ω and the microphysical properties of the reactant molecules X_1 and X_2, let us derive an explicit formula for c_μ when the molecules of species i are hard spheres of masses m_i and radii r_i. Our derivation begins by noting that the stipulation that the system be "in thermal equilibrium inside Ω at temperature T" means two things: (i) the *position* of a randomly selected X_i molecule will be randomly uniform inside the containing volume Ω, and (ii) the *velocity* of a randomly selected X_i molecule will be randomly distributed according to the Maxwell-Boltzmann density function,

$$P_i(\mathbf{v}_i) = \left(\frac{m_i}{2\pi k_B T} \right)^{3/2} \exp\left(- \frac{m_i v_i^2}{2 k_B T} \right), \qquad (5.3\text{-}30)$$

where k_B is Boltzmann's constant. It is not difficult to show that property (ii) implies that the speed of a randomly chosen X_2 molecule *relative* to a randomly chosen X_1 molecule will be randomly distributed according to the density function

$$P^*(v_{21}) = \left(\frac{m^*}{2\pi k_B T} \right)^{3/2} \exp\left(- \frac{m^* v_{21}^2}{2 k_B T} \right) 4\pi v_{21}^2, \qquad (5.3\text{-}31)$$

where $m^* \equiv m_1 m_2/(m_1 + m_2)$.[†]

Now let v_{21} be the speed of a randomly chosen X_2 molecule relative to a randomly chosen X_1 molecule. Then in the next infinitesimal time

---

[†] To prove Eq. (5.3-31), first observe that the joint density function of $\mathbf{v}_1$ and $\mathbf{v}_2$ is $P_1(\mathbf{v}_1) P_2(\mathbf{v}_2)$. Therefore, according to the RVT theorem (1.6-4), the density function of $v_{21} \equiv |\mathbf{v}_2 - \mathbf{v}_1|$ is

$$P^*(v_{21}) = \iiint d^3\mathbf{v}_1 \iiint d^3\mathbf{v}_2 \, P_1(\mathbf{v}_1) P_2(\mathbf{v}_2) \, \delta(v_{21} - |\mathbf{v}_2 - \mathbf{v}_1|).$$

Now change the integration variables here according to $(\mathbf{v}_1, \mathbf{v}_2) \to (\mathbf{V}, \mathbf{v})$, where

$$\mathbf{V} \equiv (m_1 \mathbf{v}_1 + m_2 \mathbf{v}_2)/(m_1 + m_2) \quad \text{and} \quad \mathbf{v} \equiv \mathbf{v}_2 - \mathbf{v}_1.$$

It is straightforward to show that the Jacobian of this transformation is unity, and also that

$$\tfrac{1}{2} m_1 v_1^2 + \tfrac{1}{2} m_2 v_2^2 = \tfrac{1}{2}(m_1 + m_2) V^2 + \tfrac{1}{2} m^* v^2.$$

The integrations over the Cartesian components of $\mathbf{V}$ and the polar and azimuthal angles of $\mathbf{v}$ are easily accomplished, and the final integration over the magnitude of $\mathbf{v}$ is trivial because of the Dirac delta function. The result of these six integrations is Eq. (5.3-31).

interval dt, the X_2 molecule will sweep out relative to the X_1 molecule a "collision cylinder" of height $v_{21}dt$ and base area $\pi(r_1+r_2)^2$, in the sense that if the center of the X_1 molecule happens to lie inside that cylinder at time t, then the two molecules will collide in time $[t,t+dt)$. Since the locations of the two molecules are uniformly random inside the container, then the *probability* that the center of the X_1 molecule will indeed lie inside that collision cylinder is just the ratio of the collision cylinder volume, $\pi(r_1+r_2)^2 \times v_{21}dt$, to the container volume, Ω. Therefore, the probability that a randomly chosen pair of reactant molecules will have a relative speed between v_{21} and $v_{21}+dv_{21}$, *and* will collide with each other in the next infinitesimal time interval $[t,t+dt)$, is

$$P^*(v_{21})dv_{21} \times \left(\frac{\pi(r_1+r_2)^2 v_{21}dt}{\Omega} \right).$$

Summing this probability over all possible relative speeds v_{21}, we conclude that the probability that a randomly chosen X_1-X_2 molecular pair will collide in the next dt is

$$\int_{v_{21}=0}^{\infty} P^*(v_{21})dv_{21} \times \left(\frac{\pi(r_1+r_2)^2 v_{21}dt}{\Omega} \right)$$

$$= \Omega^{-1} \pi(r_1+r_2)^2 \left\{ \int_0^\infty dv_{21}\, v_{21}\, P^*(v_{21}) \right\} dt$$

$$= \Omega^{-1} \pi(r_1+r_2)^2 \langle v_{21} \rangle dt,$$

where $\langle v_{21} \rangle$ is the mean of the density function P^* in Eq. (5.3-31).

But of course, not every X_1-X_2 molecular *collision* will result in an R_μ reaction. Let

$p_\mu \equiv$ probability that a *colliding* pair of R_μ reactant
       molecules will chemically react according to R_μ.   (5.3-32)

Then the probability on the right hand side of Eq. (5.3-29) can be obtained simply by multiplying the preceding collision probability by p_μ. Thus we have proved that the constant c_μ defined through Eq. (5.3-29) indeed *exists* in this case, and moreover is given explicitly by the formula

$$c_\mu = \Omega^{-1} \pi(r_1+r_2)^2 \langle v_{21} \rangle p_\mu. \qquad (5.3-33)$$

The average relative speed $\langle v_{21} \rangle$ is easily calculated from Eq. (5.3-31) to be $(8k_B T/\pi m^*)^{1/2}$, so it follows from our result (5.3-33) that c_μ depends on Ω, T and the properties r_1, r_2, m_1 and m_2 of the reactant molecules. An additional dependence upon the properties of the reactant molecules and the system temperature will usually enter through the probability p_μ. By way of illustration, suppose that a colliding pair of X_1 and X_2 molecules will react if and only if the kinetic energy associated with the component of their relative velocity along the line of centers at contact exceeds some "activation energy" ε_μ; in that case, a somewhat lengthy reworking of the foregoing derivation yields the result

$$c_\mu = \Omega^{-1} \pi (r_1 + r_2)^2 \left(\frac{8 k_B T}{\pi m^*} \right)^{1/2} \exp(-\varepsilon_\mu / k_B T). \qquad (5.3\text{-}34)$$

Comparing this formula with Eq. (5.3-33) reveals that in this particular case we have $p_\mu = \exp(-\varepsilon_\mu / k_B T)$. But we do not wish to concern ourselves here with the task of calculating p_μ for various reaction mechanism models. For our purposes it is enough to know that a bimolecular gas-phase chemical reaction R_μ always has a specific probability rate constant c_μ, which is defined according to Eq. (5.3-29), and which has a value that is determined by the system volume Ω, the system temperature T, and the microphysical properties of the reactant molecules.

Reactions of the form (5.3-28), with *two* reactant molecules, are by far the most commonly occurring kind in gas-phase chemical kinetics. But sometimes one encounters a *monomolecular* reaction of the form

$$X_1 \;\rightarrow\; X_2 + \dots. \qquad (5.3\text{-}35)$$

This reaction too can be described by a specific probability rate constant c_μ as defined in Eq. (5.3-29), but the functional form of c_μ for this reaction will look nothing like the bimolecular formula (5.3-33). This is because a monomolecular reaction by definition is not initiated by a collision of the reactant molecule with some other molecule. [If the reaction is actually of the form $C + X_1 \rightarrow C + X_2 + \dots$, with species C acting as a *catalyst*, then it should be treated as a bimolecular reaction and not a monomolecular reaction.] In particular, whereas a bimolecular c_μ is always inversely proportional to the system volume Ω [see Eq. (5.3-33)], a monomolecular reaction will generally be *independent* of Ω. A monomolecular reaction occurs as a result of some spontaneous change in the internal structure of a single reactant molecule, and the associated probability rate constant c_μ must usually be calculated from quantum mechanical considerations.

It is sometimes possible to consider *trimolecular* reactions of the form

$$X_1 + X_2 + X_3 \;\rightarrow\; X_4 + \dots, \qquad (5.3\text{-}36a)$$

but *only* in a special approximate sense. The problem with trimolecular reactions is that, when one straightforwardly calculates the probability that *three* randomly chosen hard-sphere molecules will come into *simultaneous* contact in the next infinitesimal time interval dt, one generally obtains a quantity that is proportional to $(dt)^2$ rather than dt. Such a dt-dependence is obviously not consistent with the prescription (5.3-29), and it rather suggests that trimolecular reactions do not occur with any physically sensible regularity. But it is sometimes both permissible and convenient to regard the trimolecular reaction (5.3-36a) as a *simplified approximation* to a multireaction sequence of the form

$$X_1 + X_2 \underset{c_2}{\overset{c_1}{\rightleftharpoons}} X^* \quad \text{and} \quad X^* + X_3 \overset{c_3}{\to} X_4 + \dots, \quad (5.3\text{-}36b)$$

wherein X^* is an "excited composite" molecule that has a very strong tendency to decay into its X_1-X_2 constituents. To see how this approximation comes about, let us calculate the probability that a randomly chosen X_1-X_2-X_3 molecular triplet will accomplish the transition described in (5.3-36a), via the multireaction sequence (5.3-36b), in the next *small but finite* time interval Δt. We reason as follows: The probability that a randomly selected X_1-X_2 molecular pair will form an X^* composite in the next Δt is, if Δt is sufficiently small, $c_1\Delta t$. The declared tendency for X^* to *quickly* decay back into an X_1-X_2 molecular pair will be reflected in a relatively *large* value of the specific probability rate constant c_2; indeed, since the probability for X^* to thus decay in the next infinitesimal time dt is $c_2 dt$, then $(1/c_2)$ is the *mean lifetime* of X^*, and that mean lifetime is *very small* since c_2 is very large. The probability that X^* will react with a randomly chosen X_3 molecule during its very brief average lifetime $(1/c_2)$ can therefore be plausibly estimated as $c_3(1/c_2)$. Finally, let us suppose that the mean lifetime $(1/c_2)$ of X^* is very small *even* compared to Δt. Then it follows from the multiplication law of probability that the probability for the first and third of reactions (5.3-36b) to *both* be accomplished within the time interval Δt is approximately

$$c_1\Delta t \times c_3(1/c_2) = (c_1 c_3/c_2)\,\Delta t. \quad (5.3\text{-}37)$$

This probability is evidently of the canonical form (5.3-29), with $c_\mu = c_1 c_3/c_2$, *provided* that we do not insist upon Δt being a true infinitesimal; in fact, the restriction $\Delta t \gg 1/c_2$ is the reason why the treatment of the multireaction sequence (5.3-36b) as the single reaction

(5.3-36a), with one specific probability rate constant c_μ, must always be regarded as an approximation.

We noted earlier that c_μ for a *monomolecular* reaction will always be independent of the system volume Ω, while c_μ for a *bimolecular* reaction will always be proportional to Ω^{-1}. To the extent that c_μ can be defined for a *trimolecular* reaction, we see from Eq. (5.3-37) that it will always be equal to the product of two bimolecular reaction constants divided by a monomolecular reaction constant; therefore, it will always be proportional to Ω^{-2}. So we have the general result that

$$R_\mu \text{ has } m \text{ reactant molecules} \implies c_\mu \propto \Omega^{-(m-1)}, \qquad (5.3\text{-}38)$$

where m can assume the values 1, 2 or 3.

The foregoing considerations should suffice to establish the physical and mathematical soundness of the dynamical description of gas-phase chemical reactions that is afforded by Eq. (5.3-29). Now we shall examine the consequences of that description in the context of a *specific* chemically reacting system. For that, we choose a system containing four chemical species, X, B_1, B_2 and B_3, which interact with each other through the following four reaction channels:

$$B_1 + X \underset{c_2}{\overset{c_1}{\rightleftharpoons}} X + X \quad \text{and} \quad X + X \underset{c_4}{\overset{c_3}{\rightleftharpoons}} B_2 + B_3. \qquad (5.3\text{-}39)$$

We shall assume that the values of the associated specific probability rate constants $c_1, \ldots, c_4$ are *all known*; furthermore, to keep our analysis within the purview of *univariate* processes, we shall suppose that species B_1, B_2 and B_3 are all "buffered," which means that their respective molecular populations N_1, N_2 and N_3 do not change appreciably with time. Our chief interest will be in the behavior of the variable

$$X(t) \equiv \text{number of X molecules inside the container at time } t. \qquad (5.3\text{-}40)$$

We shall begin by showing that $X(t)$ is in fact a well defined, temporally homogeneous, jump Markov process on the nonnegative integers.

It is obvious that the process X defined in Eq. (5.3-40) will always have nonnegative integer values. And the value of X will change only at those instants when one of the reactions (5.3-39) occurs somewhere inside the container; specifically, the occurrence of an individual reaction R_1, R_2, R_3 or R_4 will evidently cause the process to change by the respective amounts $+1$, -1, -2 or $+2$. If n is the (nonnegative) value of the process at time t, then in the next infinitesimal time interval $[t, t+dt)$ the process will increase to $n+1$ if and only if a reaction R_1 occurs somewhere inside the container.† Since the probability of that happening is by Eq. (5.3-29) $c_1 dt$ per B_1-X molecular pair, and since there are exactly $N_1 n$ such molecular pairs inside the container at time t, then by the addition law of probability we conclude that $(N_1 n)(c_1 dt)$ is the probability that X will increase from n to $n+1$ in the time interval $[t,t+dt)$. Writing this probability as $W(1 \mid n)dt$, in accordance with the fundamental definition (5.1-26), we conclude that $W(1 \mid n) = c_1 N_1 n$. The probability for an R_2 reaction to occur in $[t,t+dt)$ is similarly calculated as $c_2 dt$ times the number $n(n-1)/2$ of *distinct* X-X molecular pairs inside Ω at time t, and upon setting that probability equal to $W(-1 \mid n)dt$ we deduce that $W(-1 \mid n) = c_2 n(n-1)/2$. If we reason analogously for reaction R_3, which changes X by -2 through the reaction of a pair of X molecules, and for reaction R_4, which changes X by $+2$ through the reaction of a B_1-B_2 molecular pair, we conclude that

$$
W(v \mid n) = \begin{cases}
c_1 N_1 n & \text{, if } v = +1 \text{ and } n \geq 0, \\
(c_2/2)\, n(n-1) & \text{, if } v = -1 \text{ and } n \geq 0, \\
(c_3/2)\, n(n-1) & \text{, if } v = -2 \text{ and } n \geq 0, \\
c_4 N_2 N_3 & \text{, if } v = +2 \text{ and } n \geq 0, \\
0 & \text{, for all other } v \text{ and } n.
\end{cases}
\tag{5.3-41}
$$

Having thus established for $X(t)$ the existence of a consolidated characterizing function $W(v \mid n)$, which we note does *not* allow jumps to negative states, we conclude that $X(t)$ is indeed a temporally homogeneous jump Markov process on the nonnegative integers.

We now have only to incorporate the above formula for $W(v \mid n)$ into the equations of Subsection 5.3.A to deduce the temporal behavior of $X(t)$. But that is easier said than done. Although the forward master equation (5.3-25a) has a finite number of terms on its right hand side,

$$
\frac{\partial}{\partial t} P(n,t \mid n_0,t_0) = \sum_{v=-2}^{\text{Min}(2,n)} \left[W(v \mid n-v)\, P(n-v,t \mid n_0,t_0) \right.
$$
$$
\left. - W(-v \mid n)\, P(n,t \mid n_0,t_0) \right],
\tag{5.3-42}
$$

---

† An increase from n to $n+1$ in $[t,t+dt)$ could also be effected by the occurrence of one R_2 reaction *and* one R_4 reaction in $[t,t+dt)$; however, the probability for that to happen is proportional to $(dt)^2$, and is therefore negligibly small compared to the probability for the occurrence of a single R_1 reaction, which is proportional to dt. We can generally ignore the possibility that *more than one* reaction will occur in any *infinitesimal* time interval.

it is not at all clear how to go about obtaining an analytical solution to that equation. Nor is a numerical solution likely to come easily, because that would seem to require that we allocate one computer memory location to store the current value of $P(n,t \mid n_0,t_0)$ for each and every nonnegative value of n. Perhaps we can instead do something with the time-evolution equations for the mean and variance of $X(t)$. We have from Eqs. (5.3-14) and (5.3-15) that

$$\frac{d}{dt}\langle X(t)\rangle = \langle W_1(X(t))\rangle, \tag{5.3-43}$$

$$\frac{d}{dt}\text{var}\{X(t)\} = 2\Big(\langle X(t)\,W_1(X(t))\rangle - \langle X(t)\rangle\langle W_1(X(t))\rangle\Big) + \langle W_2(X(t))\rangle, \tag{5.3-44}$$

where, by Eqs. (5.3-10) and (5.3-41), W_1 and W_2 are given by

$$W_k(n) = (1)^k c_1 N_1\, n + (-1)^k (c_2/2)\, n(n-1)$$
$$+ (-2)^k (c_3/2)\, n(n-1) + (2)^k c_4 N_2 N_3. \tag{5.3-45}$$

But because $W_1(n)$ is a polynomial in n of degree > 1, none of the moment evolution equations will be closed (we are assuming here that c_2 and c_3 are not zero). For example, putting $k=1$ in Eq. (5.3-45) and then substituting the result into Eq. (5.3-43), we obtain

$$\frac{d}{dt}\langle X(t)\rangle = c_1 N_1 \langle X(t)\rangle - (c_2/2)\langle X(t)[X(t)-1]\rangle$$
$$- c_3\langle X(t)[X(t)-1]\rangle + 2c_4 N_2 N_3. \tag{5.3-46}$$

Since this equation for the time derivative of $\langle X(t)\rangle$ involves not only $\langle X(t)\rangle$ but also $\langle X^2(t)\rangle$, then it is not solvable without making some sort of approximation.

The most reliable way to obtain accurate numerical approximations to $\langle X(t)\rangle$ and $\text{var}\{X(t)\}$ is to carry out the procedure described in Appendix C, replacing $B_k(x)$ there by $W_k(n)$. What is usually done, though, is something that is very much easier, and very much cruder: The usual ploy is to make the *two-stage approximation*

$$\langle X(t)[X(t)-1]\rangle \doteq \langle X^2(t)\rangle \doteq \langle X(t)\rangle^2. \tag{5.3-47}$$

The first of these approximations would appear to be fairly harmless in those common situations where $X(t)$ is practically always very large compared to 1. But the second approximation seems to be rather drastic, at least so far as estimating $\text{var}\{X(t)\}$ is concerned; because it evidently amounts to asserting that $\text{var}\{X(t)\} \doteq 0$, and hence that $X(t)$ is a *deterministic* process; i.e., $X(t) \doteq \langle X(t) \rangle$. Indeed, as a result of the approximation (5.3-47), the exact-but-open evolution equation (5.3-46) for $\langle X(t) \rangle$ becomes the following *approximate-but-closed* evolution equation for the *sure* function $X(t)$:

$$\frac{d}{dt} X(t) \doteq c_1 N_1 X(t) - (c_2/2) X^2(t) - c_3 X^2(t) + 2c_4 N_2 N_3. \quad (5.3\text{-}48)$$

The ordinary differential equation (5.3-48) is called the *reaction-rate equation* for $X(t)$. The four terms on the right hand side of this equation can evidently be regarded as the contributions to the time-rate of change of $X(t)$ from the individual reaction channels (5.3-39). Recognizing that channels R_1 and R_2 each change $X(t)$ by *one* molecule per reaction, while channels R_3 and R_4 each change $X(t)$ by *two* molecules per reaction, then it follows that in this deterministic approximation we may say that

$$\left.\begin{array}{l} c_1 N_1 X(t)\, dt \doteq \text{number of } R_1 \text{ reactions occurring in } [t, t+dt) \\[4pt] (c_2/2) X^2(t)\, dt \doteq \text{number of } R_2 \text{ reactions occurring in } [t, t+dt) \\[4pt] (c_3/2) X^2(t)\, dt \doteq \text{number of } R_3 \text{ reactions occurring in } [t, t+dt) \\[4pt] c_4 N_2 N_3\, dt \doteq \text{number of } R_4 \text{ reactions occurring in } [t, t+dt) \end{array}\right\} \quad (5.3\text{-}49)$$

In truth of course, $c_1 N_1 X(t) dt$ is really the *probability* that an R_1 reaction will occur in $[t, t+dt)$, and similarly for the other three expressions.

In practice, chemists usually prefer to write the reaction rate equation (5.3-48) in terms of the species *concentrations*,

$$x(t) \equiv X(t)/\Omega, \quad b_i \equiv N_i/\Omega \quad (i=1,2,3). \quad (5.3\text{-}50)$$

By expressing $X(t)$ and N_i in Eq. (5.3-48) in terms of $x(t)$ and b_i, we can easily bring that equation into the form

$$\frac{d}{dt} x(t) \doteq k_1 b_1 x(t) - k_2 x^2(t) - 2k_3 x^2(t) + 2k_4 b_2 b_3, \quad (5.3\text{-}51)$$

where we have defined the *reaction rate constants* k_μ as follows:

$$k_1 \equiv c_1 \Omega, \quad k_2 \equiv c_2 \Omega/2, \quad k_3 \equiv c_3 \Omega/2, \quad k_4 \equiv c_4 \Omega. \quad (5.3\text{-}52)$$

Equation (5.3-51) is the time-evolution equation that any chemist would immediately write down "by inspection" for reactions (5.3-39). Since the specific probability rate constants $c_1, \dots, c_4$, being bimolecular, are all inversely proportional to the system volume Ω [see Eq. (5.3-33)], then the reaction rate constants $k_1,...,k_4$ defined in Eqs. (5.3-52) will all be *independent* of Ω; in fact, the property (5.3-38) ensures that the reaction rate constant k_μ will be independent of Ω for virtually *any* type of chemical reaction. It follows that Eq. (5.3-51) is independent of the volume of the system. Although Eq. (5.3-51) is the more commonly used form of the reaction rate equation, it will be more instructive for our purposes here to focus on the equivalent form (5.3-48).

The reaction rate equation (5.3-48) can be straightforwardly solved by separation of variables. The solution $X(t)$, which we shall not write out here but will show plots of later, turns out to have the following property: For any initial condition $X(0) \geq 0$,

$$X(t \to \infty) = X_e, \tag{5.3-53}$$

where the "equilibrium" population X_e is the solution of the algebraic equation obtained by replacing $X(t)$ in Eq. (5.3-48) by the constant X_e:

$$0 = c_1 N_1 X_e - (c_2/2) X_e^2 - c_3 X_e^2 + 2c_4 N_2 N_3. \tag{5.3-54}$$

Solving this quadratic equation for X_e is straightforward, and yields

$$X_e = \frac{c_1 N_1 + [(c_1 N_1)^2 + 8(c_2/2 + c_3)c_4 N_2 N_3]^{1/2}}{2(c_2/2 + c_3)}. \tag{5.3-55}$$

Quite often, the equilibrium population X_e is all that one needs for practical purposes. By definition, X_e makes the right hand side of Eq. (5.3-48) vanish. But suppose it happens that the combined contribution to the right hand side of that equation from each forward-backward pair of reactions (5.3-39) *separately* vanishes; i.e., suppose it happens that

$$c_1 N_1 X_e = (c_2/2) X_e^2 \quad \text{and} \quad c_3 X_e^2 = 2c_4 N_2 N_3.$$

Then it follows from Eqs. (5.3-49) that, when $X(t)$ has the equilibrium value X_e, the rate at which R_1 reactions occur will equal the rate at which R_2 reactions occur, and the rate at which R_3 reactions occur will equal the rate at which R_4 reactions occur. This state of affairs is referred to as *detailed balancing*, and a considerations of reactions (5.3-39) shows that in a detailed balance situation there will be no tendency for the B_1, B_2 and B_3 population levels to change. But it is important to understand that detailed balancing is a very special situation: By eliminating X_e between the above two equations, we obtain the relation

$$\frac{N_1^2}{N_2 N_3} = \frac{c_2^2 c_4}{2 c_1^2 c_3} \quad \text{(detailed balancing)}, \quad (5.3\text{-}56)$$

and this represents a *condition* that the system parameters must satisfy in order for detailed balancing to be realized. But if, for example, $N_1^2/N_2 N_3$ happened to be *larger* than the right hand side of Eq. (5.3-56), then when $X(t) = X_e$ reaction R_1 would be occurring more frequently than reaction R_2, and reaction R_3 would be occurring more frequently than reaction R_4; the X-molecule population would *still* remain constant at the value X_e, but in order to keep N_1, N_2 and N_3 constant we would have to continuously *add* B_1 molecules to the system and *remove* B_2 and/or B_3 molecules from the system. Notice from the formula (5.3-55) that if the values of N_1, N_2 and N_3 are all simultaneously changed by a common factor f, then X_e will be changed by the same factor f. This means that it is always possible to achieve a detailed balance solution X_e of *any* value by suitably picking the populations of the B_i species: We first pick any three values N_1, N_2 and N_3 that satisfy the detailed balance condition (5.3-56), and we then scale those three values by a common factor f in order to bring X_e to the desired value.

We have now seen how the traditional reaction rate equation (5.3-48), or (5.3-51), arises as an *approximation* to a more rigorous Markov process description. The crucial approximation made in deriving the reaction rate equation is the *neglect of the fluctuations in* $X(t)$ [see Eq. (5.3-47)]. To determine the conditions under which that approximation is really justified, we must return to our earlier description of $X(t)$ as a jump Markov process with the consolidated characterizing function $W(v \mid n)$ in Eq. (5.3-41), and try to directly calculate the effects of the fluctuations. Probably the easiest way to do that is through a series of Monte Carlo simulations, since they can be carried out *exactly* in spite of the intractability of both the master equation (5.3-42) and the mean and variance evolution equations (5.3-43) and (5.3-44). We shall examine the results of some Monte Carlo simulations in a moment. But first we want to show how the mean and variance evolution equations (5.3-43) and (5.3-44) can be used to estimate the behavior of $X(t)$ in the limit $t \to \infty$. We shall have to make an approximation in order to do this, but that approximation will be much less heavy-handed than the approximation (5.3-47) made in deriving the reaction rate equation.

We begin by making the assumption (which is undoubtedly true although we do not formally prove it here) that, as $t \to \infty$, $X(t)$ approaches a *time-independent* random variable which we shall call X_∞. The mean and variance time-evolution equations (5.3-43) and (5.3-44) therefore become in the limit $t\to\infty$,

$$0 = \langle W_1(X_\infty)\rangle, \tag{5.3-57a}$$

$$0 = 2\langle X_\infty W_1(X_\infty)\rangle + \langle W_2(X_\infty)\rangle, \tag{5.3-57b}$$

where the second equation has been slightly simplified by invoking the first equation. Now, because $W_1(n)$ and $W_2(n)$ in this instance are both quadratic in n [see Eq. (5.3-45)], the first of the above two equations will contain the moments $\langle X_\infty\rangle$ and $\langle X_\infty{}^2\rangle$, while the second equation will contain those two moments plus the moment $\langle X_\infty{}^3\rangle$; so we have two equations in three unknowns, posing the usual dilemma of "open" equations. But now we make an *approximating assumption*: We assume that, so far as Eqs. (5.3-57) are concerned, we can regard X_∞ as a *normal* random variable,

$$X_\infty \doteq N(m,\sigma^2), \tag{5.3-58}$$

whose mean m and variance σ^2 remain to be determined. This assumption allows us to use the normal moment formula (1.4-10) to write

$$\langle X_\infty\rangle \doteq m, \quad \langle X_\infty{}^2\rangle \doteq m^2 + \sigma^2, \quad \langle X_\infty{}^3\rangle \doteq m^3 + 3m\sigma^2. \tag{5.3-59}$$

If these expressions for the first three moments of X_∞ are used in Eqs. (5.3-57), then those two equations will have only the *two* unknowns m and σ^2, and we will then be in a position to solve those equations simultaneously for m and σ^2. We may expect m to be a more reliable estimate of $\langle X(t\to\infty)\rangle$ than X_e, because in computing X_e we ignored fluctuations altogether. And we may also expect that σ will provide us with a reasonably good estimate of $sdev\{X(t\to\infty)\}$, and that in turn will tell us how much error is made in treating $X(t\to\infty)$ as a sure variable instead of a random variable.

By calculating $W_1(n)$ and $W_2(n)$ explicitly from Eq. (5.3-45), substituting the results into Eqs. (5.3-57), and then expressing the first three moments of X_∞ through the approximating formulas (5.3-59), we finally obtain the following two simultaneous equations in the two unknowns m and σ^2:

$$0 = -(c_2/2+c_3)(m^2+\sigma^2) + (c_1N_1+c_2/2+c_3)m + 2c_4N_2N_3, \tag{5.3-60a}$$

$$0 = -(c_2+2c_3)(m^3+3m\sigma^2) + (2c_1N_1+3c_2/2+4c_3)(m^2+\sigma^2)$$
$$+ (c_1N_1-c_2/2+2c_3+4c_4N_2N_3)m + 4c_4N_2N_3. \tag{5.3-60b}$$

Equation (5.3-60a) is easily solved for σ^2 in terms of m. When that expression for σ^2 is substituted into Eq. (5.3-60b), the result is a *cubic* equation in m. Corresponding to the three roots m_1, m_2 and m_3 of that cubic equation, there will three *solution pairs* (m_1,σ_1), (m_2,σ_2) and (m_3,σ_3) to the simultaneous equations (5.3-60). But *only one* of those three solution pairs (m_i,σ_i) will be such that $[m_i-\sigma_i, m_i+\sigma_i]$ is an interval *wholly* contained on the *positive real* axis, and for obvious physical reasons *that* pair must be regarded as our solution for m and σ.

It is instructive to carry out some numerical calculations of m and σ for a few specific cases. Let us suppose that the specific probability rate constants for reactions (5.3-39) have the values

$$c_1 = 0.025, \quad c_2 = 20.0, \quad c_3 = 80.0, \quad c_4 = 0.00025. \qquad (5.3\text{-}61)$$

For these values, the *right* hand side of the detailed balancing condition (5.3-56) has the value 1. The table below shows results for various values of N_1, N_2 and N_3. In half of the cases, namely those with $N_1 = 1 \times 10^n$, $N_1{}^2/N_2N_3$ has the value 1, so that detailed balancing obtains. In the other cases, those with $N_1 = 4 \times 10^n$, $N_1{}^2/N_2N_3$ has the value 4, and detailed balancing does not obtain; in those cases B_1 molecules must be continually added, and B_2 and B_3 molecules continually removed, in order to keep N_1, N_2 and N_3 constant at equilibrium. For each combination of N_1, N_2 and N_3 values, the table shows the value of X_e, the *deterministic* X-molecule asymptotic population as calculated from formula (5.3-55), and also the values of m and σ, the mean and standard deviation of the *stochastic* X-molecule asymptotic population as calculated from equations (5.3-60).

Some Numerical Results for Reactions (5.3-39)

N_1	$N_2 = N_3$	X_e	m	σ	$\sigma/\sqrt{m}$
1×10^4	1×10^4	25.00	24.97	5.151	1.031
4×10^4	1×10^4	29.77	29.68	5.836	1.071
1×10^6	1×10^6	25.00×10^2	25.00×10^2	50.02	1.000
4×10^6	1×10^6	29.77×10^2	29.77×10^2	56.74	1.040
1×10^8	1×10^8	25.00×10^4	25.00×10^4	500.0	1.000
4×10^8	1×10^8	29.77×10^4	29.77×10^4	567.2	1.040

Two features of these numerical results should be noted. First, X_e constitutes a remarkably good approximation to m. This unexpectedly good agreement between X_e and m suggests that the two approximations (5.3-47) that were made in deriving the formula for X_e fortuitously tend to cancel each other; more precisely, in the double approximation (5.3-47), the quantity in the middle overestimates the two quantities on the left and right by roughly the same amount, with the consequence that $\langle X_\infty \rangle^2$ is actually a better approximation to $\langle X_\infty(X_\infty - 1) \rangle$ than is $\langle X_\infty^2 \rangle$. The second noteworthy feature of the above numerical results is the fact, made clear by the figures in the last column, that the standard deviation σ of X_∞ is always very close to the square root of its mean m. This has the consequence that the *relative* standard deviation of X_∞, namely σ/m, is approximately equal to $1/\sqrt{m}$, and hence will be very small whenever the average number of X molecules in the system is very large.

These results for the asymptotic behavior of the simple chemical system (5.3-39) are rather typical of what happens more generally for almost any chemical process $X(t)$: The traditional reaction rate equation, obtained by essentially ignoring all fluctuations in $X(t)$, has a solution that approximates $\langle X(t) \rangle$ rather well, and the fluctuations in $X(t)$ about $\langle X(t) \rangle$ are roughly on the order of $\langle X(t) \rangle^{1/2}$. Since a typical *macroscopic* chemical system will have $\langle X(t) \rangle \sim 10^{18}$, then the *relative* fluctuations in $X(t)$ will usually be exceedingly small,

$$\frac{\text{sdev}\{X(t)\}}{\langle X(t) \rangle} \sim \frac{\langle X(t) \rangle^{1/2}}{\langle X(t) \rangle} \sim \frac{1}{\langle X(t) \rangle^{1/2}} \sim 10^{-9}.$$

In that case, the solution of the deterministic reaction rate equation will by itself constitute an adequate representation of the process $X(t)$ for most practical purposes. But it is important to recognize that there are exceptions to these rules: For many *microscopic* systems, and also for some *macroscopic* systems that are *sufficiently nonlinear*, the intrinsic fluctuations in $X(t)$ can lead to behavior that cannot be adequately described by the reaction rate equation solution together with the "square-root fluctuation" rule. One example of such a chemical system will be described in Subsection 6.8.C.

Despite the analytical difficulties associated with solving the forward master equation (5.3-42) and the moment evolution equations (5.3-43) and (5.3-44), it is easy to construct *exact numerical realizations* of $X(t)$ by applying the Monte Carlo simulation algorithm of Fig. 5-1. Remembering that the characterizing functions $a(n)$ and $w(v \mid n)$ are related to the consolidated characterizing function $W(v \mid n)$ in Eq. (5.3-41) by

$$a(n) = \sum_{v=-2}^{2} W(v \mid n) \quad \text{and} \quad w(v \mid n) = \frac{W(v \mid n)}{a(n)},$$

then we see from Fig. 5-1 that a simulation of this chemical process $X(t)$ can be accomplished as follows:

1° *Initialize*: Put $t \leftarrow t_0$ and $n \leftarrow n_0$.

2° *Determine the time to the next reaction*: With $W(v \mid n)$ as given by Eq. (5.3-41), compute $a(n)$ as the sum

$$a(n) = W(-2 \mid n) + W(-1 \mid n) + W(1 \mid n) + W(2 \mid n);$$

then draw a unit uniform random number r and take

$$\tau = [1/a(n)] \ln (1/r).$$

3° *Determine the state change in the next reaction*: Draw another unit uniform random number r', and take v to be the smallest nonzero integer ≥ -2 for which

$$\sum_{v'=-2}^{v} W(v' \mid n) > a(n) r'.$$

4° *Actualize the next reaction*: Put $t \leftarrow t+\tau$ and $n \leftarrow n+v$.

5° *Record and loop*: Record $X(t)=n$ as required. If the process is to continue then return to Step 2°; otherwise, stop.

It should be emphasized that the above simulation procedure is *exact*; in particular, it is *not* premised on the approximation (5.3-47) that underlies the traditional reaction rate equation.

We show in Fig. 5-4 some simulation results for the chemical system (5.3-39), using the c_μ values (5.3-61). Figure 5-4a shows two simulations using the buffered molecule populations $N_1 = N_2 = N_3 = 10^4$, with one simulation starting at $n_0 = 0$ and the other starting at $n_0 = 70$. The two dashed curves in the figure are the corresponding solutions of the deterministic reaction rate equation (5.3-48). The dotted horizontal lines show the asymptotic one-standard deviation envelopes $[m-\sigma, m+\sigma]$ predicted by Eqs. (5.3-60). Figure 5-4b plots the same quantities, but now calculated using the buffered molecule populations $N_1 = N_2 = N_3 = 10^6$ and the respective initial conditions $n_0 = 0$ and $n_0 = 7000$. The two simulation trajectories in Fig. 5-4b were plotted by simply recording the state with a dot every fifth reaction; the up-going trajectory there

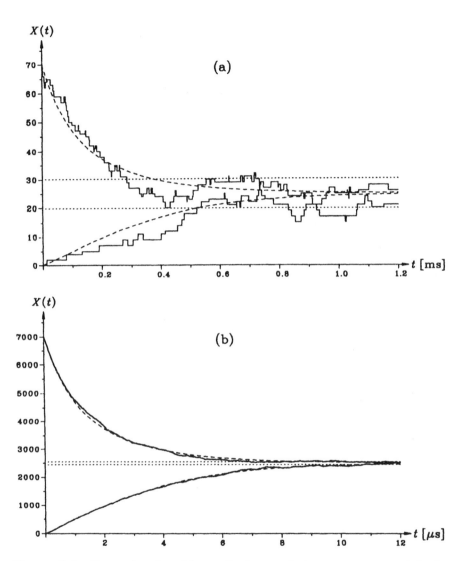

Figure 5-4. Comparing four Monte Carlo simulation trajectories with the reaction rate equation trajectories (dashed curves) for the X-molecule population in reaction set (5.3-39). In all cases the reaction constants c_μ were assigned the values (5.3-61), but the population of each buffered species B_i was kept at 10^4 for the two runs in (a) and at 10^6 for the two runs in (b). Each of the simulation trajectories in (b) consists of dots plotted every fifth reaction. The dotted horizontal lines show the asymptotic one-standard deviation envelopes that follow from assuming a time-independent *normal* density function for $X(\infty)$.

contains nearly six thousand reactions, and the down-going trajectory contains over ten thousand reactions. Notice that the time scale in Fig. 5-4a is measured in milliseconds, while the time scale in Fig. 5-4b is measured in microseconds.

A comparison of the plots in Figs. 5-4a and 5-4b shows that the relative difference between the deterministic and stochastic trajectories decreases as the number of molecules in the system increases, just as our analysis predicted. But it is important to understand that whenever discrepancies *do* occur, they should *not* be interpreted as "errors" in the Monte Carlo trajectories; rather, they must be taken as an indication that the deterministic reaction rate equation is simply not capable of describing the true behavior of the intrinsically stochastic system under consideration.

The major strength of the Monte Carlo simulation approach to chemical kinetics is also its major weakness: It not only *allows* us to simulate in a statistically correct way each and every chemical reaction event that occurs in the system, but it *demands* that we do so. For *macroscopic* systems, those reactions will usually be so closely spaced in time that a prohibitively large amount of computer effort will be required to simulate the system over time intervals of practical interest. But fortunately, for those macroscopic systems the fluctuation effects will usually be so small that the deterministic reaction rate equation can be safely applied. The two computational methods are therefore seen to complement each other.

Most chemical systems of practical interest contain *more than one* time-varying species. The consequent multivariate versions of the forward master equation and moment evolution equations are very much more unwieldy than their univariate versions. But the multivariate version of the Monte Carlo simulation algorithm remains remarkably simple and easy to implement.† In particular, whereas the corresponding set of coupled reaction rate equations must usually be solved numerically using a discrete time-step method, the intrinsically more rigorous Monte Carlo simulation algorithm never requires us to approximate an infinitesimal time-step dt by a finite time step Δt.

Finally, we should emphasize that the foregoing analysis of chemically reacting systems is valid *only* if the system is kept *well-stirred*, so that the key assumptions of randomly uniform molecular

---

† For a description of the Monte Carlo simulation algorithm for chemically reacting systems with many time-varying species, see the author's article "Exact stochastic simulation of coupled chemical reactions" in the *Journal of Physical Chemistry*, vol. 81 (1977), pp. 2340–61.

position distributions, and constant temperature molecular velocity distributions, are always valid. Our analysis does *not* apply, for example, to chemical explosions, in which reaction events occur so rapidly that pronounced spatial nonuniformities can develop. The "ideal" well-stirred system would be one in which *nonreactive* molecular collisions occur much more frequently than *reactive* molecular collisions, since that would ensure an effective thermal equilibrium with respect to the reactive collisions. Such a "self-stirring" system can in principle be realized for any set of chemical reactions by simply diluting the mixture with molecules of some nonreacting species.

- 6 -

TEMPORALLY HOMOGENEOUS BIRTH-DEATH MARKOV PROCESSES

In this final chapter we shall continue our examination, begun in Section 5.3, of temporally homogeneous jump Markov processes on the nonnegative integers. But now we shall focus on the *subclass* of those processes for which only jumps of size $+1$ or -1 are allowed. Such processes are called temporally homogeneous "one-step" or "birth-death" Markov processes. The latter appellation comes from the frequent use of these processes for modeling animal populations, where a jump from state n to state $n+1$ signals the birth of an animal, while a jump from state n to state $n-1$ signals the death of an animal. As a class, temporally homogeneous birth-death Markov processes are simple enough to be intuitively transparent and analytically tractable, yet substantive enough to provide serviceable mathematical models for many real world dynamical systems.

In Section 6.1 we shall adapt the basic results of Chapter 5 to this particular kind of discrete state jump Markov process. In Section 6.2 we shall show that a birth-death Markov process can, in certain special circumstances, be approximated by a continuous Markov process. In Section 6.3 we shall examine four simple examples of temporally homogeneous birth-death Markov processes. In the remaining sections of the chapter we shall develop a number of results that primarily concern "stable" birth-death Markov processes — results that are strikingly similar to those obtained in Chapter 3 for continuous Markov processes. In the course of our work in this chapter, we shall see that the theory of temporally homogeneous birth-death Markov processes finds applications in areas as diverse as molecular physics and business administration. We shall close the chapter with a detailed analysis of the fluctuations about and transitions between the stable states of a model bistable chemical system.

6.1 FOUNDATIONAL ELEMENTS

6.1.A THE DEFINING FUNCTIONS

Formally, a temporally homogeneous birth-death Markov process $X(t)$ can be defined as any discrete state jump Markov process whose characterizing functions $a(n,t)$ and $w(v \mid n,t)$ have the respective forms

$$a(n,t) = a(n) \quad (n \geq 0), \tag{6.1-1}$$

and

$$w(v \mid n,t) = \begin{cases} w_+(n), & \text{if } v = +1 \text{ and } n \geq 0, \\ w_-(n), & \text{if } v = -1 \text{ and } n \geq 0, \\ 0, & \text{otherwise.} \end{cases} \tag{6.1-2}$$

Here, $a(n)$ can be any nonnegative function, while $w_+(n)$ and $w_-(n)$ can be any functions that range between 0 and 1 inclusively and satisfy the normalization condition

$$w_+(n) + w_-(n) = 1 \quad (n \geq 0). \tag{6.1-3}$$

It follows from the definitions of the characterizing functions [see Eqs. (5.1-22) and (5.1-18)] that the three functions $a(n)$, $w_+(n)$ and $w_-(n)$ appearing in the above equations have the following physical interpretations:

$$a(n)dt \equiv \text{probability that the process, in state } n$$
$$\text{at time } t, \text{ will jump away in the next}$$
$$\text{infinitesimal time interval } [t,t+dt), \tag{6.1-4}$$

and

$$w_\pm(n) \equiv \text{probability that the process, upon}$$
$$\text{jumping away from state } n \text{ at time } t,$$
$$\text{will land in state } n \pm 1. \tag{6.1-5}$$

Confinement of the process to the nonnegative integers is ensured by requiring that the initial value $X(t_0) \equiv n_0$ be nonnegative, and that

$$w_-(0) = 0 \quad \text{and} \quad w_+(0) = 1. \tag{6.1-6}$$

The propagator density function $\Pi(v \mid dt; n,t)$, being generally defined for a discrete state jump Markov process as

$$\Pi(v \mid dt; n,t) \equiv \text{Prob}\{X(t+dt) = n+v, \text{ given } X(t) = n\},$$

is of course explicitly independent of t in this temporally homogeneous case. By inserting the above formulas for $a(n,t)$ and $w(v \mid n,t)$ into the general discrete state formula (5.1-23), we find that the propagator density function for a temporally homogeneous birth-death Markov process is given by

$$\Pi(v \,|\, dt; n) = \begin{cases} a(n)dt\, w_\pm(n), & \text{if } v = \pm 1 \text{ and } n \geq 0, \\ 1 - a(n)\, dt, & \text{if } v = 0 \text{ and } n \geq 0, \\ 0, & \text{otherwise.} \end{cases} \qquad (6.1\text{-}7)$$

The consolidated characterizing function $W(v \,|\, n,t)$, being by definition the product of $a(n,t)$ and $w(v \,|\, n,t)$, will evidently have the form

$$W(v \,|\, n,t) \equiv W(v \,|\, n) = \begin{cases} W_+(n), & \text{if } v = +1 \text{ and } n \geq 0, \\ W_-(n), & \text{if } v = -1 \text{ and } n \geq 0, \\ 0, & \text{otherwise,} \end{cases} \qquad (6.1\text{-}8)$$

where we have defined

$$W_\pm(n) \equiv a(n)\, w_\pm(n). \qquad (6.1\text{-}9)$$

The functions $W_+(n)$ and $W_-(n)$ will be called the **stepping functions** of the process. It follows from Eqs. (6.1-4) and (6.1-5) that the stepping functions have the following physical interpretation:

$W_\pm(n)dt$ = probability that the process, in state n at time t, will jump to state $n \pm 1$ in the next infinitesimal time interval $[t, t + dt)$.   (6.1-10)

Comparing Eqs. (6.1-7), (6.1-9) and (6.1-10) reveals that

$$\Pi(\pm 1 \,|\, dt; n) = W_\pm(n)dt \qquad (n \geq 0). \qquad (6.1\text{-}11)$$

And comparing the definition (6.1-9) with Eqs. (6.1-6) shows that

$$W_-(0) = 0 \quad \text{and} \quad W_+(0) = a(0). \qquad (6.1\text{-}12)$$

From Eqs. (6.1-9) and (6.1-3), we see that the *sum* of the stepping functions is just $a(n)$:

$$a(n) = W_+(n) + W_-(n). \qquad (6.1\text{-}13a)$$

It will later prove convenient to denote the *difference* between the stepping functions by $v(n)$:

$$v(n) \equiv W_+(n) - W_-(n). \qquad (6.1\text{-}13b)$$

Since the characterizing functions of a temporally homogeneous birth-death Markov process are completely determined by the three functions $a(n)$, $w_+(n)$ and $w_-(n)$, and since if either $w_+(n)$ or $w_-(n)$ is specified then the other will be completely determined by the normalization condition (6.1-3), then it is clear that a temporally homogeneous birth-death Markov process $X(t)$ is completely defined by *two functions* of the integer variable n. Indeed, we may select as the "defining functions" of $X(t)$ any one of the following function pairs:

$$[a(n), w_{\pm}(n)], \quad [W_+(n), W_-(n)], \quad [a(n), v(n)].$$

But it is important to recognize that the functions in these defining pairs are *not* entirely arbitrary. To be precise, the function $a(n)$ must satisfy

$$a(n) \geq 0; \tag{6.1-14a}$$

the functions $w_+(n)$ and $w_-(n)$ must satisfy

$$w_{\pm}(n) \geq 0 \quad \text{and} \quad w_+(n) + w_-(n) = 1; \tag{6.1-14b}$$

the functions $W_+(n)$ and $W_-(n)$ must, as a consequence of the preceding two inequalities, satisfy

$$W_{\pm}(n) \geq 0; \tag{6.1-14c}$$

and finally, the function $v(n)$ must satisfy

$$-a(n) \leq v(n) \leq a(n). \tag{6.1-14d}$$

The reason for this last inequality can be seen by writing the right hand side of the definition (6.1-13b) as $a(n)[w_+(n) - w_-(n)]$, and then observing that $[w_+(n) - w_-(n)]$ is confined to the interval $[-1, 1]$ on account of conditions (6.1-14b).

In the course of our analysis of the temporally homogeneous birth-death Markov process $X(t)$ described above, we shall discover that this process has a great deal in common with the temporally homogeneous *continuous* Markov process with characterizing functions $A(x) = v(x)$ and $D(x) = a(x)$. Specifically, we shall discover that the formulas for many dynamical properties of these two Markov processes are nearly identical, and that in certain special circumstances the two processes themselves can be regarded as being practically the same.

6.1.B THE NEXT-JUMP DENSITY FUNCTION AND ITS SIMULATION ALGORITHM

The next-jump density function for any temporally homogeneous discrete state Markov process is given by Eq. (5.1-41). When the birth-death characterizing functions (6.1-1) and (6.1-2) are substituted into that equation, it becomes

$$p(\tau,v \mid n,t) = \begin{cases} a(n)\exp(-a(n)\tau)\, w_+(n), & \text{if } \tau \geq 0, \ v = +1 \ \text{and} \ n \geq 0, \\ a(n)\exp(-a(n)\tau)\, w_-(n), & \text{if } \tau \geq 0, \ v = -1 \ \text{and} \ n \geq 0, \\ 0, & \text{otherwise.} \end{cases}$$

$$(6.1\text{-}15)$$

A direct derivation of this result is not difficult, and perhaps is worth reviewing here: Letting $p_0(\tau)$ denote the probability that the process, in state n at time t, will *not* leave that state in the time interval $[t, t+\tau)$, then the laws of probability and Eq. (6.1-4) allow us to write

$$p_0(\tau+d\tau) = p_0(\tau) \times [1 - a(n)d\tau].$$

This relation implies the differential equation $dp_0(\tau)/d\tau = -a(n)p_0(\tau)$ and the solution to that differential equation for the required initial condition $p_0(0) = 1$ is

$$p_0(\tau) = \exp(-a(n)\tau).$$

Combining this result with Eqs. (6.1-4) and (6.1-5) using the multiplication law of probability, we conclude that the probability that the process, in state n at time t, will remain in that state until time $t+\tau$, then jump away in the time interval $[t+\tau, t+\tau+d\tau)$, and finally land in state $n \pm 1$, is

$$\exp(-a(n)\tau) \times a(n)d\tau \times w_\pm(n) \equiv p(\tau, \pm 1 \mid n,t)\, d\tau.$$

This expression, when coupled with the fact that only jumps with $v = \pm 1$ are possible for a birth-death process, establishes the formula (6.1-15).

If we sum Eq. (6.1-15) over v using Eq. (6.1-3), we find that the marginal density function for τ is the density function of the exponential random variable with decay constant $a(n)$. Thus, the pausing time in state n is the random variable $\mathbf{E}(a(n))$, and in particular,

$$1/a(n) = \text{the mean pausing time in state } n. \qquad (6.1\text{-}16)$$

We also find from Eq. (6.1-15) that the conditional density function for the jump vector from state n, i.e., the density function for v conditioned on τ (as well as on n and t), is $w_\pm(n)$; this of course is just as we should expect on the basis of Eq. (6.1-5).

The Monte Carlo simulation algorithm for a temporally homogeneous birth-death Markov process is based wholly upon the next-jump density function. The simulation algorithm may be most easily obtained here by substituting formulas (6.1-1) and (6.1-2) for $a(n,t)$ and $w(v \mid n,t)$ into the general discrete state simulation algorithm of Fig. 5-1. Because the process $X(t)$ in this case is temporally homogeneous and its

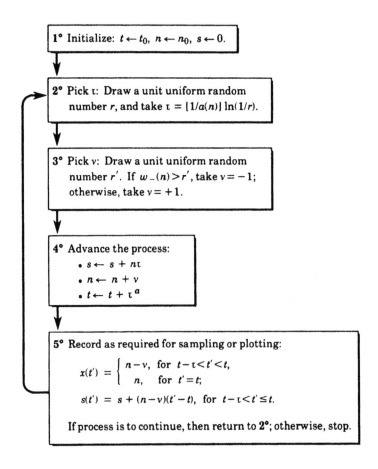

1° Initialize: $t \leftarrow t_0$, $n \leftarrow n_0$, $s \leftarrow 0$.

2° Pick ι: Draw a unit uniform random number r, and take $\iota = [1/a(n)] \ln(1/r)$.

3° Pick v: Draw a unit uniform random number r'. If $w_-(n) > r'$, take $v = -1$; otherwise, take $v = +1$.

4° Advance the process:
- $s \leftarrow s + n\iota$
- $n \leftarrow n + v$
- $t \leftarrow t + \iota^a$

5° Record as required for sampling or plotting:

$$x(t') = \begin{cases} n-v, & \text{for } t-\iota < t' < t, \\ n, & \text{for } t' = t; \end{cases}$$

$$s(t') = s + (n-v)(t'-t), \quad \text{for } t-\iota < t' \le t.$$

If process is to continue, then return to **2°**; otherwise, stop.

Note:
[a] In a simulation run containing $\sim 10^K$ jumps, the sum $t+\iota$ should be computed with at least $K+3$ digits of precision.

Figure 6-1. Exact Monte Carlo simulation algorithm for a temporally homogeneous birth-death Markov process. The algorithm produces exact sample values $x(t)$ and $s(t)$ of the process $X(t)$ and its time-integral $S(t)$ for all $t > t_0$.

jump variable v is confined to the two values ± 1, steps $2°$ and $3°$ of the algorithm of Fig. 5-1 simplify considerably along the lines of figure notes a and b. The resulting temporally homogeneous birth-death simulation algorithm is displayed in Fig. 6-1. It produces *exact* realizations $x(t)$ and $s(t)$ of the process $X(t)$ and its time-integral $S(t)$.

6.1.C THE MASTER EQUATIONS

The time-evolution of the Markov state density function $P(n,t\,|\,n_0,t_0)$ of any discrete state jump Markov process $X(t)$ is governed by the forward and backward master equations. We found in Section 5.1 that the forward master equation can be written in terms of the consolidated characterizing function $W(v\,|\,n,t)$ as in Eq. (5.1-50b). Therefore, by using the formula (6.1-8) for $W(v\,|\,n,t)$, we straightaway deduce that the *forward master equation* for a temporally homogeneous birth-death Markov process is

$$\frac{\partial}{\partial t} P(n,t\,|\,n_0,t_0)$$

$$= \lfloor\, W_-(n+1)\,P(n+1,t\,|\,n_0,t_0)\, -\, W_+(n)\,P(n,t\,|\,n_0,t_0)\,\rfloor$$

$$+\, \lfloor\, W_+(n-1)\,P(n-1,t\,|\,n_0,t_0)\, -\, W_-(n)\,P(n,t\,|\,n_0,t_0)\,\rfloor.$$

$$(6.1\text{-}17)$$

Probably the easiest way to recollect this equation is to simply derive it from the probability statement

$$P(n,t+dt\,|\,n_0,t_0) = P(n+1,t\,|\,n_0,t_0)\times W_-(n+1)dt$$

$$+\, P(n-1,t\,|\,n_0,t_0)\times W_+(n-1)dt$$

$$+\, P(n,t\,|\,n_0,t_0)\times[1 - W_-(n)dt - W_+(n)dt]. \quad (6.1\text{-}18)$$

Recalling Eq. (6.1-10), we see that this statement merely expresses the probability of finding $X(t+dt)=n$ given $X(t_0)=n_0$ as the sum of the probabilities for that to occur via each of the three time t states $n+1$, $n-1$ and n. [All other possible states at time t are two or more jumps removed from the time $t+dt$ state n, and so will contribute only terms $o(dt)$ to the right hand side of Eq. (6.1-18)]. By algebraically rearranging Eq. (6.1-18) and then passing to the limit $dt\to 0$, we easily arrive at the forward master equation (6.1-17). That equation is evidently a differential-difference equation for $P(n,t \mid n_0,t_0)$ for fixed n_0 and t_0, and is to be solved subject to the initial condition

$$P(n,t{=}t_0 \mid n_0,t_0) \; = \; \delta(n,n_0). \tag{6.1-19}$$

The backward master equation for any discrete state jump Markov process is given in terms of the consolidated characterizing function $W(v \mid n,t)$ by Eq. (5.1-53b). By substituting into that equation the formula (6.1-8) for $W(v \mid n,t)$, we straightaway deduce that the *backward master equation* for a temporally homogeneous birth-death Markov process is

$$-\frac{\partial}{\partial t_0} P(n,t \mid n_0,t_0) \; = \; W_-(n_0) [\, P(n,t \mid n_0{-}1,t_0) - P(n,t \mid n_0,t_0) \,]$$
$$+ \; W_+(n_0) [\, P(n,t \mid n_0{+}1,t_0) - P(n,t \mid n_0,t_0) \,]. \tag{6.1-20}$$

The easiest way to recollect this equation is to derive it from the probability statement

$$P(n,t \mid n_0,t_0) \; = \; W_-(n_0) dt_0 \times P(n,t \mid n_0{-}1,t_0{+}dt_0)$$
$$+ \; W_+(n_0) dt_0 \times P(n,t \mid n_0{+}1,t_0{+}dt_0)$$
$$+ \; [1 - W_-(n_0) dt_0 - W_+(n_0) dt_0] \times P(n,t \mid n_0,t_0{+}dt_0). \tag{6.1-21}$$

Recalling Eq. (6.1-10), we see that this statement merely expresses the probability of finding $X(t) = n$ given $X(t_0) = n_0$ as the sum of the probabilities for that to occur via each of the three states $n_0 - 1$, $n_0 + 1$ and n_0 at time $t_0 + dt_0$. [All other possible states at time $t_0 + dt_0$ are two or more jumps removed from the time t_0 state n_0, and so will contribute only terms $o(dt_0)$ to the right hand side of Eq. (6.1-21)]. By algebraically rearranging Eq. (6.1-21) and then passing to the limit $dt_0{\to}0$, we easily arrive at the backward master equation (6.1-20). That equation is evidently a differential-difference equation for $P(n,t \mid n_0,t_0)$ for fixed n and t, and is to be solved subject to the final condition

$$P(n,t \mid n_0,t_0{=}t) \; = \; \delta(n,n_0). \tag{6.1-22}$$

In accordance with the result (5.3-27), which holds for *any* temporally homogeneous discrete state Markov process, *the left hand sides of the forward and backward master equations* (6.1-17) *and* (6.1-20) *are interchangeable.*

An interesting variation on the forward master equation (6.1-17) can be obtained by rearranging its right hand side to read

$$\frac{\partial}{\partial t} P(n,t|n_0,t_0)$$

$$= -[\, W_+(n)\, P(n,t|n_0,t_0) - W_-(n+1)\, P(n+1,t|n_0,t_0)\,]$$

$$- \{-[\, W_+(n-1)\, P(n-1,t|n_0,t_0) - W_-(n)\, P(n,t|n_0,t_0)\,]\},$$

If we define the function $J(n,t;\, n_0,t_0)$ by

$$J(n,t;\, n_0,t_0) \equiv W_+(n-1)\, P(n-1,t|n_0,t_0) - W_-(n)\, P(n,t|n_0,t_0),$$

$$(6.1\text{-}23)$$

then the preceding equation can evidently be written

$$\frac{\partial}{\partial t} P(n,t|n_0,t_0) = -[\, J(n+1,t;\, n_0,t_0) - J(n,t;\, n_0,t_0)\,]. \qquad (6.1\text{-}24)$$

Since the quantity in brackets on the right hand side of Eq. (6.1-24) is the integer variable analog of the partial derivative of $J(n,t;\, n_0,t_0)$ with respect to n, then this version of the forward master equation is analogous to the continuity equation (3.2-5) for a *continuous* Markov process. That $J(n,t;\, n_0,t_0)$ is indeed a *probability current* at state n at time t, given $X(t_0) = n_0$, can be seen by noting from the definition (6.1-23) that $J(n,t;\, n_0,t_0)dt$ is equal to

$$P(n-1,t|n_0,t_0) \times W_+(n-1)dt - P(n,t|n_0,t_0) \times W_-(n)dt.$$

Because of Eq. (6.1-10), this is just {the probability that the process will jump from state $n-1$ to state n in time $[t,t+dt)$} *minus* {the probability that the process will jump from state n to state $n-1$ in time $[t,t+dt)$}. Thus, $J(n,t;n_0,t_0)$ indeed represents a "net probability flux" from state $n-1$ to state n at time t.

6.1.D THE MOMENT EVOLUTION EQUATIONS

The time-evolution equations for the moments of any temporally homogeneous nonnegative integer Markov process $X(t)$ are given in Eqs. (5.3-11) – (5.3-24). To apply those equations to the birth-death case, we need only calculate the propagator moment functions $W_k(n)$ defined in Eq. (5.3-10) for the birth-death function $W(v|n)$ in Eq. (6.1-8). That calculation is easy:

$$
\begin{aligned}
W_k(n) &= (-1)^k W(-1\,|\,n) + (+1)^k W(+1\,|\,n) \\
&= (-1)^k W_-(n) + (+1)^k W_+(n) \\
&= W_+(n) + (-1)^k W_-(n).
\end{aligned}
$$

Therefore, recalling Eqs. (6.1-13), we conclude that the propagator moment functions for a temporally homogeneous birth-death Markov process are given by

$$
W_k(n) = \begin{cases} v(n), & \text{if } k=1,3,5,\dots, \\ a(n), & \text{if } k=2,4,6,\dots. \end{cases} \tag{6.1-25}
$$

Upon substituting this formula for $W_k(n)$ into Eqs. (5.3-11) – (5.3-24), we obtain the following moment evolution equations for a temporally homogeneous birth-death Markov process $X(t)$:

The time-evolution equations for the moments of $X(t)$ are

$$
\begin{aligned}
\frac{d}{dt}\langle X^k(t)\rangle &= \sum_{l(\text{odd})=1}^{k} \binom{k}{l} \langle X^{k-l}(t)\, v(X(t))\rangle \\
&\quad + \sum_{l(\text{even})=2}^{k} \binom{k}{l} \langle X^{k-l}(t)\, a(X(t))\rangle
\end{aligned}
$$
$$
(t_0 \le t;\ k=1,2,\dots), \quad (6.1\text{-}26)
$$

with the initial conditions $\langle X^k(t_0)\rangle = n_0^{\ k}$. The time-evolution equations for the moments of $S(t)$, the time-integral of $X(t)$, are

$$
\frac{d}{dt}\langle S^m(t)\rangle = m\langle S^{m-1}(t)\, X(t)\rangle \quad (t_0 \le t;\ m=1,2,\dots), \quad (6.1\text{-}27)
$$

with the initial conditions $\langle S^m(t_0)\rangle = 0$, together with the auxiliary equations

$$
\begin{aligned}
\frac{d}{dt}\langle S^m(t)\, X^k(t)\rangle &= m\langle S^{m-1}(t)\, X^{k+1}(t)\rangle \\
&\quad + \sum_{l(\text{odd})=1}^{k} \binom{k}{l} \langle S^m(t)\, X^{k-l}(t)\, v(X(t))\rangle \\
&\quad + \sum_{l(\text{even})=1}^{k} \binom{k}{l} \langle S^m(t)\, X^{k-l}(t)\, a(X(t))\rangle
\end{aligned}
$$
$$
(t_0 \le t;\ m \ge 1;\ k \ge 1), \quad (6.1\text{-}28)
$$

and their initial conditions $\langle S^m(t_0)\, X^k(t_0)\rangle = 0$. These moment evolution equations will be closed if and only if $v(n)$ is at most linear in n and $a(n)$ is at most quadratic in n. If $v(n)$ and $a(n)$ do not satisfy those rather stringent conditions, then the most we can hope for is *approximate* solutions, obtained either by suitably approximating $v(n)$ and $a(n)$, or else by applying the tactics described in Appendix C.

For the mean, variance and covariance of $X(t)$ we have the differential equations

$$\frac{d}{dt}\langle X(t)\rangle = \langle v(X(t))\rangle \qquad (t_0 \leq t), \qquad (6.1\text{-}29)$$

$$\frac{d}{dt}\text{var}\{X(t)\} = 2\Big(\langle X(t)\, v(X(t))\rangle - \langle X(t)\rangle\langle v(X(t))\rangle\Big) + \langle a(X(t))\rangle$$

$$(t_0 \leq t), \qquad (6.1\text{-}30)$$

$$\frac{d}{dt_2}\text{cov}\{X(t_1),X(t_2)\} = \langle X(t_1)\, v(X(t_2))\rangle - \langle X(t_1)\rangle\langle v(X(t_2))\rangle$$

$$(t_0 \leq t_1 \leq t_2), \qquad (6.1\text{-}31)$$

with the respective initial conditions

$$\langle X(t=t_0)\rangle = n_0, \qquad (6.1\text{-}32)$$

$$\text{var}\{X(t=t_0)\} = 0, \qquad (6.1\text{-}33)$$

$$\text{cov}\{X(t_1),X(t_2=t_1)\} = \text{var}\{X(t_1)\}. \qquad (6.1\text{-}34)$$

For the mean of $S(t)$ we have

$$\langle S(t)\rangle = \int_{t_0}^{t}\langle X(t')\rangle\, dt', \qquad (6.1\text{-}35)$$

wherein the integrand is the solution to Eq. (6.1-29). For the variance of $S(t)$ we have

$$\text{var}\{S(t)\} = 2\int_{t_0}^{t}\text{cov}\{S(t'),X(t')\}\, dt', \qquad (6.1\text{-}36)$$

wherein the integrand is the solution of the differential equation

$$\frac{d}{dt}\text{cov}\{S(t),X(t)\} = \text{var}\{X(t)\} + \langle S(t)\, v(X(t))\rangle - \langle S(t)\rangle\langle v(X(t))\rangle$$

$$(t_0 \leq t) \qquad (6.1\text{-}37\text{a})$$

that satisfies the initial condition

$$\text{cov}\{S(t=t_0), X(t=t_0)\} = 0. \tag{6.1-37b}$$

And finally, for the covariance of $S(t)$ we have

$$\text{cov}\{S(t_1), S(t_2)\} = \text{var}\{S(t_1)\} + \int_{t_1}^{t_2} \text{cov}\{S(t_1), X(t_2')\}\, dt_2'$$

$$(t_0 \leq t_1 \leq t_2), \tag{6.1-38}$$

wherein the integrand is the solution of the differential equation

$$\frac{d}{dt_2}\text{cov}\{S(t_1), X(t_2)\} = \langle S(t_1)\, v(X(t_2)) \rangle - \langle S(t_1) \rangle \langle v(X(t_2)) \rangle$$

$$(t_0 \leq t_1 \leq t_2) \tag{6.1-39a}$$

that satisfies the initial condition

$$\text{cov}\{S(t_1), X(t_2=t_1)\} = \text{cov}\{S(t_1), X(t_1)\}, \tag{6.1-39b}$$

which in turn is known from the solution to Eq. (6.1-37a).

Application of the foregoing formulas will be illustrated for some specific birth-death Markov processes in Section 6.3.

6.2 THE CONTINUOUS APPROXIMATION FOR BIRTH-DEATH MARKOV PROCESSES

In Section 3.2 we found that the mean, variance and covariance of a *continuous* Markov process satisfied the time-evolution equations (3.2-12) – (3.2-14). We now observe that if the drift function $A(x,t)$ and diffusion function $D(x,t)$ in those equations are taken to be $v(x)$ and $a(x)$ respectively, then those equations become identical to the *birth-death* equations (6.1-29) – (6.1-31). In this section we shall elaborate upon this observation by showing that, under certain circumstances which are often realized in practice, a temporally homogeneous *birth-death* Markov process can be *approximated* by a temporally homogeneous *continuous* Markov process with characterizing functions

$$A(x) = v(x) \quad \text{and} \quad D(x) = a(x). \tag{6.2-1}$$

6.2.A THE FOKKER-PLANCK LIMIT OF THE BIRTH-DEATH MASTER EQUATION

That the above noted similarity between a birth-death Markov process and a continuous Markov process with characterizing functions related according to Eqs. (6.2-1) does *not* imply an *identity* between the two processes can readily be seen by comparing the time-evolution equations for their higher moments. For the temporally homogeneous *birth-death* Markov process, we have from Eq. (6.1-26) that

$$\frac{d}{dt}\langle X^k(t)\rangle = \sum_{l(\text{odd})=1}^{k} \binom{k}{l} \langle X^{k-l}(t)\, v(X(t))\rangle$$

$$+ \sum_{l(\text{even})=2}^{k} \binom{k}{l} \langle X^{k-l}(t)\, a(X(t))\rangle \quad (k \geq 1), \quad (6.2-2)$$

whereas for the temporally homogeneous *continuous* Markov process, we have from Eq. (3.2-8) that

$$\frac{d}{dt}\langle X^k(t)\rangle = \binom{k}{1}\langle X^{k-1}(t)\, A(X(t))\rangle + \binom{k}{2}\langle X^{k-2}(t)\, D(X(t))\rangle$$

$$(k \geq 1). \quad (6.2-3)$$

If $A(x) \equiv v(x)$ and $D(x) \equiv a(x)$, then these two equations will indeed be identical for $k=1$ and $k=2$. But they will *not* be identical for $k \geq 3$. So all moments of these two processes higher than the second will differ. In fact, if $v(x)$ happens to be more than linear in x, or $a(x)$ more than quadratic in x, then even the first two moments of these processes will differ; because in such cases the first two moment evolution equations, although identical in form, will necessarily involve one or more of the *differing* higher order moments. Only in the special circumstance that $v(x)$ is at most linear in x and $a(x)$ is at most quadratic in x will the first and second moments of a continuous Markov process and a birth-death Markov process be identical under the condition (6.2-1). But again, as the third and higher order moments of these two processes will *always* be different, the processes themselves can never be identical.

Suppose, however, that the characterizing functions of our birth-death process are such that $X(t)$ is virtually always *very large* compared to unity. In that commonly occurring case, each of the two sums in Eq. (6.2-2) should, by virtue of the factor $X^{k-l}(t)$, be dominated by its *lowest* l-value term. We should then be justified in *approximating* that equation by

$$\frac{d}{dt}\langle X^k(t)\rangle \approx \left(\begin{array}{c} k \\ 1 \end{array}\right)\langle X^{k-1}(t)\,v(X(t))\rangle + \left(\begin{array}{c} k \\ 2 \end{array}\right)\langle X^{k-2}(t)\,a(X(t))\rangle$$

$$(k \geq 1). \quad (6.2\text{-}4)$$

This approximated birth-death kth moment equation is evidently the same as the continuous kth moment equation (6.2-3) if the correspondence (6.2-1) holds. So we conclude that *any temporally homogeneous birth-death Markov process* $X(t)$ *that is "very large" compared to unity can be approximated by a temporally homogeneous continuous Markov process with characterizing functions* (6.2-1). Of course, the acceptability of that approximation will depend on the specifics of the situation at hand.

The *converse* of the foregoing result — namely that a large-number continuous Markov process with drift function $A(x)$ and diffusion function $D(x)$ can be approximated by the birth-death Markov process with $v(n) = A(n)$ and $a(n) = D(n)$ — has a more restricted range of validity. As was noted in Eq. (6.1-14d), birth-death Markov processes necessarily have $|v(n)| \leq a(n)$; consequently, *only* those continuous Markov processes that satisfy, in addition to the "large number" property, the condition

$$D(x) \geq |A(x)| \quad (6.2\text{-}5)$$

can have an approximating birth-death process. Roughly speaking, a continuous Markov process must be "sufficiently noisy" in order to possess an approximating birth-death Markov process. A Wiener process [see Eqs. (3.3-8)] would satisfy condition (6.2-5) if $D \geq |A|$, but not otherwise. An Ornstein-Uhlenbeck process [see Eqs. (3.3-19)] will satisfy condition (6.2-5) *only* when it happens to lie inside the asymptotic two-standard deviation interval $[-D/k, D/k]$. A Liouville process [see Eq. (3.3-2)] would never satisfy condition (6.2-5).

Our principal conclusion here, that a temporally homogeneous birth-death Markov process "on the large integers" can usually be approximated by a continuous Markov process with characterizing functions (6.2-1), was arrived at on the basis of the time-evolution equations for the *moments* of the two processes. But we can also reach this conclusion on the basis of the time-evolution equations for the *density functions* of the two processes: For the birth-death process that equation is the forward master equation (6.1-17). Since the equations (6.1-13) for $a(n)$ and $v(n)$ imply that

$$W_+(n) = [a(n) + v(n)]/2 \quad (6.2\text{-}6a)$$

and

$$W_-(n) = [a(n) - v(n)]/2, \quad (6.2\text{-}6b)$$

then we can write the forward master equation (6.1-17) as

$$\frac{\partial}{\partial t} P(n,t \mid n_0, t_0)$$

$$= \left(\frac{a(n+1) - v(n+1)}{2} \right) P(n+1, t \mid n_0, t_0) - \left(\frac{a(n) + v(n)}{2} \right) P(n, t \mid n_0, t_0)$$

$$+ \left(\frac{a(n-1) + v(n-1)}{2} \right) P(n-1, t \mid n_0, t_0) - \left(\frac{a(n) - v(n)}{2} \right) P(n, t \mid n_0, t_0).$$

A simple rearrangement of terms on the right hand side then yields

$$\frac{\partial}{\partial t} P(n,t \mid n_0, t_0)$$

$$= -\left[\frac{v(n+1) P(n+1, t \mid n_0, t_0) - v(n-1) P(n-1, t \mid n_0, t_0)}{2} \right]$$

$$+ \frac{1}{2} \Big[a(n+1) P(n+1, t \mid n_0, t_0) - 2a(n) P(n, t \mid n_0, t_0)$$

$$+ a(n-1) P(n-1, t \mid n_0, t_0) \Big]. \quad (6.2\text{-}7)$$

To interpret Eq. (6.2-7), which we emphasize is just the forward master equation for our temporally homogeneous birth-death Markov process expressed in terms of the function pair $a(n)$ and $v(n)$, let us recall from numerical analysis that a mutually consistent approximation scheme for the first two derivatives of any differentiable function $f(x)$ is

$$\frac{d}{dx} f(x) \approx \frac{f(x+\varepsilon) - f(x-\varepsilon)}{2\varepsilon} \qquad (6.2\text{-}8a)$$

and

$$\frac{d^2}{dx^2} f(x) \approx \frac{f(x+\varepsilon) - 2f(x) + f(x-\varepsilon)}{\varepsilon^2}, \qquad (6.2\text{-}8b)$$

provided ε is "sufficiently small." Now, most functions $f(x)$ that we encounter in physical modeling applications are such that these derivative approximations also hold for $\varepsilon = 1$ *if* x is sufficiently large compared to unity. For such functions, we can thus write

$$\frac{d}{dx} f(x) \approx \frac{f(x+1) - f(x-1)}{2} \qquad (\text{if } x \gg 1), \quad (6.2\text{-}9a)$$

and

$$\frac{d^2}{dx^2} f(x) \approx f(x+1) - 2f(x) + f(x-1), \quad \text{(if } x \gg 1\text{).} \quad (6.2\text{-}9b)$$

Of course it is easy to think of functions, such as $f(x) = \cos(2\pi x)$, for which approximations (6.2-9) are not viable. But for most *physical applications* of birth-death Markov process theory, it will not be unreasonable to expect that the two functions $[v(x)P(x,t \mid n_0,t_0)]$ and $[a(x)P(x,t \mid n_0,t_0)]$ will be amenable to these large-x approximations. In such instances, the forward master equation (6.2-7) of our birth-death Markov process can then be *approximated* by

$$\frac{\partial}{\partial t} P(x,t \mid n_0,t_0) \approx -\frac{\partial}{\partial x}\left[v(x) P(x,t \mid n_0,t_0) \right] + \frac{1}{2}\frac{\partial^2}{\partial x^2}\left[a(x) P(x,t \mid n_0,t_0) \right]$$

$$\text{(if } x \gg 1\text{).} \quad (6.2\text{-}10)$$

This approximated forward master equation evidently has the form of the *forward Fokker-Planck equation* (3.2-2) for the *continuous* Markov process with drift function $v(x)$ and diffusion function $a(x)$. So this "Fokker-Planck limit" of the birth-death master equation provides yet another way to see that a "large number" birth-death Markov process $X(t)$ can usually be approximated as a continuous Markov process with drift function $v(x)$ and diffusion function $a(x)$.[†]

The possibility of thus approximating a birth-death Markov process at large population levels by a continuous Markov process often turns out to be very convenient from a computational standpoint. For, although the formulas for a birth-death process are often conceptually simpler than those for a continuous process, they may actually turn out to be more tedious to work with when very many discrete states are involved (just as a discrete sum with many terms is often harder to evaluate than an approximating definite integral). But it is important to keep in mind that the continuous approximation of a birth-death Markov process is viable *only* if the effective range of the process is confined to the "very large integers." The foregoing analysis provides no warrant for such an approximation when the process $X(t)$ takes state values of order unity.

[†] Of all discrete state jump Markov processes, it is only for the *one-step* variety that the forward master equation can be written as a properly discretized *second order* partial differential equation of the Fokker-Planck type. Emphasis is laid here on the discretization being done "properly," with all x-derivatives being approximated to the same order of accuracy. It turns out that for an n-step jump Markov process, the right hand side of the forward master equation takes the form of a properly discretized partial differential equation of order $2n$.

Apart from providing alternative computational strategies, the correspondence (6.2-1) highlights an interesting point concerning the *Markov process modeling* of physical systems. To appreciate this point, we first note that the correspondence (6.2-1) is by definition the same as

$$A(x) = W_+(x) - W_-(x) \quad \text{and} \quad D(x) = W_+(x) + W_-(x). \quad (6.2\text{-}11)$$

Now, if one chooses to mathematically model some physical system as a *continuous* Markov process, then one must posit forms for the two functions $A(x)$ and $D(x)$. Those functions respectively describe, as we found in our discussion of the Langevin equation in Subsection 3.4.A, the *deterministic component* and the *randomly fluctuating component* of the forces that drive the process. On the other hand, if one chooses to mathematically model a physical system as a *birth-death* Markov process, then one must posit forms for the two functions $W_+(n)$ and $W_-(n)$. Those two functions respectively describe the tendencies for the process to *step up* or *step down* from state n. Now it turns out that, for most *real* systems, it is not possible to make a rigorous *a priori* resolution of the forces driving the system into "deterministic" and "stochastic" components. A good illustration of this point is provided by the velocity $V(t)$ of a Brownian particle, which we discussed in Chapters 3 and 4: On the *microscopic* level, it is simply not possible to separate the particle-molecule collisions into one class that gives rise to a "steady retardation" in $V(t)$ and another class that gives rise to "random fluctuations" in $V(t)$. All we can assert at the microscopic level are the propensities for various collisions to occur, and the specific kinematic outcomes of those collisions when they do occur. Indeed, these were the kinds of considerations that formed the basis for our analysis of Brownian motion in a one-dimensional gas in Section 4.5. There we derived, from a few self-evident premises, a function $W(\xi \mid v)$ that is the continuum state equivalent of the birth-death stepping functions $W_\pm(n)$ [see Eq. (4.5-16)]. It was only *after* we applied the mathematical machinery of Markov process theory that "drag effects" and "fluctuation effects" could be quantitatively inferred. But in order to have distinguished those effects at the *outset*, by positing specific forms for the drift function $A(x)$ and the diffusion function $D(x)$, we would have had to make some *ad hoc*, phenomenological assumptions. And that in fact is just what we did in our original treatment of Brownian motion in Subsection 3.4.C.

So one reason why the correspondence (6.2-11) commands our attention is that it shows quantitatively that the microscopic forces which drive a birth-death Markov process, and which are usually best described physically by the stepping functions $W_+(n)$ and $W_-(n)$, do in fact give rise to effects that can in most macroscopic circumstances be plausibly interpreted as "deterministic drift" and "stochastic diffusion."

6.2.B APPLICATION: CHEMICAL CONCENTRATION FLUCTUATIONS

To illustrate the utility of approximating a birth-death Markov process by a continuous Markov process through the correspondence (6.2-1), we consider the chemical system

$$B + X \underset{c_2}{\overset{c_1}{\rightleftharpoons}} 2X, \qquad (6.2\text{-}12)$$

for which the population of species B molecules is "buffered" to an effectively constant value N, and the population of species X molecules at time t is given by $X(t)$. The molecules of this system are assumed to be confined to a container of constant volume Ω, and to be kept "well stirred" at some constant temperature.

As discussed in Subsection 5.3.B [see Eq. (5.3-29)], the specific probability rate constants c_1 and c_2 for the above two reactions are defined so that, if $X(t) = n$, then

$$(c_1 dt) \times (Nn) = \text{Prob}\{ \text{reaction 1 will occur in } [t, t+dt) \}$$

and

$$(c_2 dt) \times [n(n-1)/2] = \text{Prob}\{ \text{reaction 2 will occur in } [t, t+dt) \}.$$

Since the occurrence of reaction 1 would increase $X(t)$ by 1, while the occurrence of reaction 2 would decrease $X(t)$ by 1, then it follows that $X(t)$ is a temporally homogeneous birth-death Markov process with stepping functions

$$W_+(n) = c_1 Nn \quad \text{and} \quad W_-(n) = (c_2/2)\, n(n-1). \qquad (6.2\text{-}13)$$

Therefore, for this birth-death Markov process the functions $v(n)$ and $a(n)$ are given by

$$v(n) \equiv W_+(n) - W_-(n) = c_1 Nn - (c_2/2)\, n(n-1), \qquad (6.2\text{-}14a)$$

and

$$a(n) \equiv W_+(n) + W_-(n) = c_1 Nn + (c_2/2)\, n(n-1). \qquad (6.2\text{-}14b)$$

Now suppose that our chemical system is "macroscopic" in the sense that the numbers of B and X molecules inside Ω are very large compared to unity. Then by our analysis in the preceding subsection, we can *approximate* $X(t)$ as the *continuous* Markov process with drift function

$$v(x) = c_1 Nx - (c_2/2) x^2 \qquad (x \gg 1), \qquad (6.2\text{-}15a)$$

and diffusion function

$$a(x) = c_1 Nx + (c_2/2) x^2 \qquad (x \gg 1). \qquad (6.2\text{-}15b)$$

In this approximation, the Markov state density function for $X(t)$ will satisfy the Fokker-Planck equation (6.2-10), and the process itself will satisfy the Langevin equation [see Eq. (3.4-8)]

$$X(t+dt) = X(t) + a^{1/2}(X(t)) \, dW(dt) + v(X(t)) \, dt, \qquad (6.2\text{-}16)$$

where $dW(dt)$ is $N(0, dt)$, the normal random variable with mean 0 and variance dt.

One benefit of thus approximating $X(t)$ is that it allows us to take advantage of the ease with which continuous Markov process theory describes *intensive* processes and *weak-noise* processes. By contrast, it is awkward to treat the intensive process

$$Z(t) \equiv X(t)/\Omega \qquad (6.2\text{-}17)$$

as a birth-death process, because $Z(t)$ will not be confined to the integers. Furthermore, it seems impossible to develop a weak-noise theory for birth-death Markov processes, as we did for continuous Markov processes in Section 3.8, because the inequality (6.1-14d) implies that birth-death Markov process are intrinsically very noisy. But having now approximated our *large-number* birth-death Markov process $X(t)$ as a *continuous* Markov process, let us see what we can learn about the behavior of the X-molecule "concentration" $Z(t)$ defined in Eq. (6.2-17).

Since $Z(t)$ is a scalar multiple of the continuous Markov process $X(t)$, then it will likewise be a continuous Markov process. By dividing the Langevin equation (6.2-16) through by Ω and then invoking Eq. (6.2-17), we can see that the Langevin equation for $Z(t)$ is

$$Z(t+dt) = Z(t) + [\Omega^{-2}a(\Omega Z(t))]^{1/2} dW(dt) + \Omega^{-1}v(\Omega Z(t)) \, dt. \quad (6.2\text{-}18)$$

From the form of this equation we immediately deduce that the continuous Markov process $Z(t)$ has drift function $\Omega^{-1}v(\Omega z)$ and diffusion function $\Omega^{-2}a(\Omega z)$. We can simplify those two characterizing functions by introducing the X and B *concentration variables*

$$z \equiv x/\Omega \quad \text{and} \quad b \equiv N/\Omega, \qquad (6.2\text{-}19)$$

and also the conventional *reaction-rate constants*

$$k_1 \equiv \Omega c_1 \quad \text{and} \quad k_2 \equiv \Omega c_2/2. \tag{6.2-20}$$

We recall from our discussion in Subsection 5.3.B [see Eqs. (5.3-38) and (5.3-52)] that these reaction-rate constants are in fact *independent* of Ω. It now follows from formulas (6.2-15) that

$$v(\Omega z) = (k_1/\Omega)(\Omega b)(\Omega z) - (k_2/\Omega)(\Omega z)^2 = \Omega (k_1 bz - k_2 z^2),$$

and similarly

$$a(\Omega z) = \Omega (k_1 bz + k_2 z^2).$$

Therefore, defining the functions v^* and a^* by

$$v^*(z) \equiv k_1 bz - k_2 z^2 \tag{6.2-21a}$$

and

$$a^*(z) \equiv k_1 bz + k_2 z^2, \tag{6.2-21b}$$

we conclude that the previously identified drift and diffusion functions of the continuous Markov process $Z(t)$ are

$$\Omega^{-1} v(\Omega z) = v^*(z) \quad \text{and} \quad \Omega^{-2} a(\Omega z) = \Omega^{-1} a^*(z).$$

Therefore, the Langevin equation (6.2-18) for $Z(t)$ can be written more compactly as

$$Z(t+dt) = Z(t) + [\Omega^{-1} a^*(Z(t))]^{1/2} dW(dt) + v^*(Z(t)) dt. \tag{6.2-22}$$

Now suppose that we approach the so-called "thermodynamic limit," in which n, N and Ω are all taken to be very large. Then the largeness of Ω implies, through this last Langevin equation, that $Z(t)$ will be a *weak-noise* continuous Markov process. More specifically, Eq. (6.2-22) has the standard weak-noise form (3.8-5) with the replacements

$$X(t) \to Z(t), \quad \varepsilon \to \Omega^{-1}, \quad D_1(x,t) \to a^*(z), \quad A(x,t) \to v^*(z).$$

So the weak-noise results of Subsection 3.8.A allow us to draw the following conclusions concerning the concentration $Z(t)$ of X molecules in the chemical system (6.2-12):

If the system volume Ω and the numbers of X and B molecules n and N are all sufficiently large, then the X-molecule concentration can be *approximately* written [see Eq. (3.8-6)]

$$Z(t) = z^*(t) + \Omega^{-1/2} Y(t). \tag{6.2-23}$$

Here, $z^*(t)$ is the *sure* function defined by the *reaction-rate equation* [see Eq. (3.8-2b)]

$$\frac{dz^*}{dt} = v^*(z^*) \equiv k_1 b z^* - k_2 z^{*2} \qquad (6.2\text{-}24a)$$

with initial condition

$$z^*(t_0) = z_0 \equiv n_0/\Omega. \qquad (6.2\text{-}24b)$$

And $Y(t)$ is the continuous Markov process with drift function $y\, v^{*\prime}(z^*(t))$ and diffusion function $a^*(z^*(t))$; i.e., $Y(t)$ is defined by the Langevin equation [see Eq. (3.3-8)]

$$Y(t+dt) = Y(t) + a^{*1/2}(z^*(t))\, dW(dt) + Y(t)\, v^{*\prime}(z^*(t))\, dt \qquad (6.2\text{-}25)$$

with initial condition $Y(t_0) = 0$, or equivalently by the Fokker-Planck equation [see Eq. (3.8-11)]

$$\frac{\partial}{\partial t} R(y,t\,|\,0,t_0) = -v^{*\prime}(z^*(t))\frac{\partial}{\partial y}\Big[y\, R(y,t\,|\,0,t_0)\Big] + \frac{1}{2}\,a^*(z^*(t))\frac{\partial^2}{\partial y^2} R(y,t\,|\,0,t_0)$$

$$(6.2\text{-}26)$$

with initial condition $R(y,t_0\,|\,0,t_0) = \delta(y)$. Furthermore, it is easily shown that for any $z_0 > 0$ the solution $z^*(t)$ of the differential equation (6.2-24a) approaches, as $t\to\infty$, an *equilibrium value* z_∞, which is the positive value of z^* that makes the right hand side of Eq. (6.2-24a) vanish:

$$z^*(t\to\infty) = z_\infty \equiv k_1 b/k_2. \qquad (6.2\text{-}27)$$

It therefore follows from the analysis presented in conjunction with Eqs. (3.8-12) – (3.8-16) that $Z(t\to\infty)$ is the *normal* random variable with mean z_∞ and variance

$$\Omega^{-1}\left|\frac{a^*(z_\infty)}{2v^{*\prime}(z_\infty)}\right| = \Omega^{-1}\left|\frac{k_1 b z_\infty + k_2 z_\infty^2}{2(k_1 b - 2k_2 z_\infty)}\right| = \Omega^{-1}\left|\frac{z_\infty}{-1}\right|;$$

that is,

$$Z(t\to\infty) = N(z_\infty, z_\infty/\Omega). \qquad (6.2\text{-}28)$$

Equations (6.2-23) – (6.2-25) provide a rigorous basis for asserting that, near the thermodynamic limit, the concentration of chemical species X follows the solution of the deterministic reaction-rate equation, but with superimposed fluctuations whose typical size is inversely proportional to the square root of the system volume Ω. In particular, Eq. (6.2-28) implies that the asymptotic concentration fluctuations about the conventional equilibrium value z_∞ are *normally* distributed with standard deviation $(z_\infty/\Omega)^{1/2}$. Evidently, the approximation of our original *birth-death* Markov process by a *continuous* Markov process has provided us with some physically useful insights into the behavior of the chemical system (6.2-12).

6.3 SOME SIMPLE BIRTH-DEATH MARKOV PROCESSES

In this section we shall examine in some detail four elementary examples of temporally homogeneous birth-death Markov processes: the Poisson process, the radioactive decay process, the random telegraph process, and the payroll process.

6.3.A THE POISSON PROCESS

The **Poisson process** is the temporally homogeneous birth-death Markov process defined by the characterizing functions

$$a(n) = a, \quad w_+(n) = 1, \quad w_-(n) = 0, \tag{6.3-1}$$

or equivalently by the stepping functions [see Eq. (6.1-9)]

$$W_+(n) = a \quad \text{and} \quad W_-(n) = 0, \tag{6.3-2}$$

where a is any positive constant. It follows from Eq. (6.1-16) that the Poisson process has a pausing time in any state n that is exponentially distributed with mean $1/a$. And it follows from Eq. (6.1-5) that this process, upon jumping away from state n, will necessarily jump to state $n+1$. The Poisson process is thus an example of a "birth-only" Markov process.

The Poisson process holds the distinction of being the one and only *completely homogeneous* birth-death Markov process. We actually encountered it earlier in Subsection 5.2.B as a special case of the completely homogeneous "one-step" Markov process defined in Eq. (5.2-13): the value $b=1$ is evidently the only b-value that is compatible with the birth-death injunction (6.1-6) against passage to negative states. In that earlier consideration of the Poisson process, we found that its Markov state density function $P(n,t \mid n_0,t_0)$ is given by the explicit formula (5.2-19). And upon comparing that formula with Eqs. (1.7-14), we concluded that the Poisson process $X(t)$ is such that $X(t) - n_0$ is the *Poisson random variable* with mean (and variance) $a(t-t_0)$. Our derivation of Eq. (5.2-19) proceeded from the general quadrature formula (5.2-9) for completely homogeneous discrete state Markov processes. We shall now show how Eq. (5.2-19) can also be derived from the birth-death forward master equation.

The birth-death forward master equation (6.1-17) reads for the stepping functions in Eqs. (6.3-2),

$$\frac{\partial}{\partial t} P(n,t \mid n_0,t_0) = -a\,P(n,t \mid n_0,t_0) + a\,P(n-1,t \mid n_0,t_0). \qquad (6.3\text{-}3)$$

We may look upon this equation as a set of coupled, ordinary differential equations in t for the infinite set of functions $P_n(t) \equiv P(n,t \mid n_0,t_0)$, where the index $n = 0, 1, 2, \dots$. This set of differential equations is amenable to solution because of the "one-way" nature of the Poisson process: Since down-going transitions can never occur in this process, then $P(n_0-1,t \mid n_0,t_0) \equiv 0$ for all $t \geq t_0$, so Eq. (6.3-3) reads for $n = n_0$:

$$\frac{\partial}{\partial t} P_{n_0}(t) = -a\,P_{n_0}(t).$$

The solution to this equation for the required initial condition $P_{n_0}(t_0) = 1$ is clearly

$$P_{n_0}(t) = e^{-a(t-t_0)}.$$

With this result, we then find from Eq. (6.3-3) the following differential equation for the function $P_{n_0+1}(t)$:

$$\frac{\partial}{\partial t} P_{n_0+1}(t) = -a\,P_{n_0+1}(t) + a\,e^{-a(t-t_0)}.$$

This differential equation is of the standard form (A-11a), so since $P_{n_0+1}(t_0) = 0$ we get from Eq. (A-11b) the solution

$$P_{n_0+1}(t) = e^{-a(t-t_0)} \int_{t_0}^{t} dt'\, a\,e^{-a(t'-t_0)}\, e^{a(t'-t_0)}$$

$$= a\,(t-t_0)\,e^{-a(t-t_0)}.$$

With this result, the master equation (6.3-3) reads for $n = n_0+2$,

$$\frac{\partial}{\partial t} P_{n_0+2}(t) = -a\,P_{n_0+2}(t) + a^2\,(t-t_0)\,e^{-a(t-t_0)},$$

and again invoking Eqs. (A-11) and the initial condition $P_{n_0+2}(t_0) = 0$, we obtain

$$P_{n_0+2}(t) = e^{-a(t-t_0)} \int_{t_0}^{t} dt'\, a^2(t'-t_0)\, e^{-a(t'-t_0)}\, e^{a(t'-t_0)}$$

$$= \frac{a^2(t-t_0)^2}{2!}\, e^{-a(t-t_0)}.$$

Continuing in this iterative way, we find by induction the general formula

$$P_{n_0+n}(t) \equiv P(n_0+n,t\,|\,n_0,t_0) = \frac{e^{-a(t-t_0)}\,[a\,(t-t_0)]^n}{n!}$$

$$(t_0 \le t,\, n = 0,1,2,\ldots), \quad (6.3\text{-}4)$$

which evidently agrees with our earlier result (5.2-19).

As is the case with virtually any temporally homogeneous birth-death Markov process, the Monte Carlo simulation of a Poisson process $X(t)$ and its time-integral $S(t)$ using the exact algorithm of Fig. 6-1 is very easy. We have in fact already exhibited in Fig. 5-2 the results of simulating a Poisson process with $a=1$, $n_0=0$ and $t_0=0$.

Although all the moments of the Poisson process $X(t)$ could be written down by simply appealing to the known moment formulas for the Poisson random variable, taking $a(t-t_0)$ as the mean, those moments can also be deduced from the general birth-death moment evolution equations (6.1-26) – (6.1-39). Those general time-evolution equations, which also enable us to compute the moments of the associated time-integral process $S(t)$, are expressed in terms of the two functions $a(n)$ and $v(n)$. The former function is of course the constant a, and the latter is easily calculated from Eqs. (6.3-2) to be

$$v(n) \equiv W_+(n) - W_-(n) = a. \qquad (6.3\text{-}5)$$

So for the Poisson process we have $a(n)=v(n)=a$. It follows from Eq. (6.1-26) that the moments of the Poisson process satisfy the differential equations

$$\frac{d}{dt}\langle X^k(t)\rangle = a \sum_{l=1}^{k} \binom{k}{l} \langle X^{k-l}(t)\rangle \quad (t_0 \le t,\, k=1,2,\ldots) \quad (6.3\text{-}6)$$

and the initial condition $\langle X^k(t_0) \rangle = n_0^k$. And it follows from Eqs. (6.1-27) and (6.1-28) that the moments of the time-integral of the Poisson process are given by

$$\langle S^m(t) \rangle = m \int_{t_0}^{t} \langle S^{m-1}(t') X(t') \rangle dt' \quad (t_0 \le t, m=1,2,...), \quad (6.3\text{-}7)$$

wherein the integrand is obtained from the solutions to the set of differential equations

$$\frac{d}{dt} \langle S^m(t) X^k(t) \rangle = m \langle S^{m-1}(t) X^{k+1}(t) \rangle + a \sum_{l=1}^{k} \binom{k}{l} \langle S^m(t) X^{k-l}(t) \rangle$$

$$(t_0 \le t; m \ge 1; k \ge 1) \quad (6.3\text{-}8)$$

for the initial conditions $\langle S^m(t_0) X^k(t_0) \rangle = 0$. Since in this case $v(n)$ is not more than linear in n and $a(n)$ is not more than quadratic in n, then all these moment evolution equations are closed, and can be solved exactly to any order. We shall not exhibit the solutions here, but we shall write down the explicit formulas for the means, variances and covariances of $X(t)$ and $S(t)$. Actually, the easiest way for us to obtain those formulas at this point is to just observe that, when we substitute $v(n)=a$ and $a(n)=a$ into the general birth-death time-evolution equations (6.1-29) – (6.1-39), those equations become identical to the time-evolution equations for the Wiener process with $A=a$ and $D=a$ [see Eqs. (3.2-12) – (3.2-19)]. So from the Wiener formulas (3.3-10) and (3.3-11), we can immediately write down the following formulas for the mean, variance and covariance of the Poisson process $X(t)$,

$$\langle X(t) \rangle = n_0 + a(t-t_0) \qquad (t_0 \le t), \qquad (6.3\text{-}9a)$$

$$\text{var}\{X(t)\} = a(t-t_0) \qquad (t_0 \le t), \qquad (6.3\text{-}9b)$$

$$\text{cov}\{X(t_1),X(t_2)\} = a(t_1-t_0) \qquad (t_0 \le t_1 \le t_2), \qquad (6.3\text{-}9c)$$

and the following formulas for the mean, variance and covariance of its time-integral $S(t)$,

$$\langle S(t) \rangle = n_0(t-t_0) + (a/2)(t-t_0)^2 \qquad (t_0 \le t), \quad (6.3\text{-}10a)$$

$$\text{var}\{S(t)\} = (a/3)(t-t_0)^3 \qquad (t_0 \le t), \quad (6.3\text{-}10b)$$

$$\text{cov}\{S(t_1),S(t_2)\} = a(t_1-t_0)^2 [(t_1-t_0)/3 + (t_2-t_1)/2]$$

$$(t_0 \le t_1 \le t_2). \quad (6.3\text{-}10c)$$

These formulas could also have been obtained by simply putting $b=1$ in Eqs. (5.2-15) and (5.2-16). Notice that formulas (6.3-9a) and (6.3-9b) are indeed consistent with our earlier conclusion that $X(t)-n_0$ is a random variable with mean and variance $a(t-t_0)$.

It should be emphasized that, although the means, variances and covariances of the Poisson process are identical to those of the Wiener process with $A=D=a$, the third and higher order moments of these two processes will all be *different*, for reasons already discussed in connection with Eqs. (6.2-2) and (6.2-3). Nevertheless, we may expect on the basis of our analysis in Subsection 6.2.A that the differences between those two processes should be minor whenever the processes are in "large number" states. In fact, it is easy to prove directly that, *if* $a(t-t_0)\gg1$, *then the Poisson process can be well approximated by the Wiener process with* $A=D=a$. For, we know from Eq. (6.3-4) that the Poisson process $X(t)$ is such that $X(t)-n_0$ is the *Poisson random variable* with mean and variance $a(t-t_0)$. And we know from Eq. (3.3-14) that the Wiener process with $A=D=a$ is such that $X(t)-n_0$ is the *normal random variable* with mean and variance $a(t-t_0)$. It then follows from the result (1.7-16) that these two random variables will be indistinguishable if their common mean $a(t-t_0)$ is very large, and that establishes our assertion. So, for example, the Poisson process with $v=a=1$, which was simulated in Fig. 5-2, will be for $at\gg1$ practically indistinguishable from the Wiener process with $A=D=1$, which was simulated in Figs. 3-10 and 3-11.

6.3.B THE RADIOACTIVE DECAY PROCESS

A radioactive atom is an atom whose nucleus is subject to the occurrence of a spontaneous internal reaction, called a "decay," that transforms it into the nucleus of another atomic species. It is known that for each radioactive species there exists a constant τ^*, which depends only on the nuclear structure of that species, such that

$$(1/\tau^*)dt = \text{probability that any radioactive atom will decay in the next infinitesimal time interval } [t,t+dt). \quad (6.3-11)$$

In consequence of this property, the "lifetime" of a radioactive atom is an exponential random variable with mean (and standard deviation) τ^*. The proof of this fact is mathematically the same as the argument we have used several times before in other contexts: Letting $p_0(\tau)$ denote the probability that a radioactive atom will *not* decay in the next time interval τ, then the laws of probability and Eq. (6.3-11) allow us to write

$$p_0(\tau+d\tau) = p_0(\tau)\times[1-(1/\tau^*)d\tau].$$

This relation implies the differential equation $dp_0(\tau)/d\tau = -(1/\tau^*)p_0(\tau)$, and the solution to that differential equation for the obvious initial condition $p_0(0)=1$ is

$$p_0(\tau) = \exp(-\tau/\tau^*).$$

Combining this result with Eq. (6.3-11), we deduce that the probability that the radioactive atom will decay between times τ and $\tau+d\tau$ is

$$p_0(\tau)\times(1/\tau^*)d\tau = (1/\tau^*)\exp(-\tau/\tau^*)\,d\tau.$$

Comparing this expression with the form of the exponential density function (1.4-5), we conclude that the lifetime of the radioactive atom is indeed the random variable $E(1/\tau^*)$.

Against this background, we now define the **radioactive decay process** $X(t)$ by

$$X(t) = \text{number of undecayed radioactive atoms at time } t \geq t_0,$$
$$\text{given } n_0 \text{ undecayed radioactive atoms at time } t_0. \qquad (6.3\text{-}12)$$

To show that $X(t)$ as thus defined is in fact a temporally homogeneous birth-death Markov process, we begin by noting that the decays of individual radioactive atoms occur *independently* of each other; this is because τ^* in Eq. (6.3-11) is independent of anything happening outside the nucleus of each radioactive atom. So from Eq. (6.3-11) and the laws of probability we may infer that, of n radioactive atoms at time t, the probability that exactly k of them will decay in $[t,t+dt)$ is

$$\binom{n}{k}\left(dt/\tau^*\right)^k\left(1-dt/\tau^*\right)^{n-k} = \begin{cases} 1-n(1/\tau^*)dt) + o(dt), & \text{if } k=0, \\ n(1/\tau^*)dt) + o(dt), & \text{if } k=1, \\ o(dt), & \text{if } k\geq 2. \end{cases}$$

On the left side here, the second factor is the probability that k specified radioactive atoms *will* decay in $[t,t+dt)$, the third factor is the probability that the remaining $n-k$ atoms will *not* decay in $[t,t+dt)$, and the first factor is the number of distinct ways of selecting k out of n atoms. The right hand side follows by simply expanding the left hand side to first order in dt. So we see that, to first order in dt, either some *one* of the n radioactive atoms will decay in the next dt, this with probability $n(1/\tau^*)dt$, or else *none* of the n radioactive atoms will decay in the next dt, this with probability $1-n(1/\tau^*)dt$. Since a single decay would reduce the population of the radioactive atoms from n to $n-1$, then we conclude from

Eqs. (6.1-4) and (6.1-5) that the radioactive decay process is the temporally homogeneous birth-death Markov process with

$$a(n) = (1/\tau^*)n, \quad w_+(n) = 0, \quad w_-(n) = 1. \tag{6.3-13}$$

Notice that the first of Eqs. (6.3-13) asserts that $a(0)=0$, and hence by Eq. (6.1-16) that the mean pausing time in state $n=0$ is infinite. Once the process $X(t)$ reaches the state $n=0$, it will stay there forever. The physical basis for this is of course that, once all n_0 radioactive atoms have decayed, nothing else is going to happen. The radioactive decay process is thus an example of a "death-only" Markov process with an "absorbing state" at $n=0$.

It follows from Eqs. (6.3-13) and the definition (6.1-9) that the stepping functions for the radioactive decay process are given by

$$W_+(n) = 0 \quad \text{and} \quad W_-(n) = (1/\tau^*)n. \tag{6.3-14}$$

By substituting these formulas into the general birth-death forward master equation (6.1-17), we get

$$\frac{\partial}{\partial t} P(n,t \mid n_0,t_0) = \frac{n+1}{\tau^*} P(n+1,t \mid n_0,t_0) - \frac{n}{\tau^*} P(n,t \mid n_0,t_0).$$
$$\tag{6.3-15}$$

Because of the "one-way" nature of the radioactive decay process, we can effect a solution to this master equation in much the same way that we obtained the solution (6.3-4) for the Poisson process: Since up-going transitions never occur, then $P(n_0+1,t \mid n_0,t_0) \equiv 0$ for all $t \geq t_0$; so, abbreviating $P(n,t \mid n_0,t_0) \equiv P_n(t)$, we see that Eq. (6.3-15) reads for $n = n_0$,

$$\frac{\partial}{\partial t} P_{n_0}(t) = -(n_0/\tau^*) P_{n_0}(t).$$

The solution to this equation for the required initial condition $P_{n_0}(t_0) = 1$ is obviously

$$P_{n_0}(t) = e^{-n_0(t-t_0)/\tau^*}$$

With this result, we have from Eq. (6.3-15) the following differential equation for $P_{n_0-1}(t)$:

$$\frac{\partial}{\partial t} P_{n_0-1}(t) = -\frac{n_0-1}{\tau^*} P_{n_0-1}(t) + \frac{n_0}{\tau^*} e^{-n_0(t-t_0)/\tau^*}.$$

This differential equation is of the form (A-11a), so since $P_{n_0-1}(t_0) = 0$ we get from Eq. (A-11b) the solution

$$P_{n_0-1}(t) = e^{-(n_0-1)(t-t_0)/\tau^*} \int_{t_0}^{t} dt' \frac{n_0}{\tau^*} e^{-n_0(t'-t_0)/\tau^*} e^{(n_0-1)(t'-t_0)/\tau^*}$$

$$= n_0 e^{-(n_0-1)(t-t_0)/\tau^*} \int_0^{(t-t_0)/\tau^*} d\tau\, e^{-\tau}$$

$$= n_0 \left[e^{-(t-t_0)/\tau^*} \right]^{n_0-1} \left[1 - e^{-(t-t_0)/\tau^*} \right].$$

Substituting this result into Eq. (6.3-15) for $n = n_0 - 2$ and again invoking Eqs. (A-11), we obtain the formula for $P(n_0-2, t \mid n_0, t_0)$. A straightforward induction argument leads to the general formula

$$P(n,t \mid n_0, t_0) = \frac{n_0!}{n!\,(n_0-n)!} \left[e^{-(t-t_0)/\tau^*} \right]^n \left[1 - e^{-(t-t_0)/\tau^*} \right]^{n_0-n}$$

$$(t \geq t_0;\ n = 0, 1, ..., n_0). \quad (6.3\text{-}16)$$

Comparing Eq. (6.3-16) with Eqs. (1.7-13), we conclude that $X(t)$ is the *binomial random variable* with mean $n_0 \exp[-(t-t_0)/\tau^*]$ and variance $n_0 \exp[-(t-t_0)/\tau^*]\{1 - \exp[-(t-t_0)/\tau^*]\}$. Furthermore, we can easily see in retrospect the correctness of Eq. (6.3-16): The second factor on the right is the probability that a particular set of n of the original n_0 radioactive atoms will *not* decay in $[t_0,t)$; the third factor is the probability that the other (n_0-n) radioactive atoms *will* decay in $[t_0,t)$; and the first factor is the number of distinct ways of dividing n_0 atoms into two groups of n and (n_0-n) atoms.

The time-integral $S(t)$ of the radioactive decay process $X(t)$ would seem to have no physical significance in the context of a population of radioactive atoms. However, if each radioactive atom represented an entity (such as an animal) that consumed some resource (such as oxygen) at a constant rate of 1 unit of that resource per unit of time, then $S(t)$ would represent the total amount of the resource consumed by all entities between times t_0 and t. In any case, time-evolution equations for the moments of the radioactive decay process and its time-integral are obtained by substituting into the general birth-death equations (6.1-26) – (6.1-39) the radioactive decay formulas for $a(n)$ and $v(n)$. The function $a(n)$ is as given in Eqs. (6.3-13). And the function $v(n)$ is, from Eqs. (6.3-14),

$$v(n) \equiv W_+(n) - W_-(n) = -(1/\tau^*)n. \qquad (6.3\text{-}17)$$

So for the radioactive decay process, we have $v(n) = -a(n) = -(1/\tau^*)n$. It then follows from Eq. (6.1-26) that the moments of the radioactive decay process satisfy the coupled differential equations

$$\frac{d}{dt}\langle X^k(t)\rangle = \frac{1}{\tau^*}\sum_{l=1}^{k}(-1)^l\binom{k}{l}\langle X^{k-l+1}(t)\rangle$$

$$(t_0 \le t,\ k=1,2,\ldots) \quad (6.3\text{-}18)$$

and the initial condition $\langle X^k(t_0)\rangle = n_0^k$. And it follows from Eqs. (6.1-27) and (6.1-28) that the moments of the time-integral of the radioactive decay process are given by

$$\langle S^m(t)\rangle = m\int_{t_0}^{t}\langle S^{m-1}(t')\,X(t')\rangle\,dt' \quad (t_0 \le t,\ m=1,2,\ldots), \quad (6.3\text{-}19)$$

wherein the integrand is obtained from the solutions to the set of differential equations

$$\frac{d}{dt}\langle S^m(t)\,X^k(t)\rangle = m\langle S^{m-1}(t)\,X^{k+1}(t)\rangle$$

$$+ \frac{1}{\tau^*}\sum_{l=1}^{k}(-1)^l\binom{k}{l}\langle S^m(t)\,X^{k-l+1}(t)\rangle$$

$$(t_0 \le t,\ m \ge 1;\ k \ge 1) \quad (6.3\text{-}20)$$

for the initial conditions $\langle S^m(t_0)\,X^k(t_0)\rangle = 0$. An inspection of the above equations will show them all to be closed, and hence solvable exactly to all orders. We shall not bother to work out all the solutions here, but as an illustrative exercise we shall derive the formulas for the means, variances and covariances of $X(t)$ and $S(t)$.

By substituting $v(n) = -a(n) = -(1/\tau^*)n$ into Eqs. (6.1-29) – (6.1-31), we obtain the following explicit time-evolution equations for the mean, variance and covariance of the radioactive decay process $X(t)$:

$$\frac{d}{dt}\langle X(t)\rangle = -(1/\tau^*)\langle X(t)\rangle \qquad (t_0 \le t), \quad (6.3\text{-}21a)$$

$$\frac{d}{dt}\text{var}\{X(t)\} = -2(1/\tau^*)\,\text{var}\{X(t)\} + (1/\tau^*)\langle X(t)\rangle \qquad (t_0 \le t), \quad (6.3\text{-}21b)$$

$$\frac{d}{dt_2}\text{cov}\{X(t_1),X(t_2)\} = -(1/\tau^*)\,\text{cov}\{X(t_1),X(t_2)\} \quad (t_0 \le t_1 \le t_2). \quad (6.3\text{-}21c)$$

The solution to Eq. (6.3-21a) subject to the initial condition (6.1-32) is obviously

$$\langle X(t)\rangle = n_0 e^{-(t-t_0)/\tau^*} \qquad (t_0 \le t). \qquad (6.3\text{-}22)$$

Substituting this result into the right hand side of Eq. (6.3-21b), and taking note of the null initial condition (6.1-33), we find with the aid of Eqs. (A-11) that

$$\text{var}\{X(t)\} = e^{-2(t-t_0)/\tau^*} \int_{t_0}^{t} dt'\, (1/\tau^*)\, n_0 e^{-(t'-t_0)/\tau^*}\, e^{2(t'-t_0)/\tau^*}$$

$$= n_0 e^{-2(t-t_0)/\tau^*} \int_{0}^{(t-t_0)/\tau^*} e^{\tau}\, d\tau$$

$$= n_0 e^{-2(t-t_0)/\tau^*} \left[e^{(t-t_0)/\tau^*} - 1 \right],$$

so

$$\text{var}\{X(t)\} = n_0 e^{-(t-t_0)/\tau^*} \left| 1 - e^{-(t-t_0)/\tau^*} \right| \qquad (t_0 \le t). \qquad (6.3\text{-}23)$$

We note that the formulas (6.3-22) and (6.3-23) for the mean and variance of $X(t)$ are consistent with the binomial density function formula for $P(n,t\,|\,n_0,t_0)$ in Eq. (6.3-16). In particular, both the formula pair (6.3-22) and (6.3-23) and the single formula (6.3-16) imply the physically obvious result that $X(t)$ approaches the sure number zero as $(t-t_0)\to\infty$. Finally, the solution to the covariance evolution equation (6.3-21c) subject to the initial condition (6.1-34) is clearly

$$\text{cov}\{X(t_1),X(t_2)\} = \text{var}\{X(t_1)\}\, e^{-(t_2-t_1)/\tau^*}.$$

Invoking the result (6.3-23), we thus have

$$\text{cov}\{X(t_1),X(t_2)\} = n_0 e^{-(t_2-t_1)/\tau^*}\, e^{-(t_1-t_0)/\tau^*} \left| 1 - e^{-(t_1-t_0)/\tau^*} \right|$$

$$(t_0 \le t_1 \le t_2). \qquad (6.3\text{-}24)$$

For the time-integral $S(t)$ of the radioactive decay process, the mean is calculated, according to Eq. (6.1-35), by simply integrating $\langle X(t)\rangle$ in Eq. (6.3-22):

$$\langle S(t)\rangle = \int_{t_0}^{t} n_0 e^{-(t'-t_0)/\tau^*}\, dt' = n_0 \tau^* \int_{0}^{(t-t_0)/\tau^*} e^{-\tau}\, d\tau;$$

thus,

$$\langle S(t)\rangle = n_0 \tau^* \left[1 - e^{-(t-t_0)/\tau^*}\right] \qquad (t_0 \leq t). \qquad (6.3\text{-}25)$$

The variance of $S(t)$ is obtained by evaluating the integral (6.1-36). But to do that, we must first calculate its integrand by solving the differential equation (6.1-37a) subject to the null initial condition (6.1-37b). That differential equation reads, for $v(n) = -(1/\tau^*)n$,

$$\frac{d}{dt}\,\text{cov}\{S(t),X(t)\} = -(1/\tau^*)\,\text{cov}\{S(t),X(t)\} + \text{var}\{X(t)\}.$$

With the null initial condition (6.1-37b), we have from Eqs. (A-11) the solution

$$\text{cov}\{S(t),X(t)\} = e^{-(t-t_0)/\tau^*} \int_{t_0}^{t} \text{var}\{X(t')\}\, e^{(t'-t_0)/\tau^*}\, dt'$$

$$= e^{-(t-t_0)/\tau^*} \int_{t_0}^{t} n_0 \left[1 - e^{-(t'-t_0)/\tau^*}\right] dt',$$

where in the last step we have invoked Eq. (6.3-23). The t'-integration here is straightforwardly accomplished, and gives

$$\text{cov}\{S(t),X(t)\} = n_0 e^{-(t-t_0)/\tau^*} \left[(t-t_0) - \tau^* + \tau^* e^{-(t-t_0)/\tau^*}\right].$$

So it follows from Eq. (6.1-36) that

$$\text{var}\{S(t)\} = 2n_0 \int_{t_0}^{t} e^{-(t'-t_0)/\tau^*} \left[(t'-t_0) - \tau^* + \tau^* e^{-(t'-t_0)/\tau^*}\right] dt'$$

The integral here is straightforwardly evaluated, with the result

$$\text{var}\{S(t)\} = n_0 \tau^{*\,2} \left[1 - e^{-2(t-t_0)/\tau^*} - \frac{2(t-t_0)}{\tau^*} e^{-(t-t_0)/\tau^*}\right]$$

$$(t_0 \leq t). \qquad (6.3\text{-}26)$$

Notice from Eqs. (6.3-25) and (6.3-26) that

$$\lim_{(t-t_0)\to\infty} \langle S(t)\rangle = n_0 \tau^* \quad \text{and} \quad \lim_{(t-t_0)\to\infty} \text{var}\{S(t)\} = n_0 \tau^{*\,2}. \qquad (6.3\text{-}27a)$$

It follows that the *relative* standard deviation in $S(t)$ satisfies

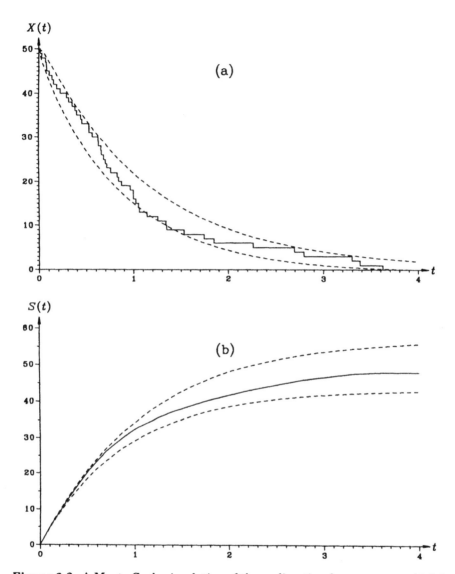

Figure 6-2. A Monte Carlo simulation of the *radioactive decay process*, which is the temporally homogeneous birth-death Markov process with stepping functions $W_+(n) = 0$ and $W_-(n) = -(1/\tau^*)n$. We have taken $\tau^* = 1$, $n_0 = 50$ and $t_0 = 0$. The realization of the process $X(t)$ is shown in (a), along with the one-standard deviation envelope predicted by Eqs. (6.3-22) and (6.3-23). The corresponding realization of the integral process $S(t)$ is shown in (b), along with the one-standard deviation envelope predicted by Eqs. (6.3-25) and (6.3-26).

$$\lim_{(t-t_0)\to\infty} \frac{\mathrm{sdev}\{S(t)\}}{\langle S(t)\rangle} = \frac{1}{\sqrt{n_0}}. \tag{6.3-27b}$$

Finally, the covariance of $S(t)$ is obtained by evaluating the integral (6.1-38). But first we must calculate the integrand there by solving the differential equation (6.1-39a) subject to the initial condition (6.1-39b). That differential equation reads, for $v(n) = -(1/\tau^*)n$,

$$\frac{d}{dt_2} \mathrm{cov}\{S(t_1),X(t_2)\} = -(1/\tau^*)\,\mathrm{cov}\{S(t_1),X(t_2)\}.$$

The solution to this equation for the initial condition (6.1-39b) is clearly

$$\mathrm{cov}\{S(t_1),X(t_2)\} = \mathrm{cov}\{S(t_1),X(t_1)\}e^{-(t_2-t_1)/\tau^*},$$

or, recalling the formula for $\mathrm{cov}\{S(t),X(t)\}$ obtained in the course of deriving formula (6.3-26),

$$\mathrm{cov}\{S(t_1),X(t_2)\} = n_0 e^{-(t_2-t_1)/\tau^*}\, e^{-(t_1-t_0)/\tau^*}$$
$$\times\left[(t_1-t_0) - \tau^* + \tau^* e^{-(t_1-t_0)/\tau^*}\right].$$

Substituting this expression into Eq. (6.1-38), we find that the t_2'-integration there can be straightforwardly performed, with the result

$$\mathrm{cov}\{S(t_1),S(t_2)\} = \mathrm{var}\{S(t_1)\} + n_0\tau^*\left[1 - e^{-(t_2-t_1)/\tau^*}\right]e^{-(t_1-t_0)/\tau^*}$$
$$\times\left[(t_1-t_0) - \tau^* + \tau^* e^{-(t_1-t_0)/\tau^*}\right]$$
$$(t_0\le t_1\le t_2). \tag{6.3-28}$$

Figure 6-2 shows the results of a Monte Carlo simulation of the radioactive decay process for $\tau^*=1$, $n_0=50$ and $t_0=0$. The simulation was carried out using the exact birth-death algorithm outlined in Fig. 6-1. In Fig. 6-2a, the solid curve is the realization of $X(t)$, and the dashed curves show the one-standard deviation envelope $\langle X(t)\rangle\pm\mathrm{sdev}\{X(t)\}$ as calculated from formulas (6.3-22) and (6.3-23). In Fig. 6-2b, the solid curve is the realization of $S(t)$, and the dashed curves show the one-standard deviation envelope $\langle S(t)\rangle\pm\mathrm{sdev}\{S(t)\}$ as calculated from formulas (6.3-25) and (6.3-26).

6.3.C THE RANDOM TELEGRAPH PROCESS

Consider a telegraph key, which when open is said to be in state $n = 0$, and when closed is said to be in state $n = 1$. Suppose this key opens and closes randomly in accordance with the following rule: There exists two positive constants a_0 and a_1 such that

$$a_n dt = \text{probability that the key, if in state } n \, (= 0 \text{ or } 1)$$
$$\text{at time } t, \text{ will jump to the other state in the}$$
$$\text{next infinitesimal time interval } [t, t + dt). \qquad (6.3\text{-}29)$$

In other words, the key jumps back and forth between its open $(n = 0)$ state and its closed $(n = 1)$ state, and its dwell time in state n is an exponential random variable with mean $1/a_n$. We let

$$X(t) = \text{state of the key at time } t \geq t_0, \text{ given that the}$$
$$\text{key was in state } n_0 \, (= 0 \text{ or } 1) \text{ at time } t_0, \qquad (6.3\text{-}30)$$

and we call $X(t)$ the **random telegraph process**. By comparing Eq. (6.3-29) with Eq. (6.1-10), we can immediately conclude that $X(t)$ is a temporally homogeneous birth-death Markov process with stepping functions

$$W_+(n) = a_0 \delta(n,0) \quad \text{and} \quad W_-(n) = a_1 \delta(n,1). \qquad (6.3\text{-}31)$$

It then follows from Eqs. (6.1-13) that the functions $a(n)$ and $v(n)$ for this birth-death Markov process are

$$a(n) = \begin{cases} a_0, & \text{if } n = 0 \\ a_1, & \text{if } n = 1 \\ 0, & \text{otherwise} \end{cases} \quad \text{and} \quad v(n) = \begin{cases} a_0, & \text{if } n = 0 \\ -a_1, & \text{if } n = 1 \\ 0, & \text{otherwise} \end{cases}. \qquad (6.3\text{-}32)$$

And it follows from these formulas and the definition (6.1-9) that the functions $w_+(n)$ and $w_-(n)$ for this birth-death Markov process are

$$w_+(n) = \delta(n,0) \quad \text{and} \quad w_-(n) = \delta(n,1). \qquad (6.3\text{-}33)$$

When the formulas for $W_\pm(n)$ in Eqs. (6.3-31) are substituted into the birth-death forward master equation (6.1-17), the result is the following pair of coupled differential equations for $P(0,t \mid n_0, t_0)$ and $P(1,t \mid n_0, t_0)$:

$$\begin{cases} \dfrac{d}{dt} P(0,t \mid n_0, t_0) = a_1 P(1,t \mid n_0, t_0) - a_0 P(0,t \mid n_0, t_0), & (6.3\text{-}34a) \\[2ex] \dfrac{d}{dt} P(1,t \mid n_0, t_0) = a_0 P(0,t \mid n_0, t_0) - a_1 P(1,t \mid n_0, t_0). & (6.3\text{-}34b) \end{cases}$$

The coupling between these two differential equations is easily handled, because we know in advance that for all $t \geq t_0$,

$$P(0,t \mid n_0,t_0) + P(1,t \mid n_0,t_0) = 1,$$

a relation that can be formally deduced by simply adding Eqs. (6.3-34) together and recalling the initial condition $P(n,t_0 \mid n_0,t_0) = \delta(n,n_0)$. So by substituting

$$P(1,t \mid n_0,t_0) = 1 - P(0,t \mid n_0,t_0)$$

into the right hand side of Eq. (6.3-34a), we get the following differential equation for $P(0,t \mid n_0,t_0)$:

$$\frac{d}{dt} P(0,t \mid n_0,t_0) = -(a_0 + a_1) P(0,t \mid n_0,t_0) + a_1.$$

This differential equation is seen to be of the form (A-11a), so by Eq. (A-11b) its solution subject to the required initial condition $P(0,t_0 \mid n_0,t_0) = \delta(0,n_0)$ is

$$P(0,t \mid n_0,t_0) = e^{-(a_0 + a_1)(t - t_0)} \left\{ \delta(0,n_0) + \int_{t_0}^{t} dt' \, a_1 \, e^{(a_0 + a_1)(t' - t_0)} \right\}$$

$$= e^{-(a_0 + a_1)(t - t_0)} \left\{ \delta(0,n_0) + \frac{a_1}{a_0 + a_1} \left[e^{(a_0 + a_1)(t - t_0)} - 1 \right] \right\}$$

or

$$P(0,t \mid n_0,t_0) = \frac{a_1}{a_0 + a_1} + \left(\delta(0,n_0) - \frac{a_1}{a_0 + a_1} \right) e^{-(a_0 + a_1)(t - t_0)}.$$

$$(6.3\text{-}35a)$$

And by simply subtracting this result from unity, we deduce that $P(1,t \mid n_0,t_0)$ is given by

$$P(1,t \mid n_0,t_0) = \frac{a_0}{a_0 + a_1} + \left(\delta(1,n_0) - \frac{a_0}{a_0 + a_1} \right) e^{-(a_0 + a_1)(t - t_0)}.$$

$$(6.3\text{-}35b)$$

Equations (6.3-35) completely specify the Markov state density function of the random telegraph process. Notice that these formulas imply that

$$\lim_{(t - t_0) \to \infty} P(n,t \mid n_0,t_0) = \begin{cases} a_1/(a_0 + a_1), & \text{if } n=0, \\ a_0/(a_0 + a_1), & \text{if } n=1. \end{cases} \qquad (6.3\text{-}36)$$

To see that these limit results are plausible, observe that if $a_1 > a_0$ then the telegraph key will always be more eager to leave the 1-state than the 0-state; so after a very long time, the key should be more likely to be found in the 0-state than in the 1-state, just as Eq. (6.3-36) implies.

To find the moments $\langle X^k(t) \rangle$ of the random telegraph process, we could proceed as usual by way of the time-evolution equations (6.1-26). The trick to doing that is to first recognize that the two functions $a(n)$ and $v(n)$ in Eqs. (6.3-32) can be expressed *analytically* as

$$\left. \begin{array}{l} a(n) = a_0 - (a_0 - a_1)n \\[2mm] v(n) = a_0 - (a_0 + a_1)n \end{array} \right\} \quad (n = 0,1). \qquad (6.3\text{-}37)$$

Since these two functions are linear in n, then the set of differential equations for the moments of $X(t)$ will be closed, and hence solvable to all orders. However, because this process has access to only the two states 0 and 1, it turns out to be much easier to calculate $\langle X^k(t) \rangle$ directly from its definition:

$$\langle X^k(t) \rangle \equiv \sum_{n=-\infty}^{\infty} n^k \, P(n,t \mid n_0,t_0)$$

$$= (0)^k \, P(0,t \mid n_0,t_0) + (1)^k \, P(1,t \mid n_0,t_0).$$

Therefore, for the random telegraph process we have *for all $k \geq 1$*,

$$\langle X^k(t) \rangle = P(1,t \mid n_0,t_0) = \frac{a_0}{a_0 + a_1} + \left(\delta(1,n_0) - \frac{a_0}{a_0 + a_1} \right) e^{-(a_0 + a_1)(t - t_0)}.$$

$$(6.3\text{-}38)$$

In particular, the mean of $X(t)$ is

$$\langle X(t) \rangle = P(1,t \mid n_0,t_0). \qquad (6.3\text{-}39)$$

And the variance of $X(t)$ is

$$\text{var}\{X(t)\} = \langle X^2(t) \rangle - \langle X(t) \rangle^2 = P(1,t \mid n_0,t_0) - P^2(1,t \mid n_0,t_0)$$

$$= P(1,t \mid n_0,t_0) \, [1 - P(1,t \mid n_0,t_0)]$$

or

$$\text{var}\{X(t)\} = P(1,t \mid n_0,t_0) \, P(0,t \mid n_0,t_0). \qquad (6.3\text{-}40)$$

Formulas (6.3-39) and (6.3-40) can of course be made more explicit by substituting from Eqs. (6.3-35). By appealing to the limit results (6.3-36), it is easy to see from the above formulas for $\langle X(t) \rangle$ and $\text{var}\{X(t)\}$ that

$$\lim_{(t-t_0)\to\infty} \langle X(t)\rangle = \frac{a_0}{a_0+a_1} \quad \text{and} \quad \lim_{(t-t_0)\to\infty} \text{var}\{X(t)\} = \frac{a_0 a_1}{(a_0+a_1)^2}.$$

$$(6.3\text{-}41)$$

The covariance of $X(t)$ can likewise be calculated directly from its definition: For $t_0 \le t_1 \le t_2$, we have

$$\text{cov}\{X(t_1),X(t_2)\} \equiv \langle X(t_1)X(t_2)\rangle - \langle X(t_1)\rangle\langle X(t_2)\rangle$$

$$= \sum_{n_1=0}^{1}\sum_{n_2=0}^{1} n_1 n_2 \, P(n_2,t_2|n_1,t_1)\,P(n_1,t_1|n_0,t_0)$$

$$- P(1,t_1|n_0,t_0)P(1,t_2|n_0,t_0)$$

$$= P(1,t_2|1,t_1)P(1,t_1|n_0,t_0)$$

$$- P(1,t_1|n_0,t_0)P(1,t_2|n_0,t_0),$$

so

$$\text{cov}\{X(t_1),X(t_2)\} = P(1,t_1|n_0,t_0)\Big[P(1,t_2|1,t_1) - P(1,t_2|n_0,t_0)\Big]$$

$$(t_0 \le t_1 \le t_2). \quad (6.3\text{-}42)$$

Again, this formula can be made more explicit by substituting from Eqs. (6.3-35). We might note that, upon making those substitutions and then passing to the limit $t_0 \to -\infty$, we discover that

$$\lim_{t_0\to-\infty} \text{cov}\{X(t_1),X(t_2)\} = \frac{a_0 a_1}{(a_0+a_1)^2}\, e^{-(a_0+a_1)(t_2-t_1)}$$

$$(t_1 \le t_2); \quad (6.3\text{-}43)$$

this implies that $(a_0+a_1)^{-1}$ is the asymptotic "decorrelation time" for the process $X(t)$.

Now let us suppose that, when the telegraph key is in its closed $(n=1)$ state, it allows passage of a unit amount of electrical energy per unit of time. Then the time-integral $S(t)$ of the random telegraph process $X(t)$ can be interpreted as the total amount of electrical energy passed by the key between times t_0 and t. We can calculate the moments of $S(t)$ by solving the time-evolution equations (6.1-27) and (6.1-28), using for $a(n)$ and $v(n)$ the functions in Eqs. (6.3-37). But we shall content ourselves here with calculating only the mean, variance and covariance of $S(t)$.

For the mean of $S(t)$, we have from Eqs. (6.1-35) and (6.3-39),

$$\langle S(t) \rangle = \int_{t_0}^{t} \langle X(t') \rangle \, dt' = \int_{t_0}^{t} P(1,t' \mid n_0, t_0) \, dt'.$$

Substituting from Eq. (6.3-35b) and then performing the t'-integration, we get the result

$$\langle S(t) \rangle = \frac{1}{a_0 + a_1} \left[a_0(t - t_0) + \left(\delta(1, n_0) - \frac{a_0}{a_0 + a_1} \right) \left(1 - e^{-(a_0 + a_1)(t - t_0)} \right) \right]$$

$$(t_0 \le t). \quad (6.3\text{-}44)$$

For the variance of $S(t)$, we must evaluate the integral (6.1-36), wherein the integrand is the solution of the differential equation (6.1-37a) for the initial condition (6.1-37b). That differential equation reads, for the $v(n)$ function in Eqs. (6.3-37),

$$\frac{d}{dt} \text{cov}\{S(t), X(t)\} = \text{var}\{X(t)\} + \langle S(t) [a_0 - (a_0 + a_1) X(t)] \rangle$$

$$- \langle S(t) \rangle \langle a_0 - (a_0 + a_1) X(t) \rangle$$

$$= -(a_0 + a_1) \text{cov}\{S(t), X(t)\} + \text{var}\{X(t)\}.$$

This equation is of the form (A-11a), so its solution for the null initial condition (6.1-37b) is, by Eq. (A-11b),

$$\text{cov}\{S(t), X(t)\} = e^{-(a_0 + a_1)(t - t_0)} \int_{t_0}^{t} \text{var}\{X(t')\} e^{(a_0 + a_1)(t' - t_0)} \, dt'.$$

Substituting into the integrand the formulas (6.3-40) and (6.3-35), we find after a straightforward but somewhat tedious integration that

$$\text{cov}\{S(t), X(t)\}$$

$$= \frac{a_0 a_1}{(a_0 + a_1)^3} \left\{ 1 + \left(\frac{a_{n_0}^2}{a_0 a_1} - 1 \right) (a_0 + a_1)(t - t_0) e^{-(a_0 + a_1)(t - t_0)} \right.$$

$$\left. - \left(\frac{a_{n_0}^2}{a_0 a_1} + 1 \right) e^{-(a_0 + a_1)(t - t_0)} + \frac{a_{n_0}^2}{a_0 a_1} e^{-2(a_0 + a_1)(t - t_0)} \right\}.$$

$$(6.3\text{-}45)$$

Now substituting this into Eq. (6.1-36) and integrating, we conclude that the variance of the integral of the random telegraph process is

$$\text{var}\{S(t)\} = \frac{2a_0 a_1}{(a_0 + a_1)^4} \left\{ (a_0 + a_1)(t - t_0) \right.$$

$$+ \left(1 - \frac{a_{n_0}^2}{a_0 a_1} \right) (a_0 + a_1)(t - t_0) e^{-(a_0 + a_1)(t - t_0)}$$

$$- 2 \left(1 - e^{-(a_0 + a_1)(t - t_0)} \right)$$

$$\left. + \frac{a_{n_0}^2}{2a_0 a_1} \left(1 - e^{-2(a_0 + a_1)(t - t_0)} \right) \right\}. \qquad (t_0 \le t) \qquad (6.3\text{-}46)$$

An inspection of Eqs. (6.3-44) and (6.3-46) shows that, in the long-time limit, the mean and variance of $S(t)$ satisfy

$$\lim_{(t - t_0) \to \infty} \langle S(t) \rangle = \frac{a_0}{a_0 + a_1} (t - t_0), \qquad \lim_{(t - t_0) \to \infty} \text{var}\{S(t)\} = \frac{2a_0 a_1}{(a_0 + a_1)^3} (t - t_0).$$

$$(6.3\text{-}47)$$

The first of these two asymptotic formulas says, quite reasonably, that $\langle S(t \to \infty) \rangle$ is the product of the asymptotic fraction of the time that the key spends in the closed state 1 [see Eqs. (6.3-36)] multiplied by the total elapsed time. And the second formula implies that sdev$\{S(t \to \infty)\}$ is proportional to the square root of the total elapsed time.

Finally, the covariance of $S(t)$ is obtained by evaluating the integral in Eq. (6.1-38), wherein the integrand is the solution of the differential equation (6.1-39a) for the initial condition (6.1-39b). That differential equation reads, for $v(n)$ as given in Eqs. (6.3-37),

$$\frac{d}{dt_2} \text{cov}\{S(t_1), X(t_2)\} = \langle S(t_1)[a_0 - (a_0 + a_1) X(t_2)] \rangle$$

$$- \langle S(t_1) \rangle \langle a_0 - (a_0 + a_1) X(t_2) \rangle$$

$$= -(a_0 + a_1) \text{cov}\{S(t_1), X(t_2)\}.$$

The solution to this differential equation for the initial condition (6.1-39b) is obviously

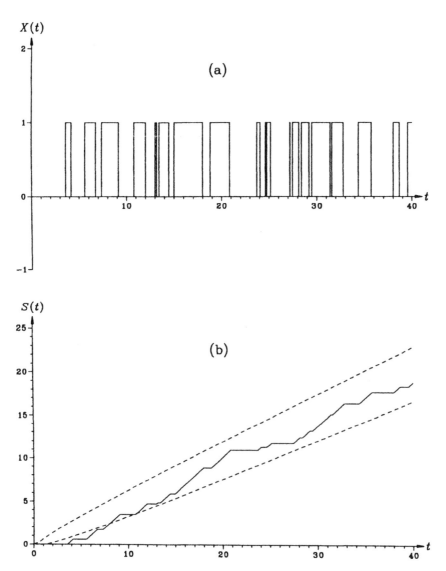

Figure 6-3. A Monte Carlo simulation of the *random telegraph process* with stepping functions $W_+(n) = \delta(n,0)$ and $W_-(n) = \delta(n,1)$. The realization of the process $X(t)$ is shown in (a); since its one-standard deviation envelope very quickly approaches the unit interval, it has not been plotted for the sake of clarity. The realization of the integral process $S(t)$ is the solid curve in (b), and the dashed curves there are the corresponding one-standard deviation envelope predicted by Eqs. (6.3-44) and (6.3-46) with $t_0 = n_0 = 0$ and $a_0 = a_1 = 1$.

$$\text{cov}\{S(t_1),X(t_2)\} = \text{cov}\{S(t_1),X(t_1)\}e^{-(a_0+a_1)(t_2-t_1)}.$$

Substituting this into Eq. (6.1-38) and then performing the t_2'-integration, we conclude that the covariance of the integral of the random telegraph process is given by

$$\text{cov}\{S(t_1),S(t_2)\} = \text{var}\{S(t_1)\} + \text{cov}\{S(t_1),X(t_1)\}\left(\frac{1 - e^{-(a_0+a_1)(t_2-t_1)}}{a_0+a_1}\right)$$

$$(t_0 \le t_1 \le t_2). \quad (6.3\text{-}48)$$

This formula can of course be made fully explicit by substituting from Eqs. (6.3-46) and (6.3-45).

Figure 6-3 shows the results of a Monte Carlo simulation of the random telegraph process for $a_0 = a_1 = 1$, $n_0 = 0$ and $t_0 = 0$. The simulation was carried out using the exact birth-death algorithm of Fig. 6-1. Figure 6-3a shows the realization of $X(t)$. The theoretical one-standard deviation envelope $\langle X(t)\rangle \pm \text{sdev}\{X(t)\}$, as calculated from Eqs. (6.3-39), (6.3-40) and (6.3-35), achieves its asymptotic form $1/2 \pm 1/2$ [see Eqs. (6.3-41)] in just a few mean jumping times. But we have not plotted that envelope in order to avoid confusion with the $X(t)$ curve, which jumps back and forth between those same limits. In Fig. 6-3b, the solid curve shows the realization of the corresponding time-integral process $S(t)$, and the dashed curves show the one-standard deviation envelope $\langle S(t)\rangle \pm \text{sdev}\{S(t)\}$ as calculated from Eqs. (6.3-44) and (6.3-46).

6.3.D THE PAYROLL PROCESS

Consider a business organization in which the hiring and leaving of employees is governed by the following probabilistic rules: There exists two positive constants h and l such that, if n is the number of employees on the company payroll at time t, then

$$hdt + o(dt) = \text{probability that a new employee} \atop \text{will be } hired \text{ in } [t,t+dt), \qquad (6.3\text{-}49a)$$

$$nldt + o(dt) = \text{probability that one of the } n \text{ current} \atop \text{employees will } leave \text{ in } [t,t+dt). \qquad (6.3\text{-}49b)$$

Against this background, we define the **payroll process** to be

$$X(t) = \text{number of employees on the company payroll at time } t. \tag{6.3-50}$$

As a consequence of the rules (6.3-49), we have

$$\text{Prob}\{X(t+dt)=m \mid X(t)=n\} = \begin{cases} h\,dt + o(dt), & \text{if } m=n+1, \\ nl\,dt + o(dt), & \text{if } m=n-1, \\ 1 - h\,dt - nl\,dt + o(dt), & \text{if } m=n, \\ o(dt), & \text{if } m \neq n-1 \text{ or } n \text{ or } n+1. \end{cases}$$

Comparing this statement with Eq. (6.1-10), we conclude that $X(t)$ is the temporally homogeneous birth-death Markov process with stepping functions

$$W_+(n) = h \quad \text{and} \quad W_-(n) = ln. \tag{6.3-51}$$

So by Eqs. (6.1-13), the associated functions $a(n)$ and $v(n)$ are

$$a(n) = h + ln \quad \text{and} \quad v(n) = h - ln. \tag{6.3-52}$$

And by the definition (6.1-9), the functions $w_\pm(n)$ are

$$w_+(n) = \frac{h}{h + ln} \quad \text{and} \quad w_-(n) = \frac{ln}{h + ln}. \tag{6.3-53}$$

Comparing Eqs. (6.3-51) with Eqs. (6.3-2) shows that if $l=0$ then the payroll process reduces to the *Poisson* process with $a=h$. And comparing Eqs. (6.3-51) with Eqs. (6.3-14) shows that if $h=0$ then the payroll process reduces to the *radioactive decay* process with $\tau^* = l^{-1}$. These observations suggest that h^{-1} can be physically interpreted as the *average time between hirings*, and l^{-1} as the *average individual employment time*.

Substituting the above formulas for the stepping functions into the birth-death forward master equation (6.1-17) gives

$$\frac{\partial}{\partial t} P(n,t \mid n_0,t_0) = l(n+1) P(n+1,t \mid n_0,t_0)$$

$$+ h P(n-1,t \mid n_0,t_0) - (h+ln) P(n,t \mid n_0,t_0). \tag{6.3-54}$$

With effort, one can obtain an explicit analytical solution to this differential-difference equation for the required initial condition $P(n,t_0 \mid n_0,t_0) = \delta(n,n_0)$, but the algebraic structure of that solution is so complicated that it is of little practical use.† In Subsection 6.4.A we shall

---

† Equation (6.3-54) is the same as Gardiner's *Handbook* equation (7.1.26) with $k_1 = l$ and $k_2 a = h$. For that equation, Gardiner gives the exact, fully analytic solution (7.1.37).

undertake a direct calculation of the *asymptotic* solution $P(n,\infty \mid n_0,t_0)$. For now, we shall content ourselves with calculating only the mean, variance and covariance of $X(t)$. But we should note that, since the functions $v(n)$ and $a(n)$ in Eqs. (6.3-52) are both linear in n, then the time-evolution equations (6.1-26) for the moments $\langle X^k(t)\rangle$ will be closed, and hence solvable exactly for all $k \geq 1$.

To calculate the mean of $X(t)$, we begin by substituting into Eq. (6.1-29) the $v(n)$ function in Eqs. (6.3-52):

$$\frac{d}{dt}\langle X(t)\rangle = \langle h - l\,X(t)\rangle = -l\langle X(t)\rangle + h.$$

This equation is seen to be of the form (A-11a), so by Eq. (A-11b) its solution for the initial condition (6.1-32) is

$$\langle X(t)\rangle = e^{-l(t-t_0)}\left\{ n_0 + \int_{t_0}^{t} h\,e^{l(t'-t_0)}\,dt' \right\}$$

$$= e^{-l(t-t_0)}\left\{ n_0 + h\left[\frac{e^{l(t-t_0)}-1}{l}\right]\right\},$$

or

$$\langle X(t)\rangle = \frac{h}{l} + \left(n_0 - \frac{h}{l}\right)e^{-l(t-t_0)} \qquad (t_0 \leq t). \qquad (6.3\text{-}55)$$

To calculate the variance of $X(t)$, we begin by substituting into Eq. (6.1-30) the $v(n)$ and $a(n)$ functions in Eqs. (6.3-52). That gives

$$\frac{d}{dt}\mathrm{var}\{X(t)\} = 2\Big(\langle X(t)[h-lX(t)]\rangle - \langle X(t)\rangle\langle h-lX(t)\rangle\Big) + \langle h+lX(t)\rangle$$

$$= -2l\,\mathrm{var}\{X(t)\} + h + l\langle X(t)\rangle$$

$$= -2l\,\mathrm{var}\{X(t)\} + 2h + (ln_0 - h)\,e^{-l(t-t_0)},$$

where the last step has invoked the result (6.3-55). Again, this differential equation is of the form (A-11a), so by Eq. (A-11b) its solution subject to the null initial condition (6.1-33) is

$$\mathrm{var}\{X(t)\}$$

$$= e^{-2l(t-t_0)}\int_{t_0}^{t}\left[2h + (ln_0 - h)\,e^{-l(t'-t_0)}\right]e^{2l(t'-t_0)}\,dt'$$

$$= e^{-2l(t-t_0)} \frac{2h}{2h} \left[\frac{e^{2l(t-t_0)} - 1}{2l} \right] + e^{-2l(t-t_0)} (ln_0 - h) \left[\frac{e^{l(t-t_0)} - 1}{l} \right]$$

or

$$\text{var}\{X(t)\} = \frac{h}{l} \left(1 - e^{-l(t-t_0)} \right) \left(1 + \frac{ln_0}{h} e^{-l(t-t_0)} \right)$$

$$(t_0 \le t). \qquad (6.3\text{-}56)$$

It follows from Eqs. (6.3-55) and (6.3-56) that

$$\lim_{(t-t_0) \to \infty} \langle X(t) \rangle = \frac{h}{l} \quad \text{and} \quad \lim_{(t-t_0) \to \infty} \text{var}\{X(t)\} = \frac{h}{l}. \qquad (6.3\text{-}57)$$

Thus, the asymptotic fluctuations of $X(t)$ about its mean will be of the order of the square root of that mean.

To calculate the covariance of $X(t)$, we begin by substituting into Eq. (6.1-31) the formula for $v(n)$ in Eqs. (6.3-52):

$$\frac{d}{dt_2} \text{cov}\{X(t_1), X(t_2)\} = \langle X(t_1)[h - lX(t_2)] \rangle - \langle X(t_1) \rangle \langle h - lX(t_2) \rangle$$

$$= -l \, \text{cov}\{X(t_1), X(t_2)\}.$$

The solution to this differential equation for the initial condition (6.1-34) is clearly

$$\text{cov}\{X(t_1), X(t_2)\} = \text{var}\{X(t_1)\} e^{-l(t_2 - t_1)},$$

or, by invoking the result (6.3-56),

$$\text{cov}\{X(t_1), X(t_2)\} = \frac{h}{l} e^{-l(t_2 - t_1)} \left(1 - e^{-l(t_1 - t_0)} \right) \left(1 + \frac{ln_0}{h} e^{-l(t_1 - t_0)} \right)$$

$$(t_0 \le t_1 \le t_2) \qquad (6.3\text{-}58)$$

Notice that this result implies that

$$\lim_{t_0 \to -\infty} \text{cov}\{X(t_1), X(t_2)\} = \frac{h}{l} e^{-l(t_2 - t_1)} \qquad (t_1 \le t_2). \qquad (6.3\text{-}59)$$

This says that the average employment time l^{-1} is also the "asymptotic decorrelation time" for the process $X(t)$.

Now let us assume that every employee on the company payroll earns wages at the *same rate*. Then by adopting a monetary unit in which that common wage rate is unity, we can interpret the time-integral $S(t)$ of the payroll process $X(t)$ according to

$$S(t) = \text{total employee wages paid out between times } t_0 \text{ and } t. \qquad (6.3\text{-}60)$$

It would obviously be useful for the administrators of the company to be able to estimate this process $S(t)$. We shall content ourselves here with calculating its mean and variance. For the mean, we substitute our result (6.3-55) for $\langle X(t) \rangle$ into the general formula (6.1-35) to get

$$\langle S(t) \rangle = \int_{t_0}^{t} \langle S(t') \rangle \, dt' = \int_{t_0}^{t} \left[\frac{h}{l} + \left(n_0 - \frac{h}{l} \right) e^{-l(t'-t_0)} \right] dt'.$$

This integral is easily evaluated, with the result

$$\langle S(t) \rangle = \frac{h}{l}(t - t_0) + \frac{1}{l}\left(n_0 - \frac{h}{l} \right)\left[1 - e^{-l(t-t_0)} \right]$$

$$(t_0 \le t). \qquad (6.3\text{-}61)$$

To calculate the variance of $S(t)$, we must evaluate the integral (6.1-36), wherein the integrand is the solution of the differential equation (6.1-37a) for the initial condition (6.1-37b). That differential equation reads, for the $v(n)$ function in Eqs. (6.3-52),

$$\frac{d}{dt}\text{cov}\{S(t), X(t)\} = \text{var}\{X(t)\} + \langle S(t)[h - lX(t)] \rangle - \langle S(t) \rangle \langle h - lX(t) \rangle$$

$$= -l\,\text{cov}\{S(t), X(t)\} + \text{var}\{X(t)\}.$$

Once again this equation is of the form (A-11a), so its solution for the null initial condition (6.1-37b) is, by Eq. (A-11b),

$$\text{cov}\{S(t), X(t)\} = e^{-l(t-t_0)} \int_{t_0}^{t} \text{var}\{X(t')\} e^{l(t'-t_0)} \, dt'$$

$$= \frac{h}{l} e^{-l(t-t_0)} \int_{t_0}^{t} \left(e^{l(t'-t_0)} - 1 \right)\left(1 + \frac{ln_0}{h} e^{-l(t'-t_0)} \right) dt',$$

where the second line has invoked the formula (6.3-56) for $\text{var}\{X(t)\}$. This last integral can be straightforwardly evaluated, with the result

$$\text{cov}\{S(t),X(t)\} = \frac{h}{l^2}\left\{1 + \left(\frac{ln_0}{h} - 1\right)l(t-t_0)e^{-l(t-t_0)}\right.$$
$$\left. - \left(\frac{ln_0}{h} + 1\right)e^{-l(t-t_0)} + \frac{ln_0}{h}e^{-2l(t-t_0)}\right\}.$$

Inserting this expression into the general formula (6.1-36) and performing the integration, we conclude that

$$\text{var}\{S(t)\} = \frac{2h}{l^3}\left\{l(t-t_0) + \left(1 - \frac{ln_0}{h}\right)l(t-t_0)e^{-l(t-t_0)}\right.$$
$$\left. - 2\left(1 - e^{-l(t-t_0)}\right) + \frac{ln_0}{2h}\left(1 - e^{-2l(t-t_0)}\right)\right\}$$

$$(t_0 \leq t). \quad (6.3\text{-}62)$$

An inspection of Eqs. (6.3-61) and (6.3-62) shows that, in the long-time limit, the mean and variance of $S(t)$ are given by

$$\lim_{(t-t_0)\to\infty} \langle S(t)\rangle = \frac{h}{l}(t-t_0) \quad \text{and} \quad \lim_{(t-t_0)\to\infty} \text{var}\{S(t)\} = \frac{2h}{l^2}(t-t_0).$$

$$(6.3\text{-}63)$$

The asymptotic formula for $\langle S(t)\rangle$ is seen, in light of the first of Eqs. (6.3-57), to be fairly obvious: it is simply the product of the asymptotic average number of payrolled employees multiplied by the elapsed time. The asymptotic formula for var$\{S(t)\}$ is not so obvious. It implies that the asymptotic fluctuations in $S(t)$ about its mean will be of the order of $(2/l)^{1/2}$ times the square root of the mean.

Figure 6-4 shows the results of a Monte Carlo simulation of the payroll process for $h=0.2$ and $l=0.005$. If the unit of time is *weeks*, then these parameter values describe a situation in which one new employee is hired on the average every $h^{-1}=5$ weeks, and each employee stays with the company for an average of $l^{-1}=200$ weeks. For these parameter values, the asymptotic mean number of employees, as predicted by the first of Eqs. (6.3-57), is $h/l=40$. We have assumed in the simulation that at time $t_0=0$ there were $n_0=20$ employees. The simulation was carried out using the exact birth-death algorithm of Fig. 6-1. In Fig. 6-4a, the solid line is the realization of the employee population, and the dashed curves show the one-standard deviation envelope $\langle X(t)\rangle \pm \text{sdev}\{X(t)\}$ predicted by Eqs. (6.3-55) and (6.3-56). We see that, as predicted by Eqs.

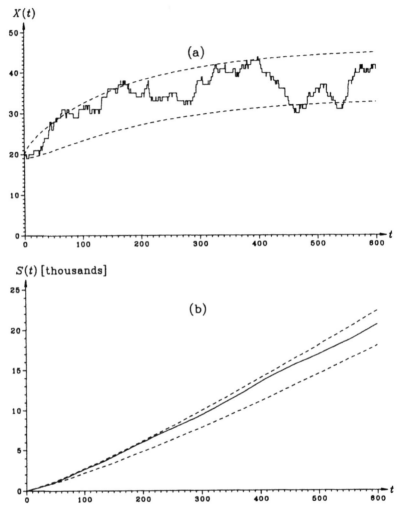

Figure 6-4. A Monte Carlo simulation of the *payroll process* defined by the stepping functions $W_+(n) = h$ and $W_-(n) = ln$, with $h = 0.2$, $l = 0.005$, and $n_0 = 20$. In the physical model, an average of one new employee is added to the company payroll every 5 time units, and a typical employee stays with the company for 200 time units. In (a) the solid curve shows the realization of the process $X(t)$, the number of employees on the payroll at time t, and the dashed curves show the one-standard deviation envelope as calculated from Eqs. (6.3-55) and (6.3-56). In (b) the solid curve shows the realization of the time-integral process $S(t)$, which can be regarded as the total wages paid out to time t, and the dashed curves show its one-standard deviation envelope as calculated from Eqs. (6.3-61) and (6.3-62).

(6.3-57), the employee population approaches the "equilibrium band" $h/l \pm (h/l)^{1/2} = 40 \pm 6.3$ in a time of the order of $l^{-1} = 200$ weeks. In Fig. 6-4b, the solid curve is the corresponding realization of the total wages paid (measured in units of the common employee weekly salary), and the dashed curves show the one-standard deviation envelope $\langle S(t) \rangle \pm \mathrm{sdev}\{S(t)\}$ predicted by Eqs. (6.3-61) and (6.3-62).

Finally, we should mention that by suitably reinterpreting the two parameters h and l, the payroll process $X(t)$ can be regarded as the instantaneous population of X molecules in the chemical system

$$X \underset{c_2}{\overset{c_1}{\rightleftharpoons}} B, \qquad (6.3\text{-}64a)$$

where the population of molecular species B is "buffered" to some effectively constant value N. To see how this comes about, let n be the number of X molecules in the system at time t. Then according to the definition (5.3-29) of the specific reaction probability rate constant, the probability of the forward reaction R_1 occurring in $[t, t + dt)$ is $(c_1 dt)(n)$, and this in turn is equal to $W_-(n)dt$ since reaction R_1 would *decrease* the X molecule population by one. And the probability of the reverse reaction R_2 occurring in $[t, t + dt)$ is $(c_2 dt)(N)$, and this in turn is equal to $W_+(n)dt$ since reaction R_2 would *increase* the X molecule population by one. Thus we have $W_-(n) = c_1 n$ and $W_+(n) = c_2 N$. So, recalling Eqs. (6.3-51), we conclude that the instantaneous X molecule population is the payroll process with

$$l = c_1 \quad \text{and} \quad h = c_2 N. \qquad (6.3\text{-}64b)$$

6.4 STABLE BIRTH-DEATH MARKOV PROCESSES

For *some* temporally homogeneous birth-death Markov processes, the Markov state density function $P(n, t \mid n_0, t_0)$ approaches, as $(t - t_0) \to \infty$, a well behaved function $P_s(n)$ that is independent of t, t_0 and n_0. Processes for which this happens are said to be **stable**, and the limit function $P_s(n)$ is called the **stationary Markov state density function** of the process. In this section we shall investigate the conditions that give rise to stability in a birth-death Markov process, and we shall examine some of the interesting attributes of stable birth-death Markov processes.

6.4.A THE STATIONARY MARKOV STATE DENSITY FUNCTION

Our investigation into the existence and form of the stationary Markov state density function $P_s(n)$ will, for heuristic reasons, proceed in a somewhat indirect manner: We shall begin by *assuming* the existence of $P_s(n)$ and deriving an explicit formula for it from the forward master equation. We shall then examine the mathematical structure of that formula and infer some specific conditions that the stepping functions of the process must satisfy if that formula is to make sense. Finally, we shall rigorously prove (in Appendix F) that *if* the stepping functions of the process *do* satisfy those specific conditions, then the function $P(n,t \mid n_0,t_0)$ will indeed always approach the asserted formula for $P_s(n)$ as $(t-t_0) \to \infty$.

We shall however circumscribe our discussion somewhat by requiring, firstly, that the process $X(t)$ be confined to a contiguous range of integer states $0 \le n \le N$, where N is allowed to be infinite, and secondly that the process be able to reach any state in $[0,N]$ from any other state in $[0,N]$. These two restrictions can be ensured by simply requiring that, for every pair of adjacent states n and $n-1$ contained in the range $[0,N]$, both of the transitions $n-1 \to n$ and $n \to n-1$ shall be possible. So we impose now the requirement that the stepping functions $W_+(n)$ and $W_-(n)$ that define the process according to Eq. (6.1-10) satisfy the condition

$$W_+(n-1) \text{ and } W_-(n) \text{ are both} \begin{cases} >0, & \text{if } n \in [1,N], \\ =0, & \text{if } n \notin [1,N]. \end{cases} \qquad (6.4\text{-}1)$$

In imposing this requirement, we are evidently ruling out the possibility that any state in $[0,N]$ is "absorbing," and also the possibility that any state in $(0,N)$ is "reflecting." The stipulation in (6.4-1) that $W_-(0)=0$ implies, however, that state 0 is a "forward-reflecting" state. Also, if N is finite then the stipulation in (6.4-1) that $W_+(N)=0$ means that state N is a "backward-reflecting" state. Naturally, we shall assume that the initial state n_0 always lies in the accessible range $[0,N]$.

The Markov state density function $P(n,t \mid n_0,t_0)$ of our process of course satisfies the forward master equation (6.1-17). Let us now *assume* it to be the case that

$$\lim_{(t-t_0) \to \infty} P(n,t \mid n_0,t_0) = P_s(n).$$

Then it would naturally follow that

$$\lim_{(t-t_0)\to\infty} \frac{\partial}{\partial t} P(n,t\,|\,n_0,t_0) = 0.$$

So in the limit $(t-t_0)\to\infty$, the forward master equation (6.1-17) takes the form

$$0 = [W_-(n+1)\,P_s(n+1) - W_+(n)\,P_s(n)]$$
$$+ [W_+(n-1)\,P_s(n-1) - W_-(n)\,P_s(n)].$$

Rearranging this equation to read

$$W_-(n)\,P_s(n) - W_+(n-1)\,P_s(n-1)$$
$$= W_-(n+1)\,P_s(n+1) - W_+(n)\,P_s(n), \qquad (6.4\text{-}2)$$

we observe that the function on the right hand side is just the function on the left hand side with n replaced by $n+1$. So we conclude that

$$W_-(n)\,P_s(n) - W_+(n-1)\,P_s(n-1) = \text{constant (w.r.t. } n).$$

Taking $n=0$, and noting from condition (6.4-1) that $W_+(-1)=W_-(0)=0$, we conclude that the constant in question here must be zero. Hence,

$$W_-(n)\,P_s(n) = W_+(n-1)\,P_s(n-1) \qquad (1\le n\le N). \qquad (6.4\text{-}3)$$

For reasons to be elaborated shortly, this relation is often referred to as the **detailed balancing condition**.

Since $W_-(n)$ in Eq. (6.4-3) is never zero [by condition (6.4-1)], then we can write that equation

$$P_s(n) = \frac{W_+(n-1)}{W_-(n)}\,P_s(n-1) \qquad (1\le n\le N). \qquad (6.4\text{-}4)$$

This is evidently a recursion relation for $P_s(n)$. A first iteration of this recursion relation gives

$$P_s(n) = \frac{W_+(n-1)}{W_-(n)}\left(\frac{W_+(n-2)}{W_-(n-1)}\,P_s(n-2)\right),$$

and $n-1$ iterations gives

$$P_{\mathrm{s}}(n) = P_{\mathrm{s}}(0) \prod_{j=1}^{n} \frac{W_{+}(j-1)}{W_{-}(j)} \qquad (1 \le n \le N).$$

The value of $P_{\mathrm{s}}(0)$ can now be deduced from the requirement that $P_{\mathrm{s}}(n)$ be properly normalized:

$$1 = \sum_{n=0}^{N} P_{\mathrm{s}}(n) = P_{\mathrm{s}}(0) + \sum_{n=1}^{N} P_{\mathrm{s}}(0) \prod_{j=1}^{n} \frac{W_{+}(j-1)}{W_{-}(j)}$$

$$= P_{\mathrm{s}}(0)\left(1 + \sum_{n=1}^{N} \prod_{j=1}^{n} \frac{W_{+}(j-1)}{W_{-}(j)} \right).$$

So, defining the **normalization constant** K by

$$K \equiv \left(1 + \sum_{n=1}^{N} \prod_{j=1}^{n} \frac{W_{+}(j-1)}{W_{-}(j)} \right)^{-1}, \qquad (6.4\text{-}5)$$

we conclude that $P_{\mathrm{s}}(n)$, if it exists at all, must be given by the formula

$$P_{\mathrm{s}}(n) = \begin{cases} K & , \text{ if } n=0, \\ K \prod_{j=1}^{n} \dfrac{W_{+}(j-1)}{W_{-}(j)} & , \text{ if } n=1,2,\ldots,N. \end{cases} \qquad (6.4\text{-}6)$$

Now let us consider under what circumstances the above formula for $P_{\mathrm{s}}(n)$ will represent a legitimate density function on the interval $[0,N]$. Since condition (6.4-1) ensures that the ratio $W_{+}(j-1)/W_{-}(j)$ in Eq. (6.4-6) is always a finite, positive number, then all that is required for the legitimacy of the expression (6.4-6) is that K be *nonzero*. That $K \equiv P_{\mathrm{s}}(0)$ is properly less than or equal to one is obvious from the definition (6.4-5), but that same definition shows that K will be greater than zero if and only if

$$\sum_{n=1}^{N} \prod_{j=1}^{n} \frac{W_{+}(j-1)}{W_{-}(j)} < \infty. \qquad (6.4\text{-}7)$$

This inequality will surely be satisfied if N is *finite*. But if $N=\infty$ then this condition represents a genuine *constraint* on the stepping functions, essentially requiring the ratio $W_{+}(j-1)/W_{-}(j)$ to approach zero with sufficient alacrity as $j \to \infty$. Roughly speaking, the process should not be able to wander off forever to infinity. In any case, by simply requiring the functions $W_{\pm}(n)$ to satisfy condition (6.4-7), we can ensure that K in Eq. (6.4-5) will satisfy the crucial nonzero condition. Note that the strict positivity of K and of $W_{+}(j-1)$ for $j\in[1,N]$ implies, through Eq. (6.4-6), that $P_s(n)$ too will be strictly positive on $[1,N]$. So we have

$$P_s(n) > 0 \text{ for all } n\in[0,N]. \tag{6.4-8}$$

What we have done to this point is to show that, if the stepping functions of the process satisfy conditions (6.4-1) and (6.4-7), then the formulas (6.4-5) and (6.4-6) define a strictly positive, time-independent solution $P_s(n)$ of the forward master equation. We have *not* as yet proved that $P(n,t\,|\,n_0,t_0)$ always *approaches* that time-independent solution as $(t-t_0)\to\infty$. But we are now in a position to do so. And in Appendix F, we give a formal proof of the following result:

Birth-Death Stability Theorem. If the stepping functions $W_{+}(n)$ and $W_{-}(n)$ of a temporally homogeneous birth-death Markov process satisfy conditions (6.4-1) and (6.4-7), then

$$\lim_{(t-t_0)\to\infty} P(n,t\,|\,n_0,t_0) = P_s(n) \text{ for all } n_0\in[0,N], \tag{6.4-9}$$

where $P_s(n)$ is the function defined through Eqs. (6.4-5) and (6.4-6).

It would of course be quite incorrect to infer from the fact that $P(n,t\,|\,n_0,t_0)$ eventually stops changing with time that $X(t)$ likewise eventually stops changing with time. In fact, condition (6.4-1) effectively guarantees that $X(t)$ will wander about *forever* over all the states 0, 1, ..., N. The import of the result (6.4-9) is rather this: If we observe the process to be in state n_0 at time t_0 and then let the process evolve *unobserved* until any time t_1 that is sufficiently later than t_0, then the probability that an observation of the process at time t_1 will find it to be in state n is $P_s(n)$. In this same vein, with the process running *unobserved* from time t_0 to some time $t_1 \gg t_0$, we also note that the probability of finding the process making the state transition $n\to n-1$ in the infinitesimal time interval $[t_1,t_1+dt_1)$ is equal to the product $P_s(n)\times W_{-}(n)dt_1$. But Eq. (6.4-3) tells us that

$$P_s(n)\times W_{-}(n)dt_1 = P_s(n-1)\times W_{+}(n-1)dt_1.$$

Therefore, if the process runs unobserved from time t_0 to time $t_1 \gg t_0$, then the probability of finding it making the transition $n\to n-1$ in $[t_1,t_1+dt_1)$ is *equal* to the probability of finding it making the transition $n-1\to n$ in $[t_1,t_1+dt_1)$. So in the long run, the system will make on the average the same number of $n\to n-1$ transitions as $n-1\to n$ transitions. This feature of a *stable* birth-death Markov process is often referred to as "detailed balancing," and is why we call Eq. (6.4-3) the detailed balancing condition. This condition is closely related to, but not quite as restrictive as, the *chemical* detailed balancing condition mentioned in connection with Eq. (5.3-56).

To illustrate the birth-death stability theorem, let us use it to decide which if any of the four birth-death Markov processes considered in Section 6.3 are stable. The *Poisson process* is defined by the stepping functions (6.3-2). Since $W_-(n) \equiv 0$, then neither of the two stability conditions (6.4-1) and (6.4-7) are satisfied, so we should not expect the Poisson process to have a stationary Markov state density function $P_s(n)$. And indeed, the unbounded asymptotic time dependencies of $\langle X(t) \rangle$ and $\text{var}\{X(t)\}$ implied by Eqs. (6.3-9) shows that this must be the case. From a broader point of view, the Poisson process is a *completely homogeneous* Markov process, and as we have noted several times in earlier chapters, completely homogeneous Markov processes cannot be stable.

The *radioactive decay process* is defined by the stepping functions (6.3-14). Since $W_+(n) \equiv 0$, then condition (6.4-1) is not satisfied, so the birth-death stability theorem technically cannot be invoked. However, we note that condition (6.4-7) *is* satisfied, since the left hand side of that inequality is exactly zero for this process. And if we just go ahead and substitute $W_+(j) = 0$ and $W_-(j) = (1/\tau^*)j$ into Eqs. (6.4-5) and (6.4-6), we get $P_s(n) = \delta(n,0)$. That this prediction is in fact correct can be seen by taking the $(t - t_0) \to \infty$ limit of the formula (6.3-16) for $P(n,t \mid n_0,t_0)$. So the radioactive decay process is stable, in a trivial sort of way, but our birth-death stability theorem technically does not apply because we did not formulate it to accommodate absorbing states.

The *random telegraph process* is defined by the stepping functions (6.3-31). Those functions evidently satisfy condition (6.4-1) with $N = 1$; moreover, because N is finite, the stepping functions also satisfy condition (6.4-7). Therefore, the birth-death stability theorem implies that the random telegraph process is stable, and has a stationary Markov state density function $P_s(n)$ that is given by Eqs. (6.4-5) and (6.4-6). The calculation of $P_s(n)$ from those formulas is easy and results in

$$P_s(0) = \frac{a_1}{a_0 + a_1} \quad \text{and} \quad P_s(1) = \frac{a_0}{a_0 + a_1}, \qquad (6.4\text{-}10)$$

which agrees exactly with our earlier result (6.3-36).

Finally, the *payroll process* is defined by the stepping functions (6.3-51). Those functions clearly satisfy condition (6.4-1) with $N = \infty$. And since $W_+(j-1)/W_-(j)=h/(lj)$ for this process, then the left hand side of condition (6.4-7) is

$$\sum_{n=1}^{N}\prod_{j=1}^{n}\frac{h}{lj} = \sum_{n=1}^{N}\frac{(h/l)^n}{n!} = \sum_{n=0}^{N}\frac{(h/l)^n}{n!} - 1 = e^{h/l} - 1,$$

which means that condition (6.4-7) is also satisfied. Therefore, the birth-death stability theorem implies that the payroll process is stable, and moreover that its stationary Markov state density function $P_s(n)$ is given by formulas (6.4-5) and (6.4-6). The equation just above together with formula (6.4-5) show that $K = e^{-h/l}$, and formula (6.4-6) then gives

$$P_s(n) = \frac{e^{-h/l}(h/l)^n}{n!} \qquad (n=0,1,\dots). \tag{6.4-11}$$

This we recognize [compare Eqs. (1.7-14)] as the density function of the Poisson random variable with mean and variance h/l, a finding that is entirely consistent with our earlier results for $\langle X(t\to\infty)\rangle$ and sdev$\{X(t\to\infty)\}$ in Eqs. (6.3-57). In the next subsection we shall exhibit (in Fig. 6-5) a plot of the function (6.4-11) for specific values of the parameters h and l. For now, we simply underscore the fact that the birth-death stability theorem has allowed us to easily deduce both the *existence* and *form* of $P_s(n)$ without having to first solve the forward master equation.

6.4.B THE POTENTIAL AND BARRIER FUNCTIONS

The formula for $P_s(n)$ in Eq. (6.4-6), being essentially the recursion relation (6.4-4) supplemented by normalization, is actually quite convenient for effecting *numerical calculations* of $P_s(n)$. But for many *analytical* purposes it turns out to be convenient to write $P_s(n)$ in another way. To derive that alternate formula for $P_s(n)$, we proceed from the formula (6.4-6), for $n \geq 1$, as follows:

$$P_s(n) = K \prod_{j=1}^{n}\frac{W_+(j-1)}{W_-(j)} = \frac{K W_+(0)}{W_+(n)}\prod_{j=1}^{n}\frac{W_+(j)}{W_-(j)}$$

$$= \frac{K W_+(0)}{W_+(n)}\exp\left(\ln\prod_{j=1}^{n}\frac{W_+(j)}{W_-(j)}\right) = \frac{K W_+(0)}{W_+(n)}\exp\left(\sum_{j=1}^{n}\ln\frac{W_+(j)}{W_-(j)}\right),$$

whence,

$$P_s(n) = \frac{K\,W_+(0)}{W_+(n)} \exp\left(-\sum_{j=1}^{n} \ln \frac{W_-(j)}{W_+(j)}\right).$$

Now we define the **potential function** $\phi(n)$ of the process $X(t)$ to be

$$\phi(n) \equiv \begin{cases} 0, & \text{if } n=0, \\[2mm] \sum_{j=1}^{n} \ln \dfrac{W_-(j)}{W_+(j)} \equiv \sum_{j=1}^{n} \ln \dfrac{w_-(j)}{w_+(j)}, & \text{if } n\geq 1, \end{cases} \qquad (6.4\text{-}12)$$

where the last equality follows from the fact that $W_\pm(n) \equiv a(n)w_\pm(n)$. With this definition, we see that the above formula for $P_s(n)$ can be written more compactly as

$$P_s(n) = \frac{K\,W_+(0)}{W_+(n)} \exp(-\phi(n)) \quad (n=0,...,N). \quad (6.4\text{-}13)$$

Notice that this formula properly gives $P_s(0)=K$, as required by Eq. (6.4-6), because of the way we have chosen to define $\phi(0)$.

The formula (6.4-13) for the stationary Markov state density function $P_s(n)$ of a birth-death Markov process is reminiscent of the formula (3.5-3) for the stationary Markov state density function $P_s(x)$ of a *continuous* Markov process. We shall find in our subsequent work that the function $\phi(n)$ defined in Eq. (6.4-12) is indeed quite analogous to the continuous potential function $\phi(x)$ defined in Eq. (3.5-4), which of course is why we have elected to use the same symbol ϕ to represent those two functions. In particular, we showed at the end of Subsection 3.5.A that the potential function $\phi(x)$ of a continuous Markov process has the property that the process always has a probabilistic tendency to move in the direction of *decreasing* $\phi(x)$, this tendency being greater the greater the magnitude of the local slope $\phi'(x)$. To see that this property also holds for the function $\phi(n)$ defined in Eq. (6.4-12), we need only examine the formula for the "derivative" of that function:

$$\phi'(n) \equiv \frac{\phi(n)-\phi(n-1)}{1} = \ln \frac{W_-(n)}{W_+(n)} = \ln \frac{w_-(n)}{w_+(n)}. \quad (6.4\text{-}14)$$

This equation shows that if $\phi'(n)>0$, so that ϕ is sloping *upward* at n, then it must be true that $w_-(n)/w_+(n)>1$, or $w_-(n)>w_+(n)$. This implies that the process in state n is more likely to step next *downward* to state $n-1$, rather than upward to state $n+1$; moreover, the strength of this downward stepping bias is evidently measured by the strength of the inequality $\phi'(n) > 0$. Similarly, if ϕ were sloping *downward* at n, then the aforementioned inequalities would all be reversed, and the process in state n would be more likely to step next *upward* to state $n+1$. If $\phi'(n) = 0$, then we would have $w_-(n)/w_+(n) = 1$, and the process would be equally likely to step away in either direction from state n. So we see that the function $\phi(n)$ defined in Eq. (6.4-12), like the function $\phi(x)$ defined in Eq. (3.5-4), has strong similarities to the potential energy function of classical mechanics. But again, we should be aware that this classical mechanics analogy cannot be pushed too far: Our Markov system lacks the "inertia" of a classical system, and the classical system lacks the "stochasticity" of our Markov system.

Notice from the slope formula (6.4-14) that if $w_+(n) = 0$, and hence $w_-(n) = 1$, then $\phi'(n) = \ln(1/0) = \infty$; thus, at any point n where it is *impossible* for the process to step up, the potential function has an infinitely steep up-slope that "reflects the process backward." Similarly, if $w_-(n) = 0$, and hence $w_+(n) = 1$, then $\phi'(n) = \ln(0/1) = -\infty$, and the potential function at n has an infinitely steep down-slope that "reflects the process forward." Of course, our condition (6.4-1) specifically excludes reflecting states from the *open* interval $(0, N)$. But since $w_-(0) = 0$, then the state 0 will always be a forward-reflecting state.[†] Furthermore, if N is *finite*, then condition (6.4-1) requires that $w_+(N) = 0$, so that state N will be a backward-reflecting state.

The slope formula (6.4-14) also allows us to make somewhat clearer the connection between $\phi(n)$ and its continuous process counterpart $\phi(x)$. Writing (6.4-14) as

$$\phi'(n) = \ln z \quad \text{where} \quad z \equiv W_-(n)/W_+(n),$$

then we can invoke an established series formula for $\ln z$ to get [‡]

$$\phi'(n) = \ln z \equiv 2\left(\frac{z-1}{z+1}\right)\left[1 + \frac{1}{3}\left(\frac{z-1}{z+1}\right)^2 + \frac{1}{5}\left(\frac{z-1}{z+1}\right)^4 + \ldots\right],$$

which is valid for all $z > 0$. But since $z \equiv W_-(n)/W_+(n)$, then

---

[†] The prediction of Eq. (6.4-14) that $\phi'(0) = -\infty$ formally requires that we extend our definition of $\phi(n)$ in Eq. (6.4-12) by stipulating that $\phi(-1) = +\infty$. This is a contrivance that seems to have no other practical consequence.

[‡] This series formula for $\ln z$ is Eq. (4.1.27) of M. Abramowitz and I. Stegun's *Handbook of Mathematical Functions*, National Bureau of Standards (1964).

$$\frac{z-1}{z+1} = \frac{W_-(n)/W_+(n) - 1}{W_-(n)/W_+(n) + 1} = -\frac{W_+(n) - W_-(n)}{W_+(n) + W_-(n)} = -\frac{v(n)}{a(n)},$$

where the last equality follows from Eqs. (6.1-13). Therefore, the derivative formula (6.4-14) of the birth-death potential function can be written in the series form

$$\phi'(n) = -\frac{2v(n)}{a(n)}\left[1 + \frac{1}{3}\left(\frac{v(n)}{a(n)}\right)^2 + \frac{1}{5}\left(\frac{v(n)}{a(n)}\right)^4 + \dots\right]. \quad (6.4\text{-}15)$$

Comparing this formula with the formula for the derivative of the *continuous* potential function in Eq. (3.5-8), namely $\phi'(x) = -2A(x)/D(x)$, we see that, under the usual correspondence $v \leftrightarrow A$ and $a \leftrightarrow D$, the two formulas are approximately the same in regions where $v(n)/a(n) \ll 1$ and the bracketed expression in Eq. (6.4-15) is approximately unity. In this connection, let us recall from Eq. (6.1-14d) that $|v(n)/a(n)|$ never exceeds unity; indeed, we have by definition

$$\frac{v(n)}{a(n)} = \frac{W_+(n) - W_-(n)}{W_+(n) + W_-(n)} = \frac{a(n)[w_+(n) - w_-(n)]}{a(n)[w_+(n) + w_-(n)]} = w_+(n) - w_-(n),$$

which clearly always lies between -1 and $+1$. Only in regions where either of the functions $w_+(n)$ or $w_-(n)$ approaches unity (and the other consequently approaches zero), will the higher order terms in brackets in Eq. (6.4-15) come into play and spoil the formal correspondence with the continuous potential function formula (3.5-8).

A function that is closely related to the potential function $\phi(n)$ and the stationary Markov state density function $P_s(n)$ is the **barrier function** $\Psi(n)$. It is defined by [compare Eq. (3.7-17)]

$$\Psi(n) \equiv \frac{1}{K W_+(0)} \exp(\phi(n)), \quad (6.4\text{-}16)$$

where K is the normalization constant (6.4-5). It is obvious from this definition that the maximums and minimums of $\Psi(n)$ will coincide exactly with the maximums and minimums of $\phi(n)$; however, because of the exponential in the definition, the peaks and valleys in $\Psi(n)$ will be greatly exaggerated in comparison to those in $\phi(n)$. If we eliminate $\phi(n)$ between Eqs. (6.4-16) and (6.4-13), we find that the barrier function can also be written in terms of the stationary Markov state density function $P_s(n)$ as

$$\Psi(n) = \frac{1}{W_+(n)\,P_{\mathrm{s}}(n)}. \qquad (6.4\text{-}17\mathrm{a})$$

And because of the detailed balancing condition (6.4-3), we can also write this as

$$\Psi(n) = \frac{1}{W_-(n+1)\,P_{\mathrm{s}}(n+1)}. \qquad (6.4\text{-}17\mathrm{b})$$

Equations (6.4-17) afford an interesting physical interpretation of the barrier function: We know that if we allow the process $X(t)$ to run *unobserved* from time t_0 to any time t_1 that is sufficiently later than t_0, then the probability of observing the process to jump from state n to state $n+1$ in the infinitesimal time interval $[t_1,t_1+dt_1)$ is

$$P_{\mathrm{s}}(n) \times W_+(n) dt_1 = [1/\Psi(n)]\,dt_1,$$

where the last step has invoked Eq. (6.4-17a). Thus, by the same reasoning that leads from Eq. (6.1-4) to Eq. (6.1-16), we may conclude that

$\Psi(n)$ = mean waiting time, after the process has
            gone unobserved for a very long time, for
            the occurrence of the transition $n \rightarrow n+1$. (6.4-18a)

Similarly, from Eq. (6.4-17b) we may deduce that

$\Psi(n)$ = mean waiting time, after the process has
            gone unobserved for a very long time, for
            the occurrence of the transition $n+1 \rightarrow n$. (6.4-18b)

So if $\Psi(n)$ happens to be relatively large, then after the process has gone unobserved for a very long period of time we should expect to have to wait a relatively long time before witnessing a transition between the particular states n and $n+1$. This, as we shall see more clearly later on, is the rationale for the barrier function's name.

For the Poisson process, the radioactive decay process and the random telegraph process, the potential and barrier functions are all ill-defined, and hence of no use. The reason for that is that at least one of the two stepping functions for each of those processes vanishes for every state n, with the consequence that Eq. (6.4-12) is mathematically undefined. However, for the *payroll process*, for which the stepping functions are as given in Eqs. (6.3-51), we can readily calculate the potential function from the definition (6.4-12):

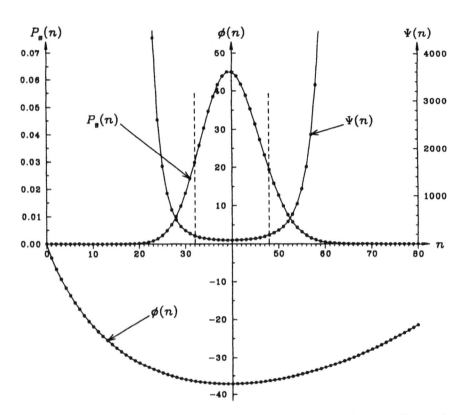

Figure 6-5. Plots of the stationary Markov state density function $P_s(n)$, the potential function $\phi(n)$, and the barrier function $\Psi(n)$ for the *payroll process* defined by the stepping functions $W_+(n) = h = 0.2$ and $W_-(n) = ln = 0.005n$, these parameter values being the same as those used in the payroll process simulation of Fig. 6-4. The dots that give the values of each plotted integer-variable function have been connected with straight line segments for the sake of clarity. The function $P_s(n)$ is the Poisson density function (6.4-11), and the functions $\phi(n)$ and $\Psi(n)$ are given by formulas (6.4-19). The two dashed vertical lines show the nominal "stable state region" $[x_i - \sigma_i/2, x_i + \sigma_i/2]$, as calculated according to the rules of Eqs. (6.4-21a), (6.4-22) and (6.4-25).

$$\phi(n) = \sum_{j=1}^{n} \ln\left(\frac{lj}{h}\right) = \ln \prod_{j=1}^{n} \left(\frac{lj}{h}\right) = \ln\left((l/h)^n\, n!\right),$$

Thus, for the payroll process the potential function is given by

$$\phi(n) = \ln(n!\,(h/l)^{-n}). \tag{6.4-19a}$$

[Although this calculation of $\phi(n)$ assumed that $n \geq 1$, the resulting formula is seen to be valid for $n = 0$ as well.] To calculate the associated barrier function, we can either use the definition (6.4-16), after first calculating $K = e^{-h/l}$ from Eq. (6.4-5), or we can use Eq. (6.4-17a), recalling that $P_s(n)$ is as given in Eq. (6.4-11). Either way, we easily find that the barrier function for the payroll process is given by

$$\Psi(n) = \frac{e^{h/l}}{h} \, n! \, (h/l)^{-n}. \tag{6.4-19b}$$

In Fig. 6-5 we show plots of the payroll process potential and barrier functions (6.4-19), along with a plot of the corresponding stationary Markov state density function (6.4-11). For these plots we have used the parameter values $h = 0.2$ and $l = 0.005$, which are the same values that were used in the payroll process simulation of Fig. 6-4. It will be observed from these plots that $P_s(n)$ has a relative maximum, and $\phi(n)$ and $\Psi(n)$ both have a relative minimum, at $n = h/l = 40$. The existence and implications of such extremums in $P_s(n)$ and $\phi(n)$ *in general* is the topic to which we turn next.

6.4.C STABLE STATES

We saw in the preceding subsection that a birth-death Markov process $X(t)$ has a stochastic tendency to move toward the nearest local minimum of its potential function $\phi(n)$. The consequent statistical favoring of states n where $\phi(n)$ is relatively small is in the main consistent with formula (6.4-13), which implies that such states are usually where $P_s(n)$ is relatively large; i.e., those are the states in which we would most likely find the process $X(t)$ a long time after it is initialized.

Because $X(t)$ tends to move toward states that are local minimums of $\phi(n)$, and tends to be found near states that are local maximums of $P_s(n)$, we call such states the **stable states** of the process. If $W_+(n)$ happens to be constant with respect to n, then it is clear from Eq. (6.4-13) that every local minimum of $\phi(n)$ will be a local maximum of $P_s(n)$, and vice versa. If, however, $W_+(n)$ is not constant, then the local minimums of $\phi(n)$ and the local maximums of $P_s(n)$ will usually not coincide, and our definition of stable state becomes somewhat ambiguous. But in virtually all cases where the notion of stable state is of *practical use*, the distance between the center of a valley in $\phi(n)$ and the center of the nearest peak in $P_s(n)$

will be small compared to the widths of those two structures. In those cases, each valley of $\phi(n)$ overlies one and only one peak of $P_s(n)$, and $X(t)$ is said to be "in" the corresponding stable state whenever it is *anywhere* inside that $\phi(n)$-valley or that $P_s(n)$-peak.

Although the relative minimums of $\phi(n)$ and the relative maximums of $P_s(n)$ can always be found by simply inspecting the graphs of those two functions, it is useful to be able to calculate those extremums directly from the stepping functions $W_+(n)$ and $W_-(n)$. A relative minimum n_i of $\phi(n)$ is by definition such that

$$\phi(n_i) < \phi(n_i - 1) \quad \text{and} \quad \phi(n_i) < \phi(n_i + 1).$$

But it follows from the definition of $\phi(n)$ in Eq. (6.4-12) that

$$\phi(n_i) = \phi(n_i - 1) + \ln\left(\frac{W_-(n_i)}{W_+(n_i)}\right)$$

and

$$\phi(n_i + 1) = \phi(n_i) + \ln\left(\frac{W_-(n_i + 1)}{W_+(n_i + 1)}\right).$$

So the preceding two inequalities imply that, for n_i any relative minimum of $\phi(n)$,

$$\frac{W_-(n_i)}{W_+(n_i)} < 1 \quad \text{and} \quad \frac{W_-(n_i + 1)}{W_+(n_i + 1)} > 1,$$

or equivalently,

$$W_+(n_i) - W_-(n_i) > 0 \quad \text{and} \quad W_+(n_i + 1) - W_-(n_i + 1) < 0.$$

Recalling the definition (6.1-13b) of the function $v(n)$, we see that these last two inequalities can be written more simply as

$$v(n_i) > 0 \quad \text{and} \quad v(n_i + 1) < 0.$$

So if $v(x)$ is any *real*-variable function that smoothly interpolates the values of the *integer*-variable function $v(n)$, then $v(x)$ must pass through zero with a negative slope at some point $x_i \in (n_i, n_i + 1)$. Therefore, if x_i is a *down-going root* of $v(x)$, then $[x_i]$, the "greatest integer in x_i," will be a relative minimum of $\phi(n)$:

$$v(x_i) = 0 \quad \text{and} \quad v'(x_i) < 0$$

$$\Rightarrow [x_i] \text{ is a relative minimum of } \phi(n). \qquad (6.4\text{-}20)$$

Now, a case could be made for letting condition (6.4-20) define the "nominal values" of the stable states of the process. Certainly that criterion is simple, and it may even apply in circumstances where $P_s(n)$ does not exist — i.e., when conditions (6.4-1) and (6.4-17) are not satisfied. However, when $P_s(n)$ *does* exist, we shall prefer to define the nominal values of the stable states to the be relative maximums of $P_s(n)$.

To deduce a criterion analogous to (6.4-20) for a relative maximum n_i of $P_s(n)$, we begin by noting that such a point would have to satisfy

$$P_s(n_i) > P_s(n_i - 1) \quad \text{and} \quad P_s(n_i) > P_s(n_i + 1).$$

But it follows from the formula for $P_s(n)$ in Eq. (6.4-6) [or more clearly from the detailed balancing condition (6.4-3)] that

$$P_s(n_i) = P_s(n_i - 1) \, \frac{W_+(n_i - 1)}{W_-(n_i)}$$

and

$$P_s(n_i + 1) = P_s(n_i) \, \frac{W_+(n_i)}{W_-(n_i + 1)} .$$

So the preceding two inequalities imply that, for n_i any relative maximum of $P_s(n)$,

$$\frac{W_+(n_i - 1)}{W_-(n_i)} > 1 \quad \text{and} \quad \frac{W_+(n_i)}{W_-(n_i + 1)} < 1,$$

or equivalently,

$$W_+(n_i - 1) - W_-(n_i) > 0 \quad \text{and} \quad W_+(n_i) - W_-(n_i + 1) < 0.$$

Now we define a new function $a(n)$ by

$$a(n) \equiv W_+(n-1) - W_-(n) \tag{6.4-21a}$$

$$\equiv v(n) - [W_+(n) - W_+(n-1)], \tag{6.4-21b}$$

where the second expression follows from the first by simply eliminating $W_-(n)$ with the help of Eq. (6.1-13b). With the definition (6.4-21a), we see that the preceding two inequalities can be written more simply as

$$a(n_i) > 0 \quad \text{and} \quad a(n_i + 1) < 0.$$

So if $a(x)$ is any *real*-variable function that smoothly interpolates the values of the *integer*-variable function $a(n)$, then $a(x)$ must pass through zero with a negative slope at some point $x_i \in (n_i, n_i + 1)$. Therefore, if x_i is a *down-going root* of $a(x)$, then $[x_i]$, the "greatest integer in x_i," will be a relative maximum of $P_s(n)$:

$$a(x_i) = 0 \quad \text{and} \quad a'(x_i) < 0$$

$$\Rightarrow [x_i] \text{ is a relative maximum of } P_s(n). \quad (6.4\text{-}22)$$

Notice from Eq. (6.4-21b) that if the function $W_+(n)$ is independent of n, then the two functions $a(n)$ and $v(n)$ will coincide with each other, in which case the criterion (6.4-22) for a relative maximum of $P_s(n)$ will be identical with the criterion (6.4-20) for a relative minimum of $\phi(n)$. So all the foregoing formulas are consistent with the fact, deduced earlier from the formula (6.4-13) for $P_s(n)$, that the maximums of $P_s(n)$ and the minimums of $\phi(n)$ will coincide with each other whenever $W_+(n)$ is constant with respect to n.

In addition to assigning to each stable state a *nominal position* n_i, which we usually take to be the associated relative maximum of $P_s(n)$, it will also be convenient to assign to each stable state a *nominal width* σ_i. We shall define σ_i to be the "effective width" of the normal or Gaussian-shaped curve that best fits $P_s(n)$ in the immediate neighborhood of the peak point n_i. The exact procedure for calculating this width is described in Appendix D. To apply that procedure here, we must first *replace* the integer-variable function $P_s(n)$ in the neighborhood of the peak by a smoothly interpolating real-variable function $P_s(x)$, and then estimate the second derivative of $P_s(x)$ at the peak. To that end, we begin by estimating the first derivative of $P_s(x)$ as

$$P_s'(x) \approx \frac{P_s(x) - P_s(x-1)}{1} = \frac{W_+(x-1)}{W_-(x)} P_s(x-1) - P_s(x-1),$$

where the second equality has invoked the recursion formula (6.4-4). So

$$P_s'(x) \approx \frac{P_s(x-1)}{W_-(x)} [W_+(x-1) - W_-(x)] \equiv \frac{P_s(x-1)}{W_-(x)} a(x),$$

where the last step has invoked the definition (6.4-21a). A final mild approximation yields the first derivative estimate

$$P_s'(x) \approx \frac{P_s(x)}{W_-(x)} a(x). \quad (6.4\text{-}23a)$$

The second derivative can now be estimated as the derivative of this formula:

$$P_s''(x) \approx \frac{P_s(x)}{W_-(x)} a'(x) + a(x) \frac{d}{dx}\left(\frac{P_s(x)}{W_-(x)} \right). \qquad (6.4\text{-}23b)$$

Now, we know from condition (6.4-22) that if x_i is a relative maximum of $P_s(x)$, then $a(x_i)=0$ and $a'(x_i)<0$. So, setting $x=x_i$ in the preceding two derivative formulas, we find from the first formula that $P_s'(x_i)=0$, as expected, and we find from the second formula that

$$P_s''(x_i) \approx - P_s(x_i) \left| \frac{a'(x_i)}{W_-(x_i)} \right|. \qquad (6.4\text{-}24)$$

Now we observe that this formula for the second derivative of $P_s(x)$ at its relative maximum x_i is precisely of the form indicated in Eqs. (D-1), provided we identify the constant c there with $|a'(x_i)/W_-(x_i)|$. And so it follows from Eq. (D-7) that the effective Gaussian width of this peak — i.e., the **nominal width** of the stable state $\lfloor x_i \rfloor$ — is

$$\sigma_i = \left| \frac{2\pi\, W_-(x_i)}{a'(x_i)} \right|^{1/2}. \qquad (6.4\text{-}25)$$

To the extent that the peak in the function $P_s(n)$ at $n=\lfloor x_i \rfloor$ is roughly symmetric and does not significantly overlap any adjacent peak, we shall say that

"$X(t)$ is in the stable state $\lfloor x_i \rfloor$"

$$\Leftrightarrow \quad X(t) \in [x_i - \sigma_i/2,\, x_i + \sigma_i/2]. \qquad (6.4\text{-}26)$$

To illustrate the foregoing results, consider the payroll process, for which the stepping functions are given by [see Eqs. (6.3-51)] $W_+(n)=h$ and $W_-(n)=ln$, with h and l both positive. Since the function $W_+(n)$ in this case is independent of n, then Eq. (6.4-21b) tells us that the functions $a(n)$ and $v(n)$ will be identical to each other; indeed, the definitions (6.4-21a) and (6.1-13b) give in this case

$$a(x) = v(x) = h - lx \quad \text{and} \quad a'(x) = v'(x) = -l.$$

Since the derivative of this function is everywhere negative, then by criteria (6.4-20) and (6.4-22) any root x_i will be at once a relative minimum of $\phi(n)$ and a relative maximum of $P_s(n)$. There is obviously only one root, namely $x_i = h/l$, so it is the one and only stable state of the payroll process. And using Eq. (6.4-25), we easily calculate the nominal width of this stable state to be

$$
\sigma_i = \left| \frac{2\pi l x_i}{-l} \right|^{1/2} = (2\pi x_i)^{1/2} = \left(\frac{2\pi h}{l} \right)^{1/2} = \left(\frac{\pi}{2} \right)^{1/2} 2 \left(\frac{h}{l} \right)^{1/2}.
$$

We found earlier that the stationary Markov state density function for the payroll process is in fact the Poisson density function with mean and variance h/l [see Eq. (6.4-11)]. The results just obtained are obviously consistent with this fact: They give the nominal value x_i of the stable state to be the mean of $P_s(n)$, and the nominal width σ_i of that stable state to be slightly more than twice the standard deviation of $P_s(n)$. In particular, for the parameter values used in Figs. 6-4 and 6-5, the formulas here predict a stable state at $x_i = 40$ with a width of $\sigma_i = 15.85$; so we would say that this process is "in" its stable state whenever it is within ± 7.93 units of the value 40, a region that is indicated by the vertical dashed lines in Fig. 6-5.

6.5 APPLICATION: THE FUNDAMENTAL POSTULATE OF STATISTICAL MECHANICS†

The birth-death stability theorem, which we stated at Eq. (6.4-9) and proved in Appendix F, allows us to gain some insight into the origins of what is often called the *fundamental postulate of statistical mechanics*:

> All possible "microstates" of an isolated physical
> system in equilibrium are equally likely.   (6.5-1)

This postulate forms the keystone in the bridge that links the phenomenological laws of thermodynamics to the microscopic laws of mechanics. In this section we shall use the birth-death stability theorem, along with a few facts of quantum mechanics, to *derive* this postulate for a very simple class of physical systems.

We consider an isolated physical system whose possible states can, in the spirit of quantum mechanics, be enumerated by the integers 0, 1, 2, ..., N. The number N will typically be enormously large for systems of macroscopic size, but quantum mechanics assures us that N will be finite if the system is spatially bounded. These $N+1$ quantum states are called

---

† The concepts and results developed in this section will not be required in subsequent sections.

the **microstates** of the system. If we know which microstate the system is currently "in," then we know the current condition of the system to the finest detail permitted by quantum mechanics. This level of description is, however, so minute that the microstates cannot all be distinguished from one another by practical, macroscopic observations. But we find that we can form "groupings" of these microstates such that it is possible to determine to which group the system's current microstate belongs. These macroscopically distinguishable groups of microstates are called the **macrostates** of the system, and they are typically far less numerous than the microstates. Figure 6-6 illustrates a hypothetical system that has twenty microstates 0, 1, ..., 19, and three macrostates 1, 2 and 3; this system is said to be in macrostate 1 whenever it is in either of microstates 0 and 1, or in macrostate 2 whenever it is in any of microstates 2 through 15, or in macrostate 3 whenever it is in any of microstates 16 through 19.

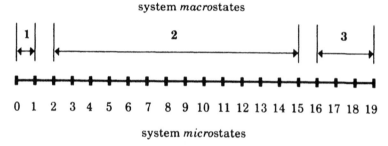

Figure 6-6. Graphical representation of the *microstates*, and their groupings into *macrostates*, for an idealized physical system.

Now we make the assumption that our system evolves with time by stochastically jumping about over its microstates. Such dynamical behavior might be thought to be a straightforward, general consequence of the laws of quantum mechanics, but the truth is that the quantum justification for this assumption is neither general nor straightforward. For in quantum mechanics, the instantaneous state of an isolated physical system generally evolves with time in a continuous, deterministic fashion. But if one takes care to specify the set of quantum microstates in a very particular way, and if one also takes due account of the quantum description of the measurement process, then quantum mechanics will usually admit the following *approximate* dynamical model: There exists a set of nonnegative numbers $\omega_{m,n}$ that can be physically interpreted according to

$\omega_{m,n}\,dt$ = probability that the system, in microstate
n at time t, will jump to microstate m in
the infinitesimal time interval $[t, t+dt)$;   (6.5-2a)

furthermore, these quantities satisfy the symmetry condition

$$\omega_{n,m} = \omega_{m,n} \quad (\text{all } m,n = 0,...,N). \qquad (6.5\text{-}2b)$$

The numbers $\omega_{m,n}$ are called the system's *transition probability
rates*.[†] Notice that the symmetry condition (6.5-2b) does *not* imply that
$\omega_{m,n}$ is constant with respect to either m or n. The practical implication
of Eq. (6.5-2a) is that the motion of the system over its microstates can be
regarded as a temporally homogeneous discrete state Markov process $X(t)$
with the consolidated characterizing function $W(m-n\mid n) = \omega_{m,n}$. For the
purposes of our discussion here, we shall make the simplifying
assumption that jumps of the system can occur only to *immediately
adjacent* microstates — i.e., that $\omega_{m,n}$ vanishes unless $m = n \pm 1$.[‡] So now
the motion of our system over its microstates can be viewed as a
temporally homogeneous *birth-death* Markov process, with stepping
functions

$$W_+(n) = \omega_{n+1,n} \quad \text{and} \quad W_-(n) = \omega_{n-1,n}. \qquad (6.5\text{-}3)$$

We are also going to assume for our analysis here that *every adjacent
microstate transition is possible*. This means that $W_+(n-1)$ and $W_-(n)$
are both strictly positive for all $n \in [1,N]$, and hence that condition (6.4-1)
of the birth-death stability theorem is satisfied. Furthermore, since the
symmetry condition (6.5-2b) implies that

$$\frac{W_+(j-1)}{W_-(j)} = \frac{\omega_{j,j-1}}{\omega_{j-1,j}} = 1 \quad \text{for all } j \in [1,N], \qquad (6.5\text{-}4)$$

[†] The justification for and limitations on the dynamical model provided by Eqs. (6.5-2)
are to be found in the *time-dependent perturbation theory* of quantum mechanics. In that
theory, the microstates of the system are taken to be the eigenstates of an "unperturbed"
energy operator that differs from the system's exact Hamiltonian operator by a very weak
"interaction" term. The true state of the system at any instant will actually be a *quantum
superposition* of the unperturbed eigenstates (microstates), rather than a *particular*
microstate as our model would seem to require; this difficulty is finessed away using the
rules of quantum measurement theory. Another difficulty with this approach is that
quantum perturbation theory requires dt in Eq. (6.5-2a) to satisfy $dt \gg \hbar/E$, where E is the
energy of the system; however, for *macroscopic* systems, $\hbar/E$ is usually so small that dt can
actually satisfy this requirement and *still* be regarded as an "infinitesimal."

[‡] But it is not hard to generalize our analysis here and in Appendix F to eliminate this
one-step restriction.

then we have

$$\sum_{n=1}^{N} \prod_{j=1}^{n} \frac{W_+(j-1)}{W_-(j)} = \sum_{n=1}^{N} \prod_{j=1}^{n} (1) = \sum_{n=1}^{N} (1) = N. \qquad (6.5\text{-}5)$$

This shows that our process also satisfies condition (6.4-7). So, since *both* conditions (6.4-1) and (6.4-7) are satisfied, we can invoke the birth-death stability theorem: The Markov state density function $P(n,t \mid n_0,t_0)$ of the instantaneous microstate $X(t)$ of our physical system approaches, as $(t-t_0)\to\infty$, a stationary form $P_s(n)$, which can be explicitly calculated from formulas (6.4-5) and (6.4-6). Formula (6.4-5), in conjunction with Eq. (6.5-5), gives $K = (1+N)^{-1}$. And formula (6.4-6), in conjunction with Eq. (6.5-4), then gives $P_s(n) = K$ for all $n \in [0,N]$. Thus we conclude that a stationary Markov state density function for this process exists and is given by

$$P_s(n) = \frac{1}{N+1} \qquad (n=0,1,...,N). \qquad (6.5\text{-}6)$$

Thus we have proved that, regardless of which microstate the system is initially in, *if we just wait long enough the system will be in any of its $N+1$ possible microstates with equal probability.* This is, for the simple idealized system under discussion, precisely the fundamental postulate of statistical mechanics.

The foregoing derivation of the fundamental postulate of statistical mechanics evidently hinges on two crucial properties of the stepping functions $W_+(n)$ and $W_-(n)$. The first is that

$$W_+(n-1) > 0 \quad \text{and} \quad W_-(n) > 0 \quad \text{for all } n \in [1,N]. \qquad (6.5\text{-}7a)$$

This property ensures that it is possible for the system to reach any microstate from any other microstate, and it is often referred to as the *ergodic property*. The second crucial property is the inverse transition symmetry relation

$$W_+(n-1) = W_-(n) \quad \text{for all } n \in [1,N]. \qquad (6.5\text{-}7b)$$

It is tempting to call this property "detailed balancing," but we reserve that term for Eq. (6.4-3), which holds for *all* stable birth-death Markov processes. Indeed, it is the *joining* of the symmetry relation (6.5-7b) and the detailed balancing relation (6.4-3) that leads to the conclusion that $P_s(n) = P_s(n-1)$, i.e., that all microstates are equally probable in the long-time limit.

The importance of the fundamental postulate of statistical mechanics is that it allows us to determine the long-time or "equilibrium" probability of any *macrostate* by simply "counting microstates." More precisely, if we let

$n(i) \equiv$ the number of microstates belonging to macrostate i,   (6.5-8)

then since all $(N+1)$ microstates are equally likely once the system reaches "equilibrium," we have

$n(i)/(N+1)$ = probability that the equilibrized system
will be found in macrostate i.   (6.5-9)

So, for example, if we make an observation on the system of Fig. 6-6 after allowing that system to evolve *unobserved* for a very long time, then the probabilities for finding the system to be in macrostates 1, 2 and 3 are, respectively, 2/20, 14/20 and 4/20. For most macroscopic physical systems N will be very large, and some *one* macrostate will own an overwhelming majority of the microstates. That macrostate will therefore be the one in which the system ultimately spends nearly all of its time, and it is called the *equilibrium macrostate*. On those rare but inevitable occasions when the system's instantaneous microstate wanders into the domain of some *other* macrostate, we say that the system has undergone a "macrostate fluctuation."

It was the nineteenth century physicist Ludwig Boltzmann who first recognized that, if we simply assert that the **entropy** $S(i)$ of the system in macrostate i is directly proportional to the logarithm of $n(i)$,

$$S(i) = k_B \ln n(i),$$ (6.5-10)

where the constant of proportionality k_B is nowadays called *Boltzmann's constant*, then we can rigorously deduce all the laws of classical thermodynamics. For example, the thermodynamic law concerning the tendency of a system to evolve toward states of higher entropy can be understood as a consequence of the inevitable tendency of the system to spend more time in macrostates that have a larger number of microstates. More quantitatively, since Eq. (6.5-10) implies that $n(i) = \exp(S(i)/k_B)$, then it follows from Eq. (6.5-9) that the relative equilibrium probability of two macrostates i and j is given by

$$\frac{\text{Prob}\{X(\infty) \in i\}}{\text{Prob}\{X(\infty) \in j\}} = \frac{n(i)}{n(j)} = \exp\left(\frac{S(i) - S(j)}{k_B}\right).$$ (6.5-11)

So if the entropy of macrostate i is even *slightly* larger than the entropy of macrostate j, then the probability of finding the equilibrized system in macrostate i will be *substantially* larger than the probability of finding it in macrostate j.

But notice that it is not only possible, but actually *inevitable*, for the entropy of the system to occasionally *decrease*: Since the process $X(t)$ whose states are illustrated in Fig. 6-6 satisfies the ergodic condition (6.5-7a), then $X(t)$ will wander forever over *all* its twenty microstates 0, 1, ..., 19. And the result (6.5-6) implies that $X(t)$ will ultimately show the same affinity for each of those microstates, and will pay no attention to the way in which they have been grouped into the three macrostates 1, 2 and 3. In the course of these eternal wanderings, $X(t)$ will occasionally make the microstate transitions 2→1 and 15→16 [see Fig. 6-6], and in so doing it will evidently move from the higher-entropy macrostate 2 to the respective lower-entropy macrostates 1 and 3. On those specific occasions, the entropy of the system will thus decrease. This is contrary to an earlier, incorrect thermodynamic dictum that the entropy of an isolated system, if it changes at all, can only increase.†

But there is still a seeming paradox here: On the one hand, since Eq. (6.5-7b) implies that $W_+(15) = W_-(16)$, then it would seem that the microstate transition 15→16 should be just as likely to occur as the microstate transition 16→15. On the other hand, since macrostate 2 has a higher entropy than macrostate 3, then we rather expect that the macrostate transition 2→3 should be less likely to occur than the macrostate transition 3→2. But how can *both* of these expectations be realized in light of the fact that the transitions 15→16 and 2→3 always occur *together*, as do also the transitions 16→15 and 3→2? The answer to this question, as to most "paradoxes" involving probability, lies in paying proper attention to *conditioning*. In particular, the microstate transition

---

† The notion of entropy as a fluctuating quantity that is subject to occasional decreases, and the concomitant notion of a fluid as an assembly of tiny particles in random motion, were not accepted by most of the scientific establishment at the very beginning of the Twentieth Century. Ludwig Boltzmann was a leading figure in the vanguard of these new ideas. The stubborn resistance he encountered from many prominent scientists of that era is widely regarded as being a contributing factor to his suicide in 1906. As a testimony to the ultimate importance of Boltzmann's insights, his important formula (6.5-10) — although in a different notation and without any quantum mechanical overtones — was engraved on his tombstone. The sad irony is that Boltzmann died apparently unaware of the fact that a relatively unknown Swiss patent officer named Albert Einstein had, about a year earlier, published a paper showing how the observed phenomenon of Brownian motion can be quantitatively understood in terms of — and apparently *only* in terms of — the molecular hypothesis. So Einstein's 1905 paper on Brownian motion, which we mentioned earlier in Subsection 3.4.C and Section 4.5, set the stage not only for the subsequent active development of Markov process theory, but also for the vindication of Boltzmann's approach to statistical mechanics.

$15 \rightarrow 16$ and the macrostate transition $2 \rightarrow 3$ are indeed the same physical event, but if we wish to calculate the probability that that event will occur in $[t,t+dt)$ then we must be careful to say whether we assume the prior conditioning $X(-\infty) = n_0$, or the prior conditioning $X(t) \in 2$. For the *former* prior conditioning we have

Prob$\{ 15 \rightarrow 16 \text{ in } [t,t+dt) \mid X(-\infty) = n_0 \}$

$= \text{Prob}\{ X(t+dt) = 16 \mid X(t) = 15 \} \times \text{Prob}\{ X(t) = 15 \mid X(-\infty) = n_0 \}$

$= W_+(15) \, dt \times 1/(N+1). \hspace{3cm} (6.5\text{-}12a)$

But for the *latter* prior conditioning we have

Prob$\{ 2 \rightarrow 3 \text{ in } [t,t+dt) \mid X(t) \in 2 \}$

$= \text{Prob}\{ X(t+dt) = 16 \mid X(t) = 15 \} \times \text{Prob}\{ X(t) = 15 \mid X(t) \in 2 \}$

$= W_+(15) \, dt \times 1/n(2). \hspace{3cm} (6.5\text{-}12b)$

Similarly, for the $16 \rightarrow 15$ or $3 \rightarrow 2$ transition we have

Prob$\{ 16 \rightarrow 15 \text{ in } [t,t+dt) \mid X(-\infty) = n_0 \}$

$= \text{Prob}\{ X(t+dt) = 15 \mid X(t) = 16 \} \times \text{Prob}\{ X(t) = 16 \mid X(-\infty) = n_0 \}$

$= W_-(16) \, dt \times 1/(N+1), \hspace{2.5cm} (6.5\text{-}13a)$

and

Prob$\{ 3 \rightarrow 2 \text{ in } [t,t+dt) \mid X(t) \in 3 \}$

$= \text{Prob}\{ X(t+dt) = 15 \mid X(t) = 16 \} \times \text{Prob}\{ X(t) = 16 \mid X(t) \in 3 \}$

$= W_-(16) \, dt \times 1/n(3). \hspace{3cm} (6.5\text{-}13b)$

Now, since $W_+(15) = W_-(16)$, then Eqs. (6.5-12a) and (6.5-13a) together imply that

$$\frac{\text{Prob}\{ 15 \rightarrow 16 \text{ in } [t,t+d.) \mid X(-\infty) = n_0 \}}{\text{Prob}\{ 16 \rightarrow 15 \text{ in } [t,t+dt) \mid X(-\infty) = n_0 \}} = 1, \hspace{1cm} (6.5\text{-}14)$$

and Eqs. (6.5-12b) and (6.5-13b) together imply that

$$\frac{\text{Prob}\{ 2 \rightarrow 3 \text{ in } [t,t+dt) \mid X(t) \in 2 \}}{\text{Prob}\{ 3 \rightarrow 2 \text{ in } [t,t+dt) \mid X(t) \in 3 \}} = \frac{n(3)}{n(2)} \ll 1. \hspace{1cm} (6.5\text{-}15)$$

Equation (6.5-14) expresses the *dynamical democracy* that obtains when the system is viewed on the *microscopic* level: It essentially says that "inverse transitions are equally likely to occur." Equation (6.5-15) on the other hand expresses the *dynamical asymmetry* that obtains on the *macroscopic* level: It essentially says that "transitions that decrease the system's entropy are less likely to occur than transitions that increase the system's entropy." These two seemingly inconsistent features are therefore seen, on close examination, to be perfectly harmonious with each other.

6.6 THE FIRST PASSAGE TIME

In this section we shall address the following general question, which often arises in a variety of practical contexts: If a given temporally homogeneous birth-death Markov process $X(t)$ is placed in some specified state n_0 at time 0, what will be the time of its first arrival in another specified state n_1? To answer this question is to discover the statistics of the **first passage time**,

$$T(n_0 \rightarrow n_1) \equiv \text{ time at which } X(t) \text{ first arrives in}$$
$$\text{state } n_1, \text{ given that } X(0) = n_0. \qquad (6.6\text{-}1)$$

Obviously, $T(n_0 \rightarrow n_1)$ is a random variable. We shall denote its density function by $Q(t; n_0 \rightarrow n_1)$, and its kth moment by $T_k(n_0 \rightarrow n_1)$:

$$T_k(n_0 \rightarrow n_1) = \int_0^{\infty} dt \, t^k Q(t; n_0 \rightarrow n_1). \qquad (6.6\text{-}2)$$

Our analysis in this section of the first passage time for temporally homogeneous birth-death Markov processes will be divided into three stages of increasing complexity and detail. In Subsection 6.6.A we shall use a relatively simple "pedestrian" approach to derive explicit formulas for the **mean first passage time** $T_1(n_0 \rightarrow n_1)$. In most situations a knowledge of this first moment of $T(n_0 \rightarrow n_1)$ will be all that one requires; indeed, the mean first passage time formulas that we shall derive through this pedestrian approach will be sufficient for nearly all of our later work in Sections 6.7 and 6.8. Nevertheless, in Subsection 6.6.B we shall undertake a more sophisticated analysis that leads to formulas for calculating $T_k(n_0 \rightarrow n_1)$ for *all* $k \geq 1$. This "moments" approach utilizes the backward master equation, and it closely parallels the analysis of the first exit time for *continuous* Markov processes given in Chapter 3. By allowing us to calculate $T_2(n_0 \rightarrow n_1)$ as well as $T_1(n_0 \rightarrow n_1)$, this approach evidently enables us to determine the standard deviation of $T(n_0 \rightarrow n_1)$. Finally, in Subsection 6.6.C, we shall show how one can, given sufficient computational resources, calculate the first passage time density function $Q(t; n_0 \rightarrow n_1)$; that, of course, would provide us with a complete characterization of the first passage time $T(n_0 \rightarrow n_1)$. But before developing these *systematic* ways of addressing the first passage time problem, let us consider that problem in the relatively simple contexts of the *random telegraph, radioactive decay* and *Poisson* processes.

The random telegraph process, which we discussed in Subsection 6.3.C, has only the two states $n = 0$ and 1. It follows immediately from the specification of this process in Eq. (6.3-29) that the pausing time in each of these states n is $E(a_n)$. Therefore, the density functions for the first passage times $T(0 \rightarrow 1)$ and $T(1 \rightarrow 0)$ in the *random telegraph process* are the simple exponential forms

$$Q(t; 0 \rightarrow 1) = a_0 \exp(-a_0 t) \quad \text{and} \quad Q(t; 1 \rightarrow 0) = a_1 \exp(-a_1 t). \quad (6.6\text{-}3\text{a})$$

It follows that the mean first passage times for this particular process are given by

$$T_1(0 \rightarrow 1) = 1/a_0 \quad \text{and} \quad T_1(1 \rightarrow 0) = 1/a_1. \quad (6.6\text{-}3\text{b})$$

The radioactive decay process, which we discussed in Subsection 6.3.B, is defined by the characterizing functions (6.3-13). This process gives the number of undecayed radioactive atoms in a population where each such atom has a lifetime against decay that is $E(1/\tau^*)$, independently of what happens to any other atom. Since this process cannot step to higher valued states, then obviously $T(n_0 \rightarrow n_1) = \infty$ if $n_1 > n_0$. For $n_1 < n_0$, we can derive a formula for the density function of $T(n_0 \rightarrow n_1)$ by reasoning as follows: Imagine the n_0 radioactive atoms at time 0 to be divided into two groups of $n_1 + 1$ atoms and $n_0 - (n_1 + 1)$ atoms. The probability that all of the atoms in the first group will *not* have decayed by time t is equal to $[\exp(-t/\tau^*)]^{n_1 + 1}$. The probability that all of the atoms in the second group *will* have decayed by time t is equal to $[1 - \exp(-t/\tau^*)]^{n_0 - n_1 - 1}$. And finally, the probability that some one of the $n_1 + 1$ atoms in the first group, undecayed at time t, *will* decay in the infinitesimal time interval $[t, t + dt)$ is equal to $(n_1 + 1)(dt/\tau^*)$. So, taking account of the number of ways of dividing n_0 atoms into two groups of $n_1 + 1$ atoms and $n_0 - n_1 - 1$ atoms, we conclude that the probability that $X(t)$ will first arrive in state n_1 in $[t, t + dt)$, given that $X(0) = n_0 > n_1$, is

$$Q(t; n_0 \rightarrow n_1)\,dt = \frac{n_0!}{(n_1 + 1)!\,(n_0 - n_1 - 1)!}$$

$$\times \left[e^{-t/\tau^*} \right]^{n_1 + 1} \times \left[1 - e^{-t/\tau^*} \right]^{n_0 - n_1 - 1} \times (n_1 + 1)\frac{dt}{\tau^*}.$$

Therefore, the density function of $T(n_0 \to n_1)$ for the *radioactive decay process* is, for $n_1 < n_0$,

$$Q(t; n_0 \to n_1) = \frac{1}{\tau^*} \frac{n_0!}{n_1!(n_0 - n_1 - 1)!} \left[e^{-t/\tau^*} \right]^{n_1 + 1} \left[1 - e^{-t/\tau^*} \right]^{n_0 - n_1 - 1}$$

$$(0 \le n_1 < n_0). \quad (6.6\text{-}4\text{a})$$

By using this density function to evaluate the integral (6.6-2) for $k = 1$, one can show that the average time for the radioactive decay process to go from n_0 to n_1 is †

$$T_1(n_0 \to n_1) = \tau^* \sum_{n = n_1 + 1}^{n_0} \frac{1}{n} \quad (0 \le n_1 < n_0). \quad (6.6\text{-}4\text{b})$$

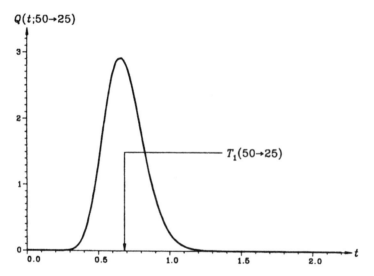

Figure 6-7. The density function of the first passage time $T(50 \to 25)$ for the *radioactive decay process* with $\tau^* = 1$, as computed from Eq. (6.6-4a). The arrow locates the corresponding mean $T_1(50 \to 25)$, as computed from Eq. (6.6-4b). Compare this plot with the radioactive decay process simulation shown in Fig. 6-2, which has the same parameter values.

† The derivation of Eq. (6.6-4b) from Eq. (6.6-4a) proceeds rather indirectly: Using Eq. (6.6-2) with $k = 1$, one shows firstly that $T_1(n_0 \to n_0 - 1) = \tau^*/n_0$, and secondly, through an integration by parts, that $T_1(n_0 \to n_1) = T_1(n_0 \to n_1 + 1) + \tau^*/(n_1 + 1)$. By combining these two results using a straightforward induction argument, one finally obtains the explicit formula for $T_1(n_0 \to n_1)$ quoted in Eq. (6.6-4b).

In Fig. 6-7 we have plotted the density function (6.6-4a) for $\tau^* = 1$, $n_0 = 50$ and $n_1 = 25$. For these parameter values, formula (6.6-4b) yields a mean first passage time of 0.6832. It is instructive to compare these first passage time results with the radioactive decay process simulation plotted in Fig. 6-2, for which we also had $\tau^* = 1$ and $n_0 = 50$. In that simulation, the process is seen to arrive in state 25 at approximately 0.67 time units, which is evidently a quite reasonable "sampling" of the density function plotted in Fig. 6-7.

Finally, we consider the Poisson process, which is defined by the characterizing functions (6.3-1). This process moves in steps of $+1$ only, with the pausing time in each state being $\mathbf{E}(a)$. Since this process cannot step to lower valued states, then obviously $T(n_0 \to n_1) = \infty$ if $n_1 < n_0$. To calculate the density function for $T(n_0 \to n_1)$ for $n_1 > n_0$, we begin by observing that the time required to go from state n_0 to state n_1 is just the sum of the individual pausing times in the $n_1 - n_0$ states n_0, $n_0 + 1$, ..., $n_1 - 1$. But the pausing time in each one of those $n_1 - n_0$ states is $\mathbf{E}(a)$. So, with $m \equiv n_1 - n_0$, we let T_1, T_2, ..., T_m be m statistically independent exponential random variables with the common decay constant a, and we write the first passage time $T(n_0 \to n_1)$ for $n_1 > n_0$ as

$$T(n_0 \to n_1) = \sum_{j=1}^{m} T_j \qquad (m \equiv n_1 - n_0).$$

The *joint* density function of the m random variables T_1, ..., T_m is clearly

$$\prod_{j=1}^{m} \left[a \exp(-at_j) \right] = a^m \exp\left(-a \sum_{j=1}^{m} t_j \right).$$

So it follows from the RVT theorem (1.6-4) that the density function of the *sum* of these m random variables is given by

$$Q(t; n_0 \to n_1) = \int_0^{\infty} dt_1 \cdots \int_0^{\infty} dt_m \left[a^m \exp\left(-a \sum_{j=1}^{m} t_j \right) \right] \delta\left(t - \sum_{j=1}^{m} t_j \right).$$

Defining the theta function $\theta(x)$ to be unity if $x > 0$ and zero if $x \leq 0$, we proceed to reduce this integral as follows:

$$Q(t; n_0 \to n_1) = a^m \int_0^{\infty} dt_1 \cdots \int_0^{\infty} dt_{m-1} \exp\left(-a \sum_{j=1}^{m-1} t_j \right)$$

$$\times \int_{-\infty}^{\infty} dt_m \, \theta(t_m) \exp\left(-at_m \right) \delta\left(t_m - \left[t - \sum_{j=1}^{m-1} t_j \right] \right)$$

$$= a^m \int_0^\infty dt_1 \dots \int_0^\infty dt_{m-1} \exp\left(-a \sum_{j=1}^{m-1} t_j\right)$$

$$\times \theta\left(t - \sum_{j=1}^{m-1} t_j\right) \exp\left(-a\left[t - \sum_{j=1}^{m-1} t_j\right]\right)$$

$$= a^m e^{-at} \int_0^t dt_1 \int_0^{t-t_1} dt_2$$

$$\times \int_0^{t-t_1-t_2} dt_3 \dots \int_0^{t-t_1-\dots-t_{m-2}} dt_{m-1}.$$

Changing integration variables from $\{t_i\}$ to $\{u_i\}$, where $u_1 \equiv t - t_1$, $u_2 \equiv t - t_1 - t_2$, etc., this becomes

$$Q(t, n_0 \to n_1) = a^m e^{-at} \int_0^t du_1 \int_0^{u_1} du_2 \int_0^{u_2} du_3 \dots \int_0^{u_{m-2}} du_{m-1}.$$

The multiple integral can now be straightforwardly evaluated, starting from the right and proceeding to the left. The result is found to be $t^{m-1}/(m-1)!$. So, recalling that $n_1 \equiv n_0 + m$, we conclude that the density function for the first passage time $T(n_0 \to n_0 + m)$ for a *Poisson process* is[†]

$$Q(t, n_0 \to n_0 + m) = a e^{-at} \frac{(at)^{m-1}}{(m-1)!} \qquad (m > 0). \qquad (6.6\text{-}5a)$$

By using this density function to evaluate the integral (6.6-2) for $k = 1$, it is easily shown that the average time for the Poisson process to go from state n_0 to state $n_0 + m$ is

$$T_1(n_0 \to n_0 + m) = m/a \qquad (m > 0). \qquad (6.6\text{-}5b)$$

This formula can be interpreted simply as the average pausing time per step $(1/a)$ multiplied by the number of required steps (m). In Fig. 6-8 we have plotted the function (6.6-5a) for $a = 1$, $n_0 = 0$ and $m = 25$. For these parameter values, Eq. (6.6-5b) yields a mean first passage time of 25.0. It is instructive to compare these first passage time results with the Poisson

---

[†] If a random variable T is equal to the sum of m statistically independent random variables T_1, T_2, ..., T_m, each one of which is $E(a)$, then T is said to be "*gamma* distributed with parameters a and m," and we write $T = \Gamma(a, m)$. The right side of Eq. (6.6-5a) is thus the density function of the random variable $\Gamma(a, m)$. Notice that Eq. (6.6-5a) is consistent with the expected result $\Gamma(a, 1) = E(a)$.

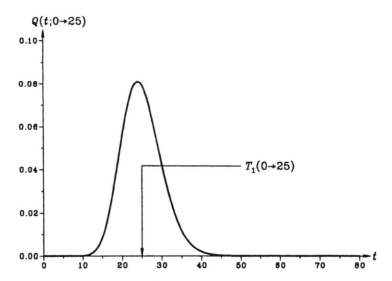

Figure 6-8. The density function of the first passage time $T(0 \rightarrow 25)$ for the *Poisson process* with $a = 1$, as computed from Eq. (6.6-5a). The arrow locates the corresponding mean $T_1(0 \rightarrow 25)$, as computed from Eq. (6.6-5b). Compare this plot with the Poisson process simulation shown in Fig. 5-2, which has the same parameter values.

process simulation plotted in Fig. 5-2, for which we also had $a = 1$ and $n_0 = 0$. In that simulation, the process is seen to arrive in state 25 at approximately 21.7 time units, which is evidently a quite reasonable "sampling" of the density function plotted in Fig. 6-8.

The foregoing calculations of the statistics of the first passage time $T(n_0 \rightarrow n_1)$ for three comparatively simple birth-death Markov processes serve to illustrate the nature and significance of that random variable. But those calculations evidently exploited specific, unique characteristics of the three processes (the "two-state" nature of the random telegraph process and the "one-way" natures of the radioactive decay and Poisson processes), and consequently they cannot be adapted to handle multistate, two-way processes. We now proceed to develop three quite general methods for analyzing the statistics of the first passage time random variable $T(n_0 \rightarrow n_1)$. The details of these general methods turn out to depend upon whether n_1 is greater than or less than n_0. For the sake of brevity we shall exhibit derivations only for the case $n_1 > n_0$, and merely quote results for the case $n_1 < n_0$. We begin by developing a relatively simple general method for calculating $T_1(n_0 \rightarrow n_1)$, the mean first passage time from state n_0 to state n_1.

6.6.A THE PEDESTRIAN APPROACH

Let us regard the given Markov process $X(t)$ as a "walker" on the nonnegative integers who starts in state n_0 and walks around, in the one-step manner dictated by its characterizing functions, until it first arrives in state n_1. With this "pedestrian" picture in mind, we define the following set of variables:

$$t(n; n_0 \to n_1) \equiv \text{average time the walker spends in state } n \text{ in the course of a first passage from state } n_0 \text{ to state } n_1; \qquad (6.6\text{-}6a)$$

$$v(n; n_0 \to n_1) \equiv \text{average number of visits by the walker to state } n \text{ in the course of a first passage from state } n_0 \text{ to state } n_1; \qquad (6.6\text{-}6b)$$

$$v_\pm(n; n_0 \to n_1) \equiv \text{average number of } n \to n \pm 1 \text{ steps taken by the walker in the course of a first passage from state } n_0 \text{ to state } n_1. \qquad (6.6\text{-}6c)$$

Since the average time the walker spends in state n on any one visit to that state us $1/a(n)$ [see Eq. (6.1-16)], then we have

$$t(n; n_0 \to n_1) / v(n; n_0 \to n_1) = 1/a(n). \qquad (6.6\text{-}7a)$$

And since the probability that the walker, upon stepping away from state n, will step to state $n \pm 1$ is $w_\pm(n) \equiv W_\pm(n)/a(n)$ [see Eqs. (6.1-5) and (6.1-9)], then we also have

$$v_\pm(n; n_0 \to n_1) / v(n; n_0 \to n_1) = W_\pm(n)/a(n). \qquad (6.6\text{-}7b)$$

Taking the ratio of Eqs. (6.6-7) gives

$$t(n; n_0 \to n_1) = v_\pm(n; n_0 \to n_1) / W_\pm(n), \qquad (6.6\text{-}8)$$

a relation that we shall make use of momentarily.

Consider now the specific case $n_1 > n_0$. It is obvious from Fig. 6-9 that, in any first passage from n_0 to n_1, the walker must make exactly one step from state $n_1 - 1$ to state n_1; therefore,

Figure 6-9. Deducing the first-passage stepping relations (6.6-9) for $n_0 < n_1$.

$$v_+(n_1-1; n_0 \to n_1) = 1 \qquad (n_0 < n_1). \qquad (6.6\text{-}9a)$$

A brief inspection of Fig. 6-9 will also reveal that, in any first passage from n_0 to n_1, the *difference* {number of $n \to n+1$ steps} minus {number of $n+1 \to n$ steps} must be exactly *one* if $n \in [n_0, n_1-2]$, and exactly *zero* if $n \in [0, n_0-1]$; therefore,

$$v_+(n; n_0 \to n_1) - v_-(n+1; n_0 \to n_1) = \theta(n+1-n_0)$$

$$(n_0 < n_1; 0 \le n \le n_1-2), \qquad (6.6\text{-}9b)$$

where we have used the customary notation for the unit step function,

$$\theta(n) \equiv \begin{cases} 0, & \text{if } n \le 0, \\ 1, & \text{if } n > 0. \end{cases} \qquad (6.6\text{-}10)$$

A little algebraic manipulation brings Eqs. (6.6-9) into the respective forms

$$\frac{v_+(n_1-1; n_0 \to n_1)}{W_+(n_1-1)} = \frac{1}{W_+(n_1-1)},$$

$$\frac{v_+(n; n_0 \to n_1)}{W_+(n)} = \frac{\theta(n+1-n_0)}{W_+(n)} + \frac{W_-(n+1)}{W_+(n)} \frac{v_-(n+1; n_0 \to n_1)}{W_-(n+1)}.$$

Now invoking the relation (6.6-8), we conclude that

$$\begin{cases} t(n_1-1; n_0 \to n_1) = \dfrac{1}{W_+(n_1-1)} & (n_0 < n_1), \qquad (6.6\text{-}11a) \\[2em] t(n; n_0 \to n_1) = \dfrac{\theta(n+1-n_0)}{W_+(n)} + \dfrac{W_-(n+1)}{W_+(n)} t(n+1; n_0 \to n_1) \end{cases}$$

$$(n_0 < n_1; \ n = n_1-2, n_1-3, ..., 1, 0). \qquad (6.6\text{-}11b)$$

Equations (6.6-11) evidently allow us to calculate $t(n; n_0 \to n_1)$ *recursively* for $n = n_1-1, n_1-2, n_1-3, ..., 1, 0$. Having thus calculated the average time spent by the walker in every accessible state n in the course of its first passage from n_0 to $n_1 > n_0$, we can then obtain the *total* average time for the first passage by simply summing those individual average times:

$$T_1(n_0 \to n_1) = \sum_{n=0}^{n_1-1} t(n; n_0 \to n_1) \qquad (n_0 < n_1). \qquad (6.6\text{-}12)$$

Equations (6.6-11) and (6.6-12) suffice to calculate the mean first passage time $T_1(n_0 \to n_1)$ for the case $n_1 > n_0$. Companion equations for the case $n_1 < n_0$ can be derived by using analogous reasoning. In that case, we must replace Eq.(6.6-9a) with an equation expressing the fact that any first passage from n_0 to $n_1 < n_0$ will entail exactly one transition $n_1 + 1 \to n_1$. And we must replace Eq. (6.6-9b) with an equation expressing the fact that, in any first passage from n_0 to $n_1 < n_0$, the *difference* {number of $n \to n - 1$ steps} minus {number of $n - 1 \to n$ steps} must be exactly *one* if $n \in [n_1 + 2, n_0]$, and exactly *zero* if $n \geq n_0 + 1$. Upon applying to those two new equations the relation (6.6-8), we arrive at the following algorithm for calculating $T_1(n_0 \to n_1)$ for the case $n_1 < n_0$: First compute $t(n; n_0 \to n_1)$ for $n = n_1 + 1,\ n_1 + 2,\ \ldots$ through the recursion relations

$$
\begin{cases}
t(n_1 + 1; n_0 \to n_1) = \dfrac{1}{W_-(n_1 + 1)} & (n_1 < n_0), \quad \text{(6.6-13a)} \\[4mm]
t(n; n_0 \to n_1) = \dfrac{\theta(n_0 + 1 - n)}{W_-(n)} + \dfrac{W_+(n-1)}{W_-(n)}\, t(n-1; n_0 \to n_1) & \\[4mm]
\qquad (n_1 < n_0;\ n = n_1 + 2, n_1 + 3, \ldots); & \text{(6.6-13b)}
\end{cases}
$$

then compute the sum

$$
T_1(n_0 \to n_1) = \sum_{n = n_1 + 1}^{\infty} t(n; n_0 \to n_1) \qquad (n_1 < n_0). \quad \text{(6.6-14)}
$$

If the process has a reflecting state N somewhere above n_0, then the summation in Eq. (6.6-14) would terminate at $n = N$. In the absence of such an upper reflecting state we must pay attention to whether or not that infinite sum converges, and if so how many terms in the sum are required for acceptable accuracy in numerical calculations. We shall see shortly that convergence of the infinite sum (6.6-14) is always assured if the process $X(t)$ is *stable* — i.e., if it has a well defined stationary Markov state density function $P_s(n)$.

Let us verify that the formulas for $T_1(n_0 \to n_1)$ in Eqs. (6.6-11) – (6.6-14) give for the Poisson and radioactive decay processes the same results that we found at the beginning of this section. For the Poisson process we have [see Eqs. (6.3-2)] $W_+(n) = a$ and $W_-(n) = 0$. In the case $n_1 < n_0$ Eqs. (6.6-13) and (6.6-14) give $T_1(n_0 \to n_1) = \infty$, as we should expect. And in the case $n_1 > n_0$ Eqs. (6.6-11) give

$$t(n; n_0 \to n_1) = \begin{cases} 1/a, & \text{for } n=n_1-1, n_1-2,...,n_0 \\ 0, & \text{for } n<n_0, \end{cases}$$

whereupon the sum in Eq. (6.6-12) gives $T_1(n_0 \to n_1)=(n_0-n_1)(1/a)$, in agreement with our earlier result (6.6-5b).

For the radioactive decay process we have [see Eqs. (6.3-14)] $W_+(n)=0$ and $W_-(n)=n/\tau^*$. In the case $n_1 > n_0$ Eqs. (6.6-11) and (6.6-12) give $T_1(n_0 \to n_1)=\infty$, as we should expect. And in the case $n_1 < n_0$ Eqs. (6.6-13) give

$$t(n; n_0 \to n_1) = \begin{cases} \tau^*/n, & \text{for } n=n_1+1, n_1+2,...,n_0 \\ 0, & \text{for } n>n_0, \end{cases}$$

whereupon Eq. (6.6-14) gives for $T_1(n_0 \to n_1)$ precisely our earlier result (6.6-4b).

The recursive nature of the formulas (6.6-11) – (6.6-14) for $T_1(n_0 \to n_1)$ makes them especially well suited for numerical calculations on a digital computer. But for some purposes it is useful to have a fully explicit formula for $T_1(n_0 \to n_1)$. To obtain such a formula, we must evidently carry out the required iterations and summations in Eqs. (6.6-11) – (6.6-14) analytically. The iteration of Eq. (6.6-11b) for $n>n_1-1$ is accomplished thusly:

$$t(n; n_0 \to n_1) = \frac{1}{W_+(n)} \left(\theta(n+1-n_0) + W_-(n+1)\, t(n+1; n_0 \to n_1) \right)$$

$$= \frac{1}{W_+(n)} \left(\theta(n+1-n_0) + \frac{W_-(n+1)}{W_+(n+1)} \left[\theta(n+2-n_0) \right. \right.$$

$$\left. \left. + W_-(n+2)\, t(n+2; n_0 \to n_1) \right] \right)$$

$$= \frac{1}{W_+(n)} \left(\theta(n+1-n_0) + \frac{W_-(n+1)}{W_+(n+1)} \left[\theta(n+2-n_0) \right. \right.$$

$$\left. \left. + \frac{W_-(n+2)}{W_+(n+2)} \left\{ \theta(n+3-n_0) + \ldots \left(\frac{W_-(n_1-1)}{W_+(n_1-1)} \right) \ldots \right\} \right] \right).$$

By carefully collecting terms in this last equation, we get the following fully explicit formula:

$t(n; n_0 \to n_1)$

$$= \frac{1}{W_+(n)} \left(\theta(n+1-n_0) + \sum_{m=n+1}^{n_1-1} \theta(m+1-n_0) \prod_{j=n+1}^{m} \frac{W_-(j)}{W_+(j)} \right)$$

$$(n_0 < n_1; \ 0 \le n \le n_1 - 2). \qquad (6.6\text{-}15)$$

If the potential function $\phi(n)$ defined in Eq.(6.4-12) exists for the process, then the j-product in Eq. (6.6-15) can be reduced as follows:

$$\prod_{j=n+1}^{m} \frac{W_-(j)}{W_+(j)} = \frac{\prod_{j=1}^{m} \left[W_-(j) / W_+(j) \right]}{\prod_{j=1}^{n} \left[W_-(j) / W_+(j) \right]} = \frac{\exp\left(\ln \prod_{j=1}^{m} \left[W_-(j) / W_+(j) \right] \right)}{\exp\left(\ln \prod_{j=1}^{n} \left[W_-(j) / W_+(j) \right] \right)}$$

$$= \frac{\exp\left(\sum_{j=1}^{m} \ln \left[W_-(j) / W_+(j) \right] \right)}{\exp\left(\sum_{j=1}^{n} \ln \left[W_-(j) / W_+(j) \right] \right)} = \frac{e^{\phi(m)}}{e^{\phi(n)}}.$$

Upon substituting this into Eq. (6.6-15), we get

$t(n; n_0 \to n_1)$

$$= \frac{1}{W_+(n)} \left(\theta(n+1-n_0) + \sum_{m=n+1}^{n_1-1} \theta(m+1-n_0) e^{\phi(m)} e^{-\phi(n)} \right)$$

$$= \frac{1}{W_+(n)} \left(\sum_{m=n}^{n_1-1} \theta(m+1-n_0) e^{\phi(m)} e^{-\phi(n)} \right),$$

a formula that Eq. (6.6-11a) shows is valid also for $n = n_1 - 1$. So if the potential function $\phi(n)$ exists for the process, then our analytical formula (6.6-15) for $t(n; n_0 \to n_1)$ can be written more compactly as

$$t(n; n_0 \to n_1) = \frac{1}{W_+(n)} e^{-\phi(n)} \sum_{m=\text{Max}(n,n_0)}^{n_1-1} e^{\phi(m)}$$

$$(n_0 < n_1; \ 0 \le n \le n_1 - 1). \qquad (6.6\text{-}16a)$$

And if the process is *stable*, so that the constant K defined in Eq. (6.4-5) is nonzero, then by multiplying and dividing the right hand side of Eq. (6.6-16a) by $KW_+(0)$ we obtain by virtue of formulas (6.4-13) and (6.4-16),

$$t(n; n_0 \to n_1) = P_s(n) \sum_{m = \text{Max}(n, n_0)}^{n_1 - 1} \Psi(m)$$

$$(n_0 < n_1;\ 0 \le n \le n_1 - 1). \quad (6.6\text{-}16b)$$

Of the three explicit formulas (6.6-15), (6.6-16a) and (6.6-16b) for $t(n; n_0 \to n_1)$, the first is evidently valid irrespective of whether the functions $\phi(n)$, $P_s(n)$ and $\Psi(n)$ exist for the process. Usually though, at least one of these three functions will exist, and one or both of the two formulas (6.6-16) will be more convenient than formula (6.6-15). By summing Eqs. (6.6-16) over n in accordance with Eq. (6.6-12), we obtain the following two explicit formulas for $T_1(n_0 \to n_1)$:

$$T_1(n_0 \to n_1) = \sum_{n=0}^{n_1 - 1} \frac{1}{W_+(n)} e^{-\phi(n)} \sum_{m = \text{Max}(n, n_0)}^{n_1 - 1} e^{\phi(m)}$$

$$(n_0 < n_1), \quad (6.6\text{-}17a)$$

$$T_1(n_0 \to n_1) = \sum_{n=0}^{n_1 - 1} P_s(n) \sum_{m = \text{Max}(n, n_0)}^{n_1 - 1} \Psi(m) \quad (n_0 < n_1). \quad (6.6\text{-}17b)$$

Formula (6.6-17a) applies whenever the function $\phi(n)$ exists, while formula (6.6-17b) applies whenever the function $P_s(n)$ exists. The common summation domain in the m-n integer lattice space in formulas (6.6-17) is evidently the region defined by the heavy dots in Fig. 6-10a. It can be seen from that figure that, by simply changing the *order* of the n and m summations, we get the following fully equivalent formulas:

$$T_1(n_0 \to n_1) = \sum_{m=n_0}^{n_1 - 1} e^{\phi(m)} \sum_{n=0}^{m} \frac{1}{W_+(n)} e^{-\phi(n)}$$

$$(n_0 < n_1), \quad (6.6\text{-}18a)$$

$$T_1(n_0 \to n_1) = \sum_{m=n_0}^{n_1 - 1} \Psi(m) \sum_{n=0}^{m} P_s(n) \quad (n_0 < n_1). \quad (6.6\text{-}18b)$$

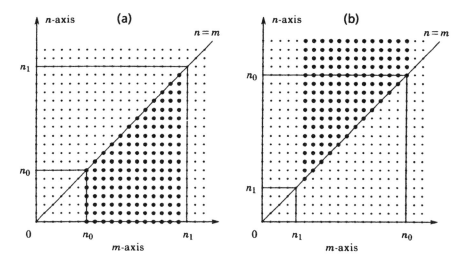

Figure 6-10. Summation regions for: (a) the $n_0 < n_1$ double-sum formulas (6.6-17) and (6.6-18); and (b) the $n_1 < n_0$ double-sum formulas (6.6-22) and (6.6-23). In each case, the summation domain is the set of heavy dots.

We know from our derivation of formulas (6.6-17) that the outer (n) summands in those two formulas can be interpreted as the average total time spent in state n during a first passage from n_0 to n_1. Interestingly, the outer (m) summands in formulas (6.6-18) also have a sensible interpretation: By putting $n_1 = n_0 + 1$ in the latter two formulas, we get

$$T_1(n_0 \to n_0 + 1) = e^{\phi(n_0)} \sum_{n=0}^{n_0} \frac{1}{W_+(n)} e^{-\phi(n)} = \Psi(n_0) \sum_{n=0}^{n_0} P_s(n).$$
(6.6-19)

From these equations we can see that the outer (m) summand in each of formulas (6.6-18) is just $T_1(m \to m + 1)$; therefore,

$$T_1(n_0 \to n_1) = \sum_{m=n_0}^{n_1 - 1} T_1(m \to m + 1) \quad (n_0 < n_1). \quad (6.6\text{-}20a)$$

It follows from this that the mean first passage time for a temporally homogeneous birth-death Markov process satisfies the following simple composition rule: For any three states $n_0 < n_1 < n_2$,

$$T_1(n_0 \to n_2) = T_1(n_0 \to n_1) + T_1(n_1 \to n_2). \quad (6.6\text{-}20b)$$

An analytical iteration of the $n_1 < n_0$ formula (6.6-13b) can be carried out using arguments very similar to the foregoing, and it yields the following results: The average total time spent in state n during a first passage from n_0 to $n_1 < n_0$ is given by the explicit formula

$$t(n; n_0 \to n_1) = \frac{1}{W_-(n)} e^{-\phi(n-1)} \sum_{m=n_1+1}^{\text{Min}(n,n_0)} e^{\phi(m-1)}$$

$$(n_1 < n_0; \; n \geq n_1 + 1) \quad (6.6\text{-}21a)$$

if the potential function is well defined, and by the explicit formula

$$t(n; n_0 \to n_1) = P_s(n) \sum_{m=n_1+1}^{\text{Min}(n,n_0)} \Psi(m-1)$$

$$(n_1 < n_0; \; n \geq n_1 + 1) \quad (6.6\text{-}21b)$$

if the stationary Markov state density function is well defined.[†] It then follows from Eq. (6.6-14) that the mean first passage time from n_0 to $n_1 < n_0$ can be written in either of the explicit representations

$$T_1(n_0 \to n_1) = \sum_{n=n_1+1}^{\infty} \frac{1}{W_-(n)} e^{-\phi(n-1)} \sum_{m=n_1+1}^{\text{Min}(n,n_0)} e^{\phi(m-1)}$$

$$(n_1 < n_0), \quad (6.6\text{-}22a)$$

$$T_1(n_0 \to n_1) = \sum_{n=n_1+1}^{\infty} P_s(n) \sum_{m=n_1+1}^{\text{Min}(n,n_0)} \Psi(m-1) \quad (n_1 < n_0). \quad (6.6\text{-}22b)$$

Formula (6.6-22a) applies whenever the function $\phi(n)$ exists, while formula (6.6-22b) applies whenever the function $P_s(n)$ exists. We observe that the double summations in these last two formulas are taken over the region of the m-n integer lattice space defined by the heavy dots in Fig. 6-10b; an inspection of that figure shows that if we simply interchange the order of the n and m summations in formulas (6.6-22), we will obtain the following equivalent formulas:

---

† The derivation of Eq. (6.6-21b) from Eq. (6.6-21a) invokes the identity

$$[1/W_-(n)] \exp(-\phi(n-1)) \equiv [1/W_+(n)] \exp(-\phi(n)).$$

This identity is an easily proved consequence of the definition of $\phi(n)$ in Eq. (6.4-12).

$$T_1(n_0 \to n_1) = \sum_{m=n_1+1}^{n_0} e^{\phi(m-1)} \sum_{n=m}^{\infty} \frac{1}{W_-(n)} e^{-\phi(n-1)}$$

$$(n_1 < n_0), \quad (6.6\text{-}23a)$$

$$T_1(n_0 \to n_1) = \sum_{m=n_1+1}^{n_0} \Psi(m-1) \sum_{n=m}^{\infty} P_s(n) \quad (n_1 < n_0). \quad (6.6\text{-}23b)$$

From the latter pair of formulas, we find by taking $n_1 = n_0 - 1$ that

$$T_1(n_0 \to n_0 - 1) = e^{\phi(n_0-1)} \sum_{n=n_0}^{\infty} \frac{1}{W_-(n)} e^{-\phi(n-1)} = \Psi(n_0-1) \sum_{n=n_0}^{\infty} P_s(n).$$

$$(6.6\text{-}24)$$

These equations reveal that the outer (m) summands in Eqs. (6.6-23) are just $T_1(m \to m - 1)$; consequently, Eqs. (6.6-23) can also be written

$$T_1(n_0 \to n_1) = \sum_{m=n_1+1}^{n_0} T_1(m \to m-1) \quad (n_1 < n_0). \quad (6.6\text{-}25a)$$

From this formula we can immediately deduce the result (6.6-20b), but now for the "down-going" state ordering $n_2 < n_1 < n_0$:

$$T_1(n_0 \to n_2) = T_1(n_0 \to n_1) + T_1(n_1 \to n_2). \quad (6.6\text{-}25b)$$

Incidently, we are now in a position to prove our earlier assertion that the infinite sum over n in Eq. (6.6-14) always converges whenever the process has a well defined stationary Markov state density function: Because of the normalization property of $P_s(n)$, it follows from Eq. (6.6-23b) that

$$T_1(n_0 \to n_1) \le \sum_{m=n_1+1}^{n_0} \Psi(m-1) \sum_{n=0}^{\infty} P_s(n) = \sum_{m=n_1+1}^{n_0} \Psi(m-1) < \infty.$$

It is both interesting and satisfying to observe the strong formal correspondence between the pair of formulas for $T_1(n_0 \to n_1)$ in Eqs. (6.6-18b) and (6.6-23b) and the pair of formulas for the mean first exit time $T_1(x_0; a,b)$ of a *continuous* Markov process in Eqs. (3.7-16a) and (3.7-16b). Specifically, if $n_0 < n_1$, then the formula for the birth-death Markov process mean first passage time $T_1(n_0 \to n_1)$ replicates the formula for the continuous Markov process mean first exit time $T_1(n_0; 0, n_1)$ with 0 a reflecting state and n_1 an absorbing state. And if $n_1 < n_0$, then the formula for the birth-death Markov process mean first passage time $T_1(n_0 \rightarrow n_1)$ replicates the formula for the continuous Markov process mean first exit time $T_1(n_0; n_1, \infty)$ with n_1 an absorbing state and ∞ a reflecting state.

The elegancies of the foregoing explicit mean first passage time formulas not withstanding, the original *recursive* formulas (6.6-11) – (6.6-14) will usually be more convenient for making numerical calculations with a digital computer. To illustrate this point, consider the payroll process discussed in Subsection 6.3.D, for which $W_+(n) = h$, $W_-(n) = ln$ and $\phi(n) = \ln(n! \, (h/l)^{-n})$ [see Eqs. (6.3-51) and (6.4-19a)]. For this process with $n_0 < n_1$, we have from Eq. (6.6-18a) that

$$T_1(n_0 \rightarrow n_1) = \frac{1}{h} \sum_{m=n_0}^{n_1 - 1} \sum_{n=0}^{m} \frac{m!}{n!} (h/l)^{n-m}, \qquad (6.6\text{-}26a)$$

and we have from Eqs. (6.6-11) that

$$t(n; n_0 \rightarrow n_1) = \begin{cases} (1/h), & \text{if } n = n_1 - 1, \\ (1/h)\,[\theta(n+1-n_0) + l(n+1)\,t(n+1; n_0 \rightarrow n_1)], \\ & \text{if } n = n_1 - 2, \ldots, 1, 0. \end{cases}$$

$$(6.6\text{-}26b)$$

If one is using a digital computer, it would clearly be easier to sum the n_1 recursively calculated values of $t(n; n_0 \rightarrow n_1)$ in Eq. (6.6-26b) than it would be to perform the explicit *double* sum in Eq. (6.6-26a).

Finally, we should note that the pedestrian approach described in this subsection also allows us to calculate the *average total number of steps* the process takes in the course of a first passage from n_0 to n_1: Equation (6.6-7a) tells us that $v(n; n_0 \rightarrow n_1)$, which can evidently be regarded as the average total number of steps away from state n that are taken in the course of the passage, is equal to $a(n)\,t(n; n_0 \rightarrow n_1)$. It follows that if we simply multiply the summands $t(n; n_0 \rightarrow n_1)$ on the right hand sides of Eqs. (6.6-12) and (6.6-14) by $a(n)$, those sums become the average total number of steps taken in the course of the passage. Similarly, the explicit formulas (6.6-17), (6.6-18), (6.6-22) and (6.6-23) for the average total *time* required for a first passage can all be converted into formulas for the average total *number of steps* required for a first passage by simply multiplying their respective summands by $a(n)$.

6.6.B THE MOMENTS APPROACH

The "pedestrian" analysis discussed in the preceding subsection evidently provides a relatively simple, intuitively straightforward method for calculating the *first moment* of the first passage time $T(n_0 \to n_1)$. Although it is possible to generalize the pedestrian approach to yield formulas for computing the second moment of $T(n_0 \to n_1)$, the analysis becomes rather complicated in the generalization and loses much of its directness and intuitive appeal. In this subsection we shall describe an approach based on the backward master equation which, although not especially direct or intuitive, does lead expeditiously to a compact set of recursion relations from which one can compute *all* the moments of $T(n_0 \to n_1)$. This "moments" approach closely parallels the first exit time analysis for continuous Markov processes that was presented in Subsection 3.7.A.

We shall again give a detailed derivation only for the case $n_1 > n_0$, and merely quote results for the case $n_1 < n_0$. We begin by defining, in analogy with Eqs. (3.7-3), the function

$$G(n_0, t; n_1) \equiv \sum_{n=0}^{n_1-1} P(n, t \mid n_0, 0), \qquad (6.6\text{-}27\text{a})$$

$$= \text{Prob}\{X(t) \in [0, n_1-1] \mid X(0) = n_0\}. \qquad (6.6\text{-}27\text{b})$$

A time-evolution equation for this function can be obtained from the *backward* master equation (6.1-20), which we rewrite here using the temporally homogeneous property (5.3-27):

$$\frac{\partial}{\partial t} P(n, t \mid n_0, 0) = W_-(n_0) [P(n, t \mid n_0-1, 0) - P(n, t \mid n_0, 0)]$$

$$+ W_+(n_0) [P(n, t \mid n_0+1, 0) - P(n, t \mid n_0, 0)]. \qquad (6.6\text{-}28)$$

By summing this equation over n from 0 to $n_1 - 1$, we obtain on account of the definition (6.6-27a),

$$\frac{\partial}{\partial t} G(n_0, t; n_1) = W_-(n_0) [G(n_0-1, t; n_1) - G(n_0, t; n_1)]$$

$$+ W_+(n_0) [G(n_0+1, t; n_1) - G(n_0, t; n_1)], \qquad (6.6\text{-}29)$$

an equation that we shall put to use shortly.

Since $0 \leq n_0 < n_1$, then our process starts out at time 0 inside the interval $[0, n_1 - 1]$, so Eq. (6.6-27b) implies the initial condition

$$G(n_0, 0; n_1) = 1 \text{ for all } n_0 \in [0, n_1 - 1]. \tag{6.6-30}$$

And since $W_-(0) = 0$, then the process can leave the interval $[0, n_1 - 1]$ only by making the transition $n_1 - 1 \to n_1$, at which point the first passage from n_0 to n_1 will be accomplished. We now *modify* the system under consideration by imposing the condition

$$W_-(n_1) = 0, \tag{6.6-31}$$

so that the once the process *does* leave the interval $[0, n_1 - 1]$ it will *never again return* to that interval. Clearly, this modification will have no effect on the process prior to its first passage to state n_1, so the modified process will exhibit the same statistics for $T(n_0 \to n_1)$ as our original process. The reason for the modification (6.6-31) is that it allows us to expand our interpretation (6.6-27b) of $G(n_0, t; n_1)$ according to

$$G(n_0, t; n_1) = \begin{array}{l} \text{probability that the process, in state} \\ n_0 \in [0, n_1 - 1] \text{ at time 0, will } \textit{not} \text{ have} \\ \text{reached state } n_1 \text{ by time t.} \end{array} \tag{6.6-32}$$

In other words, by imposing condition (6.6-31), the function $G(n_0, t; n_1)$, which is defined by the time-evolution equation (6.6-29) and the initial condition (6.6-30), becomes the probability that $T(n_0 \to n_1)$ will be *greater* than t. It then follows that

$$1 - G(n_0, t; n_1) = \text{Prob}\{ T(n_0 \to n_1) \leq t \} \equiv \int_0^t Q(t'; n_0 \to n_1) \, dt'.$$

where $Q(t; n_0 \to n_1)$ is the density function of the random variable $T(n_0 \to n_1)$. Differentiating this relation with respect to t then gives the key result

$$Q(t; n_0 \to n_1) = -\frac{\partial}{\partial t} G(n_0, t; n_1). \tag{6.6-33}$$

We now assume that $T(n_0 \to n_1)$ is a "proper" random variable, in the sense that its density function is normalized; i.e., we assume that

$$1 = \int_0^\infty dt \, Q(t; n_0 \to n_1) = \int_0^\infty dt \left[-\frac{\partial}{\partial t} G(n_0, t; n_1) \right]$$

$$= G(n_0, 0; n_1) - G(n_0, \infty; n_1)$$

$$= 1 - G(n_0, \infty; n_1),$$

where the last step follows from Eq. (6.6-30). So we see that the assumption that $T(n_0 \to n_1)$ is a proper random variable is equivalent to the assumption that

$$G(n_0, \infty; n_1) = 0. \tag{6.6-34}$$

If this assumption is *not* true, then it is evidently possible that the process will *never* arrive in state n_1, in which case the first passage time $T(n_0 \to n_1)$ will not be a well defined random variable.

With the result (6.6-33), we can evidently write the kth moment of $T(n_0 \to n_1)$ as

$$T_k(n_0 \to n_1) = \int_0^\infty dt \, t^k \left[-\frac{\partial}{\partial t} G(n_0, t; n_1) \right] \quad (k \geq 0). \tag{6.6-35}$$

In particular, since Eq. (6.6-34) ensures that $T(n_0 \to n_1)$ is a proper random variable, then the zeroth moment of $T(n_0 \to n_1)$ will be unity:

$$T_0(n_0 \to n_1) = 1. \tag{6.6-36}$$

For $k \geq 1$, an alternate formula for $T_k(n_0 \to n_1)$ follows upon first integrating Eq. (6.6-35) by parts,

$$T_k(n_0 \to n_1) = -t^k G(n_0, t; n_1) \Big|_0^\infty + \int_0^\infty G(n_0, t; n_1) \, k \, t^{k-1} \, dt,$$

and then noting that the integrated term vanishes by virtue of Eqs. (6.6-30) and (6.6-34);[†] thus,

$$k^{-1} T_k(n_0 \to n_1) = \int_0^\infty dt \, t^{k-1} G(n_0, t; n_1) \quad (k \geq 1). \tag{6.6-37}$$

Now we are in a position to convert the time-evolution equation (6.6-29) for $G(n_0, t; n_1)$ into an equation that interrelates the moments of $T(n_0 \to n_1)$. Multiplying Eq. (6.6-29) through by t^{k-1} $(k \geq 1)$ and then integrating over t from 0 to ∞, we obtain by virtue of Eqs. (6.6-35) and (6.6-37),

$$-T_{k-1}(n_0 \to n_1) = W_-(n_0) \left[k^{-1} T_k(n_0 - 1 \to n_1) - k^{-1} T_k(n_0 \to n_1) \right]$$
$$+ W_+(n_0) \left[k^{-1} T_k(n_0 + 1 \to n_1) - k^{-1} T_k(n_0 \to n_1) \right]$$
$$(0 \leq n_0 \leq n_1 - 1; k \geq 1). \tag{6.6-38}$$

[†] If the zero in Eq. (6.6-34) is not "hard enough" to secure the vanishing of $t^k G(n_0, t; n_1)$ at $t = \infty$, then we should conclude that the kth moment of $T(n_0 \to n_1)$ does not exist. So our derivation here of formulas for calculating the moments $T_k(n_0 \to n_1)$ *assumes* that those moments do in fact exist.

This equation will form the basis for our subsequent analysis. Notice that when $n_0 = n_1 - 1$, the right hand side of this equation contains a term that is proportional to $T_k(n_1 \rightarrow n_1)$. The intuitively expected fact that

$$T_k(n_1 \rightarrow n_1) = 0 \quad (k \geq 1) \tag{6.6-39}$$

is *formally* a consequence of our imposed condition (6.6-31): That condition evidently makes it impossible for the process, if initially in state n_1, to ever enter the interval $[0, n_1 - 1]$; this in turn implies through Eq. (6.6-27b) that $G(n_1, t; n_1) = 0$ for all $t \geq 0$, and that together with formula (6.6-37) gives Eq. (6.6-39). In effect, Eq. (6.6-39) provides a "boundary condition" for the set of difference equations (6.6-38).

We can rearrange Eq. (6.6-38) in two interesting ways. On the one hand, we can write it as

$$W_+(n_0) \left[T_k(n_0 \rightarrow n_1) - T_k(n_0 + 1 \rightarrow n_1) \right]$$

$$- W_-(n_0) \left[T_k(n_0 - 1 \rightarrow n_1) - T_k(n_0 \rightarrow n_1) \right] = k \, T_{k-1}(n_0 \rightarrow n_1)$$

$$(0 \leq n_0 \leq n_1 - 1; k \geq 1). \tag{6.6-40}$$

As we shall see shortly, this form of Eq. (6.6-38) will prove to be convenient for effecting a recursive solution. But it is interesting to note in passing that, if we express the functions W_+ and W_- in terms of the functions a and v through Eqs. (6.2-6), then we find after a little algebra that Eq. (6.6-38) can also be written in the form

$$v(n_0) \left[\frac{T_k(n_0 + 1 \rightarrow n_1) - T_k(n_0 - 1 \rightarrow n_1)}{2} \right] + \frac{1}{2} \, a(n_0) \left[T_k(n_0 + 1 \rightarrow n_1) \right.$$

$$\left. - 2T_k(n_0 \rightarrow n_1) + T_k(n_0 - 1 \rightarrow n_1) \right] = - k \, T_{k-1}(n_0 \rightarrow n_1)$$

$$(0 \leq n_0 \leq n_1 - 1; k \geq 1). \tag{6.6-41}$$

This set of *difference* equations will be recognized as the birth-death analog of the set of *differential* equations (3.7-12) for the moments of the first exit time of a continuous Markov process.

Either of Eqs. (6.6-40) or (6.6-41) constitutes a difference equation for T_k that is *closed* if we know T_{k-1}. And since we know that $T_0 = 1$, then we can in principle use either equation to solve successively for T_1, T_2, T_3, etc. To actually effect such a recursive solution, it is easier to use Eq. (6.6-40): We begin by introducing the auxiliary function

$$D_k(n_0; n_1) \equiv T_k(n_0 \to n_1) - T_k(n_0 + 1 \to n_1) \quad (n_0 < n_1). \quad (6.6\text{-}42)$$

With this function, Eq. (6.6-40) can evidently be written

$$W_+(n_0) D_k(n_0; n_1) - W_-(n_0) D_k(n_0 - 1; n_1) = k T_{k-1}(n_0 \to n_1),$$

whence

$$D_k(n_0; n_1) = \frac{k T_{k-1}(n_0 \to n_1)}{W_+(n_0)} + \frac{W_-(n_0)}{W_+(n_0)} D_k(n_0 - 1; n_1)$$

$$(0 \le n_0 \le n_1 - 1; \ k \ge 1).$$

If $T_{k-1}(n_0 \to n_1)$ is known as a function of n_0, then this last equation can be used to calculate $D_k(n_0; n_1)$ from $D_k(n_0 - 1; n_1)$. Such a recursive computation can be initialized by noting that, since $W_-(0) = 0$, then the formula for $D_k(0; n_1)$ is simply

$$D_k(0; n_1) = \frac{k T_{k-1}(0 \to n_1)}{W_+(0)}.$$

Once we have thus recursively calculated $D_k(n_0; n_1)$ for all $n_0 = 0, 1, \dots,$ $n_1 - 1$, we can then calculate $T_k(n_0 \to n_1)$ for all $n_0 = 0, 1, \dots, n_1 - 1$ by using Eq. (6.6-42), written now in the form

$$T_k(n_0 \to n_1) = D_k(n_0; n_1) + T_k(n_0 + 1 \to n_1) \quad (0 \le n_0 \le n_1 - 1).$$

This equation evidently allows us to calculate $T_k(n_0 \to n_1)$ from $T_k(n_0 + 1 \to n_1)$, a recursive computation that can be initialized by noting that, since $T_k(n_1 \to n_1) = 0$, then the formula for $T_k(n_1 - 1 \to n_1)$ is simply

$$T_k(n_1 - 1 \to n_1) = D_k(n_1 - 1; n_1).$$

Indeed, the last two equations taken together evidently imply that $T_k(n_0 \to n_1)$ is simply the sum of $D_k(m; n_1)$ from $m = n_0$ to $m = n_1 - 1$.

Collecting the foregoing results, we arrive at the following *algorithm for calculating all the moments of* $T(n_0 \to n_1)$ *for* $n_0 < n_1$: Starting with the (assumed) normalization condition

$$T_0(n_0 \to n_1) = 1 \quad \text{for all } n_0 \in [0, n_1 - 1], \quad (6.6\text{-}43)$$

and proceeding in turn for $k = 1, 2, 3$, etc., we calculate the values $T_k(n_0 \to n_1)$ ($n_0 = 0, 1, \dots, n_1 - 1$) from the *known* values $T_{k-1}(n_0 \to n_1)$ ($n_0 = 0, 1, \dots, n_1 - 1$) by *first* computing $D_k(n_0; n_1)$ for $n_0 = 0, 1, \dots, n_1 - 1$ through the recursion

$$D_k(n_0; n_1) = \begin{cases} \dfrac{k\,T_{k-1}(0 \to n_1)}{W_+(0)}, & \text{if } n_0 = 0, \\[2ex] \dfrac{k\,T_{k-1}(n_0 \to n_1)}{W_+(n_0)} + \dfrac{W_-(n_0)}{W_+(n_0)}D_k(n_0-1; n_1), & \\[1ex] & \text{if } n_0 = 1,2,...,n_1-1; \end{cases}$$

$$(6.6\text{-}44)$$

and *then* computing $T_k(n_0 \to n_1)$ for $n_0 = n_1 - 1,\ n_1 - 2,\ ...,\ 0$ through the recursion

$$T_k(n_0 \to n_1) = \begin{cases} D_k(n_1 - 1; n_1), & \text{if } n_0 = n_1 - 1, \\[1ex] D_k(n_0; n_1) + T_k(n_0+1 \to n_1), & \\[1ex] & \text{if } n_0 = n_1-2,\ n_1-3,...,0. \end{cases}$$

$$(6.6\text{-}45)$$

Let us verify that the algorithm described by Eqs. (6.6-43) – (6.6-45) gives for $k=1$ the same result that we obtained through the pedestrian approach of Subsection 6.6.A. By iterating the $k=1$ version of Eqs. (6.6-45), we obtain

$$T_1(n_0 \to n_1) = \sum_{m=n_0}^{n_1-1} D_1(m; n_1). \tag{6.6-46a}$$

Then by iterating the $k=1$ version of Eqs. (6.6-44), we obtain, after some fairly tedious algebra,

$$D_1(m; n_1) = e^{\phi(m)} \sum_{n=0}^{m} \frac{1}{W_+(n)} e^{-\phi(n)}, \tag{6.6-46b}$$

where ϕ is the potential function defined in Eq. (6.4-12). Upon substituting Eq. (6.6-46b) into Eq. (6.6-46a), we evidently recover the pedestrian result (6.6-18a).[†]

Companion formulas to Eqs. (6.6-43) – (6.6-45) for the case $n_1 < n_0$ can be derived by using analogous reasoning. But in that case, it is necessary to first identify the *smallest* state $N \geq n_0$ for which *either*

[†] Equation (6.6-46b) shows that $D_1(m; n_1)$ is actually independent of n_1. This fact ultimately leads to the simple additive property (6.6-20) for the first moment of $T(n_0 \to n_1)$. But for $k \geq 2$, $D_k(m; n_1)$ is *not* independent of n_1, and the additive property (6.6-20) does not hold for moments of $T(n_0 \to n_1)$ higher than the first.

$$W_+(N) = 0, \qquad (6.6\text{-}47)$$

or this condition can be *artificially imposed* without sensibly affecting the statistics of the first passage time $T(n_0 \to n_1)$. The existence of such a state N is a practical prerequisite for $T(n_0 \to n_1)$ to be a well defined random variable; certainly, the existence of such a state N is implicitly assumed in any application of the pedestrian relations (6.6-13) and (6.6-14). Given this state N, we then change the definition of G in Eqs. (6.6-27) to read

$$G(n_0, t; n_1) \equiv \sum_{n = n_1 + 1}^{N} P(n, t \,|\, n_0, 0) = \text{Prob}\{ X(t) \in [n_1 + 1, N] \,|\, X(0) = n_0 \}.$$

We also change condition (6.6-31) to read $W_+(n_1) = 0$. With these modifications $G(n_0, t; n_1)$ can be interpreted as the probability that $T(n_0 \to n_1)$ will exceed t for the case $n_1 < n_0$, and the derivation then proceeds as in the case $n_1 > n_0$. It yields once again Eq. (6.6-38), although now of course with n_0 confined to the interval $[n_1 + 1, N]$. However, to solve that equation subject to the boundary condition (6.6-39), we must now define the auxiliary function $D_k(n_0; n_1)$ by

$$D_k(n_0; n_1) \equiv T_k(n_0 \to n_1) - T_k(n_0 - 1 \to n_1) \qquad (n_1 < n_0)$$

instead of by Eq. (6.6-42). One then straightforwardly deduces the following *algorithm for calculating all the moments of* $T(n_0 \to n_1)$ *for* $n_1 < n_0$: Starting with the (assumed) normalization condition

$$T_0(n_0 \to n_1) = 1 \quad \text{for all } n_0 \in [n_1 + 1, N], \qquad (6.6\text{-}48)$$

and proceeding in turn for $k = 1$, 2, 3, etc., we calculate the values $T_k(n_0 \to n_1)$ $(n_0 = n_1 + 1, n_1 + 2, ..., N)$ from the *known* values $T_{k-1}(n_0 \to n_1)$ $(n_0 = n_1 + 1, n_1 + 2, ..., N)$ by *first* computing $D_k(n_0; n_1)$ for $n_0 = N, N-1, ..., n_1 + 1$ through the recursion

$$D_k(n_0; n_1) = \begin{cases} \dfrac{k\, T_{k-1}(N \to n_1)}{W_-(N)}, & \text{if } n_0 = N, \\[4mm] \dfrac{k\, T_{k-1}(n_0 \to n_1)}{W_-(n_0)} + \dfrac{W_+(n_0)}{W_-(n_0)} D_k(n_0 + 1; n_1), \\[2mm] & \text{if } n_0 = N-1, ..., n_1 + 1; \end{cases}$$

$$(6.6\text{-}49)$$

and *then* computing $T_k(n_0 \to n_1)$ for $n_0 = n_1 + 1$, $n_1 + 2$, ..., N through the recursion

$$T_k(n_0 \to n_1) = \begin{cases} D_k(n_1 + 1; n_1), & \text{if } n_0 = n_1 + 1, \\ D_k(n_0; n_1) + T_k(n_0 - 1 \to n_1), \\ & \text{if } n_0 = n_1 + 2, n_1 + 3, ..., N. \end{cases}$$

(6.6-50)

As a check on the $n_1 < n_0$ first passage recursion formulas (6.6-47) – (6.6-50), one can first deduce from Eqs. (6.6-50) that

$$T_1(n_0 \to n_1) = \sum_{m = n_1 + 1}^{n_0} D_1(m; n_1),$$

(6.6-51a)

and then deduce from Eqs. (6.6-49) (after a good deal of algebra) that

$$D_1(m; n_1) = e^{\phi(m-1)} \sum_{n=m}^{N} \frac{1}{W_-(n)} e^{-\phi(n-1)}.$$

(6.6-51b)

By substituting Eq. (6.6-51b) into Eq. (6.6-51a), we evidently recover (remembering that N here is "sufficiently large") the previously derived pedestrian formula (6.6-23a).

It is in principle possible to *analytically* iterate the recursion relations (6.6-43) – (6.6-45) and (6.6-48) – (6.6-50) to obtain explicit, closed formulas for *all* the moments of the first passage time $T(n_0 \to n_1)$, even as we did for the *first* moment in Subsection 6.6.A. But the higher order explicit moment formulas would most likely be so complicated as to be of little practical value. And surely, any *numerical* evaluation of $T_k(n_0 \to n_1)$ using a digital computer would be most expeditiously accomplished by using the recursion relations themselves.

In most instances we shall be satisfied with knowing only the mean and standard deviation of the first passage time. For that, we need only apply the $T_k(n_0 \to n_1)$ recursion algorithms –– namely Eqs. (6.6-43) – (6.6-45) if $n_1 > n_0$ or Eqs. (6.6-48) – (6.6-50) if $n_1 < n_0$ –– for the *two* k values 1 and 2. From the results of those calculations, we compute

$$\text{mean}\{T(n_0 \to n_1)\} = T_1(n_0 \to n_1),$$

(6.6-52a)

$$\text{sdev}\{T(n_0 \to n_1)\} = [T_2(n_0 \to n_1) - T_1^2(n_0 \to n_1)]^{1/2}.$$

(6.6-52b)

It *usually* turns out ("one-way" processes are an exception) that sdev$\{T(n_0 \to n_1)\}$ is of the same order of magnitude as $T_1(n_0 \to n_1)$. In such cases, actual first passage times that differ from $T_1(n_0 \to n_1)$ by factors ranging anywhere from $\frac{1}{2}$ to 2 will be quite commonplace, and even much larger differences should be occasionally expected.

For the random telegraph process described in Subsection 6.3.C, it is a simple exercise to show from Eqs. (6.6-43) – (6.6-45) that the mean and standard deviation of $T(0 \to 1)$ are both equal to $1/a_0$. And it is just as easy to show from Eqs. (6.6-48) – (6.6-50) that the mean and standard deviation of $T(1 \to 0)$ are both equal to $1/a_1$. These results are precisely what we should expect from the exponential first passage time density functions (6.6-3a), and they exemplify the aforementioned rule-of-thumb that sdev$\{T(n_0 \to n_1)\}$ is often of the same order of magnitude as $T_1(n_0 \to n_1)$.

A graph of the first passage time density function $Q(t; 50 \to 25)$ for the radioactive decay process with $\tau_1 = 1$ was exhibited earlier in Fig. 6-7, and a realization of that process with $n_0 = 50$ was plotted in Fig. 6-2a. By numerically iterating the recursion relations (6.6-48) – (6.6-50) for this process (using $N = 50$), we find that $T_1(50 \to 25) = 0.6832$ and $T_2(50 \to 25) = 0.4862$. From these results and Eqs. (6.6-52), we may conclude that the first passage time $T(50 \to 25)$ for this process has a mean of 0.6832 and a standard deviation of 0.1393. The figure for the mean agrees exactly with that calculated from Eq. (6.6-4b); the figure for the standard deviation is substantially smaller than the mean, as is typical of purely one-way processes, but is quite consistent with the shape of the density function curve in Fig. 6-7.

A graph of the first passage time density function $Q(t; 0 \to 25)$ for the Poisson process with $a = 1$ was exhibited earlier in Fig. 6-8, and a realization of that process with $n_0 = 0$ was plotted in Fig. 5-2a. By numerically iterating the recursion relations (6.6-43) – (6.6-45) for this process, we find that $T_1(0 \to 25) = 25$ and $T_2(0 \to 25) = 650$. From these results and Eqs. (6.6-52), we may conclude that the first passage time $T(0 \to 25)$ for this process has a mean of 25 and a standard deviation of 5. In fact, it is not difficult to prove that the density function plotted in Fig. 6-8 [the "gamma" density function of Eq. (6.6-5a) with $a = 1$ and $m = 25$] has exactly this mean and standard deviation. Again we see that the standard deviation of the first passage time for this "one-way" process is substantially smaller than the mean.

We shall apply the formulas for $T_k(n_0 \to n_1)$ derived in this subsection to some other birth-death Markov processes in Sections 6.7 and 6.8.

6.6.C THE EIGENVALUE APPROACH †

We have seen that the "pedestrian" approach of Subsection 6.6.A leads easily to an exact formula for the first moment of the first passage time $T(n_0 \to n_1)$, while the "moments" approach of Subsection 6.6.B leads, rather less easily, to exact formulas for *all* the moments of $T(n_0 \to n_1)$. In this subsection we shall raise our sights even higher by deriving an exact formula for the *density function* of $T(n_0 \to n_1)$. We shall find, however, that the price for effecting this more detailed characterization of $T(n_0 \to n_1)$ is that we have to calculate the eigenvalues and eigenvectors of an $m \times m$ matrix, where m is roughly the number of states accessible to the process in the course of a first passage from state n_0 to state n_1. So this "eigenvalue" approach will be generally feasible only if one has ready access to a moderately powerful digital computer with a reliable software package for calculating the eigenvalues and eigenvectors of large matrices. An additional practical limitation of this eigenvalue approach is that its final formulas, although mathematically exact, are *sometimes* not amenable to accurate evaluation on finite word-length computers.

To calculate the first passage time density function $Q(t; n_0 \to n_1)$ of a temporally homogeneous birth-death Markov process $X(t)$, we begin by recalling that the Markov state density function $P(n,t \mid n_0,0)$ of this process is defined by the two functions $W_+(n)$ and $W_-(n)$ through the forward master equation (6.1-17). Recalling from Eq. (6.1-13a) that

$$a(n) = W_+(n) + W_-(n), \qquad (6.6\text{-}53)$$

we can write that forward master equation as

$$\frac{\partial}{\partial t} P(n,t \mid n_0,0) = W_+(n-1) P(n-1,t \mid n_0,0) - a(n) P(n,t \mid n_0,0)$$

$$+ W_-(n+1) P(n+1,t \mid n_0,0). \qquad (6.6\text{-}54)$$

Now we introduce a *new* birth-death Markov process $X^*(t)$, with Markov state density function $P^*(n,t \mid n_0,t_0)$. The defining functions of $X^*(t)$ are the same as those of $X(t)$, except that for $X^*(t)$ we take

$$W_+(n_1) = W_-(n_1) = 0. \qquad (6.6\text{-}55)$$

The two processes $X(t)$ and $X^*(t)$ will thus be indistinguishable from each other up to the moment of their first arrival in state n_1, but thereafter

---

† The results developed in this subsection will not be required in any later section.

$X^*(t)$ will *stay* in state n_1 forever. Therefore, the probability that $X(t)$ will have made its *first passage* to state n_1 *by* time t must be equal to the probability of finding $X^*(t)$ *in* state n_1 *at* time t; in symbols,

$$\text{Prob}\{ T(n_0 \to n_1) \leq t \} = P^*(n_1, t \mid n_0, 0). \qquad (6.6\text{-}56)$$

Now, the density function $Q(t; n_0 \to n_1)$ of $T(n_0 \to n_1)$ is by definition such that

$$\int_0^t Q(t'; n_0 \to n_1) \, dt' = \text{Prob}\{ T(n_0 \to n_1) \leq t \}. \qquad (6.6\text{-}57)$$

Combining the last two equations and then differentiating with respect to t, we conclude that the density function of the first passage time of the process $X(t)$ from state n_0 to state n_1 is related to the Markov state density function of the *auxiliary* process $X^*(t)$ according to

$$Q(t; n_0 \to n_1) = \frac{\partial}{\partial t} P^*(n_1, t \mid n_0, 0). \qquad (6.6\text{-}58)$$

We can obtain an explicit formula for $P^*(n_1, t \mid n_0, 0)$ by directly solving the forward master equation for the process $X^*(t)$. That equation is of course the same as the forward master equation (6.6-54) of the process $X(t)$, but taking account of Eqs. (6.6-55). For the specific case $n_1 > n_0$, the forward master equation for $X^*(t)$ reads

$$\frac{\partial}{\partial t} P^*(n,t \mid n_0, 0) = -\Big[-W_+(n-1) P^*(n-1, t \mid n_0, 0) + a(n) P^*(n, t \mid n_0, 0)$$

$$- W_-(n+1) P^*(n+1, t \mid n_0, 0) \Big]$$

$$(n_0 < n_1; 0 \leq n \leq n_1) \quad (6.6\text{-}59)$$

wherein it is understood that Eq. (6.6-53) holds, along with the null relations

$$W_-(0) = W_+(n_1) = W_-(n_1) = 0 \qquad (6.6\text{-}60\text{a})$$

and

$$P^*(-1, t \mid n_0, 0) = P^*(n_1 + 1, t \mid n_0, 0) = 0. \qquad (6.6\text{-}60\text{b})$$

This master equation is of course to be solved subject to the initial condition

$$P^*(n, 0 \mid n_0, 0) = \delta(n, n_0). \qquad (6.6\text{-}61)$$

To effect a solution of Eq. (6.6-59), we begin by observing that it can be written in the matrix form

$$\frac{d}{dt} P^*(t) = -M \cdot P^*(t), \qquad (6.6\text{-}62)$$

where $P^*(t)$ is the $(n_1 + 1) \times 1$ matrix

$$P^*(t) \equiv \begin{bmatrix} P^*(0,t \mid n_0, 0) \\ P^*(1,t \mid n_0, 0) \\ \cdots \\ P^*(n_1, t \mid n_0, 0) \end{bmatrix} \qquad (n_0 < n_1), \qquad (6.6\text{-}63)$$

and M is the $(n_1 + 1) \times (n_1 + 1)$ matrix

$$M \equiv \begin{bmatrix} a(0) & -W_-(1) & 0 & \cdots & \cdots 0 & 0 \\ -W_+(0) & a(1) & -W_-(2) & \cdots & \cdots 0 & 0 \\ 0 & -W_+(1) & a(2) & \cdots & \cdots 0 & 0 \\ \cdots & \cdots & \cdots & \cdots & \cdots & \cdots \\ 0 & 0 & 0 & \cdots & a(n_1 - 1) & 0 \\ 0 & 0 & 0 & \cdots & -W_+(n_1 - 1) & 0 \end{bmatrix}$$
$$(n_0 < n_1). \qquad (6.6\text{-}64)$$

Notice that the matrix M has three special properties: First, it is tridiagonal. Second, the elements of each column sum to zero, owing to Eq. (6.6-53) and the null conditions (6.6-60a). And finally, excepting the last column whose elements are all zero, all elements *on* the main diagonal are positive and all elements *adjacent* to the main diagonal are negative. Now it turns out that such an $(n_1 + 1) \times (n_1 + 1)$ matrix always has $n_1 + 1$ eigenvalues, one of which is *zero*, and the other n_1 of which are *real, positive* and *distinct*.[†] So, denoting these eigenvalues by λ_i ($i = 0, 1, ..., n_1$) in order of increasing size, we have

$$0 = \lambda_0 < \lambda_1 < \lambda_2 < \ldots < \lambda_{n_1} \qquad (n_0 < n_1). \qquad (6.6\text{-}65)$$

It should however be mentioned that these eigenvalue properties do *not* hold for *some one-way* processes. The Poisson process, with $W_-(n) = 0$

[†] That one of M's eigenvalues is zero is a consequence of M having one column whose elements are all zero. That the other n_1 eigenvalues of M are real, positive and distinct can be easily proved for the cases $n_1 = 1$ and $n_1 = 2$ by direct calculation; however, to establish those properties for *any* $n_1 > 1$ requires arguments that are too lengthy to recount here.

and $W_+(n) = a(n) = a$, is one such process: Its matrix $\mathbf{M}$ will have 0's everywhere above the main diagonal, and a's everywhere *on* the main diagonal, except for the last diagonal entry which is 0. Consequently, the "characteristic polynomial" of $\mathbf{M}$, whose roots are the eigenvalues of $\mathbf{M}$, will be $(a-\lambda)^{n_1}(0-\lambda)$, and this implies that all the n_1 nonzero eigenvalues of $\mathbf{M}$ will be equal to a. It turns out that such "degeneracy" among the eigenvalues of $\mathbf{M}$ will block a crucial step in our subsequent analysis. So the eigenvalue approach described here cannot be applied to the Poisson process, nor indeed to any birth-only or death-only Markov process for which $a(n)$ has the same value for different accessible states. [Of course, for the Poisson process this failure of the eigenvalue approach is of no concern to us, because we have already calculated its first passage time density function in Eq. (6.6-5a) by another argument.]

We shall designate the (right) eigenvector of $\mathbf{M}$ corresponding to the eigenvalue λ_j by $\mathbf{v}^j$:

$$\mathbf{M} \cdot \mathbf{v}^j = \lambda_j \mathbf{v}^j \qquad (n_0 < n_1; j = 0, 1, \ldots, n_1). \tag{6.6-66a}$$

The eigenvector $\mathbf{v}^j$ is an $(n_1+1) \times 1$ matrix whose ith element we designate by v^j_i. The eigenvalue equation (6.6-66a) can thus be written in terms of the matrix elements as

$$\sum_{l=1}^{n_1+1} M_{i,l} v^j_l = \lambda_j v^j_i$$

$$(n_0 < n_1; i = 1, \ldots, n_1+1; j = 0, \ldots, n_1). \tag{6.6-66b}$$

Here the elements $M_{i,l}$ of $\mathbf{M}$ are read off from formula (6.6-64) in the usual way; in particular, $M_{i,i} = a(i-1)$, $M_{i,i+1} = -W_-(i)$, and $M_{i+1,i} = -W_+(i-1)$.

Now let us suppose that all the eigenvalues and eigenvectors of $\mathbf{M}$ have been calculated, say by using a reliable computer software package, so that the values of λ_j and v^j_i are known for all values of their indices. From these numbers we then proceed to construct two auxiliary $(n_1+1) \times (n_1+1)$ matrices, $\mathrm{e}^{-\lambda t}$ and $\mathbf{V}$, according to the following definitions:

$$(\mathrm{e}^{-\lambda t})_{i,j} \equiv \mathrm{e}^{-\lambda_{j-1} t} \delta(i,j) \qquad (n_0 < n_1; i,j = 1, \ldots, n_1+1), \tag{6.6-67a}$$

$$(\mathbf{V})_{i,j} \equiv v^{j-1}_i \qquad (n_0 < n_1; i,j = 1, \ldots, n_1+1). \tag{6.6-67b}$$

The matrix $\mathrm{e}^{-\lambda t}$ is evidently a *diagonal* matrix whose diagonal elements are formed from the eigenvalues of $\mathbf{M}$, while the matrix $\mathbf{V}$ has as its *columns* the eigenvectors of $\mathbf{M}$. Since the eigenvalues of $\mathbf{M}$ are distinct, then the eigenvectors of $\mathbf{M}$, and hence the columns of $\mathbf{V}$, will be linearly independent; this implies that $\mathbf{V}$ will have a well defined inverse, $\mathbf{V}^{-1}$. From these facts, it is not difficult to prove (as we do in Appendix G) that the requisite solution of the differential equation (6.6-62) is

$$\mathbf{P}^*(t) = \mathbf{V} \cdot e^{-\lambda t} \cdot \mathbf{V}^{-1} \cdot \mathbf{P}^*(0). \tag{6.6-68}$$

With the result (6.6-68), we are now in a position to calculate the function $P^*(n_1,t \mid n_0,0)$ that appears in Eq. (6.6-58). We have, recalling Eq. (6.6-63)

$$P^*(n_1,t \mid n_0,0) = [\mathbf{P}^*(t)]_{n_1+1}$$

$$= [\mathbf{V} \cdot e^{-\lambda t} \cdot \mathbf{V}^{-1} \cdot \mathbf{P}^*(0)]_{n_1+1}$$

$$= \sum_{i=1}^{n_1+1} \sum_{j=1}^{n_1+1} \sum_{k=1}^{n_1+1} (\mathbf{V})_{n_1+1,\,i} \, (e^{-\lambda t})_{i,\,j}$$
$$\times (\mathbf{V}^{-1})_{j,\,k} \, (\mathbf{P}^*(0))_k$$

$$= \sum_{i=1}^{n_1+1} \sum_{j=1}^{n_1+1} \sum_{k=1}^{n_1+1} (\mathbf{V})_{n_1+1,\,i} \, e^{-\lambda_{j-1} t} \delta(i,j)$$
$$\times (\mathbf{V}^{-1})_{j,\,k} \, P^*(k-1,0 \mid n_0,0)$$

$$= \sum_{i=1}^{n_1+1} \sum_{k=1}^{n_1+1} (\mathbf{V})_{n_1+1,\,i} \, e^{-\lambda_{i-1} t} (\mathbf{V}^{-1})_{i,\,k} \, \delta(k-1,n_0)$$

$$P^*(n_1,t \mid n_0,0) = \sum_{i=1}^{n_1+1} (\mathbf{V})_{n_1+1,\,i} \, e^{-\lambda_{i-1} t} (\mathbf{V}^{-1})_{i,\,n_0+1}.$$

So if we let

$$(\mathbf{V})_{n_1+1,\,i} \, (\mathbf{V}^{-1})_{i,\,n_0+1} \equiv -C_{i-1}(n_0,n_1) \quad (i=1,\ldots,n_1+1),$$

then we have

$$P^*(n_1,t \mid n_0,0) = -\sum_{i=1}^{n_1+1} C_{i-1}(n_0,n_1) e^{-\lambda_{i-1} t} \equiv -\sum_{i=0}^{n_1} C_i(n_0,n_1) e^{-\lambda_i t},$$

or, recalling that $\lambda_0 = 0$,

$$P^*(n_1, t \mid n_0, 0) = -C_0(n_0, n_1) - \sum_{i=1}^{n_1} C_i(n_0, n_1) e^{-\lambda_i t}.$$

By substituting this last formula into Eq. (6.6-58) and then performing the t-differentiation, we conclude that the density function for the first passage time $T(n_0, n_1)$ for the case $n_0 < n_1$ is given by

$$Q(t; n_0 \to n_1) = \sum_{i=1}^{n_1} C_i(n_0, n_1) \lambda_i e^{-\lambda_i t} \qquad (n_0 < n_1). \quad (6.6\text{-}69)$$

Here, $\lambda_1, \lambda_2, \ldots, \lambda_{n_1}$ are the n_1 nonzero eigenvalues of the matrix $\mathbf{M}$ in Eq. (6.6-64), arranged in order of increasing size. And the coefficients $C_i(n_0, n_1)$ are defined by

$$C_i(n_0, n_1) \equiv -(\mathbf{V})_{n_1 + 1, i + 1} (\mathbf{V}^{-1})_{i + 1, n_0 + 1}$$

$$(n_0 < n_1; i = 1, \ldots, n_1), \quad (6.6\text{-}70)$$

where $\mathbf{V}$ is the matrix whose columns are constructed from the eigenvectors of $\mathbf{M}$ according to the prescription in Eqs. (6.6-66b) and (6.6-67b).

Notice that it is easy to calculate from the density function (6.6-69) a formal expression for the *moments* of $T(n_0, n_1)$: We have for any $k \geq 0$,

$$T_k(n_0 \to n_1) = \int_0^\infty dt\, t^k\, Q(t; n_0 \to n_1)$$

$$= \sum_{i=1}^{n_1} C_i(n_0, n_1) \lambda_i \int_0^\infty dt\, t^k\, e^{-\lambda_i t}$$

$$= \sum_{i=1}^{n_1} C_i(n_0, n_1) \lambda_i \left(k! / \lambda_i^{k+1} \right),$$

the last step following from the integral identity (A-2); therefore,

$$T_k(n_0 \to n_1) = k! \sum_{i=1}^{n_1} \frac{C_i(n_0, n_1)}{\lambda_i^k} \qquad (n_0 < n_1; k \geq 0). \quad (6.6\text{-}71)$$

Formulas (6.6-69) – (6.6-71) would appear to provide a complete characterization of the first passage time $T(n_0, n_1)$ for the case $n_0 < n_1$.

But the application of those formulas will evidently entail a great deal of computation: First one must compute, for $j = 0, 1, ..., n_1$, the eigenvalues λ_j and associated eigenvectors $(v^j_1, ..., v^j_{n_1+1})$ of the $(n_1+1) \times (n_1+1)$ matrix $\mathbf{M}$ in Eq. (6.6-64). Then, after forming the $(n_1+1) \times (n_1+1)$ matrix $\mathbf{V}$ from the eigenvectors of $\mathbf{M}$ according to Eq. (6.6-67b), one must compute the elements of the inverse matrix $\mathbf{V}^{-1}$. Finally, after evaluating the coefficients $C_i(n_0,n_1)$ from the matrix elements of $\mathbf{V}$ and $\mathbf{V}^{-1}$ according to formula (6.6-70), one arrives at the formulas for the density function and moments of $T(n_0,n_1)$ in Eqs. (6.6-69) and (6.6-71), respectively.[†] The matrix eigenvalue, eigenvector and inversion calculations will require, in most circumstances, a fairly powerful and reliably programmed digital computer.

The formula for $Q(t; n_0,n_1)$ in Eq. (6.6-69) evidently has the structure of a "weighted sum" of exponential density functions $\lambda_i e^{-\lambda_i t}$ $(i = 1,...,n_1)$, with the ith such density function having the "weight" $C_i(n_0,n_1)$. These weights have the property that they sum to unity, as can be seen by evaluating the unit zeroth moment $T_0(n_0,n_1)$ from formula (6.6-71):

$$1 = T_0(n_0 \to n_1) = 0! \sum_{i=1}^{n_1} \frac{C_i(n_0,n_1)}{\lambda_i^0} = \sum_{i=1}^{n_1} C_i(n_0,n_1)$$

$$(n_0 < n_1). \qquad (6.6\text{-}72)$$

It is tempting to infer from these observations that $T(n_0,n_1)$ is a kind of "statistical mixture" of exponential random variables, but such is emphatically not the case. The reason is that some of the weights $C_i(n_0,n_1)$ will nearly always be *negative*, and this clearly rules out any interpretation of $C_i(n_0,n_1)$ as a probability density function with respect to the index i. Indeed, if all the coefficients $C_i(n_0,n_1)$ in Eq. (6.6-69) were positive, then $Q(t; n_0,n_1)$ would necessarily assume its maximum value at $t = 0$; yet we should certainly not expect $t = 0$ to be the most probable value of $T(n_0,n_1)$ in those cases where n_1 is many steps removed from n_0.

But the weighted exponential structure of formula (6.6-69) does have an interesting consequence for the *long-time behavior* of $Q(t; n_0,n_1)$. Recalling that the eigenvalues λ_i are ordered according to Eq. (6.6-65), then we see that as t increases in Eq. (6.6-69) the relative contribution of the *higher* indexed terms *decreases*. In fact, if we write Eq. (6.6-69) as

---

[†] Formula (6.6-70) shows that one actually needs to compute *only* those elements of $\mathbf{V}^{-1}$ in its (n_0+1)th column. That same formula is also seen to require only the elements in the last row of $\mathbf{V}$, but in fact *all* the elements of $\mathbf{V}$ will be needed to compute the required elements of $\mathbf{V}^{-1}$.

$$Q(t; n_0 \to n_1) = C_1(n_0, n_1) \lambda_1 e^{-\lambda_1 t} \left(1 + \sum_{i=2}^{n_1} \frac{C_i(n_0, n_1) \lambda_i}{C_1(n_0, n_1) \lambda_1} e^{-(\lambda_i - \lambda_1) t} \right),$$

then the λ_i-ordering (6.6-65) allows us to conclude that

$$Q(t; n_0 \to n_1) \approx C_1(n_0, n_1) \lambda_1 e^{-\lambda_1 t} \quad \text{if } t \gg \frac{1}{\lambda_2 - \lambda_1}. \qquad (6.6\text{-}73)$$

So after a sufficiently long time, $Q(t; n_0 \to n_1)$ dies off exponentially with a decay constant that is equal to the smallest nonzero eigenvalue of the matrix $\mathbf{M}$.

Companion formulas to Eqs. (6.6-69) – (6.6-71) for the case $n_1 < n_0$ can be derived by using analogous arguments. In that case, as was explained in connection with Eq. (6.6-47), it is necessary to identify the smallest state $N \geq n_0$ for which *either* $W_+(N) = 0$ *or* $W_+(N)$ can be *set* to zero without sensibly affecting the statistics of $T(n_0, n_1)$. That done, the arguments leading to Eqs. (6.6-58) and (6.6-59) apply equally well to the case $n_1 < n_0$, except that the variable n in Eq. (6.6-59) now ranges over the interval $[n_1, N]$. The matrix equation (6.6-62) then follows, with $\mathbf{P}^*(t)$ now the $(N - n_1 + 1) \times 1$ matrix

$$\mathbf{P}^*(t) \equiv \begin{bmatrix} P^*(n_1, t \mid n_0, 0) \\ P^*(n_1 + 1, t \mid n_0, 0) \\ \cdots \\ P^*(N, t \mid n_0, 0) \end{bmatrix} \quad (n_1 < n_0), \qquad (6.6\text{-}74)$$

and $\mathbf{M}$ the $(N - n_1 + 1) \times (N - n_1 + 1)$ tridiagonal matrix

$$\mathbf{M} \equiv \begin{bmatrix} 0 & -W_-(n_1 + 1) & 0 & \cdots & 0 & 0 \\ 0 & a(n_1 + 1) & -W_-(n_1 + 2) & \cdots & 0 & 0 \\ 0 & -W_+(n_1 + 1) & a(n_1 + 2) & \cdots & 0 & 0 \\ \cdots & \cdots & \cdots & \cdots & \cdots & \cdots \\ 0 & 0 & 0 & \cdots & a(N-1) & -W_-(N) \\ 0 & 0 & 0 & \cdots & -W_+(N-1) & a(N) \end{bmatrix}$$

$$(n_1 < n_0). \qquad (6.6\text{-}75)$$

This matrix $\mathbf{M}$ will (in most cases) be found to have $N - n_1 + 1$ eigenvalues λ_j that satisfy

$$0 = \lambda_0 < \lambda_1 < \lambda_2 < \ldots < \lambda_{N-n_1} \qquad (n_1 < n_0). \qquad (6.6\text{-}76)$$

The $N - n_1 + 1$ eigenvectors $v^j \equiv (v^j{}_1, v^j{}_2, \ldots, v^j{}_{N-n_1+1})$ of $\mathbf{M}$ are related to the eigenvalues through the set of eigenvalue equations

$$\sum_{l=1}^{N-n_1+1} M_{i,l} v^j{}_l = \lambda_j v^j{}_i$$

$$(n_1 < n_0; \ i = 1, \ldots, N - n_1 + 1; \ j = 0, \ldots, N - n_1), \qquad (6.6\text{-}77)$$

wherein the elements $M_{i,l}$ are now to be read off from Eq. (6.6-75). The definitions (6.6-67) hold as before, except that the upper limit on the indices i and j are now $N - n_1 + 1$; in particular,

$$(\mathbf{V})_{i,j} = v^{j-1}{}_i \qquad (n_1 < n_0; \ i,j = 1, \ldots, N - n_1 + 1). \qquad (6.6\text{-}78)$$

Formula (6.6-68) also holds for $n_1 < n_0$, but now we employ that formula to calculate $P^*(n_1, t \mid n_0, 0) = (\mathbf{P}^*(t))_1$, in accordance with Eq. (6.6-74).

The final formula for the density function of $T(n_0 \to n_1)$ for $n_1 < n_0$ turns out to be

$$Q(t; n_0 \to n_1) = \sum_{i=1}^{N-n_1} C_i(n_0, n_1) \lambda_i e^{-\lambda_i t} \qquad (n_1 < n_0), \qquad (6.6\text{-}79)$$

wherein

$$C_i(n_0, n_1) \equiv -(\mathbf{V})_{1,\,i+1} (\mathbf{V}^{-1})_{i+1,\,n_0-n_1+1}$$

$$(n_1 < n_0; \ i = 1, \ldots, N - n_1). \qquad (6.6\text{-}80)$$

Calculation of the kth moment of the density function (6.6-79) is straightforward, and gives

$$T_k(n_0 \to n_1) = k! \sum_{i=1}^{N-n_1} \frac{C_i(n_0, n_1)}{\lambda_i^k} \qquad (n_1 < n_0; \ k \geq 0). \qquad (6.6\text{-}81)$$

And as in the $n_0 < n_1$ case, the $k = 0$ version of this moment formula implies that

$$\sum_{i=1}^{N-n_1} C_i(n_0, n_1) = 1 \qquad (n_1 < n_0). \qquad (6.6\text{-}82)$$

But again, we cannot look upon the coefficients $C_i(n_0, n_1)$ in formula (6.6-79) as "statistical weights," because in all but the most trivial cases *some* of those coefficients will be negative. Finally, the asymptotic condition (6.6-73) holds also for $n_1 < n_0$.

With the principal formulas of the eigenvalue approach now derived, let's examine briefly the applicability of these formulas. Their simplest application would be to the *random telegraph process* [see Subsection 6.3.C], although as we noted earlier in this section [see Eqs. (6.6-3a)], the density function $Q(t; n_0 \rightarrow n_1)$ for that process can be written down essentially by inspection. For the case $n_0 = 0$ and $n_1 = 1$, the matrix $\mathbf{M}$ in Eq. (6.6-64) reads

$$\mathbf{M} = \begin{bmatrix} a_0 & 0 \\ -a_0 & 0 \end{bmatrix}.$$

The characteristic equation for this matrix is obviously $(a_0 - \lambda)(0 - \lambda) = 0$, from which we may infer that the two eigenvalues are $\lambda_0 = 0$ and $\lambda_1 = a_0$. The corresponding eigenvectors (whose normalizations are of no consequence) can be identified as

$$\mathbf{v}^0 = \begin{bmatrix} 0 \\ 1 \end{bmatrix} \quad \text{and} \quad \mathbf{v}^1 = \begin{bmatrix} 1 \\ -1 \end{bmatrix},$$

and the matrices $\mathbf{V}$ and $\mathbf{V}^{-1}$ are then found to be

$$\mathbf{V} = \begin{bmatrix} 0 & 1 \\ 1 & -1 \end{bmatrix} \quad \text{and} \quad \mathbf{V}^{-1} = \begin{bmatrix} 1 & 1 \\ 1 & 0 \end{bmatrix}.$$

From the definition (6.6-70), we then compute

$$C_1(0,1) = -(\mathbf{V})_{2,2}(\mathbf{V}^{-1})_{2,1} = -(-1)(1) = 1.$$

The formula (6.6-69) then yields the result

$$Q(t; 0 \rightarrow 1) = C_1(0,1)\lambda_1 e^{-\lambda_1 t} = a_0 e^{-a_0 t},$$

which evidently agrees with the first of Eqs. (6.6-3a). The second of Eqs. (6.6-3a) can be similarly derived from the $n_1 < n_0$ formulas (6.6-75) – (6.6-80).

As mentioned earlier, the *Poisson process* cannot be analyzed using the eigenvalue approach because of the degeneracy of its $\mathbf{M}$ eigenvalues; indeed, it seems obvious that there are no constants λ_i and $C_i(n_0, n_1)$ for which the linear combination of exponentials in Eq. (6.6-69) can produce the Poisson process first passage time density function (6.6-5a). On the other hand, the *radioactive decay process*, for $n_1 < n_0$, can easily be shown to have *distinct* M-eigenvalues; therefore, the first passage time for the radioactive decay process should be analyzable in terms of the $n_1 < n_0$ formulas (6.6-75) – (6.6-80), with $N = n_0$. We shall not explicitly go through that calculation here, though, because we can obtain the final result in a much simpler way: We have already deduced in Eq. (6.6-4a) an exact formula for $Q(t; n_0 \to n_1)$ for the radioactive decay process. If we simply expand the last factor in that formula using the binomial formula,

$$\left[1 - e^{-t/\tau^*} \right]^{n_0 - n_1 - 1}$$

$$= \sum_{j=0}^{n_0 - n_1 - 1} \binom{n_0 - n_1 - 1}{j} \left[1 \right]^{n_0 - n_1 - 1 - j} \left[-e^{-t/\tau^*} \right]^{j},$$

then upon collecting terms we find that Eq. (6.6-4a) can be brought into the form

$$Q(t; n_0, n_1) = \sum_{i=1}^{n_0 - n_1} \frac{(-1)^{i-1} n_0!}{\tau^* n_1! (i-1)! (n_0 - n_1 - i)!} \exp\left(-\frac{n_1 + i}{\tau^*} t \right).$$

$$(6.6\text{-}83)$$

This has the form of Eq. (6.6-79), with $N = n_0$, provided that we take

$$\begin{cases} \lambda_i = \dfrac{n_1 + i}{\tau^*} & (i = 1, \dots, n_0 - n_1), \quad (6.6\text{-}84a) \\[2em] C_i(n_0, n_1) = \dfrac{(-1)^{i-1} n_0!}{n_1! (n_1 + i)(i-1)!(n_0 - n_1 - i)!} \\[1em] \hspace{6em} (i = 1, \dots, n_0 - n_1). \quad (6.6\text{-}84b) \end{cases}$$

That the values λ_i in Eq. (6.6-84a) are indeed eigenvalues of M can be seen by simply examining the matrix M in Eq. (6.6-75) for the radioactive decay stepping functions $W_+(n) = 0$ and $W_-(n) = n/\tau_1$: that matrix has zeros everywhere *below* the main diagonal, and the values given by Eq. (6.6-84a) as the nonzero elements *on* the main diagonal.

The results (6.6-84) imply, through formula (6.6-81), that the mean first passage time of the radioactive decay process is given by

$$T_1(n_0 \to n_1) = \tau^* \sum_{i=1}^{n_0 - n_1} \frac{(-1)^{i-1} n_0!}{n_1! (n_1 + i)^2 (i-1)! (n_0 - n_1 - i)!}$$

$$(0 \le n_1 \le n_0). \quad \text{(6.6-85a)}$$

Let's compare this formula for the radioactive decay process mean first passage time with our earlier formula (6.6-4b). Changing the summation index in Eq. (6.6-4b) from n to $i = n - n_1$, that equation becomes

$$T_1(n_0 \to n_1) = \tau^* \sum_{i=1}^{n_0 - n_1} \frac{1}{n_1 + i} \quad (0 \le n_1 \le n_0). \quad \text{(6.6-85b)}$$

As incredible as it may seem, the two formulas (6.6-85) *are* mathematically equivalent. But to say that these formulas are "mathematically" equivalent is *not* to say that they are "computationally" equivalent. The table below shows the values of the individual terms in the two formulas for $\tau^* = 1$, $n_0 = 20$ and $n_1 = 12$:

Evaluating the sums in Eqs. (6.6-85)
for $\tau^* = 1$, $n_0 = 20$ and $n_1 = 12$.

Term index i	Term value in Eq. (6.6-85a)	Term value in Eq. (6.6-85b)
1	5963.0769	0.0769
2	−35991.4286	0.0714
3	94057.6000	0.0667
4	−137779.6875	0.0625
5	122047.0588	0.0588
6	−65317.7778	0.0556
7	19541.0526	0.0526
8	−2519.4000	0.0500
Sum	0.4945	0.4945

From the third column in the table, we see that a calculation of $T_1(20 \to 12)$ accurate to four decimal places using the simpler formula (6.6-85b) requires four decimal places of precision. In contrast, from the second column in the table we see that a calculation of $T_1(20 \to 12)$ to that same accuracy using the eigenvalue formula (6.6-85a) requires *ten* decimal places of precision. This disparity gets worse if $n_0 - n_1$ is increased: For $n_0 = 50$ and $n_1 = 25$ [see Fig. 6-7], the simpler formula (6.6-85b) gives the result $T_1(50 \to 25) = 0.6832$ in a computation that again requires only four decimal places of precision.   But the same calculation using the eigenvalue formula (6.6-85a) now requires *twenty-two* decimal places of precision, and is thus beyond the capability of even the "double-precision" mode of a standard 32-bit computer!

Formula (6.6-85a) is quite frankly a numerical analyst's nightmare. The difficulty with that formula is obviously due to the behavior of the coefficients $C_i(n_0, n_1)$, and this circumstance raises a question about the general practicality of the eigenvalue formulas (6.6-69) and (6.6-79) for $Q(t; n_0 \to n_1)$: Even though those formulas are mathematically exact, the coefficients $C_i(n_0, n_1)$ *might* vary with i over such an enormous range of oppositely signed values that the sums cannot be evaluated on an ordinary, finite word-length computer.   By contrast, the recursion formulas for $T_1(n_0 \to n_1)$ derived through the pedestrian approach, and also the recursion formulas for $T_k(n_0 \to n_1)$ derived through the moments approach, *never* require summing terms of opposite signs; consequently, the pedestrian and moments formulas are not unduly sensitive to truncations that inevitably occur in finite word-length computers.

So, despite the formal elegance of the eigenvalue approach, the practical usefulness of its formulas may be rather limited.   Perhaps the most generally useful result of the eigenvalue analysis is to be found in Eq. (6.6-73):   The long-time tail of $Q(t; n_0 \to n_1)$ for most temporally homogeneous birth-death Markov processes approaches zero like $e^{-\lambda_1 t}$, where λ_1 is the smallest nonzero eigenvalue of the appropriate matrix $\mathbf{M}$. This implies that a first passage from n_0 to n_1 will almost always be accomplished in a time less than a few multiples of $1/\lambda_1$.

6.7 FIRST EXIT FROM AN INTERVAL

In this section we address a problem that is closely related to the first passage time problem discussed in the preceding section.   We consider a temporally homogeneous birth-death Markov process $X(t)$ that starts out in state n_0, and we let n_1 and n_2 be any two states that bracket n_0 according to

$$0 \le n_1 < n_0 < n_2. \tag{6.7-1}$$

We seek to determine the following parameters that characterize the *first exit* of $X(t)$ from the interval $\lfloor n_1 + 1, n_2 - 1 \rfloor$:

$p_i \equiv p_i(n_0; n_1,n_2) \equiv$ probability that the process, starting in state n_0 inside the interval $[n_1 + 1, n_2 - 1]$, will first exit that interval via state n_i ($i = 1,2$). (6.7-2)

$T_1(n_0; n_1,n_2) \equiv$ average time for the process, starting in state n_0 inside the interval $[n_1 + 1, n_2 - 1]$, to first exit that interval. (6.7-3)

In short, p_i is the probability that the process will reach the bracketing state n_i before it reaches the *other* bracketing state, and $T_1(n_0; n_1,n_2)$ is the average time it takes the process to first reach *either* bracketing state.† The probabilities p_1 and p_2 are called the **splitting probabilities**, and $T_1(n_0; n_1,n_2)$ is called the **mean first exit time**. We shall assume in our analysis here that there are no reflecting or absorbing states in the interval $[n_1 + 1, n_2 - 1]$ that could prevent the process from reaching either of the states n_1 and n_2. This assumption guarantees that p_1 and p_2 will satisfy

$$p_1 + p_2 = 1. \qquad (6.7\text{-}4)$$

In Subsection 6.7.A we shall derive general formulas for calculating the splitting probabilities and the mean first exit time. In Subsection 6.7.B we shall illustrate the use of those formulas by applying them to the so-called "drunkard's walk" process.

6.7.A SPLITTING PROBABILITIES AND THE MEAN FIRST EXIT TIME

There is more than one way to calculate the splitting probabilities (6.7-2) and the mean first exit time (6.7-3). The procedure that we shall use here is an extension of the "pedestrian" method of Subsection 6.6.A for calculating the mean first passage time. In the spirit of that method, we begin by looking upon the process $X(t)$ as a "walker" on the nonnegative integers, who starts in state n_0 and walks around in the one-step manner dictated by its characterizing functions until it *first* arrives in *either* of the two states n_1 and n_2. Relative to this journey-of-first-exit from the interval $[n_1 + 1, n_2 - 1]$, we define for each state in that interval the following functions:

---

† Note that the subscript "1" on $T_1(n_0; n_1,n_2)$ does not refer to state n_1, but rather designates the *first moment* of the first exit time random variable $T(n_0; n_1,n_2)$.

$$t(n) \equiv \text{average total time the walker spends in state } n, \qquad (6.7\text{-}5a)$$

$$v(n) \equiv \text{average number of visits the walker makes to state } n, \qquad (6.7\text{-}5b)$$

$$v_\pm(n) \equiv \text{average number of } n \to n \pm 1 \text{ steps taken by the walker.} \qquad (6.7\text{-}5c)$$

Since the average time spent by the walker in state n per visit is $1/a(n)$, then the first two of these functions are related by

$$t(n)/v(n) = 1/a(n). \qquad (6.7\text{-}6a)$$

And since the probability that the walker, upon stepping away from state n, will step to state $n \pm 1$ is $w_\pm(n) \equiv W_\pm(n)/a(n)$, then the last two of functions (6.7-5) are related by

$$v_\pm(n)/v(n) = W_\pm(n)/a(n). \qquad (6.7\text{-}6b)$$

Eliminating $v(n)$ between Eqs. (6.7-6) gives

$$v_\pm(n) = W_\pm(n)\, t(n). \qquad (6.7\text{-}7)$$

This relation, and also the obvious relation

$$T_1(n_0; n_1, n_2) = \sum_{n=n_1+1}^{n_2-1} t(n), \qquad (6.7\text{-}8)$$

will be put to use shortly.

Imagine now that we have an infinitely large ensemble of independent journeys of our walker, each journey starting in state n_0 and ending at the first arrival in either of states n_1 and n_2. By definition, a fraction p_1 of those journeys will end in state n_1, and a fraction p_2 will end in state n_2. For any pair of states $n-1$ and n inside the interval $[n_1, n_0]$, such as the pair of states shown just to the *left* of state n_0 in Fig. 6-11, consider for each individual journey the *difference* {number of $n \to n-1$ steps taken} minus {number of $n-1 \to n$ steps taken}. It should be obvious that for any journey which ends in state n_1 that difference will be

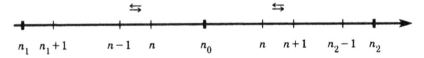

Figure 6-11. Deducing the first-exit stepping relations (6.7-9).

exactly 1, while for any journey which ends in state n_2 that difference will be exactly 0. So for all states $n \in [n_1 + 1, n_0]$,

$$\{\# \, n \to n - 1 \text{ steps}\} - \{\# \, n - 1 \to n \text{ steps}\} = \begin{cases} 1, & \text{if journey ends in } n_1 \\ 0, & \text{if journey ends in } n_2 \end{cases},$$

$$= \begin{cases} 1, & \text{with probability } p_1 \\ 0, & \text{with probability } p_2 \end{cases}.$$

Therefore, the *average* of this difference, which by definition (6.7-5c) is just $v_-(n) - v_+(n-1)$, can be computed according to the familiar rule (1.7-11) as

$$v_-(n) - v_+(n-1) = 1 \cdot p_1 + 0 \cdot p_2 = p_1,$$

valid for all $n \in [n_1 + 1, n_0]$. In particular, putting $n = n_1 + 1$ and recognizing that *none* of these journeys will have an upgoing step from state n_1, we get $v_-(n_1 + 1) - 0 = p_1$. Thus we have shown that

$$v_-(n_1 + 1) = p_1, \tag{6.7-9a}$$

$$v_-(n) = p_1 + v_+(n-1), \quad \text{if } n \in [n_1 + 2, n_0]. \tag{6.7-9b}$$

Similar reasoning applied to any pair of states n and $n+1$ inside the interval $[n_0, n_2]$, such as the pair of states shown just to the *right* of state n_0 in Fig. 6-11, leads to the equations

$$v_+(n_2 - 1) = p_2, \tag{6.7-9c}$$

$$v_+(n) = p_2 + v_-(n+1), \quad \text{if } n \in [n_0, n_2 - 2]. \tag{6.7-9d}$$

Upon eliminating the v_- and v_+ functions from Eqs. (6.7-9) through our earlier result (6.7-7), we obtain the following set of formulas:

$$t(n_1 + 1) = \frac{p_1}{W_-(n_1 + 1)}, \tag{6.7-10a}$$

$$t(n) = \frac{1}{W_-(n)} \left(p_1 + W_+(n-1) \, t(n-1) \right)$$

$$(n = n_1 + 2, ..., n_0); \tag{6.7-10b}$$

$$t(n_2 - 1) = \frac{p_2}{W_+(n_2 - 1)}, \tag{6.7-10c}$$

$$t(n) = \frac{1}{W_+(n)}\left(p_2 + W_-(n+1)\,t(n+1)\right)$$

$$(n=n_2-2,...,n_0). \quad (6.7\text{-}10\mathrm{d})$$

Equations (6.7-10a) and (6.7-10b) together evidently allow us to recursively calculate $t(n_1+1)$, $t(n_1+2)$, ..., $t(n_0)$ in terms of the given functions $W_+(n)$ and $W_-(n)$ *and* the as yet unknown number p_1. But notice that each of the $t(n)$'s thus calculated will be *directly proportional* to p_1. Similarly, Eqs. (6.7-10c) and (6.7-10d) together determine values for $t(n_2-1)$, $t(n_2-2)$, ..., $t(n_0)$, these values all being directly proportional to p_2. So if we define the auxiliary functions

$$t_1(n) \equiv t(n)/p_1, \quad \text{for } n\in[n_1+1,n_0], \quad (6.7\text{-}11\mathrm{a})$$

and

$$t_2(n) \equiv t(n)/p_2, \quad \text{for } n\in[n_0,n_2-1], \quad (6.7\text{-}11\mathrm{b})$$

then upon dividing Eqs. (6.7-10a,b) through by p_1, and Eqs. (6.7-10c,d) through by p_2, we obtain the following pair of p_i-*independent* recursion relations:

$$\left\{ \begin{array}{l} t_1(n_1+1) = \dfrac{1}{W_-(n_1+1)}, \quad\quad\quad\quad\quad\quad (6.7\text{-}12\mathrm{a}) \\[2em] t_1(n) = \dfrac{1}{W_-(n)}\left(1 + W_+(n-1)\,t_1(n-1)\right) \\[1em] \quad\quad\quad\quad\quad (n=n_1+2,...,n_0); \quad (6.7\text{-}12\mathrm{b}) \end{array} \right.$$

and

$$\left\{ \begin{array}{l} t_2(n_2-1) = \dfrac{1}{W_+(n_2-1)}, \quad\quad\quad\quad\quad\quad (6.7\text{-}13\mathrm{a}) \\[2em] t_2(n) = \dfrac{1}{W_+(n)}\left(1 + W_-(n+1)\,t_2(n+1)\right) \\[1em] \quad\quad\quad\quad\quad (n=n_2-2,...,n_0). \quad (6.7\text{-}13\mathrm{b}) \end{array} \right.$$

Now we can recursively solve Eqs. (6.7-12) for $t_1(n_1+1)$, $t_1(n_1+2)$, ..., $t_1(n_0)$, and also Eqs. (6.7-13) for $t_2(n_2-1)$, $t_2(n_2-2)$, ..., $t_2(n_0)$. Taking the ratio of Eqs. (6.7-11) with $n=n_0$, we find that $p_2/p_1 = t_1(n_0)/t_2(n_0)$; this *ratio* relation between p_1 and p_2, combined with the *sum* relation (6.7-4), allows us to conclude that p_1 and p_2 are given by

$$p_1 = \frac{t_2(n_0)}{t_1(n_0) + t_2(n_0)} \quad \text{and} \quad p_2 = \frac{t_1(n_0)}{t_1(n_0) + t_2(n_0)}. \qquad (6.7\text{-}14)$$

With the values of the splitting probabilities p_1 and p_2 thus determined, we can finally calculate the mean first exit time from Eqs. (6.7-8) and (6.7-11):

$$T_1(n_0; n_1, n_2) = p_1 \sum_{n=n_1+1}^{n_0} t_1(n) + p_2 \sum_{n=n_0+1}^{n_2-1} t_2(n). \qquad (6.7\text{-}15)$$

Formulas (6.7-12) – (6.7-15) evidently suffice to calculate the splitting probabilities and the mean first exit time. If one is concerned only with obtaining numerical results on a digital computer, then those formulas are most expeditiously utilized in precisely the manner just outlined; indeed, the recursive nature of those equations, in conjunction with the cumulative nature of the two sums in Eq. (6.7-15), allow all these computations to be performed by a computer program whose memory requirements do not increase with the difference $n_2 - n_1$. But in some situations, especially those where the process has a well defined potential function $\phi(n)$ or a well defined stationary Markov state density function $P_s(n)$, it is convenient to have fully explicit formulas for p_1, p_2 and $T_1(n_0; n_1, n_2)$. The obvious first step towards deriving such formulas is to iterate Eqs. (6.7-12) and (6.7-13) analytically to obtain explicit (nonrecursive) formulas for $t_1(n)$ and $t_2(n)$.

The analytical iteration of the recursion relations (6.7-12) proceeds as follows: For any $n \in [n_1 + 2, n_0]$, we have

$$t_1(n) = \frac{1}{W_-(n)} \left(1 + \frac{W_+(n-1)}{W_-(n-1)} \left(1 + W_+(n-2) t_1(n-2) \right) \right)$$

$$= \frac{1}{W_-(n)} \left(1 + \frac{W_+(n-1)}{W_-(n-1)} \left(1 + \frac{W_+(n-2)}{W_-(n-2)} \left(1 + \right. \right. \right.$$

$$\left. \left. \left. \dots + \frac{W_+(n_1+1)}{W_-(n_1+1)} \right) \dots \right) \right)$$

$$= \frac{1}{W_-(n)} \left(1 + \sum_{m=n_1+1}^{n-1} \prod_{j=m}^{n-1} \frac{W_+(j)}{W_-(j)} \right)$$

$$= \frac{1}{W_-(n)}\left(1 + \sum_{m=n_1+1}^{n-1} \frac{\prod_{j=1}^{n-1}\left[W_+(j)/W_-(j)\right]}{\prod_{j=1}^{m-1}\left[W_+(j)/W_-(j)\right]}\right)$$

$$= \frac{1}{W_-(n)}\left(1 + \sum_{m=n_1+1}^{n-1} \frac{e^{-\phi(n-1)}}{e^{-\phi(m-1)}}\right)$$

$$t_1(n) = \frac{1}{W_-(n)} \sum_{m=n_1+1}^{n} \frac{e^{-\phi(n-1)}}{e^{-\phi(m-1)}}, \qquad n\in[n_1+2,n_0].$$

In the penultimate step, we have assumed the existence of the potential function $\phi(n)$ as defined in Eq. (6.4-12), and we have invoked some well known properties of the exponential and logarithm functions. Observing now that if we put $n = n_1 + 1$ in this last equation we do in fact get the correct result (6.7-12a), then we have

$$t_1(n) = \frac{1}{W_-(n)} e^{-\phi(n-1)} \sum_{m=n_1+1}^{n} e^{\phi(m-1)}, \quad n\in[n_1+1,n_0].$$

Finally, by invoking the identity mentioned in the footnote on page 460, we conclude that

$$t_1(n) = \frac{1}{W_+(n)} e^{-\phi(n)} \sum_{m=n_1}^{n-1} e^{\phi(m)}, \quad n\in[n_1+1,n_0]. \qquad (6.7\text{-}16a)$$

The analytical iteration of the recursion relations (6.7-13) proceeds in a similar way, with the result

$$t_2(n) = \frac{1}{W_+(n)} e^{-\phi(n)} \sum_{m=n}^{n_2-1} e^{\phi(m)}, \quad n\in[n_0,n_2-1]. \qquad (6.7\text{-}16b)$$

If the process in question happens to be *stable*, so that the constant K in Eq. (6.4-5) is nonzero, then by multiplying and dividing both of Eqs. (6.7-16) by $K\,W_+(0)$ and invoking Eqs. (6.4-13) and (6.4-16), we obtain the alternate formulas

$$t_1(n) = P_s(n) \sum_{m=n_1}^{n-1} \Psi(m), \quad n \in [n_1+1, n_0]; \quad (6.7\text{-}17a)$$

$$t_2(n) = P_s(n) \sum_{m=n}^{n_2-1} \Psi(m), \quad n \in [n_0, n_2-1]. \quad (6.7\text{-}17b)$$

With Eqs. (6.7-12) and (6.7-13) now fully iterated, it is a simple matter to obtain explicit formulas for the splitting probabilities. Putting $n = n_0$ in Eqs. (6.6-16) gives

$$t_1(n_0) = \frac{1}{W_+(n_0)} e^{-\phi(n_0)} \sum_{m=n_1}^{n_0-1} e^{\phi(m)},$$

$$t_2(n_0) = \frac{1}{W_+(n_0)} e^{-\phi(n_0)} \sum_{m=n_0}^{n_2-1} e^{\phi(m)},$$

and

$$t_1(n_0) + t_2(n_0) = \frac{1}{W_+(n_0)} e^{-\phi(n_0)} \left[\sum_{m=n_1}^{n_0-1} e^{\phi(m)} + \sum_{m=n_0}^{n_2-1} e^{\phi(m)} \right]$$

$$= \frac{1}{W_+(n_0)} e^{-\phi(n_0)} \sum_{m=n_1}^{n_2-1} e^{\phi(m)}.$$

Inserting these last three expressions into Eqs. (6.7-14), we conclude that the splitting probabilities are given explicitly by

$$p_1(n_0; n_1, n_2) = \frac{\displaystyle\sum_{m=n_0}^{n_2-1} e^{\phi(m)}}{\displaystyle\sum_{m=n_1}^{n_2-1} e^{\phi(m)}} \quad \text{and} \quad p_2(n_0; n_1, n_2) = \frac{\displaystyle\sum_{m=n_1}^{n_0-1} e^{\phi(m)}}{\displaystyle\sum_{m=n_1}^{n_2-1} e^{\phi(m)}}.$$

$$(6.7\text{-}18a)$$

And if the process happens to be *stable*, so that the barrier function of Eq. (6.4-16) is well defined, then these formulas can evidently be written as

$$p_1(n_0; n_1, n_2) = \frac{\sum\limits_{m=n_0}^{n_2-1} \Psi(m)}{\sum\limits_{m=n_1}^{n_2-1} \Psi(m)} \quad \text{and} \quad p_2(n_0; n_1, n_2) = \frac{\sum\limits_{m=n_1}^{n_0-1} \Psi(m)}{\sum\limits_{m=n_1}^{n_2-1} \Psi(m)}.$$

$$\text{(6.7-18b)}$$

This latter pair of formulas has an interesting geometrical interpretation in terms of the "area" under the barrier function $\Psi(n)$ between n_1 and $n_2 - 1$ inclusively: We see that $p_1(n_0; n_1, n_2)$ is the fraction of that area to the *right* of n_0 (inclusively), while $p_2(n_0; n_1, n_2)$ is the fraction of that area to the *left* of $n_0 - 1$ (inclusively). We shall amplify this geometric interpretation of Eqs. (6.7-18) later in Section 6.8.

With the above formulas (6.7-16) − (6.7-18) for p_i and $t_i(n)$, we can now obtain an explicit formula for the mean first exit time from Eq. (6.7-15): In terms of the *potential function* we evidently have

$$T_1(n_0; n_1, n_2) = p_1 \sum_{n=n_1+1}^{n_0} \frac{e^{-\phi(n)}}{W_+(n)} \sum_{m=n_1}^{n-1} e^{\phi(m)}$$

$$+ p_2 \sum_{n=n_0+1}^{n_2-1} \frac{e^{-\phi(n)}}{W_+(n)} \sum_{m=n}^{n_2-1} e^{\phi(m)}, \quad \text{(6.7-19a)}$$

wherein it is understood that p_1 and p_2 are given by Eqs. (6.7-18a). And in terms of the *stationary Markov state density function and the barrier function* we have

$$T_1(n_0; n_1, n_2) = p_1 \sum_{n=n_1+1}^{n_0} P_s(n) \sum_{m=n_1}^{n-1} \Psi(m)$$

$$+ p_2 \sum_{n=n_0+1}^{n_2-1} P_s(n) \sum_{m=n}^{n_2-1} \Psi(m), \quad \text{(6.7-19b)}$$

wherein it is understood that p_1 and p_2 are given by Eqs. (6.7-18b).

Finally, we should mention that it is not difficult to also calculate the *average total number of steps* taken in the course of first exiting the interval $[n_1+1,n_2-1]$. Equation (6.7-6a) tells us that $v(n)$, which can be regarded as the average total number of steps away from state n in the first exit journey, is equal to $a(n)t(n)$; that in turn is by Eqs. (6.7-11) equal to $a(n)p_1t_1(n)$ if $n \in [n_1+1,n_0]$, and to $a(n)p_2t_2(n)$ if $n \in [n_0,n_2-1]$. Therefore, by merely inserting a factor $a(n)$ into each of the summands in Eqs. (6.7-15) and (6.7-19), we straightaway convert those formulas for the average total *time* required for the first exit into formulas for the average total *number of steps* required for the first exit.

6.7.B APPLICATION: THE DRUNKARD'S WALK PROCESS

To illustrate the first exit results derived in the preceding subsection, let us consider a process $X(t)$ for which

$$a(n) = a, \quad w_+(n) = b, \quad w_-(n) = 1 - b \quad (n=1,2,\dots), \quad (6.7\text{-}20)$$

where a and b are any two constants satisfying

$$a > 0, \quad \tfrac{1}{2} \le b < 1. \tag{6.7-21}$$

Evidently, this process has a mean pausing time in any state $n \ge 1$ of $1/a$; furthermore, upon stepping away from state n, the process will step to state $n+1$ with probability b, and to state $n-1$ with probability $1-b$. Our injunction against letting $b=1$ disallows the Poisson process, for which a first *passage* analysis would be more appropriate than a first *exit* analysis. Our injunction against letting b be less than $\tfrac{1}{2}$ can be circumvented by simply renumbering the states in the opposite direction.

We have in fact already encountered processes meeting the above requirements in Subsection 5.2.B; there, Eqs. (6.7-20) were presumed to hold for *all* $n \in (-\infty,\infty)$. For a *birth-death* process, though, we must have $w_-(0)=0$, and in that case the results than we derived in Subsection 5.2.B [specifically, Eqs. (5.2-15) – (5.2-18)] will not apply. But since all we shall be interested in here is the *first exit* of any such process from some positive interval $[n_1+1,n_2-1]$, then it is clear that the values of the characterizing functions outside of that interval will be irrelevant. In particular, the following first exit analysis will be valid for any process satisfying Eqs. (6.7-20), regardless of whether that process is allowed to enter the negative axis.

If the process has $b=\tfrac{1}{2}$, as does the one simulated in Fig. 5-3a, then it is commonly referred to as the "classic drunkard's walk process." The allusion here is to a drunkard who steps, or staggers, from his present state n to either of states $n \pm 1$ with equal probability, these steps being taken on the average once per $1/a$ units of time. The circumstance $b > \frac{1}{2}$ can be incorporated into this metaphor by imagining the drunkard to then be walking either in a wind or on a slope, so that a step of $+1$ is more likely than a step of -1.

For any $b \in [\frac{1}{2}, 1)$, we suppose our process $X(t)$ to be initially in some state $n_0 \geq 1$, and we wish to investigate the first exit of the process from some specified interval $[n_1 + 1, n_2 - 1]$, where $0 \leq n_1 < n_0 < n_2$. Now, unless this process happens to be confined by upper and lower reflecting boundaries which prevent it from wandering off forever from the interval in question, the process will *not* have a well defined stationary Markov state density function $P_s(n)$; so we shall carry out our first exit analysis by using the potential function formulas developed in the preceding subsection. The potential function $\phi(n)$ for our process is, according to the definition (6.4-12),

$$\phi(n) \equiv \sum_{j=1}^{n} \ln \frac{w_-(j)}{w_+(j)} = \sum_{j=1}^{n} \ln\left(\frac{1-b}{b}\right) = n \ln\left(\frac{1-b}{b}\right),$$

or

$$\phi(n) = \ln\left(\frac{1-b}{b}\right)^n \qquad (n \geq 1). \qquad (6.7\text{-}22a)$$

Thus it follows that

$$e^{\phi(n)} = \left(\frac{1-b}{b}\right)^n \qquad (n \geq 1). \qquad (6.7\text{-}22b)$$

At this point, we must consider separately the two cases $b = \frac{1}{2}$ and $b > \frac{1}{2}$.

For the *classic drunkard's walk* process with $b = \frac{1}{2}$, Eq. (6.7-22b) evidently simplifies to

$$e^{\phi(n)} = 1 \qquad (b = 1/2; \, n \geq 1). \qquad (6.7\text{-}23)$$

So, using Eqs. (6.7-18a), we calculate the probability p_1 of arriving at n_1 before arriving at n_2, and the probability p_2 of arriving at n_2 before arriving at n_1, as

$$p_1(n_0; n_1, n_2) = \frac{\displaystyle\sum_{m=n_0}^{n_2-1} (1)}{\displaystyle\sum_{m=n_1}^{n_2-1} (1)} = \frac{(n_2-1) - (n_0) + 1}{(n_2-1) - (n_1) + 1},$$

and

$$p_2(n_0; n_1, n_2) = \frac{\sum_{m=n_1}^{n_0-1} (1)}{\sum_{m=n_1}^{n_2-1} (1)} = \frac{(n_0-1) - (n_1) + 1}{(n_2-1) - (n_1) + 1} ;$$

therefore,

$$p_1(n_0; n_1, n_2) = \frac{n_2 - n_0}{n_2 - n_1} \quad \text{and} \quad p_2(n_0; n_1, n_2) = \frac{n_0 - n_1}{n_2 - n_1}$$

$$(b = 1/2). \quad (6.7\text{-}24)$$

These results are eminently reasonable; in particular, they imply that $p_1 \rightarrow 1$ as $n_0 \rightarrow n_1$, $p_2 \rightarrow 1$ as $n_0 \rightarrow n_2$, and $p_1 + p_2 = 1$.

To calculate the mean first exit time for the case $b = \frac{1}{2}$, we merely substitute Eqs. (6.7-23) and (6.7-24) into formula (6.7-19a). Remembering that $W_+(n) \equiv a(n)w_+(n)$, which in this case is equal to $a/2$, then we have from Eq. (6.7-19a),

$$T_1(n_0; n_1, n_2) = \left(\frac{n_2 - n_0}{n_2 - n_1} \right) \sum_{n=n_1+1}^{n_0} \frac{1}{a/2} \sum_{m=n_1}^{n-1} (1)$$

$$+ \left(\frac{n_0 - n_1}{n_2 - n_1} \right) \sum_{n=n_0+1}^{n_2-1} \frac{1}{a/2} \sum_{m=n}^{n_2-1} (1)$$

$$= \left(\frac{n_2 - n_0}{n_2 - n_1} \right) \frac{2}{a} \sum_{n=n_1+1}^{n_0} (n - n_1)$$

$$+ \left(\frac{n_0 - n_1}{n_2 - n_1} \right) \frac{2}{a} \sum_{n=n_0+1}^{n_2-1} (n_2 - n)$$

$$= \frac{2/a}{n_2 - n_1} \left((n_2 - n_0) \sum_{m=1}^{n_0 - n_1} m + (n_0 - n_1) \sum_{m=1}^{n_2 - n_0 - 1} m \right).$$

Using the algebraic identity

$$\sum_{m=1}^{N} m \equiv \frac{N(N+1)}{2},$$

it is straightforward to conclude that

$$T_1(n_0; n_1, n_2) = \frac{(n_2 - n_0)(n_0 - n_1)}{a} \qquad (b = 1/2). \quad (6.7\text{-}25)$$

An interesting special case of the classic drunkard's walk process is the case in which n_1 and n_2 are symmetrically located about n_0 at a distance ε. Equations (6.7-24) and (6.7-25) then reduce to

$$p_1(n_0; n_0 - \varepsilon, n_0 + \varepsilon) = p_2(n_0; n_0 - \varepsilon, n_0 + \varepsilon) = 1/2$$
$$(b = 1/2), \quad (6.7\text{-}26a)$$

and

$$T_1(n_0; n_0 - \varepsilon, n_0 + \varepsilon) = \frac{\varepsilon^2}{a} \qquad (b = 1/2). \quad (6.7\text{-}26b)$$

That the splitting probabilities are both equal to 1/2 in this case is to be expected from simple symmetry considerations. The result (6.7-26b) for the mean first exit time is less obvious, but more intriguing: It implies that the average time for the process to first move a distance ε away from n_0 in either direction is equal to ε^2/a. This in turn implies that, in a time t, the process will on the average move a distance $(at)^{1/2}$. So we see that the result (6.7-26b) is consistent with the formula (5.2-15b) for var$\{X(t)\}$ for the unrestricted drunkard's walk, which implies that the standard deviation of that process after a time t is $(at)^{1/2}$. Both of these results show too that the classic drunkard's walk process has much in common with Brownian motion and molecular diffusion [see Eqs. (3.4-22b) and (4.4-42)]. In fact, Brownian motion and molecular diffusion are sometimes mathematically modeled using the classic drunkard's walk process. But it should be clear from our discussion in Section 4.5 that a more rigorous modeling of those physical phenomena is obtained through *continuum* state jump Markov process theory.

Now let us consider briefly the *biased drunkard's walk process*, i.e., the case with $\frac{1}{2} < b < 1$ in which the walker has a bias for stepping to higher states. In that case we have from Eq. (6.7-22b) that $e^{\phi(n)} = \beta^n$, where for convenience we have abbreviated $\beta \equiv (1-b)/b$. So it follows from the second of Eqs. (6.7-18a) that the probability of the process reaching state n_2 before reaching state n_1 is

$$p_2(n_0; n_1, n_2) = \frac{\displaystyle\sum_{m=n_1}^{n_0-1} \beta^m}{\displaystyle\sum_{m=n_1}^{n_2-1} \beta^m} = \frac{\beta^{n_1} \displaystyle\sum_{m=n_1}^{n_0-1} \beta^{m-n_1}}{\beta^{n_1} \displaystyle\sum_{m=n_1}^{n_2-1} \beta^{m-n_1}} = \frac{\displaystyle\sum_{n=0}^{n_0-n_1-1} \beta^n}{\displaystyle\sum_{n=0}^{n_2-n_1-1} \beta^n}.$$

By appealing to the algebraic identity

$$\sum_{n=0}^{N} \beta^n = \frac{1 - \beta^{N+1}}{1 - \beta} \qquad (0 < \beta < 1),$$

noting as we do that $\beta \equiv (1-b)/b$ is indeed confined to the interval $(0,1)$ when b is confined to the interval $(\tfrac{1}{2}, 1)$, we can bring the preceding formula into the form

$$p_2(n_0; n_1, n_2) = \frac{1 - \beta^{(n_0-n_1-1)+1}}{1 - \beta^{(n_2-n_1-1)+1}};$$

thus, recalling our definition of β,

$$p_2(n_0; n_1, n_2) = \frac{1 - \left(\dfrac{1-b}{b}\right)^{n_0-n_1}}{1 - \left(\dfrac{1-b}{b}\right)^{n_2-n_1}} \qquad (1/2 < b < 1). \qquad (6.7\text{-}27)$$

And of course, the probability $p_1(n_0; n_1, n_2)$ will be equal to 1 minus this quantity.

As the correctness of the result (6.7-27) is not altogether obvious, some simple checks would seem to be in order. By taking $n_1 = n_0 - 1$ and $n_2 = n_0 + 1$, it is not hard to show that the right side of Eq. (6.7-27) reduces to simply b; this is clearly correct, since the second of Eqs. (6.7-20) implies that the probability for the process in state n_0 to step next to state $n_0 + 1$ instead of state $n_0 - 1$ is b. Another check is obtained by letting $b \to 1$, in which case the right hand side of Eq. (6.7-27) is observed to approach 1; this again is clearly correct, since in the limit $b \to 1$ only upgoing transitions are possible. Finally, we note that putting $b = \tfrac{1}{2}$ in Eq. (6.7-27) gives an indeterminate form $0/0$; however, it is not difficult to show that the right side of Eq. (6.7-27) *approaches* $(n_0 - n_1)/(n_2 - n_1)$ as b approaches $\tfrac{1}{2}$, which is just what we expect from our $b = \tfrac{1}{2}$ results (6.7-24).

We could go on from here and calculate an explicit formula for $T_1(n_0; n_1,n_2)$ for the biased drunkard's walk process, but since that calculation is rather tedious we shall not do so. We shall instead simply emphasize the fact that *numerical* calculations of the splitting probabilities and the mean first exit time are usually most easily accomplished by using a digital computer to iterate the nonexplicit relations (6.7-12) – (6.7-15).

6.8 STABLE STATE FLUCTUATIONS AND TRANSITIONS

Much of the interest in birth-death Markov process theory for physical science applications concerns processes that have one or more *well defined stable states*. Such a process will spend most of its time near some stable state, as we say, "fluctuating about" that stable state. And if the process has more than one stable state, then it will occasionally (and indeed inevitably) pass from the vicinity of one stable state to the vicinity of another, as we say, "transitioning" from the former stable state to the latter. In this section we shall use the concepts and results developed in previous sections to quantitatively describe such stable state fluctuations and transitions.

Our attention in this section will accordingly be focused on a birth-death Markov process $X(t)$ that, first of all, has a well defined stationary Markov state density function $P_s(n)$ and a well defined potential function $\phi(n)$. We recall that the stepping functions $W_+(n)$ and $W_-(n)$ determine the function $P_s(n)$ through Eqs. (6.4-5) and (6.4-6), and the function $\phi(n)$ through Eq. (6.4-12). The two functions $P_s(n)$ and $\phi(n)$ are in turn related to each other through Eq. (6.4-13). The process $X(t)$ will also have a well defined barrier function $\Psi(n)$, which may be defined either in terms of $\phi(n)$ through Eq. (6.4-16), or in terms of $P_s(n)$ through Eqs. (6.4-17).

The stable states n_i of our process are formally defined to be the relative maximums of $P_s(n)$. These relative maximums are, according to the result (6.4-22), essentially the down-going roots of the function

$$a(n) \equiv W_+(n-1) - W_-(n). \qquad (6.8-1)$$

We shall assume here that $P_s(n)$ has at each of its relative maximums n_i a *distinct peak* — one that falls away on both sides to values that are small compared to its value at n_i. Corresponding to each distinct peak in $P_s(n)$

will be a *distinct valley* in $\phi(n)$. The lowest point in such a valley will, according to the result (6.4-20), be essentially a down-going root of the function

$$v(n) \equiv W_+(n) - W_-(n). \tag{6.8-2}$$

Since in most cases the down-going roots of the two functions $v(n)$ and $a(n)$ will not exactly coincide — the general exception being for processes in which $W_+(n)$ is independent of n — then the lowest point in any $\phi(n)$-valley will usually not coincide *exactly* with the highest point in the corresponding $P_s(n)$-peak. But we shall assume for our deliberations here that the difference between a valley minimum of $\phi(n)$ and the corresponding peak maximum of $P_s(n)$ is *negligibly small* compared to the widths of that valley and peak. In other words, for the processes of interest here, any stable state n_i can be regarded as an *approximate* relative minimum of $\phi(n)$. And since it follows from Eq. (6.4-16) that the two functions $\phi(n)$ and $\Psi(n)$ have identical extremums, then any stable state n_i will also be an approximate relative minimum of $\Psi(n)$. In thus assuming that the down-going roots n_i of $a(n)$ are approximate down-going roots of $v(n)$, we are evidently assuming that

$$W_+(n_i - 1) \approx W_-(n_i) \approx W_+(n_i), \tag{6.8-3}$$

and that $v(n_i)$ and $a(n_i)$ are accordingly given approximately by

$$v(n_i) \approx 0 \quad \text{and} \quad a(n_i) \approx 2W_-(n_i) \approx 2W_+(n_i). \tag{6.8-4}$$

For any stable birth-death Markov process $X(t)$ that meets the criteria just set forth, we shall examine in Subsection 6.8.A the *fluctuations* that occur about a given stable state. And for any such process that has exactly *two* stable states, we shall examine in Subsection 6.8.B the *transitions* that occur between those stable states. We shall illustrate our results in Subsection 6.8.C by applying them to an idealized model of a bistable chemical system.

6.8.A FLUCTUATIONS ABOUT A STABLE STATE

As just stated, we shall be concerned here with fluctuations about a stable state whose nominal value n_i locates the highest point of a well resolved peak in the function $P_s(n)$, and the (approximate) lowest point of a well resolved valley in each of the functions $\phi(n)$ and $\Psi(n)$. Although we intend for our discussion to be otherwise quite general, it will be helpful to keep in mind the specific example of the payroll process. That process is defined by the stepping functions $W_+(n) = h$ and $W_-(n) = ln$, and has a single stable state $n_1 = h/l$. We found earlier [see Eqs. (6.4-11) and (6.4-19)] that the functions $P_s(n)$, $\phi(n)$ and $\Psi(n)$ for the payroll process are given explicitly by

$$P_s(n) = \frac{e^{-h/l}(h/l)^n}{n!}, \qquad (6.8\text{-}5a)$$

$$\phi(n) = \ln(n!\,(h/l)^{-n}), \qquad (6.8\text{-}5b)$$

$$\Psi(n) = \frac{e^{h/l}}{h}\,n!\,(h/l)^{-n}. \qquad (6.8\text{-}5c)$$

These three functions were plotted earlier in Fig. 6-5 for $h = 0.2$ and $l = 0.005$, for which case the stable state $n_1 = 40$. A plot of a Monte Carlo simulation of the payroll process for those same values of h and l is shown in Fig. 6-4a.

To say that $X(t)$ is "fluctuating about the stable state n_i" is essentially to say that $X(t)$ is somewhere inside the n_i-peak of the function $P_s(n)$. We showed in Subsection 6.4.C that the width of that peak can be conveniently estimated by the parameter

$$\sigma_i = \left|\frac{2\pi\,W_-(n_i)}{a'(n_i)}\right|^{1/2}, \qquad (6.8\text{-}6)$$

which we call the **nominal width** of the stable state n_i. This nominal width σ_i is, more precisely, the effective width of the Gaussian peak that best fits $P_s(n)$ in the immediate neighborhood of n_i, where by "effective width" of a peak we mean the area under the peak divided by the maximum height of the peak. So if we let

$$\{n_i\} \equiv \text{ set of all states in the interval } [n_i - \sigma_i/2,\, n_i + \sigma_i/2], \quad (6.8\text{-}7)$$

then to say that $X(t)$ is fluctuating about the stable state n_i is *roughly* to say that $X(t) \in \{n_i\}$. The nominal width σ_i evidently constitutes a reasonable measure of the *range of fluctuations* of the process about the stable state n_i.

An evaluation of the right hand side of Eq. (6.8-6) for the stable state $n_1 = h/l$ of the payroll process gives, as was noted near the end of Subsection 6.4.C, the result $\sigma_1 = (2\pi h/l)^{1/2}$. This is a little more than twice the standard deviation of the Poissonian density function $P_s(n)$ in Eq. (6.8-5a). For the h and l values used in Figs. 6-4 and 6-5, we find that

$\sigma_1 = 15.85$. The dashed vertical lines in Fig. 6-5 indicate the boundaries of the region $\{n_1\} = [n_1 - \sigma_1/2, n_1 + \sigma_1/2]$ in this particular case.

Complementing σ_i as a descriptor of the fluctuations about the stable state n_i is a parameter τ_i which is called the **fluctuation time** of the stable state n_i. We formally define τ_i to be the average time for the process, initially in state n_i, to *first* reach *either* of the two states $n_i \pm \sigma_i/2$, or in the notation of our first exit analysis in Subsection 6.7.A,

$$\tau_i \equiv T_1(n_i;\, n_i - \sigma_i/2, n_i + \sigma_i/2). \tag{6.8-8}$$

In other words, τ_i is the average time it takes for the process, starting at the center of the set of states $\{n_i\}$, to "fluctuate out" to the boundary of that set. We could of course calculate this quantity exactly by using the equations for the mean first exit time that we derived in Subsection 6.7.A; however, it will usually be more convenient to have a *simple approximate formula* for τ_i. We can derive such a formula by reasoning as follows.

According to Eq. (6.7-19b), the mean first exit time in Eq. (6.8-8) is given exactly by

$$\tau_i = P_1 \sum_{n=n_i-\sigma_i/2+1}^{n_i} \sum_{m=n_i-\sigma_i/2}^{n-1} P_s(n)\, \Psi(m)$$

$$+ P_2 \sum_{n=n_i+1}^{n_i+\sigma_i/2-1} \sum_{m=n}^{n_i+\sigma_i/2-1} P_s(n)\, \Psi(m), \tag{6.8-9}$$

where p_1 is the probability of reaching $n_i - \sigma_i/2$ before $n_i + \sigma_i/2$, and p_2 is the probability of reaching $n_i + \sigma_i/2$ before $n_i - \sigma_i/2$. Assuming that the n_i-peak in $P_s(n)$ is roughly symmetric, it is reasonable to make the approximation $p_1 \approx p_2 \approx 1/2$, in which case the above formula reduces to

$$\tau_i \approx \left(\sum_{n=n_i-\sigma_i/2+1}^{n_i} \sum_{m=n_i-\sigma_i/2}^{n-1} + \sum_{n=n_i+1}^{n_i+\sigma_i/2-1} \sum_{m=n}^{n_i+\sigma_i/2-1} \right) \frac{P_s(n)\, \Psi(m)}{2}.$$

$$\tag{6.8-10}$$

This formula expresses τ_i as a sum or "discrete integral" of the function $[P_s(n)\Psi(m)/2]$ over the disjoint domain of the m-n integer lattice space shown in heavy solid dots in Fig. 6-12. From the point of view of the m-axis, the combined summation region is roughly symmetrically centered on the interval $n_i - \sigma_i/2 \le m \le n_i + \sigma_i/2$, where the summand

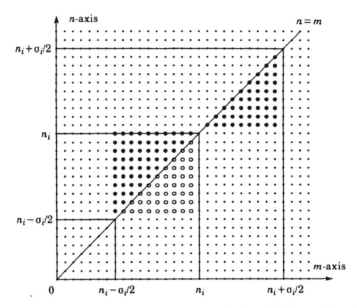

Figure 6-12. The m-n integer lattice space, with the solid heavy dots showing the summation domains in Eqs. (6.8-10) and (6.8-11). The left-most heavy-dotted region is the summation domain of the *first* double sums in those two equations, and the right-most heavy-dotted region is the summation domain of the *second* double sums. The open-dotted region is congruent to the right-most heavy-dotted region, and aids in counting the total number of solid heavy dots.

factor $\Psi(m)$ forms a *valley* whose *minimum* value is $\Psi(n_i)$. And from the point of view of the n-axis, the combined summation region is roughly symmetrically centered on the interval $n_i - \sigma_i/2 \le n \le n_i + \sigma_i/2$, where the summand factor $P_s(n)$ forms a *peak* whose *maximum* value is $P_s(n_i)$. We now propose to *approximate* the full summand $[P_s(n)\Psi(m)/2]$ everywhere in the summation region by the *constant* value $[P_s(n_i)\Psi(n_i)/2]$. In so doing, we evidently *over*estimate the factor $P_s(n)$ while *under*estimating the factor $\Psi(m)$, and we bank on the fact that these two misestimations will tend to offset each other in the *product* $P_s(n)\Psi(m)$. So we approximate the summand in Eq. (6.8-10) by

$$\frac{P_s(n)\,\Psi(m)}{2} \approx \frac{P_s(n_i)\,\Psi(n_i)}{2} = \frac{1}{2\,W_+(n_i)},$$

where the last step follows from the relation (6.4-17a). With this approximation, Eq. (6.8-10) evidently becomes

$$\tau_i \approx \frac{1}{2\,W_+(n_i)} \left(\sum_{n=n_i-\sigma_i/2+1}^{n_i} \sum_{m=n_i-\sigma_i/2}^{n-1} + \sum_{n=n_i+1}^{n_i+\sigma_i/2-1} \sum_{m=n}^{n_i+\sigma_i/2-1} \right).$$

$$(6.8\text{-}11)$$

The two double sums here can be evaluated algebraically; however, since those sums just count the total number of lattice points in the two heavy-dotted triangular regions of Fig. 6-12, it is easier to deduce the answer geometrically: Notice in Fig. 6-12 that the right-most heavy-dotted region is congruent to the open-dotted region, and further that the open-dotted region together with the left-most heavy-dotted region forms a *square array* of lattice points with exactly $\sigma_i/2$ points on a side. So the number of lattice points in that square, and hence the total value of the two double sums in Eq. (6.8-11), is $(\sigma_i/2)^2$. We conclude that the fluctuation time τ_i should be reasonably well approximated by the formula

$$\tau_i \approx \frac{(\sigma_i/2)^2}{2\,W_+(n_i)}. \qquad (6.8\text{-}12\text{a})$$

The simple formula (6.8-12a) can be made even simpler by first replacing $W_+(n_i)$ by $W_-(n_i)$, in accordance with Eq. (6.8-3), and then invoking formula (6.8-6) for σ_i. This gives the alternate formula

$$\tau_i \approx \frac{\Pi}{4\,|a'(n_i)|}, \qquad (6.8\text{-}12\text{b})$$

where the absolute value operation is necessary because $a(n)$ always has a negative slope at any stable state. Formula (6.8-12b) allows us to easily estimate the *time scale* of the fluctuations about the stable state n_i, in the same way that the formula (6.8-6) allows us to easily estimate the *distance scale* of those fluctuations.

We can also write the result (6.8-12a), by using the second of Eqs. (6.8-4), as $\tau_i \approx (\sigma_i/2)^2/a(n_i)$. We can gain some insight into this result by comparing it with formula (6.7-26b), which gives the symmetrical mean first exit time for the classic drunkard's walk process: We see that our approximation here for τ_i is equal to $T_1(n_i; n_i-\sigma_i/2, n_i+\sigma_i/2)$ for a birth-death process that has $a(n)=a(n_i)$ and $w_\pm(n)=1/2$ *everywhere* inside the interval of interest. Since this model for approximately calculating τ_i ignores the "restoring forces" that tend to bring the process back to the center n_i of the potential function valley, then we may expect that the approximate τ_i formulas (6.8-12) will generally tend to *underestimate* the mean first exit time on the right hand side of Eq. (6.8-8).

For the payroll process with $h = 0.2$ and $l = 0.005$, a calculation of the fluctuation time τ_1 for the stable state $n_1 = 40$ using either of formulas (6.8-12) yields $\tau_1 = 157$. To check this result, we have used the recursive formulas (6.7-12) – (6.7-15) to calculate *exact* mean first exit times $T_1(40; 40 - \varepsilon, 40 + \varepsilon)$ for ε ranging from 1 to 16. The results of those calculations are displayed in the second column of the table below. Since $\sigma_1/2$ here has the value $7.93 \approx 8$, then the figures in the table suggest that τ_1 should be about 213. As we anticipated, the value 157 from the simple formulas (6.8-12) underestimates the true value of $T_1(40; 40 - 8, 40 + 8)$. But notice that this estimate of 157 *exceeds* the true value of $T_1(40; 40 - 7, 40 + 7)$, so the underestimation is not as serious as it might at first appear. In fact, the figures in the table show that $T_1(40; 40 - \varepsilon, 40 + \varepsilon)$ is a very rapidly increasing function of ε; furthermore, it is reasonable to expect that these mean first exit times will be accompanied by standard deviations having comparable magnitudes. So we conclude that the value $\tau_1 = 157$ predicted by the

For the Payroll Process of Figs. 6-4 and 6-5

ε	$T_1(40; 40 - \varepsilon, 40 + \varepsilon)$	$T(40 \to 40 - \varepsilon)$ mean	std. dev.	$T(40 \to 40 + \varepsilon)$ mean	std. dev.
1	2.5	41.4	111	43.1	115
2	10.1	89.0	170	92.2	176
3	23.3	144	228	149	236
4	42.7	210	306	215	299
5	69.4	288	362	292	371
6	105	383	448	384	455
7	152	500	554	495	556
8	213	649	689	630	680
9	294	840	865	798	834
10	402	1095	1101	1008	1029
11	546	1440	1428	1275	1281
12	743	1923	1891	1621	1610
13	1016	2621	2568	2076	2048
14	1401	3662	3587	2684	2638
15	1956	5271	5174	3509	3445
16	2768	7854	7735	4649	4567

simple approximate formulas (6.8-12) is in fact a quite reasonable estimate of the time required for a "typical" fluctuation away from the stable state $n_1 = 40$.

The last two double columns of the above table show the means and standard deviations of the first *passage* times $T(40 \rightarrow 40 - \varepsilon)$ and $T(40 \rightarrow 40 + \varepsilon)$. These figures were calculated using the general recursion formulas derived in the "moments" analysis of Subsection 6.6.B, and they provide some interesting comparisons with the first *exit* time figures in the second column. The $\varepsilon = 8$ entries for instance show that, if the process starts in its stable state 40, then the average time to first reach the state $40 - 8$ is roughly the same as the average time to first reach the state $40 + 8$, but that those times are about *thrice* the average time to first reach *either* of the states 40 ± 8. Furthermore, the first passage times $T(40 \rightarrow 40 - 8)$ and $T(40 \rightarrow 40 + 8)$ both have standard deviations that are slightly larger than their means. These effects are even more pronounced for smaller values of ε. The $\varepsilon = 1$ entries for instance show that, if the process starts in state 40, then it will take on the average 2.5 time units to first reach *either* of the two states 39 and 41, the value 2.5 being just the mean pausing time $1/a(40)$ in state 40. But the mean first passage times $T_1(40 \rightarrow 39)$ and $T_1(40 \rightarrow 41)$ are both seen to be about 16 times larger than 2.5. The reason why $T_1(40 \rightarrow 39)$, for instance, is so much larger than $T_1(40; 39, 41)$ is really not hard to understand: Starting in state 40, the process will roughly half of the time step to state 39 after about 2.5 units of time. But the *other* half of the time the process will step to state 41 instead of 39, and from there it will in *some* instances wander around over states $n \geq 40$ for quite awhile before finally passing to state 39. It is these "widely wandering" sojourns that are responsible for $T_1(40 \rightarrow 39)$ being so much larger than $T_1(40; 39, 41)$. By contrast, the mean first exit time $T_1(40; 39, 70)$ turns out to equal the mean first passage time $T_1(40 \rightarrow 39)$ to three significant figures.

The data in the above table also reveal an interesting feature about the *standard deviations* of the first passage times $T(40 \rightarrow 40 - \varepsilon)$ and $T(40 \rightarrow 40 + \varepsilon)$: Those standard deviations are roughly *equal* to the associated means for $\varepsilon > 8$, but they gradually increase to nearly *thrice* the mean as ε decreases from 8 to 1.

As we have seen, the fluctuation time τ_i estimates the average time for a *typical* fluctuation away from the stable state n_i. Often, however, we are interested in the average time for an *extreme* fluctuation away from the stable state n_i. More specifically, let us consider a *monostable* process — i.e., a process such as the payroll process with a single stable state n_1 — and let us suppose that z is some state that differs from n_1 by more than two or three multiples of the half-width $\sigma_1/2$ of the stable state. How long, on the average, should it take the process to "fluctuate" from n_1 out to that relatively remote state z? We could of course answer this question *exactly* by using the mean first passage time formulas (6.6-17b) and (6.6-22b):

$$T_1(n_1 \to z) = \begin{cases} \sum_{n=0}^{z-1} P_s(n) \sum_{m=\text{Max}(n,n_1)}^{z-1} \Psi(m), & \text{for } z > n_1, \quad \text{(6.8-13a)} \\[2em] \sum_{n=z+1}^{\infty} P_s(n) \sum_{m=z+1}^{\text{Min}(n,n_1)} \Psi(m-1), & \text{for } z < n_1. \quad \text{(6.8-13b)} \end{cases}$$

But it will usually be more convenient to have *simple approximate* formulas that apply specifically when $|n_1 - z|$ is large compared to σ_1. To derive such formulas, we proceed as follows.

Consider first the case where z lies comfortably *above* the region $[n_1 - \sigma_1, n_1 + \sigma_1]$, such as the point $z = 60$ in the payroll process of Fig. 6-5. Then $T_1(n_1 \to z)$ will be given exactly by formula (6.8-13a). But in the sum over n in that formula, $P_s(n)$ will differ appreciably from zero only in the interval $[n_1 - \sigma_1, n_1 + \sigma_1]$; accordingly, we should be able to approximate Eq. (6.8-13a) by

$$T_1(n_1 \to z) \approx \sum_{n=n_1-\sigma_1}^{n_1+\sigma_1} P_s(n) \sum_{m=\text{Max}(n,n_1)}^{z-1} \Psi(m), \quad z \gg n_1 + \sigma_1.$$

Now in the sum over m in this last formula, suppose we replace the lower limit $\text{Max}(n,n_1)$ by simply n_1. The only error we make in doing that occurs when $n > n_1$, and that error amounts to erroneously accumulating terms $\Psi(m)$ for $m \in [n_1, n]$ for $n \leq n_1 + \sigma_1$. But owing to the behavior of the function $\Psi(m)$, the contribution of such terms will be *very small* compared to the contribution of terms with $m \in (n_1 + \sigma_1, z)$. So we should make only a very small error in replacing $\text{Max}(n,n_1)$ in the above formula by n_1:

$$T_1(n_1 \to z) \approx \sum_{n=n_1-\sigma_1}^{n_1+\sigma_1} P_s(n) \sum_{m=n_1}^{z-1} \Psi(m), \quad z \gg n_1 + \sigma_1.$$

The two sums over n and m are now seen to be *independent* of each other, and they can be evaluated separately. In particular, since n_1 is presumed here to be the *prominent* stable state of a *monostable* process, then the n-sum over $P_s(n)$ should be approximately unity. Thus we arrive at the following *approximate* formula for the mean time for a fluctuation of a monostable process from its stable state n_1 to any point z lying *well above* $n_1 + \sigma_1$:

$$T_1(n_1 \to z) \approx \sum_{n=n_1}^{z-1} \Psi(n) \qquad (z \gg n_1 + \sigma_1). \qquad (6.8\text{-}14a)$$

An entirely analogous argument applied to formula (6.8-13b) yields the following formula for the mean time for a fluctuation of a monostable process from its stable state n_1 to any point z lying *well below* $n_1 - \sigma_1$:

$$T_1(n_1 \to z) \approx \sum_{n=z}^{n_1} \Psi(n) \qquad (z \ll n_1 - \sigma_1). \qquad (6.8\text{-}14b)$$

Several noteworthy features of the extreme fluctuation formulas (6.8-14) can be brought out by examining the $\Psi(n)$ plot in Fig. 6-5 for the payroll process with $h = 0.2$ and $l = 0.005$. According to Eq. (6.8-14a), the average time for the process to fluctuate from the stable state n_1 out to some state $z \gg n_1 + \sigma_1$ is equal to the "area" under the function $\Psi(n)$ between $n = n_1$ and $n = z - 1$ inclusively. In the case of the plot in Fig. 6-5, we have $n_1 = 40$ and $\sigma_1 \approx 16$, so we can take z to be any number that is "significantly greater" than 56. For any such z, we observe from the figure that the major portion of the area under $\Psi(n)$ between n_1 and $z - 1$ will always be near the *upper* limit $z - 1$, where the function $\Psi(n)$ is relatively large and fast becoming larger; accordingly, we cannot change the upper summation limit in Eq. (6.8-14a), say from $z - 1$ to z, without significantly affecting the value of the sum. However, the area contribution from n-values near the lower boundary n_1 is by comparison very small, so we could shift the *lower* summation limit to any point in the interval $\{n_1\} \equiv [n_1 - \sigma_1/2, n_1 + \sigma_1/2]$ without materially altering the total area up to $z - 1$. What this means is that the formula (6.8-14a) can be looked upon as the mean first passage time to z from *any* state in the stable state region $\{n_1\}$. Finally, the fact that the area under $\Psi(n)$ between n_1 and $z - 1$ directly measures the "average waiting time" for a fluctuation from n_1 to z obviously implies that the *larger* that area is, the *less likely* it will be for such a fluctuation to occur during any given observation time. So the steep slope of the barrier function $\Psi(n)$ as n increases beyond $n_1 + \sigma_1$ essentially forms a "barrier" that increasingly inhibits the passage of the process.

Similar remarks apply to the formula (6.8-14b). It implies that for any state $z \ll n_1 - \sigma_1$, the "area" under the $\Psi(n)$ curve between z and n_1

inclusively is numerically equal to the average waiting time for a fluctuation to z from any state in the stable state region $\{n_1\}$. And the steep slope of $\Psi(n)$ for $n < n_1 - \sigma_1$ forms a "barrier" that inhibits, although does *not* ultimately prevent, the occurrence of such extreme fluctuations.

To test the accuracy of the extreme fluctuation formulas (6.8-14), we show in the table below a comparison of values calculated from those approximate formulas with values calculated from the exact formulas (6.8-13) for the payroll process with $h = 0.2$ and $l = 0.005$. These figures indicate that the approximate formulas (6.8-14) are quite accurate whenever $|z - n_1| \geq 3\sigma_1/2$. Of course, we should keep in mind that these mean first passage times will all be accompanied by standard deviations of roughly the same value; so in actual realizations of the process, we will routinely encounter extreme fluctuation times that differ from these average values by factors ranging from $\frac{1}{2}$ to 2.

For the Payroll Process of Figs. 6-4 and 6-5

ε	$T_1(40 \to 40 + \varepsilon)$		$T_1(40 \to 40 - \varepsilon)$	
	exact Eq. (6.8-13a)	approximate Eq. (6.8-14a)	exact Eq. (6.8-13b)	approximate Eq. (6.8-14b)
8	6.30×10^2	8.38×10^2	6.49×10^2	8.67×10^2
16	4.65×10^3	4.97×10^3	7.85×10^3	8.19×10^3
24	8.02×10^4	8.06×10^4	1.038×10^6	1.038×10^6
32	4.23×10^6	4.23×10^6	9.44×10^9	9.44×10^9
40	5.93×10^8	5.93×10^8	1.208×10^{18}	1.208×10^{18}

6.8.B TRANSITIONS BETWEEN STABLE STATES IN A BISTABLE PROCESS

In the last part of the preceding subsection we discussed the occurrence of "extreme fluctuations" away from a stable state in a *monostable* birth-death Markov process. Such extreme fluctuations can of course also occur in a birth-death Markov process with *more* than one stable state, and their effect then is often to carry the process into the domain of a neighboring stable state. The process is then said to have made a "spontaneous transition" from the former stable state to the latter. In this subsection we shall identify, and derive formulas for, several parameters that characterize stable state transitions in a *bistable* birth-death Markov process.

We let n_1 and n_2 denote the two stable states of the process, with $n_1 < n_2$. Two obvious transition parameters are the average time it takes the process to go from n_1 to n_2, and the average time it takes the process to go from n_2 to n_1. These **mean stable state transition times** $T_1(n_1 \rightarrow n_2)$ and $T_1(n_2 \rightarrow n_1)$ can obviously be calculated from the first passage time formulas derived in Section 6.6. If we are concerned only with obtaining *numerical* results, we can use either the simple recursion formulas (6.6-11) – (6.6-14) of the "pedestrian" approach, or else the more complicated recursion formulas (6.6-43) – (6.6-45) and (6.6-48) – (6.6-50) of the "moments" approach (the more complicated formulas allowing us to also calculate the associated standard deviations in the transition times). But let us instead see what insights can be gained from the *analytical* formulas (6.6-18b) and (6.6-23b), namely

$$T_1(n_1 \rightarrow n_2) = \sum_{m=n_1}^{n_2-1} \Psi(m) \sum_{n=0}^{m} P_s(n) \qquad (6.8\text{-}15)$$

and

$$T_1(n_2 \rightarrow n_1) = \sum_{m=n_1+1}^{n_2} \Psi(m-1) \sum_{n=m}^{\infty} P_s(n), \qquad (6.8\text{-}16)$$

in the case where n_1 and n_2 are distinct, well separated stable states.

Figure 6-13 shows schematically how the functions $P_s(n)$, $\phi(n)$ and $\Psi(n)$ might appear for the type of process that we are considering here. The stationary Markov state density function $P_s(n)$ has two prominent, well separated peaks at $n = n_1$ and n_2. The precise values of n_1 and n_2 can be calculated, according to theorem (6.4-22), as the down-going roots of the function $a(n) \equiv W_+(n-1) - W_-(n)$. Each stable state n_i has a *nominal width* σ_i, which is defined as the Gaussian effective width of the n_i-peak in $P_s(n)$, and which can be calculated from formula (6.8-6). In assuming the stable states to be "well separated," we are essentially assuming that there is a comfortable distance between the extreme upper edge $n_1 + \sigma_1$ of the lower peak and the extreme lower edge $n_2 - \sigma_2$ of the upper peak. In the region between the two peaks the function $P_s(n)$ is presumed to be very small, although Eq. (6.4-8) assures us that $P_s(n)$ will be strictly positive for all $n \geq 0$.

The potential function $\phi(n)$ will have *valleys* overlying the n_1 and n_2 peaks of the function $P_s(n)$. The true minimums of those valleys can be calculated, according to theorem (6.4-20), as the down-going roots of the function $v(n) \equiv W_+(n) - W_-(n)$; they will usually differ somewhat from

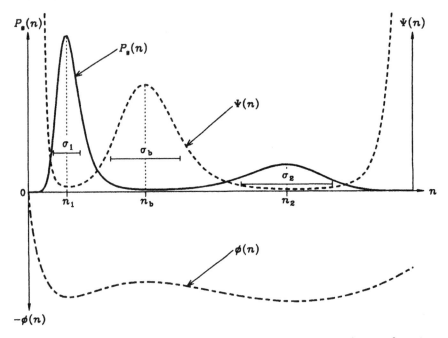

Figure 6-13. Schematic plots of the stationary Markov state density function $P_s(n)$, the potential function $\phi(n)$, and the barrier function $\Psi(n)$, for a typical bistable birth-death Markov process. The three functions are really defined only for n a nonnegative integer, and each function has its own ordinate scale. The nominal values of the two stable states, n_1 and n_2, are relative maximums of $P_s(n)$ and approximate relative minimums of $\phi(n)$ and $\Psi(n)$. The barrier state n_b is a relative maximum of $\phi(n)$ and $\Psi(n)$ and an approximate relative minimum of $P_s(n)$. The Gaussian effective widths of the n_1 and n_2 peaks in $P_s(n)$ are designated σ_1 and σ_2 respectively, and can be calculated from formula (6.8-6). The Gaussian effective width of the n_b peak in $\Psi(n)$ is designated σ_b, and can be calculated from formula (6.8-21).

n_1 and n_2, but we shall assume here that those differences are very small compared to σ_1 and σ_2. Since $\phi(n)$ has relative minimums at (approximately) n_1 and n_2, then it must have a relative maximum at some point n_b that lies somewhere *between* n_1 and n_2. We shall call n_b a **barrier state.** A simple modification in the arguments leading to the relative minimum theorem (6.4-20) reveals that the relative maximum n_b of $\phi(n)$ can be calculated as the *up-going* root of the function $v(n)$; i.e., we have

$$v(n_b) = 0 \quad \text{and} \quad v'(n_b) > 0. \qquad (6.8\text{-}17)$$

The potential function $\phi(n)$ is defined in terms of the process stepping functions according to [see Eq. (6.4-12)]

$$\phi(n) \equiv \begin{cases} 0, & \text{if } n=0, \\ \displaystyle\sum_{j=1}^{n} \ln \frac{W_-(j)}{W_+(j)}, & \text{if } n \geq 1. \end{cases} \qquad (6.8\text{-}18)$$

The stationary Markov state density function $P_s(n)$ can be expressed in terms of the function $\phi(n)$ by [see Eq. (6.4-13)]

$$P_s(n) = \frac{K W_+(0)}{W_+(n)} e^{-\phi(n)} \qquad (n \geq 0), \qquad (6.8\text{-}19)$$

where the constant K is chosen so that that this density function is properly normalized.

The barrier function $\Psi(n)$ can be expressed in terms of either $\phi(n)$ or $P_s(n)$ through the formulas [see Eqs. 6.4-16) and (6.4-17)]

$$\Psi(n) \equiv \frac{1}{K W_+(0)} e^{\phi(n)} = \frac{1}{W_+(n) P_s(n)} \qquad (n \geq 0). \quad (6.8\text{-}20)$$

These two formulas for $\Psi(n)$ have several noteworthy implications: First, $\Psi(n)$, following $\phi(n)$, will have two relative minimums at approximately n_1 and n_2, and one relative maximum at exactly n_b. Secondly, although the sign of $\phi(n)$ is unrestricted, $\Psi(n)$ will always be strictly positive. Thirdly, owing to the exponentiation in the first formula for $\Psi(n)$, the n_1 and n_2 valleys and the n_b peak in the function $\Psi(n)$ will be much more pronounced or exaggerated than in the function $\phi(n)$. Finally, the approximate inverse relationship between $\Psi(n)$ and $P_s(n)$ implied by the second formula for $\Psi(n)$ suggests that $\Psi(n)$ will be very small in the regions of the n_1 and n_2 peaks in $P_s(n)$; hence, the n_b peak in $\Psi(n)$ should not significantly overlap the n_1 and n_2 peaks in $P_s(n)$. All these features are illustrated schematically in Fig. 6-13.

The n_b peak in the barrier function $\Psi(n)$ will be called the **barrier peak**. Like the stable state peaks in $P_s(n)$, the barrier peak in $\Psi(n)$ will have a *nominal width* σ_b that can be conveniently defined as the effective width of the Gaussian curve that best fits the function $\Psi(n)$ in the immediate neighborhood of n_b. As shown in Appendix D, this definition of σ_b leads to the formula

$$\sigma_b = (2\pi/c)^{1/2} \quad \text{where} \quad c \equiv -\Psi''(n_b)/\Psi(n_b).$$

We shall later find it useful to have an explicit formula for σ_b comparable to the formula for σ_1 and σ_2 in Eq. (6.8-6), so we proceed now to evaluate the constant c.

From the first of Eqs. (6.8-20) we deduce that

$$\Psi'(n) = \Psi(n)\,\phi'(n);$$

therefore,

$$\Psi''(n) = \Psi'(n)\,\phi'(n) + \Psi(n)\,\phi''(n).$$

Since n_b is a relative maximum of $\Psi(n)$, then $\Psi'(n_b)=0$, so

$$\Psi''(n_b) = \Psi(n_b)\,\phi''(n_b).$$

Thus we conclude from the foregoing definition of c that

$$c = -\phi''(n_b),$$

which we note is positive since n_b is also a relative maximum of $\phi(n)$. To evaluate this second derivative of $\phi(n)$, we start by observing from Eq. (6.8-18) that the first derivative of $\phi(n)$ can be taken to be

$$\phi'(n) = \frac{\phi(n) - \phi(n-1)}{1} = \ln \frac{W_-(n)}{W_+(n)}.$$

It follows from this that

$$\phi''(n) = \frac{W_+(n)}{W_-(n)} \left(\frac{W_-'(n)\,W_+(n) - W_-(n)\,W_+'(n)}{W_+^2(n)} \right)$$

$$= \frac{W_-'(n)\,W_+(n) - W_-(n)\,W_+'(n)}{W_-(n)\,W_+(n)}.$$

So, putting $n = n_b$, and noting that the first of Eqs. (6.8-17) implies that $W_+(n_b) = W_-(n_b)$, we get

$$-c = \phi''(n_b) = \frac{W_-'(n_b) - W_+'(n_b)}{W_-(n_b)} = \frac{-v'(n_b)}{W_-(n_b)}.$$

Substituting this result into our earlier formula for σ_b, we conclude that

$$\sigma_b = \left(\frac{2\pi\, W_-(n_b)}{v'(n_b)} \right)^{1/2}. \qquad (6.8\text{-}21)$$

This formula evidently allows us to directly calculate the effective width of the barrier peak in the function $\Psi(n)$, in the same way that formula (6.8-6) allows us to directly calculate the effective widths of the steady state peaks in the function $P_s(n)$.

Now we are ready to see what simplifications can be made in the exact mean transition time formulas (6.8-15) and (6.8-16) in circumstances like that in Fig. 6-13 where the n_1, n_2 and n_b peaks do not significantly overlap — i.e., when

$$n_1 + \sigma_1/2 < n_b - \sigma_b/2 \quad \text{and} \quad n_b + \sigma_b/2 < n_2 - \sigma_2/2. \qquad (6.8\text{-}22)$$

Consider first the formula (6.8-15) for $T_1(n_1 \to n_2)$. Although the m-sum in that formula extends from n_1 to $n_2 - 1$, it is clear from Fig. 6-13 that the summand, being proportional to $\Psi(m)$, will be significantly different from zero only for $m \in [n_b - \sigma_b, n_b + \sigma_b]$; consequently, we may approximate Eq. (6.8-15) as

$$T_1(n_1 \to n_2) \approx \sum_{m = n_b - \sigma_b}^{n_b + \sigma_b} \Psi(m) \sum_{n=0}^{m} P_s(n).$$

Next we observe that the n-sum, which runs from 0 to some value in the interval $[n_b - \sigma_b, n_b + \sigma_b]$, can be legitimately restricted to run from $n_1 - \sigma_1$ to $n_1 + \sigma_1$, because the summand $P_s(n)$ is elsewhere very small. Thus,

$$T_1(n_1 \to n_2) \approx \sum_{m = n_b - \sigma_b}^{n_b + \sigma_b} \Psi(m) \sum_{n = n_1 - \sigma_1}^{n_1 + \sigma_1} P_s(n).$$

Similar reasoning applied to the exact formula (6.8-16) leads to the result

$$T_1(n_2 \to n_1) \approx \sum_{m = n_b - \sigma_b}^{n_b + \sigma_b} \Psi(m) \sum_{n = n_2 - \sigma_2}^{n_2 + \sigma_2} P_s(n).$$

These two results can evidently be combined into the single formula

$$T_1(n_i \to n_j) \approx \left\{ \sum_{m = n_b - \sigma_b}^{n_b + \sigma_b} \Psi(m) \right\} \left\{ \sum_{n = n_i - \sigma_i}^{n_i + \sigma_i} P_s(n) \right\}. \qquad (6.8\text{-}23)$$

The content of Eq. (6.8-23) can be expressed in geometrical terms by saying that the average waiting time for a spontaneous transition from stable state n_i to stable state n_j is approximately equal to the *product* of {the area under the *initial* stable state peak in $P_s(n)$} times {the area under the barrier peak in $\Psi(n)$}. That $T_1(n_i \rightarrow n_j)$ should be proportional to the area under the n_i peak in $P_s(n)$ is quite reasonable, since that area is roughly proportional to the time the process spends in the vicinity of stable state n_i, just "waiting" for the occurrence of a spontaneous transition to stable state n_j. That the constant of proportionality should be the area under the n_b peak in $\Psi(n)$ is not so obvious, but is nonetheless plausible in light of our results (6.8-14) for extreme fluctuations in monostable systems. Evidently, spontaneous transitions from either stable state to the other will take longer, and hence seem more inhibited, if the total area under the barrier peak in $\Psi(n)$ is made larger.

It follows from Eq. (6.8-23) that

$$\frac{T_1(n_1 \rightarrow n_2)}{T_1(n_2 \rightarrow n_1)} \approx \left\{ \sum_{n=n_1-\sigma_1}^{n_1+\sigma_1} P_s(n) \right\} \div \left\{ \sum_{n=n_2-\sigma_2}^{n_2+\sigma_2} P_s(n) \right\}. \quad (6.8\text{-}24)$$

The ratio on the left is a reasonable measure of the *stability of stable state* n_1 *relative to stable state* n_2. The formula says that this relative stability is approximately equal to the ratio of the area under the n_1-peak in $P_s(n)$ to the area under the n_2-peak in $P_s(n)$. So the stable state that has the largest $P_s(n)$ peak in terms of *total area* is the "more stable."

Another simple relationship between the two mean stable state transition times follows from the fact that, for a *bistable* system, we will typically have

$$\sum_{n=n_1-\sigma_1}^{n_1+\sigma_1} P_s(n) + \sum_{n=n_2-\sigma_2}^{n_2+\sigma_2} P_s(n) \approx \sum_{n=0}^{\infty} P_s(n) = 1.$$

By combining this with Eq. (6.8-23), we may conclude that the average time for a "round trip" between the two stable states is approximately equal to the total area under the barrier peak:

$$T_1(n_1 \rightarrow n_2) + T_1(n_2 \rightarrow n_1) \approx \sum_{m=n_b-\sigma_b}^{n_b+\sigma_b} \Psi(m). \quad (6.8\text{-}25)$$

The results (6.8-23) – (6.8-25) can be expressed in several other ways that are sometimes more convenient for computational purposes. Since the nominal widths σ_1, σ_2 and σ_b of the peaks in $P_s(n)$ and $\Psi(n)$ are all *effective* widths, then the areas under these peaks can be expressed in the simple forms

$$\sum_{m=n_b-\sigma_b}^{n_b+\sigma_b} \Psi(m) \approx \Psi(n_b)\,\sigma_b,$$

$$\sum_{n=n_i-\sigma_i}^{n_i+\sigma_i} P_s(n) \approx P_s(n_i)\,\sigma_i \quad (i=1,2).$$

Therefore, Eqs. (6.8-23) – (6.8-25) can be written in the alternate respective forms

$$T_1(n_i \to n_j) \approx \Psi(n_b)\,\sigma_b\,P_s(n_i)\,\sigma_i, \qquad (6.8\text{-}26)$$

$$\frac{T_1(n_1 \to n_2)}{T_1(n_2 \to n_1)} \approx \frac{P_s(n_1)\,\sigma_1}{P_s(n_2)\,\sigma_2}, \qquad (6.8\text{-}27)$$

and

$$T_1(n_1 \to n_2) + T_1(n_2 \to n_1) \approx \Psi(n_b)\,\sigma_b. \qquad (6.8\text{-}28)$$

Formulas (6.8-26) and (6.8-27) can in turn be made a bit more explicit by noting firstly from Eqs. (6.8-20) and (6.8-21) that

$$\Psi(n_b)\,\sigma_b = \frac{1}{K\,W_+(0)}\,e^{\phi(n_b)}\left(\frac{2\pi\,W_-(n_b)}{v'(n_b)}\right)^{1/2},$$

and secondly from Eqs. (6.8-19) and (6.8-6) that

$$P_s(n_i)\,\sigma_i = \frac{K\,W_+(0)}{W_+(n_i)}\,e^{-\phi(n_i)}\left|\frac{2\pi\,W_-(n_i)}{a'(n_i)}\right|^{1/2}$$

$$\approx K\,W_+(0)\,e^{-\phi(n_i)}\left|\frac{2\pi}{W_-(n_i)\,a'(n_i)}\right|^{1/2},$$

where the last step has made use of Eq. (6.8-3). Substituting these two expressions into Eqs. (6.8-26) and (6.8-27), we obtain the often useful formulas

$$T_1(n_i \to n_j) \approx 2\pi\left|\frac{W_-(n_b)}{W_-(n_i)\,a'(n_i)\,v'(n_b)}\right|^{1/2}\exp(\phi(n_b)-\phi(n_i)),$$

$$(6.8\text{-}29)$$

and

$$\frac{T_1(n_1 \to n_2)}{T_1(n_2 \to n_1)} \approx \left| \frac{W_-(n_2) a'(n_2)}{W_-(n_1) a'(n_1)} \right|^{1/2} \exp(\phi(n_2) - \phi(n_1)).$$

$$(6.8\text{-}30)$$

The controlling factors in these last two formulas are obviously the exponentials. Equation (6.8-29) shows that $T_1(n_i \to n_j)$ is a strongly increasing function of the potential difference $\phi(n_b) - \phi(n_i)$. And Eq. (6.8-30) shows that the relative stability of the two stable states depends very sensitively on the difference in the depths of their potential valleys: If stable state n_i has only a *slightly* deeper potential valley than stable state n_j, then stable state n_i may be *much* more stable that stable state n_j.

We can make an interesting connection at this point with our results in Subsection 3.7.B for the mean stable state transition times in a *continuous* bistable Markov process. We noted in Eq. (6.8-4) that $a(n_i) \approx 2W_-(n_i)$. And it is also true, because of the first of Eqs. (6.8-17), that $a(n_b) \approx 2W_-(n_b)$. It therefore follows that the function W_- appearing in Eqs. (6.8-29) and (6.8-30) can be everywhere replaced by the function a. That done, those two equations evidently become identical to Eqs. (3.7-22) and (3.7-23) under the usual functional correspondence $a \leftrightarrow D$ and $v \leftrightarrow A$. We thus have another example of the strong parallelism that exists between *birth-death* Markov processes and *continuous* Markov processes.

An interesting observation follows by making an approximate calculation of the mean first passage time from the stable state n_1 to the barrier state n_b: Starting with the general mean first passage time formula (6.6-18b), and using the same straightforward arguments that led to Eq. (6.8-23), we get

$$T_1(n_1 \to n_b) = \sum_{m=n_1}^{n_b - 1} \Psi(m) \sum_{n=0}^{m} P_s(n)$$

$$\approx \sum_{m=n_b - \sigma_b}^{n_b - 1} \Psi(m) \sum_{n=0}^{m} P_s(n)$$

$$\approx \left\{ \sum_{m=n_b - \sigma_b}^{n_b - 1} \Psi(m) \right\} \left\{ \sum_{n=n_1 - \sigma_1}^{n_1 + \sigma_1} P_s(n) \right\}.$$

Dividing this result by the formula for $T_1(n_1 \to n_2)$ in Eq. (6.8-23) then gives

$$\frac{T_1(n_1 \to n_b)}{T_1(n_1 \to n_2)} \approx \left\{ \sum_{m=n_b - \sigma_b}^{n_b - 1} \Psi(m) \right\} \div \left\{ \sum_{m=n_b - \sigma_b}^{n_b + \sigma_b} \Psi(m) \right\} \approx \frac{1}{2},$$

$$(6.8\text{-}31)$$

where our final approximation has assumed that the barrier peak is roughly symmetric in shape. This result implies that, on the average, it takes about *twice* as long for the process to go from n_1 to n_2 as for the process to go from n_1 to n_b. The reason for this is really not so hard to understand: Once the process has reached the barrier state n_b, where according to the first of Eqs. (6.8-17) we have $W_+(n_b) = W_-(n_b)$, it is roughly a "heads-or-tails" decision whether the process will go quickly *on* to stable state n_2, or go quickly *back* to stable state n_1.

We should emphasize again that all the results (6.8-23) – (6.8-31) are predicated on the assumption that the n_1, n_2 and n_b peaks are *cleanly separated*, in the sense of Eq. (6.8-22). If that condition is *not* satisfied, then we must use the exact mean first passage time formulas (6.8-15) and (6.8-16). When the separation condition *is* satisfied, then our analysis has painted the following picture of the behavior of the bistable process if initially placed in the stable state n_i: In a time of order τ_i, as given by either of formulas (6.8-12), the process will first fluctuate out to one of the boundaries $n_i \pm \sigma_i/2$ of the stable state region $\{n_i\}$. And in a much longer time of order $T_1(n_i \to n_j)$, as given by either of formulas (6.8-26) or (6.8-29), the process will undergo a transition to the other stable state n_j. There it will fluctuate around with a fluctuation time τ_j until roughly a time $T_1(n_j \to n_i)$ later, when it makes a transition back to the stable state n_i. This scenario then repeats itself ad infinitum.

Finally, let us consider a somewhat different question: What happens if we start the process out in some state n that lies *between* the stable states n_1 and n_2? We rather expect that the process will tend to move to stable state n_1 if $n < n_b$, or to stable state n_2 if $n > n_b$. But we must remember that the barrier presented to the process by the n_b peak in the function $\Psi(n)$ is only a *probabilistic* barrier –– one that will inevitably be overcome through the laws of chance; indeed, if that were not so, then spontaneous transitions between the stable states would never occur at all. The proper way to address the question just posed is to use the notion of *splitting probabilities* introduced in Section 6.7: From the general formulas (6.7-18b), we see that the probability that the process, starting from any state $n \in [n_1 + 1, n_2 - 1]$, will go first to stable state n_2 rather than stable state n_1 is

$$p_2(n; n_1, n_2) = \frac{\displaystyle\sum_{m = n_1}^{n-1} \Psi(m)}{\displaystyle\sum_{m = n_1}^{n_2 - 1} \Psi(m)}, \quad n \in [n_1 + 1, n_2 - 1]. \quad (6.8\text{-}32)$$

And of course, 1 minus this quantity is the probability $p_1(n; n_1, n_2)$ that the process will reach n_1 before reaching n_2. A schematic plot of the function $p_2(n; n_1, n_2)$ as given by Eq. (6.8-32) is shown in Fig. 6-14,

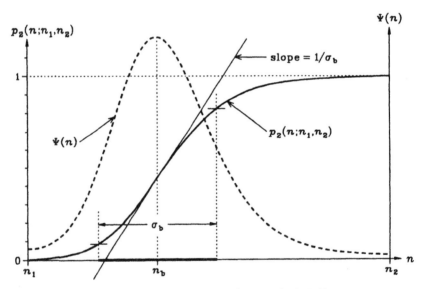

Figure 6-14. A schematic plot of the *splitting probability* $p_2(n; n_1, n_2)$, as measured against the left ordinate axis, superimposed on a plot of the barrier function $\Psi(n)$, as measured against the right ordinate axis, for a bistable birth-death Markov process with stable states n_1 and n_2 and barrier state n_b (compare Fig. 6-13). The two functions are really defined only for integer values of n. The maximum slope of the function $p_2(n; n_1, n_2)$ occurs at $n = n_b$, and is approximately equal to the reciprocal of the effective width σ_b of the barrier peak in $\Psi(n)$. As can be seen from the figure, most of the 0-to-1 increase in $p_2(n; n_1, n_2)$ takes place inside the region $[n_b - \sigma_b/2, n_b + \sigma_b/2]$, which we therefore call the *transition region*.

together with a superimposed plot of the barrier function $\Psi(n)$. As we should expect, $p_2(n; n_1,n_2)$ rises monotonically from a value near zero at its lower limit $n = n_1 + 1$ to a value near unity at its upper limit $n = n_2 - 1$. Equation (6.8-32) tells us that $p_2(n; n_1,n_2)$ is essentially the "normalized integral" of the barrier function $\Psi(n)$ for $n \in [n_1 + 1, n_2 - 1]$.

An important implication of formula (6.8-32) can be brought out by first noting that the "slope" of the function p_2 at any point n is

$$p_2'(n; n_1,n_2) = \frac{p_2(n+1; n_1,n_2) - p_2(n; n_1,n_2)}{1} = \frac{\Psi(n)}{\displaystyle\sum_{m=n_1}^{n_2-1} \Psi(m)}.$$

Since $\Psi(n)$ assumes its maximum value for $n \in [n_1 + 1, n_2 - 1]$ at the point $n = n_b$, then the *maximum slope* of the rising function $p_2(n; n_1,n_2)$ occurs at $n = n_b$:

$$p_2'(\max) = p_2'(n_b; n_1,n_2) = \frac{\Psi(n_b)}{\displaystyle\sum_{m=n_1}^{n_2-1} \Psi(m)}. \qquad (6.8\text{-}33\text{a})$$

Now, since σ_b is defined as the Gaussian effective width of the barrier peak in $\Psi(n)$, then we can approximate the area under that peak as

$$\sum_{m=n_1}^{n_2-1} \Psi(m) \approx \Psi(n_b)\,\sigma_b.$$

Substituting this into the preceding equation, we conclude that the maximum slope of the function $p_2(n; n_1,n_2)$ occurs at $n = n_b$ and has the approximate value $1/\sigma_b$:

$$p_2'(\max) = p_2'(n_b; n_1,n_2) \approx \frac{1}{\sigma_b}. \qquad (6.8\text{-}33\text{b})$$

The result (6.8-33b) can be given an interesting and useful graphical interpretation [see Fig. 6-14]: It says that the tangent line to the curve of $p_2(n; n_1,n_2)$ at the point $n = n_b$ rises from 0 to 1 in a horizontal distance of (approximately) σ_b. Now, if the barrier peak is roughly symmetric in shape, so that $p_2(n; n_1,n_2)$ assumes a value of roughly $\frac{1}{2}$ at $n = n_b$, then this tangent line will cut the horizontal lines $p_2 = 0$ and $p_2 = 1$ at roughly

$n = n_b - \sigma_b/2$ and $n = n_b + \sigma_b/2$, respectively. Figure 6-14 shows this behavior of the line passing through the point $(n_b, p_2(n_b; n_1, n_2))$ with slope $1/\sigma_b$. We note that $p_2(n; n_1, n_2)$ does *most* of its rising from 0 to 1 inside the interval

$$\{n_b\} \equiv [n_b - \sigma_b/2, \, n_b + \sigma_b/2]. \tag{6.8-34}$$

This interval $\{n_b\}$ is therefore called the **transition region** of the bistable process. If $X(t)$ initially lies *below* that region, then it is much more likely that $X(t)$ will visit stable state n_1 before stable state n_2. If $X(t)$ initially lies *above* that region, then $X(t)$ is much more likely to visit n_2 before n_1. It is only when $X(t)$ initially lies *inside* the transition region $\{n_b\}$ that the question of which stable state will be visited first becomes impossible to answer with a high degree of certainty.

6.8.C EXAMPLE OF A BISTABLE PROCESS: THE SCHLÖGL REACTIONS

At the end of Chapter 5 we showed how certain chemically reacting systems can be described using Markov process theory. We considered there as a specific example a well stirred, gas-phase mixture of chemical species B_1, B_2, B_3 and X, which interreact according to reaction channels (5.3-39). We showed that if the molecular populations of species B_1, B_2 and B_3 are "buffered" to remain essentially constant, then the instantaneous molecular population of the chemical species X can be accurately modeled as a temporally homogeneous jump Markov process $X(t)$ on the nonnegative integers. We also showed that $X(t \to \infty)$ in that case is a time-independent random variable whose mean is approximately equal to the "equilibrium" value X_e given in Eq. (5.3-55), and whose standard deviation is approximately equal to $(X_e)^{1/2}$. This asymptotic behavior is in many ways typical of most chemically reacting systems: the system eventually settles down to fluctuating about a single, well defined stable state, with the random fluctuations about that state being noticeable only if the system is microscopic in size.

An interesting exception to this common situation is provided by the following model set of coupled chemical reactions:

$$B_1 + 2X \underset{c_2}{\overset{c_1}{\rightleftharpoons}} 3X \quad \text{and} \quad B_2 \underset{c_4}{\overset{c_3}{\rightleftharpoons}} X . \tag{6.8-35}$$

These reactions are called the *Schlögl reactions*, after the scientist who first studied them.† In the Schlögl reactions, B_1 and B_2 denote buffered species whose respective molecular populations N_1 and N_2 are assumed to remain essentially constant over the time intervals of interest. Our focus of course is on the behavior of the quantity

$$X(t) \equiv \text{number of X molecules in the system at time } t. \quad (6.8\text{-}36)$$

To see that $X(t)$ in this case is a temporally homogeneous birth-death Markov process [notice that $X(t)$ for reactions (5.3-39) is *not* a birth-death process since it can take jumps of length 2], we need only take note of two facts: First, $c_\mu dt$ gives, by definition (5.3-29), the probability that a particular combination of R_μ reactant molecules will react thusly in the next infinitesimal time increment dt; so if $X(t) = n$, then the probability that an R_1 reaction will occur in $[t, t+dt)$ is equal to $c_1 dt$ times the number $N_1 n(n-1)/2!$ of distinct combinations of R_1 reactant molecules available at time t, and similarly for the other three reaction channels. Secondly, we note that reactions R_1 and R_3 of the set (6.8-35) each increase $X(t)$ by exactly 1, while reactions R_2 and R_4 each decrease $X(t)$ by exactly 1. From these two facts and the laws of probability it follows that, if $X(t) = n$, then the probability that $X(t+dt)$ will equal $n+1$ is

$$W_+(n)\, dt = (c_1 dt)\, [N_1 n(n-1)/2!] + (c_3 dt)\, (N_2),$$

while the probability that $X(t+dt)$ will equal $n-1$ is

$$W_-(n)\, dt = (c_2 dt)\, [n(n-1)(n-2)/3!] + (c_4 dt)\, (n).$$

And so we conclude that the instantaneous X-molecule population $X(t)$ is a temporally homogeneous birth-death Markov process, which we shall call the **Schlögl process**, with stepping functions

$$\begin{cases} W_+(n) = (c_1 N_1/2)n(n-1) + c_3 N_2, & (6.8\text{-}37a) \\[2mm] W_-(n) = (c_2/6)n(n-1)(n-2) + c_4 n. & (6.8\text{-}37b) \end{cases}$$

What makes the Schlögl reactions especially interesting is that, for certain ranges of values of the constant parameters c_1, c_2, c_3, c_4, N_1 and N_2, the process will be *bistable*, meaning that $X(t)$ will have not one but *two* stable or "equilibrium" states. This possibility arises from the cubic n-dependence of $W_-(n)$. As a consequence of that cubic n-dependence, the two functions

---

† F. Schlögl, "On Thermodynamics Near a Steady State", *Zeitschrift für Physik*, vol. 248 (1971), pp. 446–58.

$$a(n) \equiv W_+(n-1) - W_-(n) \tag{6.8-38a}$$

and

$$v(n) \equiv W_+(n) - W_-(n) \tag{6.8-38b}$$

will also be cubics. It then becomes possible for $a(n)$ to have *two* down-going roots, which by theorem (6.4-22) define two stable states, since those two roots can be separated by a third *up*-going root. For example, Fig. 6-15 shows plots of $a(n)$ and $v(n)$ for the parameter values

$$\left\{ \begin{array}{l} c_1 = 3 \times 10^{-7}, \ c_2 = 10^{-4}, \ c_3 = 10^{-3}, \ c_4 = 3.5; \qquad (6.8\text{-}39a) \\[2mm] N_1 = 1 \times 10^5, \ N_2 = 2 \times 10^5. \qquad\qquad\qquad\quad (6.8\text{-}39b) \end{array} \right.$$

We see from these plots that $a(n)$ has two down-going roots at 82.9 and 563.1, separated by an up-going root at 256.9. And $v(n)$ has two down-going roots at 84.8 and 570.0, separated by an up-going root at 248.4. So we expect on the basis of theorems (6.4-22) and (6.8-17) that the Schlögl process in this case will have stable states n_1 and n_2 and barrier state n_b given by

$$n_1 = 82, \quad n_b = 248, \quad n_2 = 563. \tag{6.8-40}$$

We note in passing that the roots of $a(n)$ are very close to the roots of $v(n)$. This means that we can usually put $W_+(n) \approx W_-(n)$ at $n = n_1$ and n_2.

It should however be pointed out that the Schlögl reactions (6.8-35) do *not* exactly model any known *real* chemical system. The reason is that genuinely trimolecular reactions like R_1 and R_2 are practically nonexistent in Nature. And of course, it is precisely the trimolecular character of reaction R_2 that gives rise to the crucial cubic form of $a(n)$. As discussed in Chapter 5, a trimolecular reaction of the form (5.3-36a) is at best only a rough approximation to a multireaction sequence of the form (5.3-36b), which involves an additional reacting species. It is possible to construct a bistable chemical system *without* any trimolecular reactions *if* we allow the system to have more than one time-varying species, but such a system could then not be analyzed using the well developed machinery of *univariate* Markov process theory. So the Schlögl reactions should be regarded as a highly idealized model of a bistable chemical system, whose somewhat strained physical credentials are counterbalanced by its mathematical tractability. The Schlögl reactions provide a simple model of a chemical system in which the ever present and usually inconsequential random fluctuations can actually produce *macroscopic* effects, such as a spontaneous transition from one stable state n_i to the other stable state n_j.

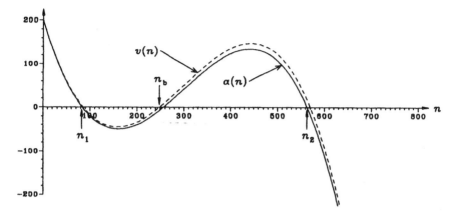

Figure 6-15. The functions $a(n)$ and $v(n)$ for the *Schlögl process* with parameter values (6.8-39). The two down-going roots of $a(n)$ locate the stable states n_1 and n_2, while the up-going root of $v(n)$ locates the barrier state n_b. See figure below.

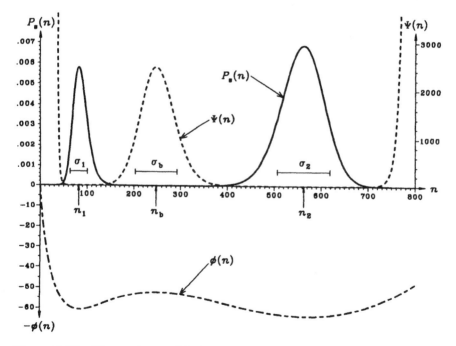

Figure 6-16. The stationary Markov state density function $P_s(n)$, the potential function $\phi(n)$, and the barrier function $\Psi(n)$ for the *Schlögl process* with stepping functions (6.8-37) and parameter values (6.8-39). The indicated peak widths σ_1, σ_2 and σ_b were computed from formulas (6.8-6) and (6.8-21).

In Fig. 6-16 we show a plot of the stationary Markov state density function $P_s(n)$, the potential function $\phi(n)$, and the barrier function $\Psi(n)$ for the Schlögl process with the parameter values (6.8-39). These functions were computed from the stepping functions (6.8-37) using the formulas developed in Section 6.4.† We observe in these plots the appearance of the stable state peaks and the barrier peak in precisely the places anticipated in Fig. 6-15. A calculation of the nominal widths of those three peaks from formulas (6.8-6) and (6.8-21) yields the values

$$\sigma_1 = 36.4, \quad \sigma_b = 89.6, \quad \sigma_2 = 112.4. \tag{6.8-41}$$

Figure 6-16 shows that these values characterize the "widths" of the respective peaks quite adequately. Despite appearances to the contrary, the functions $P_s(n)$ and $\Psi(n)$ are *strictly positive* for all $n \geq 0$; in particular, we find that $P_s(0) = 4.538 \times 10^{-29}$, $P_s(n_b) = 3.60 \times 10^{-7}$, $\Psi(n_1) = 0.578$ and $\Psi(n_2) = 0.030$.

In Fig. 6-17 we show plots of four separate Monte Carlo simulations of the Schlögl process, each simulation starting out in a different state (specifically, $n_0 = 0$, 238, 258 and 800). These simulations were carried out using the exact Monte Carlo simulation algorithm of Fig. 6-1, but omitting the calculation of the integral $S(t)$ since it has no physical significance for this process. Each $X(t)$ trajectory in Fig. 6-17 was constructed by plotting unconnected dots obtained by sampling the process every fifth reaction. This plotting technique is easy to implement, and it reveals the greater reaction frequency that obtains at higher population levels. The two solid horizontal lines in Fig. 6-17 show the stable states n_1 and n_2, and the dashed horizontal line shows the barrier state n_b, all as specified by Eqs. (6.8-40); the dotted horizontal lines show the respective $\pm\sigma/2$ bands, as specified by Eqs. (6.8-41). We note the tendency of the process to move toward the stable states n_i $(i = 1, 2)$, where the respective bands $[n_i - \sigma_i/2, n_i + \sigma_i/2]$ indeed appear to provide reasonable estimates of the *typical fluctuation ranges*. Runs with $n_0 < n_b$ usually migrate to the lower stable state n_1, while runs with $n_0 > n_b$ usually migrate to the upper stable state n_2. However, as if to

---

† Specifically, the function $\phi(n)$ was computed by using a recursive form of Eq. (6.4-12), namely $\phi(0) = 0$ and $\phi(n) = \phi(n-1) + \ln[W_-(n)/W_+(n)]$. The function $P_s(n)$ was computed rather more indirectly as follows: First we temporarily set $P_s(0) = 1$ and calculated $P_s(n) = P_s(n-1) \times [W_+(n-1)/W_-(n)]$, in accordance with Eq. (6.4-4), for n running from 1 up to some "suitably large" value; next we calculated K as the reciprocal of the sum of those temporary $P_s(n)$ values, in accordance with Eq. (6.4-5); finally, we converted the temporary $P_s(n)$ values to properly normalized values by multiplying each by K. With K known, the function $\Psi(n)$ is most easily computed from formula (6.4-16), although either of formulas (6.4-17) could be used instead.

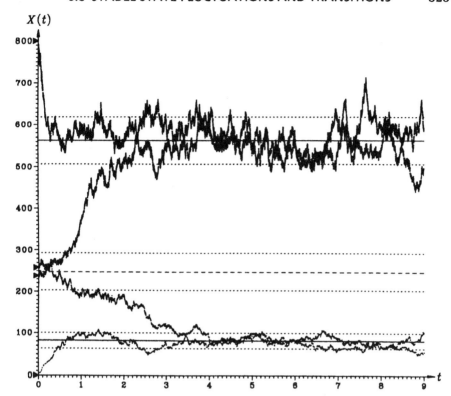

Figure 6-17. Four simulation runs of the *Schlögl process* using the same parameter values as in Fig. 6-16. The four markers on the vertical axis indicate the values of n_0 for the runs (the vertical axis in this figure is the *horizontal* axis of Fig. 6-16). The two solid horizontal lines show the stable states n_1 and n_2, the dashed line shows the barrier state n_h, and the dotted lines show the respective $\pm\sigma/2$ bands. The trajectories were generated by using the algorithm of Fig. 6-1 and marking with a simple dot the state after every five reactions.

emphasize the inherent unpredictability associated with the *transition region* $[n_b - \sigma_b/2, n_b + \sigma_b/2]$, the trajectory with $n_0 = n_b - 10$ actually migrated to the *upper* stable state, while the trajectory with $n_0 = n_b + 10$ migrated to the *lower* stable state.

Apart from their statistical jaggedness, the $X(t)$ trajectories in Fig. 6-17 correspond for the most part to what one would calculate from the traditional deterministic "reaction-rate" equation, and it is perhaps appropriate to briefly review why that is so. As was discussed in Subsection 5.3.B, the reaction-rate equation is obtained by starting with the time-evolution equation for $\langle X(t) \rangle$, in this case Eq. (6.1-29),

$$\frac{d}{dt} \langle X(t) \rangle = \langle v(X(t)) \rangle,$$

and then simplifying the right hand side by making *two approximations*. The first approximation springs from the presumption that $X(t)$ will usually be very large compared to unity. Specifically, we observe from Eqs. (6.8-37) and (6.8-38) that, if $n \gg 1$, then the two functions $a(n)$ and $v(n)$ can both be approximated by the *same function*, namely,

$$v(n) \doteq a(n) \doteq R(n) \equiv (c_1 N_1/2)n^2 + c_3 N_2 - (c_2/6)n^3 - c_4 n$$

$$(n \gg 1). \quad (6.8\text{-}42a)$$

The second simplifying approximation is to put

$$\langle X^k(t) \rangle \doteq \langle X(t) \rangle^k \text{ for all } k \geq 2. \quad (6.8\text{-}42b)$$

Since the $k = 2$ version of this approximation amounts to asserting that var$\{X(t)\} \doteq 0$, then the general result (1.3-10) implies that the Markov process $X(t)$ is now being approximated as a *deterministic* process, namely $X(t) \doteq \langle X(t) \rangle$. And the foregoing equations show that the time-evolution equation for this deterministic process is

$$\frac{dX(t)}{dt} \doteq R(X(t)), \quad (6.8\text{-}43)$$

where R is the function defined in Eq. (6.8-42a). Equation (6.8-43) is essentially the "reaction-rate equation." It implies that, as $t \to \infty$, $X(t)$ will asymptotically approach the particular stable state that lies on the same side of n_b as the initial state n_0. The proof of this statement is a simple consequence of the fact that the function $R(n)$ in Eq. (6.8-43) is, by Eq. (6.8-42a), a good approximation to *both* of the functions plotted in Fig. 6-15. So if X is initially to the *left* of the root n_1 where R is positive, then we will have $dX/dt > 0$, and X will move up toward n_1; on the other hand, if X is initially *between* the roots n_1 and n_b where R is negative, then we will have $dX/dt < 0$, and X will move down toward n_1. Similar reasoning shows that if X is initially to the *right* of the root n_b then it will always move toward n_2.

In point of fact, the simple, deterministic reaction-rate equation (6.8-43) gives a very good description of many important features of the Schlögl process. But there are *some* interesting features of the process that cannot be understood through that equation, and which require a proper Markovian analysis. Those features include: the genuinely nondeterministic behavior of the process in the transition region around

n_b; the fluctuating behavior of the process when it is "in" either of the stable states n_1 or n_2; and the occurrence of spontaneous transitions between the stable states. We shall now discuss each of these three points in turn, using the machinery developed in Subsections 6.8.A and 6.8.B.

Equation (6.8-32) is an exact formula for the probability $p_2(n; n_1, n_2)$ that the process, starting in any state $n \in [n_1 + 1, n_2 - 1]$, will go first to state n_2 instead of state n_1. Of course, the probability $p_1(n; n_1, n_2)$ that the process will go first to state n_1 will be equal to $1 - p_2(n; n_1, n_2)$. The formula for $p_2(n; n_1, n_2)$ in Eq. (6.8-32) is evidently expressed solely in terms of the barrier function $\Psi(n)$, which in turn is plotted in Fig. 6-16 for the Schlögl process with the parameter values (6.8-39). For that function $\Psi(n)$, it is a simple matter to compute from Eq. (6.8-32) the function $p_2(n; n_1, n_2)$, and we show the result of that computation in Fig. 6-18. As expected, the function $p_2(n; n_1, n_2)$ rises monotonically from near zero at $n = n_1 + 1$ to near unity at $n = n_2 - 1$. We have also indicated in Fig. 6-18 the transition region $[n_b - \sigma_b/2, n_b + \sigma_b/2]$, where σ_b is the width of the barrier peak in $\Psi(n)$ calculated earlier in Eq. (6.8-41). As expected, most of the change in $p_2(n; n_1, n_2)$ takes place inside of that region. Of the two

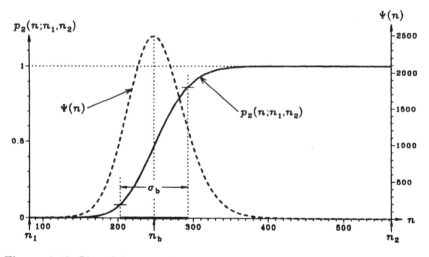

Figure 6-18. Plot of the probability $p_2(n; n_1, n_2)$ that the Schlögl process, with parameter values (6.8-39) and starting in state n, will reach stable state n_2 before stable state n_1. Superimposed as the dashed curve is the portion of the barrier function $\Psi(n)$ between n_1 and n_2 [see Fig. 6-16)], of which $p_2(n; n_1, n_2)$ is just the normalized integral. Note that most of the rise in $p_2(n; n_1, n_2)$ from 0 to 1 takes place inside the transition region, $\{n_b\} \equiv [n_b - \sigma_b/2, n_b + \sigma_b/2]$.

simulation runs plotted in Fig. 6-17 that started out inside the transition region, we can read off from Fig. 6-18 that the run with $n_0 = 238$ had a probability of 0.358 of going first to n_2 (as it in fact did), while the run with $n_0 = 258$ had a probability of 0.573 of going first to n_2 (as it in fact did not). The curve is seen to cross the value $\frac{1}{2}$ between the n values 251 and 252, a few units above n_b.

Turning next to the fluctuating behavior of $X(t)$ when it is "in" one of its stable states, we have already noted from our simulations in Fig. 6-17 that the values σ_1 and σ_2 in Eq. (6.8-41), which were calculated from formula (6.8-6), do a good job of characterizing the *ranges* of the stable state fluctuations. Another descriptive parameter in this regard is the stable state *fluctuation time* τ_i, which is defined as the average time for the process, starting in state n_i, to first reach either of the two states $n_i \pm \sigma_i/2$. By using the recursive formulas (6.7-11) – (6.7-15) for the mean first exit time, we have calculated the following *exact* values for τ_1 and τ_2 for the parameter values (6.8-39):

$$\tau_1 \equiv T_1(n_1; n_1 - \sigma_1/2, n_1 + \sigma_1/2) = 0.716, \qquad (6.8\text{-}44a)$$

$$\tau_2 \equiv T_1(n_2; n_2 - \sigma_2/2, n_2 + \sigma_2/2) = 0.420. \qquad (6.8\text{-}44b)$$

It is interesting to note that the narrower stable state n_1 actually has the longer fluctuation time, a feature that is more or less discernible from the simulations in Fig. 6-17. Now, in Subsection 6.8.A, we argued that the stable state fluctuation times could be *approximately* calculated from the simple formulas (6.8-12). In the present instance, Eq. (6.8-12a) gives

$$\tau_1 \approx \pi / 4 |a'(n_1)| = 0.564, \qquad (6.8\text{-}45a)$$

$$\tau_2 \approx \pi / 4 |a'(n_2)| = 0.320. \qquad (6.8\text{-}45b)$$

These approximate figures are seen to underestimate the exact values (6.8-44) by nearly 25%. This discrepancy is, however, insignificant from a practical point of view, owing to the relatively large fluctuations typically associated with first exit times. To illustrate this point, twenty independent simulation runs of the Schlögl process were made, each starting in the stable state $n_1 = 82$ and terminating upon the first arrival in either of the states $n_1 - \sigma_1/2 = 64$ and $n_1 + \sigma_1/2 = 100$. The elapsed times for these twenty simulation runs ranged from a low of 0.149 to a high of 1.207, and they had an arithmetic average of 0.590. Clearly, as a "typifying value" for those times, the approximate value 0.564 in Eq. (6.8-45a) is just as serviceable as the exact mean 0.716 in Eq. (6.8-44a). It may also be noted that, although the approximate τ_1 value 0.564 of Eq.

(6.8-45a) is indeed *less* than the mean first exit time $T_1(82;64,100)$ $=0.716$ of Eq. (6.8-45a), it is *greater* than the mean first exit time $T_1(82;66,98) = 0.532$ for a slightly narrower interval about n_1.

Lastly, we turn the matter of spontaneous transitions between the stable states $n_1 = 82$ and $n_2 = 563$. Using the first passage time recursion formulas (6.6-43) – (6.6-45) and (6.6-48) – (6.6-50), we have calculated the first two moments of the first passage times $T(82{\rightarrow}563)$ and $T(563{\rightarrow}82)$ for the Schlögl process with the parameter values (6.8-39). Those calculations yielded the following *exact* results:

$$T_1(n_1{\rightarrow}n_2) = 5.031\times10^4, \quad \mathrm{sdev}\{T(n_1{\rightarrow}n_2)\} = 5.030\times10^4, \quad (6.8\text{-}46\mathrm{a})$$

$$T_1(n_2{\rightarrow}n_1) = 1.781\times10^5, \quad \mathrm{sdev}\{T(n_2{\rightarrow}n_1)\} = 1.781\times10^5. \quad (6.8\text{-}46\mathrm{b})$$

Note that the standard deviations in these stable state transition times have the usual property of being practically equal to their means. In Subsection 6.8.B we argued that the means of these stable state transition times could be *approximately* calculated through the explicit formula (6.8-29). Upon evaluating that formula for the cases under consideration, we find that

$$T_1(n_1{\rightarrow}n_2) \approx 4.70\times10^4, \quad (6.8\text{-}47\mathrm{a})$$

$$T_1(n_2{\rightarrow}n_1) \approx 1.719\times10^5. \quad (6.8\text{-}47\mathrm{b})$$

Comparing these approximate values with the exact values in Eqs. (6.8-46), we see that the approximating formula (6.8-29) has underestimated $T_1(n_1{\rightarrow}n_2)$ by about 7%, and $T_1(n_2{\rightarrow}n_1)$ by about 3%. But in view of the expected 100% relative fluctuations in the transition times predicted by the standard deviation values in Eq. (6.8-46), these underestimations are practically inconsequential. Comparing these mean transition times with the fluctuation times τ_i in Eqs. (6.8-45), we see that the process will normally make very many fluctuations about a stable state before it makes a transition to the other stable state.

The relative stability of stable state n_1 compared to stable state n_2 is measured by the ratio $T_1(n_1{\rightarrow}n_2)/T_1(n_2{\rightarrow}n_1)$. Using the exact times in Eq. (6.8-46) we get 0.282 for that ratio, while using the approximate times in Eq. (6.8-47) we get 0.273. As discussed in connection with Eq. (6.8-24), this ratio should approximately be equal to the ratio of the area under the n_1-peak in $P_s(n)$ to the area under the n_2-peak in $P_s(n)$. The area ratios just quoted are clearly consistent with the plot of $P_s(n)$ in Fig. 6-16. In this case, stable state n_2 is "more stable" than stable state n_1 by a factor of approximately 4.

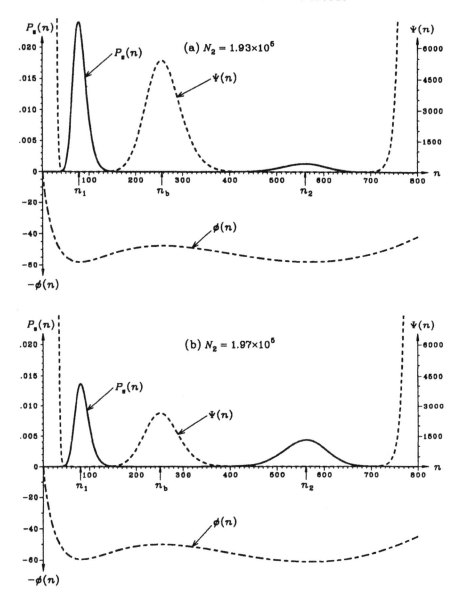

Figure 6-19. Plots of the stationary Markov state density function $P_s(n)$ (solid curve), the potential function $\phi(n)$ (dot-dash curve), and the barrier function $\Psi(n)$ (dashed curve) for the Schlögl process, using the parameter values (6.8-39) but with different values of N_2. We see that as N_2 is increased from the value 1.93×10^5 in (a) to the slightly higher value 2.05×10^5 in (d), comparably small

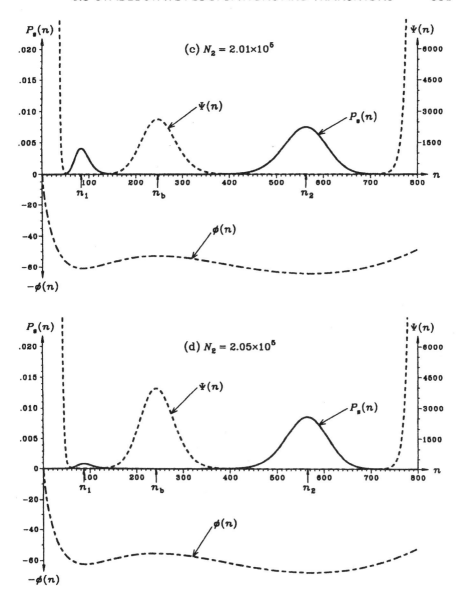

(Fig. 6-19, cont.) changes are induced in n_1, n_2, n_b, σ_1, σ_2, σ_b, τ_1 and τ_2. But the changes induced in $T_1(n_1 \rightarrow n_2)$ and $T_1(n_2 \rightarrow n_1)$, which are determined by the areas under the peaks in $P_s(n)$ and $\Psi(n)$, are evidently quite large. In short, the slight increase in N_2 shifts both stable states slightly upward, and simultaneously induces a dramatic stability shift from a strongly stable n_1 to a strongly stable n_2.

In Eq. (6.8-31), we deduced from general considerations that the average time required for the process to go from stable state n_i to the other stable state n_j should be roughly *twice* the average time required to go from n_i to the barrier state n_b. The explanation for this is that once the process has made it from n_i to n_b, it is about as likely to rush back to n_i as on to n_j. To check this prediction, the pedestrian recursion relations (6.6-11) – (6.6-14) were used to calculate the exact mean first passage times $T_1(n_1 \rightarrow n_b)$ and $T_1(n_2 \rightarrow n_b)$ for the Schlögl process under consideration. The results are that $T_1(n_1 \rightarrow n_b) = 2.339 \times 10^4$, which is 46% of $T_1(n_1 \rightarrow n_2)$ in Eq. (6.8-46a), and $T_1(n_2 \rightarrow n_b) = 0.953 \times 10^5$, which is 54% of $T_1(n_2 \rightarrow n_1)$ in Eq. (6.8-46b). So we conclude that the approximation (6.8-31) is reasonably good in this instance.

Finally, it is interesting to consider the effects of changing the value of the parameter N_2 in the Schlögl process. From the form of the Schlögl reactions (6.8-35), it would seem that an increase in the number of B_2 molecules should promote an increase in the number of X molecules. An understanding of what in fact happens can be obtained by examining the effect that an increase in N_2 would have on the functions $a(n)$ and $v(n)$ plotted in Fig. 6-15: Since N_2 enters those functions only as an additive constant through the function $W_+(n)$ [see Eqs. (6.8-37) and (6.8-38)], then increasing N_2 from its current value 2.00×10^5 would simply shift the $a(n)$ and $v(n)$ curves in Fig. 6-15 vertically upward. This would evidently cause the outside roots of $a(n)$, n_1 and n_2, to increase, and the middle root of $v(n)$, n_b, to decrease. A continuing increase in N_2 would eventually cause n_1 and n_b to approach each other, coalesce, and then disappear, leaving an upshifted n_2 as the only stable state of the system. In a similar way, if N_2 is steadily decreased from the value 2.00×10^5, the $a(n)$ and $v(n)$ curves will move vertically downward, causing n_1 and n_2 to decrease and n_b to increase; eventually, n_b and n_2 will coalesce and then disappear, leaving a down-shifted n_1 as the only stable state. So increasing (decreasing) the B_2-molecule population will indeed increase (decrease) the X-molecule population. But more interestingly, we see that a sufficiently large change either way in the B_2-molecule population will change the Schlögl process from a *bistable* process to a *monostable* process.

The plots in Fig. 6-19 shows how the functions $P_s(n)$, $\phi(n)$ and $\Psi(n)$ appear for four slightly different values of N_2, ranging from 1.93×10^5 to 2.05×10^5, with all other parameter values as given in Eqs. (6.8-39). All these cases are comfortably within the bistable regime (since a coalescence of n_b with either n_1 or n_2 is evidently not imminent in any of these plots). We see that this rather modest (6%) increase in N_2 induces comparably modest changes in the locations and widths of the stable states and the barrier state. And calculations with formula (6.8-12) will reveal similarly modest changes in the stable state fluctuation times. But the mean stable state transition times change quite dramatically; more specifically, the ratio $T_1(n_2 \to n_1)/T_1(n_1 \to n_2)$, which measures the stability of stable state n_2 relative to stable state n_1, is found from Eq. (6.8-27) to increase by a factor of about 168. This large shift in the relative stability of the two stable states is obvious from the sequence of plots in Fig. 6-19. That sequence of plots also suggests that the area under the barrier peak in $\Psi(n)$, which by Eq. (6.8-25) measures the average total time for a "round trip" between the two stable states, achieves a relative minimum when the areas under the two stable state peaks in $P_s(n)$ become approximately equal.

It should be apparent from the foregoing considerations that the Markov process model of the Schlögl reactions (6.8-35) gives us a much more complete picture of the dynamics of those reactions than does the traditional deterministic reaction rate equation (6.8-43). We conclude by remarking that the graphics used in Figs. 6-13 and 6-14 to illustrate a "generic" bistable birth-death Markov process are in fact plots for the Schlögl process with parameter values $c_1 N_1 = 110$, $c_2 = 1$, $c_3 N_2 = 1.02 \times 10^5$ and $c_4 = 4750$. Readers wishing to test their understanding of this final subsection are invited to repeat its analysis using these parameter values in place of those in Eq. (6.8-39).

Appendix A

SOME USEFUL INTEGRAL IDENTITIES

On many occasions in this book we require the values of certain definite integrals. Those definite integrals that are "well known" are listed below, without proof, for convenience in referencing. On one occasion we use an integral identity that is not well known, namely Eq. (A-10), and for that identity we have supplied a proof.

The following identities may be found in common tables of integrals:

$$\int_a^b dx\, x^n = \frac{b^{n+1} - a^{n+1}}{n+1}$$

$$\equiv \frac{b-a}{n+1} \sum_{j=0}^n a^j b^{n-j} \qquad (n=0,1,...). \quad \text{(A-1)}$$

$$\int_0^\infty dx\, x^n \exp(-ax) = \frac{n!}{a^{n+1}} \qquad (a>0;\, n=0,1,...). \quad \text{(A-2)}$$

$$\int_{-\infty}^\infty dx\, x^n \exp(-a^2 x^2) = \begin{cases} \dfrac{n!}{(n/2)!\,(2a)^n} \dfrac{\pi^{1/2}}{|a|}, & \text{if } n=0,2,4..., \\[2mm] 0, & \text{if } n=1,3,5,.... \end{cases} \quad \text{(A-3)}$$

$$\int_0^\infty dx\, x^n \exp(-a^2 x^2) = \frac{1}{2|a|^{n+1}} \left(\frac{n-1}{2} \right)! \qquad (n=1,3,5,...). \quad \text{(A-4)}$$

$$\int_{-\infty}^\infty \frac{dx}{(x-c)^2 + a^2} = \frac{\pi}{|a|}. \quad \text{(A-5)}$$

$$\int_{-\infty}^{\infty} dx\, e^{ibx} \exp(-a^2 x^2) = 2 \int_{0}^{\infty} dx \cos(bx) \exp(-a^2 x^2)$$

$$= \frac{\pi^{1/2}}{|a|} \exp(-b^2/4a^2). \qquad \text{(A-6)}$$

$$\int_{-\infty}^{\infty} dx\, e^{-iux} \frac{1}{(2\pi a^2)^{1/2}} \exp\left(-\frac{(x-m)^2}{2a^2}\right) = e^{-ium} \exp(-u^2 a^2/2).$$

(A-7)

[Note that Eq. (A-7) can be proved by substituting $x' = (x-m)/a$ on the left hand side, and then invoking Eq. (A-6).]

$$\int_{-\infty}^{\infty} dx\, \frac{e^{ibx}}{x^2 + a^2} = 2 \int_{0}^{\infty} dx\, \frac{\cos(bx)}{x^2 + a^2} = \frac{\pi}{|a|} \exp(-|ab|). \qquad \text{(A-8)}$$

$$\int_{0}^{\infty} dx \cos(bx) \exp(-ax) = \frac{a}{a^2 + b^2} \qquad (a>0). \qquad \text{(A-9)}$$

The integral identity

$$\int_{0}^{\pi} du\, \cos(nu - a \sin u) \exp(a \cos u) = \frac{a^n \pi}{n!} \qquad (n=0,1,\dots) \quad \text{(A-10)}$$

will not be found in common integral tables, but it can be proved as follows: Denoting the function of n and a on the left side of Eq. (A-10) by $f_n(a)$, then it is straightforward to show that the derivative of this function with respect to a is given by

$$f_n'(a) = \int_{0}^{\pi} du\, \cos[(n-1)u - a \sin u] \exp[a \cos u] \equiv f_{n-1}(a).$$

It follows that the kth derivative of $f_n(a)$ with respect to a is

$$f_n^{(k)}(a) = f_{n-k}(a) = \int_{0}^{\pi} du\, \cos[(n-k)u - a \sin u] \exp[a \cos u].$$

In particular, for $a=0$ we have

$$f_n^{(k)}(0) = \int_{0}^{\pi} du\, \cos[(n-k)u] = \begin{cases} \pi, & \text{if } k=n \\ 0, & \text{if } k \neq n \end{cases}.$$

Therefore, by Taylor's theorem we conclude that

$$f_n(a) = \sum_{k=0}^{\infty} \frac{1}{k!} f_n^{(k)}(0) \, a^k = \frac{1}{n!} f_n^{(n)}(0) \, a^n = \frac{1}{n!} \, n! \, a^n,$$

which evidently establishes Eq. (A-10).

We also make frequent use in this book of the fact that the first order linear differential equation

$$\frac{dy}{dt} = ky + f(t), \tag{A-11a}$$

where k is constant with respect to t and f is any function of t, has as its solution the function

$$y(t) = e^{k(t-t_0)} \left\{ y(t_0) + \int_{t_0}^{t} dt' \, f(t') \, e^{-k(t'-t_0)} \right\}. \tag{A-11b}$$

That this function satisfies the $t = t_0$ condition is obvious; that it also satisfies the differential equation (A-11a) is straightforwardly proved by direct differentiation.

In proving Eq. (A-11b) and Eq. (A-10), and also at several places in the text, we make use of the following well known rule for differentiating definite integrals:

$$\frac{d}{dx} \int_{a(x)}^{b(x)} h(x,z) \, dz = \int_{a(x)}^{b(x)} \frac{\partial h(x,z)}{\partial x} \, dz + h(x,b(x)) \, b'(x) - h(x,a(x)) \, a'(x), \tag{A-12}$$

where the primes on the right denote differentiation with respect to x.

Appendix B

INTEGRAL REPRESENTATIONS OF THE
DELTA FUNCTIONS

The **Kronecker delta function** $\delta(n_1,n_2)$ is defined for n_1 and n_2 any *integer* variables by

$$\delta(n_1,n_2) \equiv \begin{cases} 0, & \text{if } n_1 \neq n_2 \\ 1, & \text{if } n_1 = n_2 \end{cases}. \tag{B-1}$$

Using the fact that $e^{iu} = \cos u + i\sin u$, plus the fact that the integral of $\cos ku$ over the u-interval $[-\pi, \pi]$ vanishes for any nonzero integer k, it is easy to show that $\delta(n_1,n_2)$ as thus defined has the *integral representation*

$$\delta(n_1,n_2) = (2\pi)^{-1} \int_{-\pi}^{\pi} du \exp[iu(n_1 - n_2)]. \tag{B-2}$$

The real variable analog of the Kronecker delta function is the **Dirac delta function** $\delta(x_1 - x_2)$, which can be defined for x_1 and x_2 any *real* variables by the pair of equations

$$\delta(x_1 - x_2) = 0, \quad \text{if } x_1 \neq x_2, \tag{B-3a}$$

$$\int_{-\infty}^{\infty} dx_1\, \delta(x_1 - x_2) = 1. \tag{B-3b}$$

We note that these two equations imply that

$$\delta(x_2 - x_1) = \delta(x_1 - x_2). \tag{B-4}$$

We also note that these two equations leave the value of $\delta(0)$ *undefined*, although some might try to argue that Eqs. (B-3a) and (B-3b) together imply that $\delta(0)dx = 1$, and hence that $\delta(0) = 1/dx$. In any event, $\delta(x_1 - x_2)$ is of real use *only* when it appears as a factor in the integrand of an integral over x_1 or x_2.

To derive an integral representation of $\delta(x_1 - x_2)$ that is analogous to the integral representation of $\delta(n_1, n_2)$ in Eq. (B-2), it is necessary to first introduce a new kind of infinite integral. We recall that the "ordinary" infinite integral of a function $f(u)$ is defined by

$$\int_{-\infty}^{\infty} du\, f(u) \equiv \lim_{L_1, L_2 \to \infty} \int_{-L_1}^{L_2} du\, f(u), \tag{B-5}$$

where existence requires convergence *regardless* of the way in which L_1 and L_2 approach infinity. We now introduce another kind of infinite integral through the *definition*

$$\oint_{-\infty}^{\infty} du\, f(u) \equiv \lim_{\kappa \downarrow 0} \lim_{L \to \infty} \int_{-L}^{L} du\, f(u)\, e^{-\kappa|u|}, \tag{B-6}$$

wherein it is understood that the κ-limit is to be taken *after* the L-limit is taken. If the function f is such that the ordinary infinite integral (B-5) exists, then the special infinite integral (B-6) will also exist and will be equal to the ordinary infinite integral. However, the special integral (B-6) might exist even if the ordinary integral (B-5) does not. For example, although the two functions $\sin u$ and $\cos u$ do not have an infinite integral in the sense of definition (B-5), it is nevertheless the case that

$$\oint_{-\infty}^{\infty} du\, \sin u = 0 \quad \text{and} \quad \oint_{-\infty}^{\infty} du\, \cos u = 0;$$

here, the first relation is a simple consequence of the symmetric integration limits in (B-6), but the proof of the second relation must invoke the integral identity (A-9) with $b = 1$ and $a = \kappa$. Our motivation for introducing the special integral (B-6) is the following useful theorem, which is the Dirac analog of the Kronecker result (B-2):

Dirac Integral Representation Theorem. In any x_1-integral containing $\delta(x_1 - x_2)$ as a factor in the integrand, we may write

$$\delta(x_1 - x_2) = (2\pi)^{-1} \oint_{-\infty}^{\infty} du\, \exp[iu(x_1 - x_2)], \tag{B-7}$$

provided that, whenever convergence of the x_1-integral requires explicit recognition of the special nature of the u-integral definition (B-6), the associated κ-limit is taken *after* the x_1-integration is performed.

Proof. We shall prove that the expression on the right side of Eq. (B-7) satisfies both of Eqs. (B-3) when the special proviso of the theorem is taken into account.

To show that the representation (B-7) satisfies Eq. (B-3a), we take $x_1 \neq x_2$ and observe that

$$\oint_{-\infty}^{\infty} du \exp[iu(x_1 - x_2)] = \oint_{-\infty}^{\infty} du \cos[u(x_1 - x_2)] + i\oint_{-\infty}^{\infty} du \sin[u(x_1 - x_2)]$$

$$= \lim_{\kappa \downarrow 0} \lim_{L \to \infty} \int_{-L}^{L} du \cos[u(x_1 - x_2)] e^{-\kappa|u|}$$

$$+ i \lim_{\kappa \downarrow 0} \lim_{L \to \infty} \int_{-L}^{L} du \sin[u(x_1 - x_2)] e^{-\kappa|u|}.$$

Noting that $\cos[u(x_1 - x_2)]$ and $e^{-\kappa|u|}$ are both *even* functions of u, while $\sin[u(x_1 - x_2)]$ is an *odd* function of u, we see that the integral in the second term vanishes, leaving

$$\oint_{-\infty}^{\infty} du \exp[iu(x_1 - x_2)] = \lim_{\kappa \downarrow 0} \lim_{L \to \infty} 2\int_{0}^{L} du \cos[u(x_1 - x_2)] e^{-\kappa u}$$

$$= \lim_{\kappa \downarrow 0} 2\int_{0}^{\infty} du \cos[u(x_1 - x_2)] e^{-\kappa u}$$

$$= 2 \lim_{\kappa \downarrow 0} \frac{\kappa}{\kappa^2 + (x_1 - x_2)^2} \qquad \text{[by (A-9)].}$$

Since this last κ-limit is evidently zero for all $x_1 \neq x_2$, we may conclude that the representation (B-7) does indeed satisfy Eq. (B-3a).

Now we must show that the representation (B-7) also satisfies Eq. (B-3b). The demonstration proceeds as follows:

$$\int_{-\infty}^{\infty} \left[(2\pi)^{-1} \oint_{-\infty}^{\infty} du \exp[iu(x_1 - x_2)] \right] dx_1$$

$$= (2\pi)^{-1} \int_{-\infty}^{\infty} dx_1 \lim_{\kappa \downarrow 0} \lim_{L \to \infty} \int_{-L}^{L} du \exp[iu(x_1 - x_2)] e^{-\kappa|u|}$$

$$= (2\pi)^{-1} \int_{-\infty}^{\infty} dx_1 \lim_{\kappa \downarrow 0} \lim_{L \to \infty} \left[\int_{-L}^{L} du \cos[u(x_1 - x_2)] e^{-\kappa|u|} \right.$$

$$\left. + i \int_{-L}^{L} du \sin[u(x_1 - x_2)] e^{-\kappa|u|} \right]$$

$$= (2\pi)^{-1} \int_{-\infty}^{\infty} dx_1 \lim_{\kappa \downarrow 0} \lim_{L \to \infty} 2\int_{0}^{L} du \cos[u(x_1 - x_2)] e^{-\kappa u}$$

$$= \pi^{-1} \int_{-\infty}^{\infty} dx_1 \lim_{\kappa \downarrow 0} \int_{0}^{\infty} du \cos[u(x_1 - x_2)] e^{-\kappa u}$$

$$= \pi^{-1} \int_{-\infty}^{\infty} dx_1 \lim_{\kappa \downarrow 0} \frac{\kappa}{\kappa^2 + (x_1 - x_2)^2} \qquad \text{[by (A-9)]}$$

$$= \pi^{-1} \lim_{\kappa \downarrow 0} \int_{-\infty}^{\infty} dx_1 \frac{\kappa}{\kappa^2 + (x_1 - x_2)^2} \qquad \text{[by theorem proviso]}$$

$$= \pi^{-1} \lim_{\kappa \downarrow 0} \kappa \left(\frac{\pi}{\kappa} \right) \qquad \text{[by (A-5)].}$$

Since this last expression is obviously unity, we conclude that the representation (B-7) indeed satisfies Eq. (B-3b).

Having thus shown that the entity defined through Eqs. (B-7) and (B-6) satisfies both of conditions (B-3), we conclude that that entity is indeed the Dirac delta function.

Appendix C

AN APPROXIMATE SOLUTION
PROCEDURE FOR "OPEN" MOMENT
EVOLUTION EQUATIONS

The moments of a Markov process $X(t)$ with propagator moment functions $B_k(x,t)$ satisfy the infinite set of coupled, first order differential equations (2.7-6). In the commonly encountered case in which the propagator moment functions are explicitly independent of t (i.e., in the temporally homogeneous case), equations (2.7-6) read

$$\frac{d}{dt}\langle X^n(t)\rangle = \sum_{k=1}^{n}\binom{n}{k}\langle X^{n-k}(t)B_k(X(t))\rangle \quad (t_0\le t; n=1,2,...). \quad (C\text{-}1)$$

These equations are to be solved subject to the initial conditions

$$\langle X^n(t_0)\rangle = x_0^n \quad (n=1,2,...). \quad (C\text{-}2)$$

If $B_k(x)$ happens to be a polynomial in x of degree $\le k$, then the right side of Eq. (C-1) will evidently involve only moments of order $\le n$; in that case the system of equations is said to be "closed," and they can be solved successively for $n=1, 2$, etc. Otherwise, the right side of Eq. (C-1) will involve moments of $X(t)$ of order $>n$; in that case the system of equations is said to be "open," and no *exact* solution method is available.

We shall describe here a procedure for finding *approximate* solutions to Eqs. (C-1) in the open case for the two most often required moments, $\langle X(t)\rangle$ and $\langle X^2(t)\rangle$. Our approximate solution procedure is rather lengthy and tedious, although those drawbacks can be mitigated to a greater or lesser degree by judicious use of a computer. Otherwise, the procedure would appear, in principle at least, to be fairly widely applicable.

We begin by assuming that the propagator moment functions $B_k(x)$ are expandable in a well behaved Taylor series as

$$B_k(x) = \sum_{j=0}^{\infty} b_{kj} x^j,$$

where we have defined

$$b_{kj} \equiv \frac{1}{j!} \frac{d^j B_k}{dx^j} \bigg|_{x=0} \qquad (k \geq 1; j \geq 0). \qquad \text{(C-3)}$$

Substituting this expansion into Eq. (C-1), we get

$$\frac{d}{dt} \langle X^n(t) \rangle = \sum_{k=1}^{n} \binom{n}{k} \sum_{j=0}^{\infty} \langle X^{n-k}(t) \, b_{kj} \, X^j(t) \rangle$$

$$= \sum_{k=1}^{n} \sum_{j=0}^{\infty} \binom{n}{k} b_{kj} \langle X^{n-k+j}(t) \rangle$$

$$\equiv \sum_{v=0}^{\infty} \sum_{k=1}^{n} \sum_{j=0}^{\infty} \binom{n}{k} b_{kj} \, \delta(v, n-k+j) \langle X^v(t) \rangle,$$

where $\delta(i,j)$ is the Kronecker delta function. But

$$\sum_{k=1}^{n} \sum_{j=0}^{\infty} \binom{n}{k} b_{kj} \, \delta(v, n-k+j) = \sum_{j=0}^{\infty} \sum_{k=1}^{n} \delta(k, n-v+j) \binom{n}{k} b_{kj}$$

$$= \sum_{\substack{j=0 \\ 1 \leq n-v+j \leq n}}^{\infty} \binom{n}{n-v+j} b_{n-v+j,j}$$

$$= \sum_{j=\text{Max}(0, v+1-n)}^{v} \binom{n}{n-v+j} b_{n-v+j,j}.$$

Therefore, if we define the constants $\Omega_{n,v}$ by

$$\Omega_{n,v} \equiv \sum_{j=\text{Max}(0, v+1-n)}^{v} \binom{n}{n-v+j} b_{n-v+j,j} \qquad (n \geq 1; v \geq 0), \qquad \text{(C-4)}$$

then the preceding equation for the derivative of $\langle X^n(t) \rangle$ becomes

$$\frac{d}{dt} \langle X^n(t) \rangle = \sum_{v=0}^{\infty} \Omega_{n,v} \langle X^v(t) \rangle \qquad (t_0 \leq t; \, n=1,2,\ldots). \qquad \text{(C-5)}$$

Equations (C-5) are formally equivalent to Eqs. (C-1), and are of course subject to the same initial conditions (C-2).

So, having calculated the constants $b_{k,j}$ from the propagator moment functions $B_k(x)$ according to Eq. (C-3), and then combined those constants according to Eq. (C-4) to obtain the constants $\Omega_{n,\nu}$, we propose to embark on the following program: First, we truncate the *first two* of equations (C-5) at $\nu = 2$, and then solve that closed pair of coupled differential equations for $\langle X(t) \rangle$ and $\langle X^2(t) \rangle$. Next, we truncate the *first three* of equations (C-5) at $\nu = 3$, solve that closed set of coupled differential equations for $\langle X(t) \rangle$, $\langle X^2(t) \rangle$ and $\langle X^3(t) \rangle$, and then compare the new solutions for $\langle X(t) \rangle$ and $\langle X^2(t) \rangle$ with the solutions obtained previously. If the discrepancy between those solutions exceeds the desired solution accuracy, then we truncate the *first four* of equations (C-5) at $\nu = 4$, solve that closed set of coupled differential equations for the first four moments, and compare the newest solutions for $\langle X(t) \rangle$ and $\langle X^2(t) \rangle$ to those obtained in the preceding truncation. We continue this procedure until the solutions obtained for $\langle X(t) \rangle$ and $\langle X^2(t) \rangle$ in successive solution sets are, for *practical* purposes, identical. The final solution pair should then constitute an acceptable approximation to $\langle X(t) \rangle$ and $\langle X^2(t) \rangle$.

Once the constants $\Omega_{n,\nu}$ have been calculated, the feasibility of the program just outlined evidently hinges on having an efficient algorithm for solving a *closed* set of *first* order, *linear* differential equations with *constant* coefficients. Happily, well developed solution algorithms, both analytical and numerical, exist for equations of this kind. For our purposes here, a *numerical* solution procedure will usually be preferred. In fact, given a sufficiently powerful symbolic manipulation program, the entire procedure outlined here, from computing the derivatives for $b_{k,j}$ to comparatively displaying successive approximate solutions for $\langle X(t) \rangle$ and $\langle X^2(t) \rangle$, could be completely coded for automatic machine computation.

Appendix D

ESTIMATING THE WIDTH AND AREA OF A
FUNCTION PEAK

Consider a positive function $f(x)$ that has a smooth, reasonably symmetric peak at $x = x_0$. There are several possible ways of quantitatively assigning a "width" and "area" to that peak. These ways will usually give results that agree with each other to within factors of order unity, but most will break down for certain specific peak shapes. The problem of course is that one can never be sure where to place the left and right boundaries of the peak. In this appendix we shall describe one method for assigning widths and areas to peaks that is fairly convenient and robust, and in any case is quite adequate for our purposes in this book. In brief, the method first approximates $f(x)$ in the vicinity of x_0 by a Gaussian or normal-shaped function. It then defines the "area" of the peak to be the total area under the approximating Gaussian function, and the "width" of the peak to be the so-called "effective width" of the Gaussian function.

We let $y_0 > 0$ denote the height of the peak in f at x_0, and we observe that the existence of the peak implies that f will have at x_0 a vanishing first derivative and a negative second derivative; thus, we have

$$f(x_0) = y_0, \quad f'(x_0) = 0, \quad f''(x_0) = -y_0 c, \qquad \text{(D-1)}$$

where c is some positive constant. Our method seeks an estimate of the width and area of the peak solely in terms of the two parameters y_0 and c. We begin by approximating the function $f(x)$ in the vicinity of the peak by a normal-shaped or "Gaussian" function $g(x)$ which satisfies conditions (D-1); i.e.,

$$g(x_0) = y_0, \quad g'(x_0) = 0, \quad g''(x_0) = -y_0 c. \qquad \text{(D-2)}$$

A Gaussian function that obviously satisfies the *first two* of conditions (D-2) is

$$g(x) = y_0 \exp[-(x-x_0)^2/2a^2] \quad (a > 0). \tag{D-3}$$

We shall now choose the standard deviation parameter a here so that the *third* of conditions (D-2) is likewise satisfied. By straightforward differentiation with respect to x, we get

$$g'(x) = -g(x)(x-x_0)/a^2,$$

and

$$g''(x) = -g'(x)(x-x_0)/a^2 - g(x)/a^2.$$

The first of these relations shows that indeed $g'(x_0) = 0$. Using this and the fact that $g(x_0) = y_0$, the second relation then gives

$$g''(x_0) = -y_0/a^2.$$

Comparing this with the last of conditions (D-2), we conclude that we must choose a so that

$$1/a^2 = c. \tag{D-4}$$

And so we conclude that the approximating Gaussian function for the peak in f at $x = x_0$ is

$$g(x) = y_0 \exp[-c(x-x_0)^2/2]. \tag{D-5}$$

We now simply *define* the area A under the x_0-peak in $f(x)$ to be the total area under the approximating Gaussian function $g(x)$:

$$A \equiv \int_{-\infty}^{\infty} dx\, g(x) = y_0 \int_{-\infty}^{\infty} dx \exp\left[-c(x-x_0)^2/2\right].$$

Applying the integral identity (A-3), we get

$$A = y_0 (2\pi/c)^{1/2}. \tag{D-6}$$

So this is our estimate of the *area* of the x_0-peak in $f(x)$.

Finally, we define the width σ of the x_0-peak in $f(x)$ to be the "effective width" of the approximating Gaussian peak $g(x)$. The effective width of a peak is defined in general as *the width of the rectangle that has the same height and area as the peak*; i.e., the effective width of a peak is equal to the area of the peak divided by the height of the peak. In the present case, where our peak has height y_0 and area A, the result (D-6) implies that the effective width is

$$\sigma = (2\pi/c)^{1/2}. \tag{D-7}$$

So this is our estimate of the *width* of the x_0-peak in $f(x)$.

Appendix E

CAN THE ACCURACY OF THE CONTINUOUS PROCESS SIMULATION FORMULA BE IMPROVED?

Consider a temporally homogeneous continuous Markov process $X(t)$ with characterizing functions $A(x)$ and $D(x)$. The Langevin equation (3.4-4) tells us that if the value of the process at time t is $x(t)$, then the value of the process at an infinitesimally later time $t+\mathrm{d}t$ can be computed as

$$x(t+\mathrm{d}t) = x(t) + n[D(x(t))\mathrm{d}t]^{1/2} + A(x(t))\mathrm{d}t, \qquad \text{(E-1)}$$

where n is a sample value of the unit normal random variable $N \equiv N(0,1)$. The continuous process Monte Carlo simulation formula (3.9-1), which we rewrite here as

$$x(t+\Delta t) \approx x(t) + n[D(x(t))\Delta t]^{1/2} + A(x(t))\Delta t, \qquad \text{(E-2)}$$

was evidently obtained simply by replacing the infinitesimal variable $\mathrm{d}t$ in Eq. (E-1) by the *finite* variable Δt. We found in Section 3.9 that Eq. (E-2) is exact if $A(x)$ and $D(x)$ are both constants, but that otherwise Eq. (E-2) must be regarded as an approximation that becomes exact only in the limit of vanishingly small Δt. The question that we shall address in this appendix is whether we can modify the right side of Eq. (E-2) so as to improve its general accuracy for finite Δt.

To orient our thinking on this question, let us consider first the special case in which $D(x) \equiv 0$. Equation (E-1) then becomes equivalent to the ordinary differential equation

$$\frac{\mathrm{d}x}{\mathrm{d}t} = A(x) \quad [\, D(x) \equiv 0 \,], \qquad \text{(E-3)}$$

and Eq. (E-2) reduces to the well known *Euler integration formula*,

$$x(t+\Delta t) \approx x(t) + A(x(t))\Delta t \qquad [\, D(x) \equiv 0 \,]. \qquad \text{(E-4)}$$

Now, one way to improve the accuracy of Eq. (E-4) is to argue as follows:

$$x(t+\Delta t) \approx x(t) + \frac{dx}{dt}\,\Delta t + \frac{1}{2}\frac{d^2x}{dt^2}(\Delta t)^2 \qquad \text{[by Taylor's theorem]}$$

$$= x(t) + A(x(t))\,\Delta t + \frac{1}{2}\frac{d}{dt}A(x(t))\,(\Delta t)^2 \qquad \text{[by (E-3)]}$$

$$= x(t) + A(x(t))\,\Delta t + \frac{1}{2}\left[\partial_x A(x(t))\,\frac{dx}{dt}\right](\Delta t)^2 \quad \text{[by chain rule]}$$

$$= x(t) + A(x(t))\,\Delta t + \frac{1}{2}\left[\partial_x A(x(t))\right]A(x(t))\,(\Delta t)^2 \qquad \text{[by (E-3)]}.$$

Thus we deduce the formula

$$x(t+\Delta t) \approx x(t) + A(x(t))\Delta t[1 + (1/2)\,\partial_x A(x(t))\Delta t] \quad [\, D(x) \equiv 0 \,], \quad \text{(E-5)}$$

which should clearly be more accurate than the formula (E-4). Notice that Eq. (E-5) properly reduces to Eq. (E-4) if *either* $A(x)$ is constant *or* Δt is infinitesimally small.

Because the foregoing derivation of Eq. (E-5) invoked Eq. (E-3), it assumed that $x(t)$ is a *differentiable function of t*. That assumption would pose a serious problem for adapting the analysis to the case of a nonzero diffusion function $D(x)$, since the process $X(t)$ in that case does *not* have a *t*-derivative. But there is another route to Eq. (E-5) that does not overtly assume that $x(t)$ is differentiable. This is the *modified Euler hypothesis*. To explain that hypothesis most simply, we first abbreviate

$$x(t) \equiv x \quad \text{and} \quad x(t+\Delta t) - x(t) \equiv \Delta x,$$

and then note from Eq. (E-4) that the *simple* Euler hypothesis is just the statement that

$$\Delta x \approx \Delta_s x \equiv A(x)\Delta t \qquad \text{(simple Euler).} \qquad \text{(E-6a)}$$

The *modified* Euler hypothesis is then the statement that

$$\Delta x \approx \Delta_m x \equiv A(x + \Delta_s x/2)\Delta t \quad \text{(modified Euler).} \quad \text{(E-6b)}$$

In words, the modified Euler hypothesis proposes to evaluate the slope function A not at the point x, but rather at the point which the *simple* Euler hypothesis would assert is halfway between x and $x+\Delta x$. We can prove that the modified Euler hypothesis (E-6b) leads to the improved estimate (E-5) by reasoning as follows:

$$\Delta_m x \approx [A(x) + (\Delta_s x/2)\,\partial_x A(x)]\,\Delta t \quad \text{[by Taylor's theorem]}$$

$$= A(x)\Delta t + [A(x)\Delta t/2\,|\,\partial_x A(x)\,\Delta t \quad \text{[by (E-6a)]}$$

$$= A(x)\Delta t\,[1 + (1/2)\,\partial_x A(x)\,\Delta t\,],$$

which is evidently the same as Eq. (E-5). So we see that, when $D(x) \equiv 0$, the modified Euler hypothesis produces a more accurate formula than does the simple Euler hypothesis.

The adaptation of the modified Euler algorithm to the continuous Markov process case in which $D(x)$ is *not* identically zero would appear to be fairly straightforward: In our abbreviated notation, we may rewrite the simulation formula (E-2) as

$$\Delta x \approx \Delta_s x \equiv n\,[D(x)\Delta t]^{1/2} + A(x)\Delta t, \tag{E-7}$$

and it constitutes a *simple* Euler formula, analogous to Eq. (E-6a). The corresponding *modified* Euler formula analogous to Eq. (E-6b) should then be given by

$$\Delta x \approx \Delta_m x \equiv n\,[D(x + \Delta_s x/2)\Delta t]^{1/2} + A(x + \Delta_s x/2)\Delta t. \tag{E-8}$$

To mathematically simplify Eq. (E-8), let us temporarily put $D^{1/2}(x) \equiv E(x)$. Then Eq. (E-8) becomes, using Taylor's theorem and Eq. (E-7),

$$\Delta x \approx n\,E(x + \Delta_s x/2)(\Delta t)^{1/2} + A(x + \Delta_s x/2)\Delta t$$

$$\approx n\,[E(x) + (\Delta_s x/2)\,\partial_x E(x)]\,(\Delta t)^{1/2} + [A(x) + (\Delta_s x/2)\,\partial_x A(x)]\,\Delta t$$

$$= n\,E(x)(\Delta t)^{1/2} + A(x)\Delta t + \frac{1}{2}\left\{ n\,E(x)(\Delta t)^{1/2} + A(x)\Delta t \right\}$$

$$\times \left\{ n\,\partial_x E(x)(\Delta t)^{1/2} + \partial_x A(x)\Delta t \right\}$$

$$= n\,E(x)(\Delta t)^{1/2} + [\,A(x) + (n^2/2)\,E(x)\,\partial_x E(x)\,]\,\Delta t$$

$$+ (n/2)\{\,E(x)\,\partial_x A(x) + A(x)\,\partial_x E(x)\,\}(\Delta t)^{3/2}$$

$$+ (1/2)\,A(x)\,\partial_x A(x)\,(\Delta t)^2.$$

So, replacing $E(x)$ by $D^{1/2}(x)$, and hence also $E(x)\partial_x E(x)$ by $\frac{1}{2}\partial_x D(x)$, we conclude that the modified Euler formula corresponding to the simple Euler formula (E-7) is

$$\Delta x \approx \Delta_m x = n\,[D(x)\Delta t\,]^{1/2} + [A(x) + (n^2/4)\,\partial_x D(x)]\,\Delta t$$

$$+ (n/2)\{D^{1/2}(x)\,\partial_x A(x) + A(x)\,\partial_x D^{1/2}(x)\}\,(\Delta t)^{3/2}$$

$$+ (1/2)\,A(x)\,\partial_x A(x)\,(\Delta t)^2. \tag{E-9}$$

Now we must determine whether formula (E-9) really constitutes a valid improvement over Eq. (E-7). Let's consider first the case in which $D(x) \equiv D$, a constant. Equation (E-9) then becomes

$$\Delta x \approx n\,[D\,\Delta t\,]^{1/2} + A(x)\,\Delta t$$

$$+ (n/2)\,D^{1/2}\partial_x A(x)\,(\Delta t)^{3/2} + (1/2)\,A(x)\,\partial_x A(x)\,(\Delta t)^2,$$

and this in turn immediately factors to

$$\Delta x \approx \{n\,[D\,\Delta t\,]^{1/2} + A(x)\,\Delta t\}\,\{1 + (1/2)\,\partial_x A(x)\,\Delta t\}$$

$$[D(x) \equiv D]. \tag{E-10}$$

This is the "improved formula" asserted in Eq. (3.9-7b). Notice that it properly reduces to the simple Euler formula (E-7) if *either* $A(x)$ is constant *or* Δt is infinitesimally small; furthermore, it properly reduces to Eq. (E-5) in the deterministic case $D = 0$. So it would appear that the modified Euler hypothesis indeed produces a valid improvement over the simple Euler formula (E-7) when $D(x)$ is independent of x.

But now let's consider the more general case in which $D(x)$ is *not* independent of x. In that case, if we take Δt to be the infinitesimal dt, then the modified Euler formula (E-9) evidently reduces to

$$\Delta x \approx n\,[D(x)\,dt\,]^{1/2} + [A(x) + (n^2/4)\,\partial_x D(x)]\,dt. \tag{E-11}$$

This formula is obviously not correct, because it does not agree with the $\Delta t \to dt$ version of Eq. (E-7), namely, the Langevin equation (E-1). Evidently, if we were to use the modified Euler formula (E-9) for nonconstant $D(x)$, then we would be simulating a continuous Markov process whose diffusion function is indeed $D(x)$, but whose drift function is $A(x) + (N^2/4)\partial_x D(x)$ instead of $A(x)$!

So we conclude that *the modified Euler hypothesis leads to an incorrect simulation formula when the diffusion function depends explicitly on x*. This finding leads us to make the *conjecture* that the question advanced in the title of this appendix has a negative answer when $D(x)$ is not a constant.

The failure of the modified Euler approach for nonconstant $D(x)$ inevitably casts doubt on the validity of its prediction (E-10) for constant $D(x)$, and that is why our advocacy of Eq. (3.9-7b) in the text is so hesitant. But we *can* show that Eq. (E-10), or (3.9-7b), is indeed a valid second-order formula for the *Ornstein-Uhlenbeck* process, which has $A(x) = -kx$ and $D(x) \equiv D$. We have from the main Ornstein-Uhlenbeck result (3.3-24) that, if $X(t) = x$, then $X(t + \Delta t)$ will be given *exactly* by

$$X(t + \Delta t) = \mathbf{N}(xe^{-k\Delta t}, (D/2k)(1 - e^{-2k\Delta t})) \qquad \text{(E-12a)}$$

for all $\Delta t > 0$. Using theorem (1.6-7), we can rewrite this formula as

$$X(t + \Delta t) = xe^{-k\Delta t} + [(D/2k)(1 - e^{-2k\Delta t})]^{1/2}N, \qquad \text{(E-12b)}$$

where N is as usual the random variable $\mathbf{N}(0,1)$. Equation (E-12b) constitutes an *exact* stepping formula for the Ornstein-Uhlenbeck process. Now, if we simply expand the right hand side of that formula to second order in $k\Delta t$, then we obtain after collecting terms,

$$X(t + \Delta t) \approx x + \{N[D\Delta t]^{1/2} - kx\Delta t\}\{1 - (1/2)k\Delta t\}. \qquad \text{(E-13)}$$

But this formula, which is unquestionably correct to second order in Δt, is precisely the formula predicted by Eq. (E-10), or (3.9-7b). So we conclude that Eq. (3.9-7b) is *probably* okay for any continuous Markov process with $D(x) = D$.

We shall leave this matter here with the comment that the relationship between the *simple* infinitesimal Euler formula (E-1) and the *modified* infinitesimal Euler formula (E-11) has overtones of the relationship between the *Ito stochastic calculus* and the *Stratonovich stochastic calculus*. The interested reader may consult Gardiner's book [see Bibliography] for an introduction to these calculi.

Appendix F

PROOF OF THE BIRTH-DEATH
STABILITY THEOREM

In this appendix we give a formal proof of the theorem asserting Eq. (6.4-9). The function $P_s(n)$ appearing in that theorem is evidently *defined* through Eqs. (6.4-5) and (6.4-6). We know from our discussion of those equations in the text that this function $P_s(n)$ is a strictly positive density function on the integers $n = 0, 1, ..., N$, and further that it satisfies the "detailed balancing" condition

$$W_-(n)\,P_s(n) = W_+(n-1)\,P_s(n-1) \quad (1 \le n \le N). \quad \text{(F-1)}$$

We also know that the Markov state density function $P(n,t \mid n_0,t_0)$ of the process satisfies the forward master equation (6.1-17), which, abbreviating $P(n,t \mid n_0,t_0) \equiv P(n,t)$, reads

$$\frac{\partial}{\partial t} P(n,t) = [W_-(n+1)\,P(n+1,t) - W_+(n)\,P(n,t)]$$

$$+ [W_+(n-1)\,P(n-1,t) - W_-(n)\,P(n,t)]$$

$$(n = 0,1,...,N). \quad \text{(F-2)}$$

To show that Eq. (6.4-9) follows from the facts just stated, it will be convenient to introduce the notation [compare Eqs. (5.3-5) and (5.3-6)]

$$\omega_{m,n} \equiv \begin{cases} W_+(n), & \text{if } m = n+1 \text{ and } 0 \le n \le N-1 \\ W_-(n), & \text{if } m = n-1 \text{ and } 1 \le n \le N \\ 0, & \text{otherwise.} \end{cases} \quad \text{(F-3)}$$

The detailed balancing condition (F-1) can then be written as

$$\omega_{m,n}\,P_s(n) = \omega_{n,m}\,P_s(m), \quad \text{(F-4)}$$

where m and n can be *any* integers in $[0,N]$ since both sides will be zero unless $m = n \pm 1$. Also, the forward master equation (F-2) can be written

$$\frac{\partial}{\partial t} P(n,t) = [\omega_{n,n+1} P(n+1,t) - \omega_{n+1,n} P(n,t)]$$
$$+ [\omega_{n,n-1} P(n-1,t) - \omega_{n-1,n} P(n,t)],$$

or

$$\frac{\partial}{\partial t} P(n,t) = \sum_{m=0}^{N} [\omega_{n,m} P(m,t) - \omega_{m,n} P(n,t)], \qquad (F-5)$$

since all terms in the m-sum except the $m = n \pm 1$ terms will be zero because of definition (F-3).[†]

Since $P_s(m) \neq 0$, then Eq. (F-4) can be written

$$\omega_{n,m} = \frac{\omega_{m,n} P_s(n)}{P_s(m)}.$$

Substituting this into Eq. (F-5) gives

$$\frac{\partial}{\partial t} P(n,t) = \sum_{m=0}^{N} \left[\frac{\omega_{m,n} P_s(n)}{P_s(m)} P(m,t) - \omega_{m,n} P(n,t) \right],$$

or, since $P_s(n) \neq 0$,

$$\frac{\partial}{\partial t} P(n,t) = \sum_{m=0}^{N} \omega_{m,n} P_s(n) \left[\frac{P(m,t)}{P_s(m)} - \frac{P(n,t)}{P_s(n)} \right]. \qquad (F-6)$$

Equation (F-6) is the form of the forward master equation that will be convenient for our purposes here.

Now we introduce the function

$$G(t) \equiv \sum_{n=0}^{N} \frac{[P(n,t) - P_s(n)]^2}{P_s(n)}. \qquad (F-7)$$

We observe that $G(t) \geq 0$, with the condition $G(t) = 0$ obtaining if and only if $P(n,t) = P_s(n)$ for all $n \in [0,N]$. The strategy of our proof here will be to

---

[†] Our arguments from this point on actually will not require that the integers m and n be nonnegative, nor even that $\omega_{m,n}$ be zero if $m \neq n \pm 1$. All that will be required is the satisfaction of Eqs. (F-4) and (F-5), with the sum over m in the latter extending over the appropriate range.

show that $G(t) \to 0$ as $t \to \infty$. This would evidently imply that $P(n,t) \to P_s(n)$ as $t \to \infty$ for all $n \in [0,N]$, which is precisely what we want to prove.

To show that $G(t) \to 0$ as $t \to \infty$, we begin by calculating $G'(t)$, the derivative of G with respect to t. We have from the definition (F-7) that

$$G'(t) = \sum_{n=0}^{N} \frac{1}{P_s(n)} 2[P(n,t) - P_s(n)] \frac{\partial}{\partial t} P(n,t)$$

$$= 2 \sum_{n=0}^{N} \left(\frac{P(n,t)}{P_s(n)} - 1 \right) \frac{\partial}{\partial t} P(n,t).$$

Substituting for $\partial P/\partial t$ from the forward master equation (F-6), we get

$$G'(t) = 2 \sum_{n=0}^{N} \sum_{m=0}^{N} \omega_{m,n} P_s(n) \left(\frac{P(n,t)}{P_s(n)} - 1 \right) \left(\frac{P(m,t)}{P_s(m)} - \frac{P(n,t)}{P_s(n)} \right).$$

(F-8a)

If in this equation we first replace the factor $\omega_{m,n} P_s(n)$ by $\omega_{n,m} P_s(m)$ in accordance with the detailed balance condition, and then *relabel* the summation indices $n \to m$ and $m \to n$, we get

$$G'(t) = 2 \sum_{m=0}^{N} \sum_{n=0}^{N} \omega_{m,n} P_s(n) \left(\frac{P(m,t)}{P_s(m)} - 1 \right) \left(\frac{P(n,t)}{P_s(n)} - \frac{P(m,t)}{P_s(m)} \right).$$

This can easily be rearranged to read

$$G'(t) = 2 \sum_{n=0}^{N} \sum_{m=0}^{N} \omega_{m,n} P_s(n) \left(1 - \frac{P(m,t)}{P_s(m)} \right) \left(\frac{P(m,t)}{P_s(m)} - \frac{P(n,t)}{P_s(n)} \right).$$

(F-8b)

Now adding Eqs. (F-8a) and (F-8b) and dividing by 2, we conclude that

$$G'(t) = - \sum_{n=0}^{N} \sum_{m=0}^{N} \omega_{m,n} P_s(n) \left(\frac{P(m,t)}{P_s(m)} - \frac{P(n,t)}{P_s(n)} \right)^2. \qquad (F-9)$$

Equation (F-9) shows that $G'(t) \leq 0$, with equality obtaining if and only if

$$\frac{P(m,t)}{P_s(m)} = \frac{P(n,t)}{P_s(n)} \quad \text{for all } m,n \in [0,N].$$

But this would imply that $P(n,t)/P_s(n)=c$ where c is independent of n, and normalization considerations would in turn require that $c=1$. So we conclude that $G'(t)=0$ if and only if $P(n,t)=P_s(n)$ for all $n\in[0,N]$.

We have now established that $G(t)\geq 0$ and $G'(t)\leq 0$, with *both* equalities obtaining if and only if the two functions $P(n,t)$ and $P_s(n)$ coincide everywhere with each other. Since $P(n,t)$ does not coincide with $P_s(n)$ at $t=t_0$ — because $P(n,t_0)\equiv\delta(n,n_0)$ is obviously not everywhere positive on $[0,N]$ as $P_s(n)$ is — then we have $G(t_0)>0$ and $G'(t_0)<0$. The latter inequality causes $G(t)$ to decrease toward zero as t increases from t_0, and that decrease in $G(t)$ implies, by Eq. (F-7), a decrease in the overall disparity between the two functions $P(n,t)$ and $P_s(n)$. Equation (F-9) implies that this decrease in $G(t)$, and the concomitant decrease in the difference between the two functions $P(n,t)$ and $P_s(n)$, must continue for so long as $P(n,t)$ differs from $P_s(n)$ anywhere in the interval $[0,N]$. This establishes the result (6.4-9).

Appendix G

SOLUTION TO THE MATRIX
DIFFERENTIAL EQUATION (6.6-62)

In this appendix we want to establish the fact that the matrix differential equation (6.6-62), namely

$$\frac{d}{dt} \mathbf{P}^*(t) = -\mathbf{M} \cdot \mathbf{P}^*(t), \qquad (\text{G-1})$$

has as its solution the function (6.6-68), namely

$$\mathbf{P}^*(t) = \mathbf{V} \cdot e^{-\lambda t} \cdot \mathbf{V}^{-1} \cdot \mathbf{P}^*(0) \qquad (t \geq 0), \qquad (\text{G-2})$$

wherein $\mathbf{V}$ and $e^{-\lambda t}$ are defined in terms of the eigenvectors and eigenvalues of $\mathbf{M}$ through Eqs. (6.6-66) and (6.6-67). Our proof proceeds by way of two lemmas.

The first lemma is that the matrix $\mathbf{M}$ can be written as

$$\mathbf{M} = \mathbf{V} \cdot \lambda \cdot \mathbf{V}^{-1}, \qquad (\text{G-3})$$

where λ is defined to be the diagonal matrix

$$(\lambda)_{i,j} \equiv \lambda_{j-1} \delta(i,j) \qquad (i,j = 1,...,n_1 + 1). \qquad (\text{G-4})$$

To prove Eq. (G-3), we first note that the eigenvalue equation (6.6-66b) for $\mathbf{M}$ can be written

$$\sum_{l=1}^{n_1+1} M_{i,l} v^{j-1}_l = \lambda_{j-1} v^{j-1}_i \equiv \sum_{l=1}^{n_1+1} v^{l-1}_i \delta(l,j) \lambda_{j-1},$$

where the indices i and j range from 1 to $n_1 + 1$ inclusively. The outside members of this equation can, by virtue of the respective definitions of $\mathbf{V}$ and λ in Eqs. (6.6-67b) and (G-4), be expressed as

$$\sum_{l=1}^{n_1+1} (\mathbf{M})_{i,l}(\mathbf{V})_{l,j} = \sum_{l=1}^{n_1+1} (\mathbf{V})_{i,l}(\mathbf{\Lambda})_{l,j},$$

or more compactly,

$$\mathbf{M}\cdot\mathbf{V} = \mathbf{V}\cdot\mathbf{\Lambda}.$$

Multiplying this equation from the right by $\mathbf{V}^{-1}$ establishes Eq. (G-3).

The second lemma we require here is that

$$\frac{d}{dt}e^{-\mathbf{\Lambda} t} = -\mathbf{\Lambda}\cdot e^{-\mathbf{\Lambda} t}. \tag{G-5}$$

To prove this lemma we proceed as follows, using the definition of $e^{-\mathbf{\Lambda} t}$ in Eq. (6.6-67a):

$$\left(\frac{d}{dt}e^{-\mathbf{\Lambda} t}\right)_{i,j} = \frac{d}{dt}\left(e^{-\mathbf{\Lambda} t}\right)_{i,j}$$

$$= \frac{d}{dt}\left(e^{-\lambda_{j-1}t}\,\delta(i,j)\right)$$

$$= -\lambda_{j-1}e^{-\lambda_{j-1}t}\,\delta(i,j)$$

$$= -\sum_{l=1}^{n_1+1}\lambda_{l-1}\,\delta(l,j)\,e^{-\lambda_{j-1}t}\,\delta(i,l)$$

$$= -\sum_{l=1}^{n_1+1}\lambda_{l-1}\,\delta(i,l)\times e^{-\lambda_{j-1}t}\,\delta(l,j)$$

$$= -\sum_{l=1}^{n_1+1}(\mathbf{\Lambda})_{i,l}\left(e^{-\mathbf{\Lambda} t}\right)_{l,j}$$

$$\left(\frac{d}{dt}e^{-\mathbf{\Lambda} t}\right)_{i,j} = -\left(\mathbf{\Lambda}\cdot e^{-\mathbf{\Lambda} t}\right)_{i,j},$$

and this establishes Eq. (G-5).

Using lemmas (G-3) and (G-5), we can now straightforwardly evaluate the derivative of the right hand side of Eq. (G-2) as follows:

$$\frac{d}{dt}\left(\mathbf{V}\cdot e^{-\boldsymbol{\lambda} t}\cdot\mathbf{V}^{-1}\cdot\mathbf{P}^*(0)\right) = \mathbf{V}\cdot\left(\frac{d}{dt}e^{-\boldsymbol{\lambda} t}\right)\cdot\mathbf{V}^{-1}\cdot\mathbf{P}^*(0)$$

$$= \mathbf{V}\cdot\left(-\boldsymbol{\lambda}\cdot e^{-\boldsymbol{\lambda} t}\right)\cdot\mathbf{V}^{-1}\cdot\mathbf{P}^*(0) \quad \text{[by (G-5)]}$$

$$= -\mathbf{V}\cdot\boldsymbol{\lambda}\cdot\left(\mathbf{V}^{-1}\cdot\mathbf{V}\right)\cdot e^{-\boldsymbol{\lambda} t}\cdot\mathbf{V}^{-1}\cdot\mathbf{P}^*(0)$$

$$= -\left(\mathbf{V}\cdot\boldsymbol{\lambda}\cdot\mathbf{V}^{-1}\right)\cdot\left(\mathbf{V}\cdot e^{-\boldsymbol{\lambda} t}\cdot\mathbf{V}^{-1}\cdot\mathbf{P}^*(0)\right)$$

$$\frac{d}{dt}\left(\mathbf{V}\cdot e^{-\boldsymbol{\lambda} t}\cdot\mathbf{V}^{-1}\cdot\mathbf{P}^*(0)\right) = -\mathbf{M}\cdot\left(\mathbf{V}\cdot e^{-\boldsymbol{\lambda} t}\cdot\mathbf{V}^{-1}\cdot\mathbf{P}^*(0)\right). \quad \text{[by (G-3)]}$$

This last equation shows that the function $\mathbf{P}^*(t)$ specified by Eq. (G-2) does indeed satisfy the differential equation (G-1).

Finally, it is easy to see that the expression (G-2) also satisfies the requisite initial condition: The definition (6.6-67a) shows that $e^{-\boldsymbol{\lambda} t}$ for $t=0$ is the unit matrix $\mathbf{1}$; so for $t=0$ Eq. (G-2) gives

$$\mathbf{P}^*(t=0) = \mathbf{V}\cdot\mathbf{1}\cdot\mathbf{V}^{-1}\cdot\mathbf{P}^*(0) = \mathbf{V}\cdot\mathbf{V}^{-1}\cdot\mathbf{P}^*(0) = \mathbf{P}^*(0),$$

as it should.

INDEX

Printed and bound by CPI Group (UK) Ltd, Croydon, CR0 4YY

03/10/2024

01040418-0012